Data Communications

and Computer Networks:

A Business User's Approach

▶ **Curt M. White**

DePaul University

COURSE
TECHNOLOGY

Thomson Learning

ONE MAIN STREET, CAMBRIDGE, MA 02142

Australia • Canada • Denmark • Japan • Mexico • New Zealand • Philippines
Puerto Rico • Singapore • South Africa • Spain • United Kingdom • United States

Data Communications
and Computer Networks:
A Business User's Approach

Data Communications and Computer Networks: A Business User's Approach is published by Course Technology.

Credits

Managing Editor	Jennifer Normandin
Vice President, Publisher	Kristen Duerr
Production Editor	Christine Spillett
Developmental Editor	Marilyn Freedman
Associate Product Manager	Amanda Young
Marketing Manager	Susan Ogar
Text Designer	Carol Keller, Books by Design
Cover Designer	Efrat Reis
Interactive Modules	Designs with Attitude

© 2001 by Course Technology, a division of Thomson Learning
Thomson Learning is a trademark used herein under license.

Microsoft and Visual Basic are registered trademarks.

Disclaimer
Course Technology reserves the right to revise this publication and make changes from time to time in its content without notice.

The Web addresses in this book are subject to change from time to time as necessary without notice.

For more information, contact Course Technology, One Main Street, Cambridge, MA 02142; or find us on the World Wide Web at *www.course.com*.

For permission to use material from this text or product, contact us by
■ www.thomsonrights.com
■ 1-800-730-2214
■ 1-800-730-2215

ISBN 0-619-01529-2

Printed in America
1 2 3 4 5 6 7 8 9 CRKEN 04 03 02 01 00

To Kathleen, Hannah Colleen, and
Samuel Memphis—it's never boring.

Brief Contents

1 Introduction to Computer Networks and Data Communications 1

2 Fundamentals of Data and Signals 29

3 The Media: Conducted and Wireless 67

4 Making Connections 105

5 Multiplexing: Sharing a Medium 143

6 Errors, Error Detection, and Error Control 165

7 Local Area Networks: The Basics 191

8 Local Area Networks: Internetworking 231

9 Local Area Networks: Software and Support Systems 261

10 Introduction to Wide Area Networks 289

11 The Internet 317

12 Telecommunication Systems 357

13 Network Security 393

14 Network Design and Management 423

Appendix 453

Glossary 467
Index 482

Contents

Preface   **xvii-xxiv**

1 **Introduction to Computer Networks
        and Data Communications**   **1**

Introduction   **2**
The Language of Computer Networks   **3**
The Big Picture of Networks   **4**
Computer Networks—Basic Configurations   **6**
        Computer terminal-to-mainframe computer configurations   6
        Microcomputer-to-mainframe computer configuration   7
        Microcomputer-to-local area network configuration   8
        Microcomputer-to-Internet configuration   9
        Local area network-to-local area network configuration   10
        Local area network-to-wide area network configuration   10
        Sensor-to-local area network connection   11
        Satellite and microwave configurations   12
        Wireless telephone configurations   13
Network Architecture Models   **14**
        The Open Systems Interconnection (OSI) model   16
        The Internet Model   19
        Logical and physical connections   22
The Internet Model in Action   **23**
Summary   25
Key Terms   27
Review Questions   27
Exercises   27
Thinking Outside the Box   28
Projects   28

2 **Fundamentals of Data and Signals**   **29**

Introduction   **30**
Data and Signals   **30**
        Analog versus digital   31
        Fundamentals of signals   34
        Loss of signal strength   36
Converting Data into Signals   **39**
        Transmitting digital data with digital signals: digital encoding schemes 39
        Transmitting digital data with analog signals   43
        Transmitting analog data with digital signals   47
        Transmitting analog data with analog signals   52
Spread Spectrum Technology   **53**
Data Codes   **54**
        EBCDIC   55

ASCII   57

Baudot code   58

Data and Signal Conversions in Action   59

Summary   62

Key Terms   64

Review Questions   64

Exercises   64

Thinking Outside the Box   66

Projects   66

3   The Media: Conducted and Wireless   67

Introduction   68

Twisted Pair Wire   68

Coaxial Cable   74

Fiber Optic Cable   76

Introduction to Wireless Transmissions   80

Terrestrial microwave transmission   81

Satellite microwave transmission   82

Mobile telephones   86

Cellular digital packet data   89

Pagers   89

Infrared transmissions   90

Broadband wireless systems   91

Media Selection Criteria   94

Cost   94

Speed   95

Distance and expandability   95

Environment   96

Security   96

Conducted Media in Action   97

Wireless Media in Action   98

Summary   101

Key Terms   102

Review Questions   102

Exercises   103

Thinking Outside the Box   103

Projects   104

4   Making Connections   105

Introduction   106

Modems   107

Basic modem operating principles   107

Data transmission rate   107

Standard telephone operations   108

Connection negotiation   108

Compression and error correction   108

Facsimile   **109**
Security   **109**
Self-testing (loop back)   **109**
Internal versus external models   **110**
Modems for laptops   **110**

Breaking Bandwidth Limitations   **111**

Alternatives to Traditional Modems   **114**

Channel Service Unit/Data Service Unit (CSU/DSU)   **114**
Cable modems   **115**
ISDN modems   **117**
DSL modems   **118**

Modem Pools   **118**

Interfacing a Computer to Modems and Other Devices   **119**

Data terminal equipment and data circuit-terminating equipment   **120**
Interface standards   **120**

Interfacing a Computer and a Peripheral   **127**

FireWire   **127**
Universal Serial Bus (USB)   **128**

Data Link Connections   **129**

Asynchronous connections   **129**
Synchronous connections   **131**
Half Duplex, Full Duplex, and Simplex Connections   **132**

Terminal-to-Mainframe Computer Connections   **133**

Making Computer Connections In Action   **135**

Summary   **137**
Key Terms   **138**
Review Questions   **139**
Exercises   **139**
Thinking Outside the Box   **140**
Projects   **141**

5   Multiplexing: Sharing a Medium   **143**

Introduction   **144**

Frequency Division Multiplexing   **144**

Time Division Multiplexing   **147**

Synchronous time division multiplexing   **148**
Statistical time division multiplexing   **154**

Dense Wavelength Division Multiplexing   **157**

Comparison of Multiplexing Techniques   **159**

Business Multiplexing In Action   **160**

Summary   **162**
Key Terms   **163**
Review Questions   **163**
Exercises   **163**
Thinking Outside the Box   **164**
Projects   **164**

6 Errors, Error Detection, and Error Control   165

Introduction   166
Noise and Errors   166
  White Noise   167
  Impulse noise   167
  Crosstalk   169
  Echo   169
Jitter   170
  Delay distortion   171
  Attenuation   171
Error Prevention   171
Error Detection Techniques   172
  Parity checks   173
  Cyclic redundancy checksum   175
Error Control   178
  Do nothing   178
  Return a message   178
  Correct the error   184
Error Detection and Error Control in Action   186
Summary   187
Key Terms   188
Review Questions   188
Exercises   189
Thinking Outside the Box   190
Projects   190

7 Local Area Networks: The Basics   191

Introduction   192
Functions of a Local Area Network   193
Advantages and Disadvantages of Local Area Networks   195
Basic Network Topologies   196
  Bus/tree topology   197
  Star-wired bus topology   201
  Ring topology   203
  Wireless LANs   206
  Comparison of bus, star-wired bus, ring and wireless topologies 208
Medium Access Control Protocols   209
  Contention-based protocols   209
  Round robin protocols   211
  Reservation protocols   214
Medium Access Control Sublayer   215
IEEE 802 Frame Formats   216
Local Area Network Systems   218
  Ethernet   218
  IBM token ring   220

Fiber data distributed interface (FDDI) 220
100VG-AnyLAN 221
LANs in Action: A Small Office Solution 223
LANs in Action: A Home Office Local Area Network Solution 225
Summary 226
Key Terms 227
Review Questions 228
Exercises 228
Thinking Outside the Box 229
Projects 229

8 **Local Area Networks: Internetworking** 231
Introduction 232
Why Interconnect Local Area Networks? 232
Bridges 234
Transparent bridge 236
Source-routing bridges 240
Remote bridges 243
Hubs 244
Switches 244
Isolating traffic patterns and providing multiple access 247
Full duplex switches 250
Network Servers 251
Routers 252
LAN Internetworking in Action: A Small Office Revisited 254
Summary 256
Key Terms 258
Review Questions 258
Exercises 258
Thinking Outside the Box 260
Projects 260

9 **Local Area Networks: Software and Support Systems** 261
Introduction 262
Network Operating Systems 262
Client/server systems 264
Current Network Operating Systems 266
Novell NetWare 267
Microsoft Windows NT 271
Unix 273
Linux 274
IBM's OS/2 275
Summary of network operating systems 276

Network Software: Utilities, Tools, and Applications   278
    Utilities   278
    Internet server software   279
    Programming tools   280
    Application software   280
Software Licensing Agreements   281
LAN Support Devices   283
LAN Software in Action: A University Makes a Choice   284
    Primary uses of current network   284
    Changes due to Microsoft Office suite   285
    Network support staff training   285
    Cost of NT   285
Summary   286
Key Terms   286
Review Questions   287
Exercises   287
Thinking Outside the Box   288
Projects   288

10 Introduction to Wide Area Networks   289
Introduction   290
Wide Area Network Basics   291
    Types of network subnets   293
    Connection-oriented versus connectionless network applications   295
    Combinations of network applications with subnet types   297
Routing   298
    Dijkstra's least cost algorithm   301
    Flooding   302
    Centralized routing   305
    Distributed routing   306
    Isolated routing   307
    Adaptive routing versus static routing   308
    Routing examples   308
Network Congestion   309
    Preventing network congestion   309
    Handling network congestion   309
WANs in Action: Making Internet Connections   311
    A home-to-Internet connection   311
    A work-to-Internet connection   312
Summary   313
Key Terms   314
Review Questions   314
Exercises   315
Thinking Outside the Box   316
Projects   316

11 The Internet   **317**

  Introduction   318

  Internet Services   319
        File transfer protocol (FTP)   319
        Remote login (Telnet)   320
        Internet telephony   321
        Electronic mail   322
        Listservs   324
        Usenet   324
        Streaming audio and video   325

  The World Wide Web   325
        Creating web pages   327
        e-Commerce   331
        Cookies and State Information   332

  Intranets and Extranets   332

  Internet Protocols   333
        The Internet Protocol (IP)   334
        The TCP Protocol   338
        Internet Control Message Protocol (ICMP)   339
        User Datagram Protocol (UDP)   340
        Address Resolution Protocol (ARP)   340
        Tunneling protocols   341
        Locating a document on the Internet   341

  The Future of the Internet   345
        IPv6   347
        Internet2   348

  The Internet in Action: A Company Creates a VPN   349

        Summary   351
        Key Terms   352
        Review Questions   353
        Exercises   353
        Thinking Outside the Box   355
        Projects   355

12 Telecommunication Systems   **357**

  Introduction   358

  Basic Telephone Systems   358
        Limitations of telephone signals   359
        Telephone lines, trunks, and numbers   360
        The telephone network before and after 1984   361
        Telephone networks after 1996   363

  Leased Line Services   364

  Integrated Services Digital Network (ISDN)   365

  Frame Relay   366
        Frame relay setup   368
        Committed Information Rate (CIR)   370

Frame relay vs. the Internet   371
Voice over Frame Relay (VoFR)   371
Frame relay switched virtual circuits   372

Asynchronous Transfer Mode (ATM)   373
ATM classes of service   374
Advantages and disadvantages of ATM   376

Digital Subscriber Line   376
DSL basics   377
DSL formats   378
Comparison of DSL to ISDN, Frame Relay, ATM, and cable modems   379

Computer Telephony Integration   380

Telecommunications Systems in Action – A Company Makes a Service Choice   382
Current data network   382
New data network   383
New data applications   383
Prices   384
Making the choice   384

Summary   386
Key Terms   388
Review Questions   388
Exercises   389
Thinking Outside the Box   390
Projects   390

13   Network Security   393

Introduction   394

Basic Security Measures   394
External security   395
Operational security   395
Surveillance   396
Passwords and ID systems   397
Auditing   399
Access rights   400
Guarding against viruses   401

Standard System Attacks   402

Basic Encryption and Decryption Techniques   404
Monoalphabetic substitution-based ciphers   405
Polyalphabetic substitution-based ciphers   405
Transposition-based ciphers   406
Public key cryptography and secure sockets layer   407
Data Encryption Standard   408

Digital signatures 409
Pretty Good Privacy (PGP) 410
Kerberos 410
Public Key Infrastructure 411
Firewalls 413
Firewall efficacy 414
Basic firewall types 415
Security Policy Design Issues 416
Network Security in Action: Banking and PKI 418
Summary 419
Key Terms 420
Review Questions 420
Exercises 421
Thinking Outside the Box 421
Projects 422

14 Network Design and Management **423**
Introduction 424
Systems Development Life Cycle 424
Planning phase 427
Analysis phase 427
Design phase 428
Implementation phase 428
Maintenance phase 428
Network Modeling 429
Feasibility Studies 431
Capacity Planning 434
Creating a Baseline 437
Network Manager Skills 439
Generating Useable Statistics 441
Managing Operations 443
Simple Network Management Protocol (SNMP) 443
Network Diagnostic Tools 445
Tools that test and debug network hardware 445
Tools that analyze data transmitted over the network 446
Capacity Planning and Network Design in Action: BringBring Corporation 446
Summary 449
Key Terms 450
Review Questions 450
Exercises 451
Thinking Outside the Box 452
Projects 452

Appendix: Pioneering Protocols   453
BISYNC Transmission   454
Transparency   456
Synchronous Data Link Control (SDLC)   457
High Level Data Link Control (HDLC)   462
Public data networks and X.25   463

Glossary   467
Index   482

Preface

Data Communications and Computer Networks: A Business User's Approach is intended for a one-semester course in *business data communications* for students majoring in business, information systems, management information systems, and other applied fields of computer science. It is a readable resource for computer network users that draws on examples as they might occur in business environments.

As a computer network user, more than likely you will not be the one who designs, installs, and maintains the network. Instead, you will have interactions, either direct or indirect, with individuals who do. This book will help you get the most out of those interactions with network personnel.

In a generic sense, this book provides an owner's manual for the individual computer user. In a world in which computer networks are involved in nearly every facet of business and personal life, it is paramount that each of us understands the basic features, operations, and limitations of different types of computer networks. This understanding will make us better managers, better employees, and simply better computer users. Reading this book should give you a strong foundation that will enable you to work effectively with network administrators, network installers, and network designers.

Here are some of the many scenarios in which the knowledge contained in this book would be particularly useful:

▶ You work for a company and must deal directly with a network specialist. To better understand the specialist and conduct a meaningful dialog with him or her, you need a basic understanding of the many aspects of computer networks.

▶ You are in management within a company and are involved with one or more network specialists who provide you with network recommendations. You do not want to find yourself in a situation in which you blindly accept the recommendations of network professionals. So that you may make intelligent decisions regarding network resources, you need to know the basic concepts of data communications and computer networks.

▶ You work in a small company, and each employee wears many hats. Thus, you may need to perform some level of network assessment, administration, or support.

▶ You have your own business and need to fully understand the advantages of using computer networks to support your operations. To optimize those advantages, you should have a good grasp of the basic characteristics of a computer network.

▶ You have a computer at home or at work and you simply wish to learn more about computer networks. You have realized that to keep your job skills current and remain a key player in the information technology arena, you must understand how different computer networks work and their advantages and shortcomings.

In a university setting, this book can be used at practically any level above the first year. Instructors who wish to use this book at the graduate level can draw on the many advanced projects at the end of each chapter to create a more challenging environment for the advanced student.

Defining Characteristics of This Book

The major theme employed throughout this book is to introduce readers to the next level of details found within the fields of computer networks and data communications. This increased level of detail includes the network technology and standards underlying the applications and the computer and network systems that support them. This level of detail is more than just an introduction to terminology. It introduces the concepts necessary for a more in-depth understanding. Once this in-depth understanding is attained, the often complex topic of data communications becomes less intimidating. To facilitate this understanding, the book incorporates three major concepts: readability, a balance between the technical and the practical, and currency.

Readability

Great care has been taken to provide the often technical material in as readable a fashion as possible. Terminology, while unavoidable, is presented in a clear fashion with a minimal use of acronyms and an even lesser use of computer jargon.

Balance between the technical and the practical

A major goal of *Data Communications and Computer Networks* is to achieve a good balance between the more technical aspects of data communications and the everyday practical aspects. Throughout each chapter, sections entitled Details delve into the more specialized aspects of the topic at hand. If a reader is not interested in this technical information, the Details sections can be bypassed.

Current technology

Every attempt has been made to present the most current trends in data communications and computer networks. Some of these topics include:

- PCS wireless systems
- Dense wavelength division multiplexing
- Switching in local area networks
- 56K modems
- Cable modems
- Frame relay and asynchronous transfer mode

- ► Current LAN network operating systems (Novell NetWare 5 and Windows 2000)
- ► Dynamic HTML, XML, and IPv6
- ► Public key infrastructure and digital signatures
- ► Local multipoint distribution service

It is also important to remember the many older technologies still in prevalent use today. Discussions of older technologies in wide use today can be found, when appropriate, in each chapter. In the case of technologies that have diminished in use over the years, but still provide a good historical base for more modern technologies, information has been removed from the main body of the book and is provided, instead, in an appendix at the end.

Since it is impossible for a textbook to keep abreast of the quickly changing field of computer networks, the author has provided a series of web pages *(http://bach.cs.depaul.edu/cwhite)* that feature some of the more recent developments in each major area of technology.

Organization

The organization of *Data Communications and Computer Networks* roughly follows the OSI model from the physical layer to the upper layers. It has been carefully designed to consist of 14 chapters in order to fit well into a typical 15- or 16-week semester along with any required exams. The intent was to create a set of chapters each of which provides a balanced introduction to the study of computer networks. Each instructor may choose to emphasize or de-emphasize certain topics, allowing for flexibility in the week-to-week curriculum.

Chapter One, Introduction to Computer Networks and Data Communications, introduces the many types of connections found within computer networks. This chapter introduces many of the major concepts that will be discussed in the following chapters.

Chapters Two, Fundamentals of Data and Signals, and Three, The Media - Conducted and Wireless, cover basic concepts that are critical to the proper understanding of all computer networks and data communications. These chapters, in particular, establish the theme that is carried throughout the rest of the textbook—introducing the reader to the next level of details about the technology and standards underlying the applications and systems that support the applications.

Chapters Four, Connecting to the World-Interfacing and Modems, Five, Multiplexing-Sharing a Medium, and Six, Errors, Error Detection and Error Control, cover the basic technologies necessary to connect computers and networks to one another. Chapter Four discusses how a connection or interface is created between two communicating devices, such as a computer and a modem. Chapter Five describes how two or more devices can share a single medium. Chapter Six explains the actions that can take place when a data transmission produces an error.

Chapters Seven, Local Area Networks - The Basics, Eight, Local Area Networks – Internetworking, and Nine, Local Area Networks - The Software, cover the very popular topic of local area networks. Entire chapters are devoted to the basic concepts, how local area networks connect internally and to one another, and the software that supports them.

Chapters Ten, Introduction to Wide Area Networks, Eleven, The Internet, and Twelve, Telecommunication Networks, cover the many topic areas of wide area networks. Once upon a time a student studied either data communications or voice communications. With the merging of these two technologies, a student who does not understand something from both areas of study will suffer when placed into the modern work environment. Therefore, Chapter Ten provides a gentle introduction to the overall topic of wide area networks. Chapter Eleven delves into the details of the Internet including TCP, IP, DNS, and the World Wide Web. Chapter Twelve provides a detailed introduction to the area of telecommunications.

Chapter Thirteen, Network Security, covers the current trends of security, including digital signatures and public key infrastructures. Chapter Fourteen, Network Design and Management, the final chapter, introduces the systems development life cycle, feasibility studies, capacity planning, and baseline studies, and shows how they apply to the analysis and design of computer networks.

Finally, the appendix, Pioneering Protocols, introduces a number of older technologies (BISYNC, HDLC, SDLC, and X.25) that can be considered non-essential to the business understanding of data communications and computer networks. These older protocols, however, provide a good historical background for and understanding of many modern protocols that are based on them.

Features

To assist readers in better understanding the technical nature of data communications and computer networks, a number of significant features have been added to each chapter. These features include both older, well-tested pedagogical techniques as well as some newer techniques.

Opening Case

Each chapter begins with a short case or vignette that emphasizes the main concept of the chapter and sets the stage for exploration. These cases are designed to spark readers' interest and create a desire to learn more about the chapter's concepts.

Learning Objectives

Following the opening case is a list of learning objectives that should be accomplished by the end of the chapter. Each objective is tied to the main sections of the chapter. Readers can use the objectives to gain a grasp of the scope and intent of the chapter. The objectives also work in conjunction with the end of chapter summary and review questions, so that readers can assess whether they have adequately mastered the material.

Details

Many chapters contain one or more Details sections, which dig deeper into a particular topic. Readers more interested in these technical details will find the sections valuable. Since these sections are physically separate from the main text, if you are not interested in the technical details, the sections may be omitted. Omitting these sections will not affect your overall understanding of the chapter's material.

In Action

At the end of the chapter's main content presentation is an "In Action" example that demonstrates the chapter's key topic in a realistic environment. Although a number of "In Action" examples include imaginary persons and places, every attempt was made to make the examples as close as possible to what would be found in a business or home environment. Thus, the "In Action" example helps the reader visualize the concepts presented in the chapter.

End-of-Chapter Material

The end-of-chapter material is designed to help readers review the content of the chapter and assess whether they have adequately mastered the concepts. It includes:

- A bulleted summary that readers can use as a review of the key topics of the chapter and as a study guide
- A list of the key terms used within the chapter
- A list of review questions that readers can use to quickly check if they understand the chapter's key components
- A set of exercises that draw on the material presented in the chapter
- A set of more in depth "Thinking Outside the Box" exercises, which require readers to consider possible alternatives as well as their advantages and disadvantages
- A set of projects that require readers to reach beyond the material found within the text and use outside resources to compose a response

Glossary

At the end of the book you will find a glossary that includes the key terms from each chapter.

CD-ROM

An interactive CD-ROM containing a set of visual demonstrations of many key data communications and networking concepts accompanies the text. The text passages that correspond to these interactive animations are called out as such in the book. There is a visual demonstration for each of the following concepts:

- ► Chapter One: Introduction to Computer Networks and Data Communications – Layer encapsulation example
- ► Chapter Two: Fundamentals of Data and Signals – dB loss and gain example
- ► Chapter Four: Interfacing and Modems – RS232 example of two modems establishing a connection
- ► Chapter Five: Multiplexing – Packets from multiple sources coming together for synchronous TDM and a second example demonstrating statistical TDM
- ► Chapter Six: Errors, Error Detection, and Error Control – Sliding window example using ARQ error control
- ► Chapter Seven: Local Area Networks (The Basics) – CSMA/CD example with workstations sending packets and collisions happening; a second example demonstrating a token ring local area network with token and packets circling the ring
- ► Chapter Eight: Local Area Networks (Internetworking) – Two LANs with a bridge showing how bridge tables are created and packets are routed. A second example showing one LAN with a switch in place of a hub
- ► Chapter Ten: Introduction to Wide Area Networks – Datagram network sending individual packets and virtual circuit network first creating a connection then sending packets down a prescribed path.
- ► Chapter Eleven: The Internet – Domain Name System as it tries to find the dotted decimal notation for a given URL.

Teaching Supplements

The Instructor's Resource Kit, which can be requested separately from Course Technology, contains an electronic Instructor's Manual, Course Test Manager Test Engine and Test Bank, and PowerPoint presentations with graphics from the book. The Instructor's Manual includes the following information for each chapter:

- ▶ Learning objectives
- ▶ Lecture notes - a summary of all major headings and their subheadings
- ▶ Quick Quiz - A short set of questions that may be used to help determine a reader's understanding of the material
- ▶ Solutions to the exercises at the end of the chapter
- ▶ Discussion topics - A set of discussion questions that may be used to stimulate an interactive dialogue in class.

Finally, the author's Web pages (*http://bach.cs.depaul.edu/cwhite*) contain recent information on topics that were introduced in the text. Since the material in this field evolves so quickly, the author's web pages can be used to keep the reader up to date.

Acknowledgments

Producing a textbook requires the skills and dedication of many people. Unfortunately, the final product displays only the author's name on the cover, and not the names of those who provided countless hours of input and professional advice. I would first like to thank the people at Course Technology for being so vitally supportive and one of the best teams an author could hope to work with: Jennifer Normandin, Managing Editor; Amanda Young, Associate Product Manager; and Christine Spillett, Senior Production Editor. Another person who is amazingly vital to the production of a textbook is the development editor. Once again I was lucky and was awarded the opportunity to work with one of the best: Marilyn Freedman. Her support and confidence in me kept me moving forward no matter how large the number of edits necessary to turn a rough draft into a polished chapter.

I would also like to thank the reviewers who undoubtedly spent many hours reading rough drafts and providing their valuable technical expertise and insights:

Azad Azadmanesh, University of Nebraska at Omaha

France Belanger, Virginia Tech

Richard Darnell, Salt Lake Community College

Badie Farah, Eastern Michigan University

Someswar Kesh, Central Missouri State University

Judy Scheeren, Westmoreland County Community College

Claude Simpson, Northwestern State University

I must also thank a number of colleagues at DePaul University who listened to my problems, provided ideas for exercises, proofread some of the technical chapters, and provided many fresh ideas when I could think of none myself: Greg Brewster, Jacob Furst, Ruth Ter Bush, and Rosalee Wolfe.

Finally I thank my family: my wife Kathleen, my daughter Hannah, and my son Samuel. It was your love and support (and letting me use the computer) that kept me going, day after day, week after week, and month after month.

1

Introduction to Computer Networks and Data Communications

MAKING PREDICTIONS is a difficult task, and predicting the future of computing is no exception. History is filled with computer-related predictions that were so inaccurate that today they are amusing. For example, consider the following predictions:

> "I think there is a world market for maybe five computers."
> *Thomas Watson, chairman of IBM, 1943*

> "There is no reason anyone would want a computer in their home." *Ken Olson, president and founder of Digital Equipment Corporation, 1977*

> "640K ought to be enough for anybody." *Bill Gates, chairman and CEO of Microsoft, 1981*

> "I have traveled the length and breadth of this country, and talked with the best people, and I can assure you that data processing is a fad that won't last out the year."
> *Editor in charge of business books for Prentice Hall, 1957*

Apparently, it doesn't matter how famous you are or how influential your position, it is very easy to make very bad predictions. It is hard to imagine that anyone can make a prediction that is worse than any of those above. Buoyed by this false sense of optimism, let's make a few predictions of our own:

Some day we will wear a computer-like a suit of clothes, and when we shake hands with a person, data will transfer down our skin, across our shaking hands, and into the other person's computer.

To check out at the grocery, one day we will simply pass an entire cart of groceries by a sensor, and in a single moment, every item in the cart will be rung up. Our account will be debited, and we will be out the door without stopping the cart.

One day, you'll wake up in the morning and step on the bathroom scale. When you walk into the kitchen, the kiosk on the refrigerator will read "I see you are still 5 pounds overweight. Why don't you have a grapefruit and dry toast for breakfast this morning? Since you are low on grapefruit, I've ordered some for you. They will be delivered this afternoon. Signed, *your family physician.*"

Do these predictions sound far-fetched and filled with mysterious technologies that only scientists and engineers can understand? They shouldn't, because they aren't predictions. They are happening today. But they would not be happening today without computer networks and data communications.

Objectives

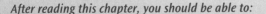

After reading this chapter, you should be able to:

▶ Define the basic terminology of computer networks.

▶ Recognize the individual components of the big picture of computer networks.

▶ Outline the basic network configurations.

▶ Cite the reasons for using a network model and how it applies to current network systems.

▶ List the layers of the OSI model and describe the duties of each layer.

▶ List the layers of the Internet model and describe the duties of each layer.

▶ Compare the OSI and Internet models and list their differences and similarities.

▐ Introduction ▶

The world of computer networks and data communications is a surprisingly vast and increasingly significant field of study. Once considered primarily the domain of communications engineers and technicians, computer networks now involve business managers, computer programmers, system designers, office managers, home computer users, and everyday citizens. It is virtually impossible for the average person to spend 24 hours without directly or indirectly using some form of computer network.

Ask any group "Who today has used a computer network?" and roughly one-half of the people may answer "yes". Then ask the people who did not answer yes, "How did you get to work, school, or the store today if you did not use a computer network?" Most transportation systems use extensive communication networks to monitor the flow of vehicles and trains. Expressways and highways have computerized systems for controlling traffic signals and limiting access during peak traffic times. Some major cities are considering placing a satellite dish atop each city bus so that the precise location of each bus is known. With this information, it would be possible to keep the buses more evenly spaced and on time.

If you become lost while driving, there is even a satellite system that will tell you precisely where your automobile is and give you directions, will unlock your car doors if you leave your keys in the ignition, and will locate your car in a crowded parking lot by beeping the horn and flashing the headlights.

Banking relies heavily on computer networks and data communications. Home banking using a telephone or a personal computer and a modem continues to grow in popularity. Twenty-four hour automatic teller machines are everywhere, and most allow transfers, inquiries, and deposits, as well as cash withdrawals.

Grocery stores, with laser scanning and barcode pricing, can record every purchase made. With grocery stores now requiring customers to present a store ID card to receive a sale price, each item purchased can be linked to a particular customer. The computer and data communications systems involved allow the grocery store to track the buying habits of each individual.

Cable television continues to expand, offering extensive programming, pay-per-view options, and in some markets, multi-megabit connectivity to the Internet. The telephone system, the oldest and most extensive network of communicating devices, continues to become more of a computer network every day. Cellular telephone systems cover virtually every major metropolitan area, including one system that allows users to be anywhere in the world and still place or receive telephone calls. Pager systems have evolved to the point that they can receive text messages and some can even return a reply.

Welcome to the amazing world of computer networks! Unless you have spent the last 24 hours in complete isolation, it is near impossible not to use some form of a computer network and data communications. Because of this growing integration of computer networks and data communications into business and life, we cannot leave this area of study to technicians. All of us—particularly information systems, business, and computer science students—need to understand the basic concepts.

The Language of Computer Networks

Over the years, numerous terms and definitions relating to computer networks and data communications have emerged. To gain insight into the many subfields of study, and to better introduce the emphasis of this textbook, let's examine the more common terms and their definitions.

A **computer network** is an interconnection of computers and computing equipment using either wires or radio waves over small or large geographic areas. Computer networks that use radio waves are termed **wireless** and can involve broadcast radio (AM and FM), microwaves, or satellite transmissions. Networks that are small in geographic size—spanning a room, a building, or a campus—are **local area networks.** Networks that serve an area of 3 to 30 miles—approximately the area of a typical city—are called **metropolitan area networks.** Large networks encompassing parts of states, multiple states, countries, and the world are **wide area networks.** Chapters Seven, Eight and Nine concentrate on local area networks, and Chapters Ten and Eleven concentrate on wide area networks. Metropolitan area networks, which are rare, contain components from both local area networks and wide area networks and will not be discussed in detail.

Data communications is the transfer of digital or analog data using digital or analog signals. The study of data communications usually begins with the introduction of two very important building blocks: data and signals. Data is information that has been translated into a form that is more conducive to storage, transmission, and calculation. As we shall see in Chapter Two, a signal is used to transmit data. Both the data and the signal can be analog or digital, allowing for four possible combinations. Analog data transmitted by analog signals, and digital data transmitted by digital signals are fairly straightforward. Digital data transmitted by analog signals requires the digital data to be modulated onto an analog signal, such as happens with a modem and the telephone system. Modems will be discussed in detail in Chapter Four. Analog data transmitted by digital signals requires the data to be sampled at specific intervals and then digitized into a digital signal, such as happens with a digitizer, or **codec.** The analog and digital signals are then transmitted over conducted media or wireless media (discussed in Chapter Three).

Usually, only one signal can be transmitted over a medium at one time. For a medium to transmit multiple signals simultaneously, something must alter the signals so that they do not interfere with one another. Transmitting multiple signals on one medium is **multiplexing,** and is covered in detail in Chapter Five.

Transmitting data and signals between a sender and a receiver, or between a computer and a modem requires interfacing, which is covered in Chapter Four. When the signals transmitted between computing devices are corrupted and errors result, error detection and error control are necessary. These topics are discussed in detail in Chapter Six.

A **voice network** transmits telephone signals, and a **data network** transmits computer data. The differences between data networks and voice networks are slowly beginning to disappear. Networks that were designed primarily for voice now carry data, and networks that were designed to carry data are now transmitting voice in real-time. Some experts (but not all) predict that one day there will be no distinction and that one network will efficiently and effectively carry all types of traffic.

Throughout this text, most of our time and energy will concentrate on computer networks and data communications with a business emphasis. However, to receive a well-rounded introduction to this exciting field of study, the text will also introduce telecommunications, network design, and network management.

Telecommunications is the study of telephones and the systems that transmit telephone signals. Although this book deals primarily with computer networks and data communications, Chapter Twelve is devoted to introducing several of the more recent systems that transmit data and voice signals over high speed telephone networks, such as asynchronous transfer mode (ATM), frame relay, synchronous optical network (SONET), integrated services digital network (ISDN), and digital subscriber line (DSL).

Network management is the design, installation, and support of a network and its hardware and software. Chapter Fourteen discusses many of the basic concepts necessary to properly support the design and improvement of network hardware and software, as well as the more common management techniques used to support a network.

The Big Picture of Networks

If you could create a picture that gives an overview of all computer networks, what might this one picture include? Figure 1-1 shows such a picture, and it includes examples of both local and wide area networks. The picture shows a number of different types of local area networks (LAN1 and LAN2). Although a full description of the different types of local area networks is not necessary at this time, most include the following hardware:

> ▶ **workstations**, which are personal computers or computer terminals, at which users reside;
> ▶ **servers**, which are the computers that store the network software and shared or private user files;
> ▶ **bridges**, which are the connecting devices between separate local area networks;
> ▶ **routers**, which are the connecting devices between local area networks and wide area networks; and
> ▶ **hubs**, which are the collection points for the wires that interconnect the workstations.

Switch →

Wide area networks can also be of many types. Although many different technologies are used to support wide area networks, all wide area networks include the following components:

> ▶ **nodes**, which are the computing devices that allow workstations to connect to the network and that make the decisions as to which route a piece of data will follow next;
> ▶ some type of **high speed telephone line**, which runs from one node to another; and
> ▶ **subnet**, which is the collection of nodes and telephone lines into a cohesive unit.

Figure 1-1

An overall view of the interconnection between local area networks and wide area networks

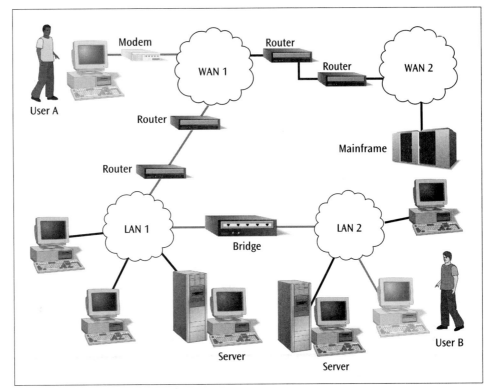

To see how the local area networks and wide area networks work together, consider User A (in the upper left corner) who wishes to send an e-mail message to User B (in the lower right corner). User A's computer requires both the necessary hardware and software to communicate with the first wide area network. Assuming that User A's computer is connected to the wide area network through a conventional telephone line, User A needs a modem. Furthermore, if the wide area network is part of the Internet, User A's computer requires software that talks the talk of the Internet: TCP/IP (Transmission Control Protocol / Internet Protocol).

Notice that there is no direct connection between WAN1, where User A resides, and LAN2, where User B resides. For User A's e-mail message to reach its intended receiver (User B), User A's software attaches the appropriate address information that WAN1 uses to route User A's message to the router that connects LAN1 to WAN1. Once the e-mail message is on LAN1, the bridge connecting LAN1 and LAN2 uses this address information to pass the message to LAN2. Further address information then routes User A's e-mail message to User B, whose software accepts and displays the message.

Under normal traffic and conditions, this procedure takes only a second or two. When you begin to understand all the steps involved and the great number of transformations that a simple e-mail message must undergo, the fact that it takes *only* a second or two to deliver an e-mail message is amazing.

Computer Networks – Basic Configurations

The beginning of this chapter described a few of the many application areas of data communications. Based on that sampling, you can see that setting out all the different types of jobs and services that use some sort of computer network and data communications would generate an enormous list. Instead, let's examine general network systems and configurations to see how extensive the uses of data communications and computer networks are. The general configurations include: terminal-to-mainframe computer configurations; microcomputer-to-mainframe computer configurations; microcomputer-to-local area network configurations; microcomputer-to-Internet configurations; local area network-to-local area network configurations; local area network-to-wide area network configurations; sensor-to-local area network systems; satellite and microwave configurations; and wireless telephone configurations.

Computer terminal-to-mainframe computer configurations

During the 1960's and 1970's, the terminal-to-mainframe connection was in virtually every office, manufacturing, and academic environment. A computer terminal is a device that is essentially a keyboard and screen with no long term storage capabilities. Computer terminals are used for entering data into a system, such as a mainframe computer, and then displaying results from the mainframe. Today many business systems still employ a terminal-to-mainframe connection, although the number of these systems in use is not what it used to be. These types of systems are used for inquiry/response applications, interactive applications, and data entry applications, such as you might find when applying for a new drivers license at the department of motor vehicles. Terminal-to-mainframe connection systems use "dumb" terminals because the end user is doing relatively simple data entry and retrieval operations, and a workstation with a lot of computing power is not necessary. Since the terminal does not possess a lot of computing power, the mainframe computer controls the sending and receiving of data to and from each terminal. This requires special types of protocols and the data is usually transmitted at relatively slow speeds, such as 9600 bits per second (bps). (Figure 1-2).

Figure 1-2
Using a terminal to perform a text-based input transaction

Microcomputer-to-mainframe computer configuration

As the microcomputer (or personal computer) began to emerge in the late 1970s and early 1980s, many of the same end users who had a terminal on their desks now also found a microcomputer on their desks (and thus had very little room for anything else). During this period, terminal-emulation cards were developed, which allowed a microcomputer to imitate the abilities of a computer terminal. As terminal-emulation cards were added to microcomputers, terminals were removed from the end users' desks and microcomputers began to serve both functions. Now, if users wished to access the mainframe computer, they could download information from the mainframe computer to their microcomputers, perform operations on the data, and then upload the information to the mainframe. The overall configuration is similar to Figure 1-2, except users are interacting with microcomputers instead of terminals.

Microcomputer-to-local area network configuration

Simplest
Type of Network we
2 Pc connected
together

Perhaps the most common network configuration today, the microcomputer-to-local area network (LAN) connection, is found in virtually every business and academic environment. The LAN, as we shall see in Chapter Seven, is an excellent tool for sharing software and peripherals. In many LANs, application software, such as word processing or spreadsheet software, resides on a central computer called a server. Using microcomputers connected to a LAN, end users can request and download an application, which then executes on their computers. If users wish to print documents on a high-quality network printer, the LAN contains the network software necessary to route their print requests to the appropriate printer. Figure 1-3 shows a diagram of this type of microcomputer-to-local area network configuration.

Figure 1-3

A microcomputer lab, showing the cabling that exits from the back of a computer and runs to a collection point of the LAN in the back of the room

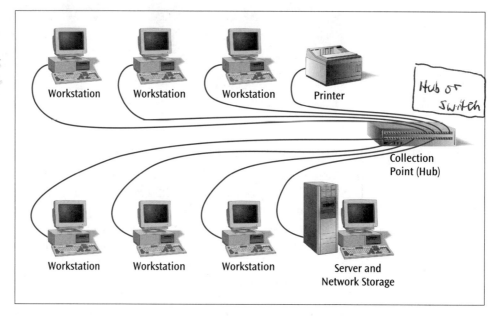

Hub or Switch

One very common form of microcomputer-to-local area network configuration in the business world is the client/server system. In a **client/server system**, a user at a microcomputer, or client machine, issues a request for some form of data or service. This request travels across the system to a server that contains a large repository of data and programs. The server fills the request and returns the results to the client, where the results are displayed on the client's monitor. Client/server systems are also found on microcomputer-to-mainframe configurations, in which the microcomputer is the client and the mainframe is the server.

Microcomputer-to-Internet configuration

With the explosive growth of the Internet, and the desire of users to dial into the Internet from home, the microcomputer-to-Internet connection is growing steadily. Most home users connect to the Internet using a modem and a dial-up telephone service, which currently limits data transfer to approximately 56,000 bits per second (56 Kbps). (Actually the connections don't achieve 56Kbps, but that is a mystery we will examine in Chapter Four.) Users who wish to connect at speeds higher than 56 Kbps use telecommunications services as ISDN or digital subscriber line, or access the Internet through cable television services. Many of these alternative telecommunication services will be examined in Chapter Twelve.

To communicate with the Internet using a dial-up modem, the user's computer must connect to another computer that is already communicating with the Internet. Two of the easiest ways to establish this connection are through the services of an Internet service provider (ISP) or by establishing an account with an information service, such as America Online. In either case, the user's computer requires software to communicate with the Internet. The Internet only "talks" TCP/IP, so users must use software that supports the TCP and IP protocols. Once the user's computer is talking TCP/IP, a connection to the Internet can be established. Figure 1-4 shows a typical microcomputer-to-Internet connection.

Figure 1-4
A microcomputer sending data over a telephone line to an Internet service provider and into the Internet

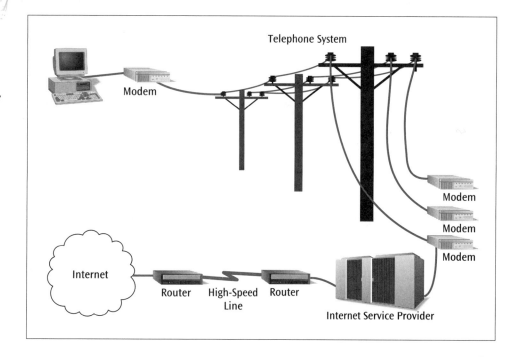

Local area network-to-local area network configuration

Since the local area network is a standard in business and academic environments, it should come as no surprise that many organizations need the services of multiple local area networks. It is possible to connect two local area networks so that they can share software and peripherals. The device that usually connects two or more LANs is the bridge. Sometimes it is more important to prevent data from flowing between networks than it is to allow data to flow from one network to another. The bridge can filter out traffic that is not intended for the neighboring network, thus minimizing the amount of traffic flowing. Figure 1-5 provides an example of two LANs connected by a bridge.

Figure 1-5
Two local area networks connected by a bridge

Isolate

Interconnect

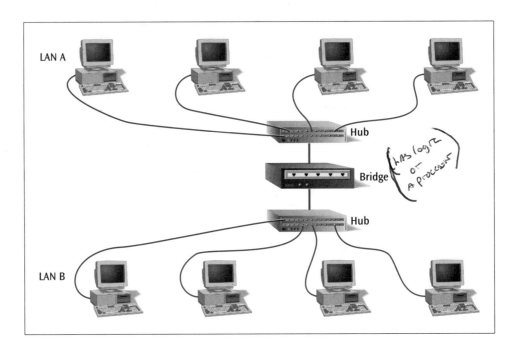

Local area network-to-wide area network configuration

You have already seen that the local area network is commonly found in business and academic environments. If a user is working at a microcomputer that is connected to a local area network, and the user wishes to access the Internet (a wide area network), the user's local area network has to have a connection to the Internet. A device called a router is employed to connect the two networks. Although bridges and routers both connect networks, as you shall see in Chapter Eight, they perform different functions. The bridge is a simpler device that can connect two similar local area networks. The router is a more elaborate device. One of the primary reasons a router is more elaborate is the simple fact that it is more difficult to convert the data from a local area network into the data bound for a wide area network. Since a bridge is converting the data on one local area network into the data on a similar local area network, less work is performed. Thus, the router has a bigger job to perform than the bridge does. Figure 1-6 shows a local area network connection to a wide area network via a router.

Figure 1-6
Local area network to a wide area network connection

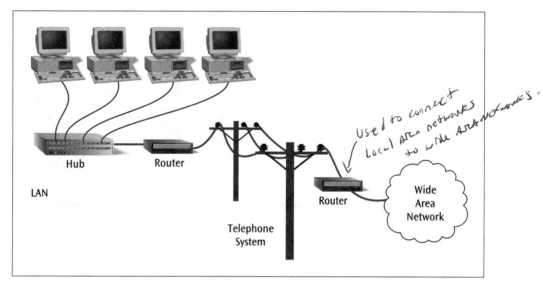

Sensor-to-local area network connection

In many network applications no human directly provides a data input, such as typing on the keyboard of a personal computer. Often, the action of a person or thing triggers a sensor—for example, a left turn light at a traffic intersection—that is connected to a network. In many left turn lanes, a separate left turn signal will appear if, and only if, there is one or more vehicles in the left turn lane. A sensor embedded in the roadway detects the movement of an automobile in the lane above and triggers the left turn mechanism in the traffic signal control box at the side of the road. If this traffic signal control box is connected to a larger traffic control system, the sensor is connected to a network.

Another example of sensor-to-local area network connections is found within manufacturing environments. Assembly lines, robotic control devices, oven temperature controls, and chemical analysis equipment often use sensors connected to data-gathering computers that control movements and operations, sound alarms, and compute experimental or quality control results. Figure 1-7 shows a diagram of a typical sensor-to-local area network connection in a manufacturing environment.

Figure 1-7
An automobile moves down an assembly line and triggers a sensor

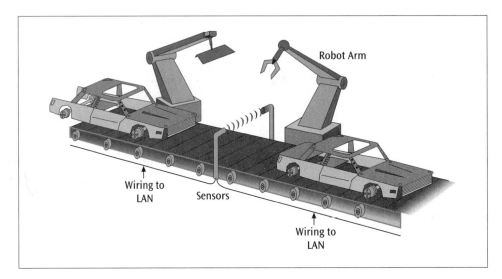

Satellite and microwave configurations

If the distance between two networks is great and it is difficult (if not impossible) to run a wire between the two networks, satellite and microwave transmission systems can be an extremely effective way to connect two networks or computer systems. Satellite and microwave configurations are continuously evolving technologies and are used in many applications. Examples of these applications include broadcast, cable, and direct TV; meteorology; intelligence operations; mobile maritime telephony; GPS-style surface navigation systems; e-mail, paging, and worldwide mobile telephone systems; and videoconferencing. Figure 1-8 shows a diagram of a typical satellite system.

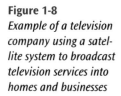

Figure 1-8
Example of a television company using a satellite system to broadcast television services into homes and businesses

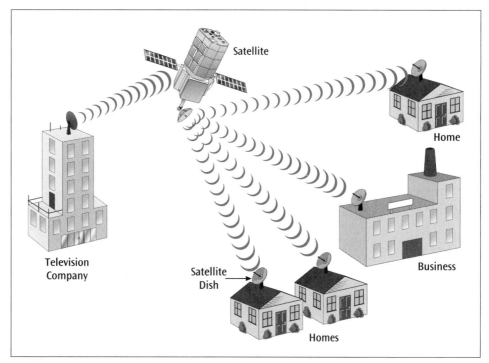

Wireless telephone configurations

One of the most explosive areas of growth in recent years has been in wireless telephone networks. The wireless telephone has become almost as commonplace as the pager, and newer wireless technologies that conduct telephone conversations with less background noise are joining the older services. Since the technology does not require wires, a wireless telephone configuration is only constrained by the size of a particular market. In addition to consumer wireless telephone services, police and emergency services, as well as business employees, use a portion of the wireless telephone frequencies for the transmission of data between laptop or mobile computers and a central computer system. One technique for the transmission of computer data using wireless telephone frequencies is called cellular digital packet data (CDPD). Figure 1-9 shows an example of a laptop computer connected to a wireless telephone system to transmit and receive data.

Figure 1-9
An example of a laptop computer connected to a wireless telephone system to transmit and receive data

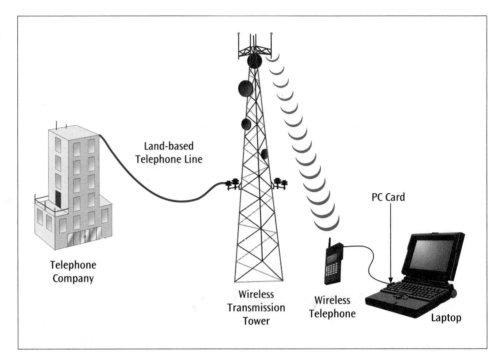

Network Architecture Models

The size and complexity of computer networks continues to grow at an amazing rate. The mere fact that a computer network contains so many different, yet intertwining, parts is proof of this level of complexity. Consider that a typical computer network within a business contains the following components:

▶ wires;

▶ printed circuit boards;

▶ wiring connectors and jacks;

▶ computers;

▶ centrally located wiring concentrators;

▶ backup power supplies;

▶ disk drives;

▶ tape drives;

▶ computer applications such as word processors, e-mail programs, accounting, marketing and electronic commerce software;

▶ computer programs to support the transfer of data;

▶ computer programs to check for errors when the data is transferred;

▶ computer programs that allow access to the network;

> ► computer programs that protect user transactions from unauthorized viewing; and

> ► computer programs that make sure any data that leaves a transmitting station eventually arrives at the receiving station.

This list could go on, but the point should be clear—when someone uses a computer network to perform an application, many pieces come together to assist in the operation.

Two big questions come from this large interaction of components. First, how do all of these pieces work together harmoniously? You don't want two pieces performing the same function, or no pieces performing a necessary function. Like a well-oiled machine, all components must work together to produce a product.

Second, does the choice of one piece depend on the choice of another piece? To make the pieces as modular as possible, you don't want the selection of one piece to affect the choice of another piece. For example, if you create a network and originally plan to use one type of wiring, but later change your mind and use a different type of wiring, will that change affect the choice of word processor? Such an interaction would seem highly unlikely. But can the choice of wiring affect the choice of the software program that checks for errors in the data sent over the wires? The answer to this question is not as obvious.

To keep the pieces working together harmoniously and to allow modularity between the pieces, it is necessary to use a **network architecture model**, or communications model, that places the appropriate network pieces in layers. Each layer in the model defines what services are provided by either hardware or software or both. For example, most organizations that produce some type of product or perform a service have a division of labor. Secretaries do the paperwork; accountants keep the books; laborers perform the manual duties; scientists design products; engineers test the products; managers control operations. Rarely is one person capable of performing all these duties. Large software applications operate the same way. Different procedures perform different tasks, and the whole would not function without the proper operation of each of its parts. Communications software is no exception. As the size of the applications grows, the need for a division of labor becomes increasingly important.

A communications application such as an e-mail system that accepts the message "Sam, how about lunch? Kathleen." has many parts. To begin, the e-mail "application worker" prompts the user to enter a message and an intended receiver. The application worker creates the appropriate data package with message contents and addresses and sends it to a "presentation worker." The presentation worker examines the data package to determine whether something like encryption or data compression is required and, if so, performs the necessary functions. The modified data package is then passed to the "session worker" that is responsible for adding backup synchronization points in case of network failures. A backup synchronization point is similar to leaving a bookmark in a book as you read ahead. If the book should fall out of your hands and you lose the page you were reading, the bookmark will show you where you were a few pages ago and allow you to catch up to the current page.

Next, the updated data package goes to a "transport worker" that is responsible for providing overall transport integrity. The transport worker may establish a connection with the intended receiver, monitor the flow between sender and receiver, and perform the necessary operations to recover lost data in the event of some data disappearing or becoming unreadable.

The "network worker" takes the data package next and may add routing information so that the data package can find its way through the network. Next to get the data package is the "data link worker" that inserts error checking information and prepares the data package for transmission. The final worker is the "physical worker," which transmits the data package over some form of wire or through the air using radio waves.

Each worker has its own job function. Figure 1-10 shows how the workers work together to create a single package for transmission.

Figure 1-10
The network workers performing their job duties at each layer in the model

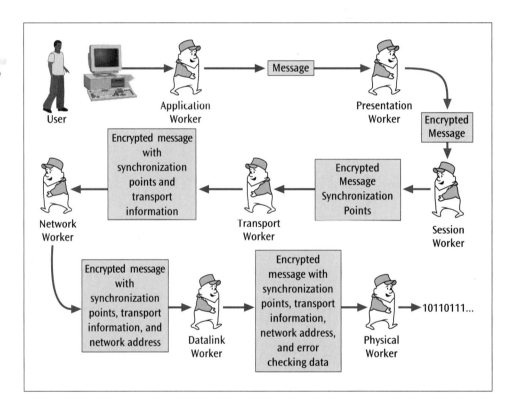

The Open Systems Interconnection (OSI) model

The scheme shown in Figure 1-10 is basically the architectural model adopted by the International Standards Organization (ISO) when its members created the **Open Systems Interconnection (OSI)** reference model. As shown in Figure 1-11, the model consists of seven layers. Note that the layers do not specify precise protocols or exact services, but rather define a *model* for the functions that need to be performed. Note also that each layer provides a service for the next layer. For example, the network layer is concerned with finding the best path for the data from one point to the next within the network, but relies on the transport layer to make sure the data received at the very end is exactly the same as the data originally transmitted. With each layer performing its designated function, the layers work together to allow an application to send its data over a network of computers.

Figure 1-11
The seven layers of the OSI model

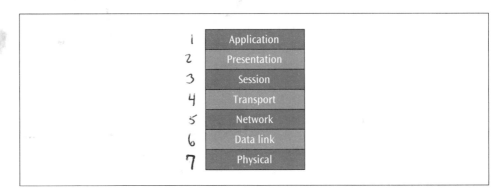

The **physical layer** handles the transmission of bits over a communications channel. To perform this transmission of bits, the physical layer handles voltage levels, plug and connector dimensions, pin configurations, and other electrical and mechanical issues. The choice of wire or wireless transmission media is usually determined at the physical layer. Any encoding or modulation technique is also determined at the physical layer because the digital or analog data is encoded or modulated onto a digital or analog signal at this point in the process.

The **data link layer** is responsible for taking the raw data and transforming it into a cohesive unit called a **frame**. This frame contains an identifier that signals the beginning and ending of the frame, as well as spaces for control information and address information. The address information identifies a particular workstation in a line of multiple workstations. The data link layer can also incorporate some form of error detection software. If there is an error, the data link layer is responsible for error control, informing the sender of the error. The data link layer must also perform flow control. In a large network in which the data hops from node to node as it makes its way across the network, flow control makes sure that one node does not overwhelm the next node with too much data.

The **network layer** is responsible for creating, maintaining, and ending network connections. As this layer sends the package of data from node to node within a network and between multiple networks, network addressing is necessary for the system to recognize the next intended receiver. To determine a path through the network, routing information is determined and applied to each packet or group of packets. The network layer also performs congestion control, which ensures that the network does not become saturated at any one point. In networks that use a broadcast distribution scheme, in which the transmitted data is sent to all other stations, the network layer may be very simple.

The first three layers of the OSI model deal with passing the data packet from node to node in the network or system. The fourth layer, the **transport layer**, is the first one concerned with an error-free, end-to-end flow of data. Thus, the transport layer performs end-to-end error and flow control. If the underlying network experiences problems such as reset or restart conditions, the transport layer will try to recover from the error and return the end-to-end connection to a known safe state. To ensure that the data arrives error-free at the final destination, the transport layer must be able to work across all kinds of reliable and unreliable networks. The transport layer is large and important, and acts as a manager who must cover for both reliable and unreliable workers.

The **session layer** is responsible for establishing sessions between users and for handling the service of **token management**, which controls who talks when during the current session by passing a software token back and forth. Additionally, the session layer establishes **synchronization points**, which are backup points in case of errors or failures. For example, while transmitting a large document such as an electronic book, the session layer may insert a synchronization point at the end of each chapter. If an error occurs during transmission, both sender and receiver can back up to the last synchronization point (for example, the beginning of a chapter) and start retransmission from there. The significance of the session layer has diminished greatly over the years. Most network applications do not use tokens to manage a conversation. If they do, the "token" is inserted by the application layer, not the session layer. Likewise, synchronization points, if used, are inserted by the application layer. For these reasons, most network applications today do not have a session layer.

Details ▶

The Internet's Request for Comment (RFC)

Network models, like communication protocols, computer hardware, and application software, continue to evolve daily. The Internet model is a good example of a large set of protocols and standards that is constantly being revised and improved. An **Internet standard** is a tested specification that is both useful and adhered to by users who work with the Internet. Let's examine the path a proposal must follow on the way to becoming an Internet standard.

All Internet standards start as an **Internet draft**, which is a preliminary work in progress. One or more internal Internet committees work on a draft, improving it until it is in an acceptable form. When the Internet authorities feel the draft is ready for the public, it is published as a **Request for Comment** (RFC), which opens the document to all interested parties. The RFC is assigned a number and it enters its first phase: **proposed standard**. A proposed standard is a proposal that is stable, of interest to the Internet community, and fairly well understood. The specification is tested and implemented by a number of different groups and the results are published. If the proposal passes at least two independent and interoperable implementations, the proposed standard is elevated to **draft standard**. After any feedback from test implementations are taken into account, and if the draft standard experiences no further problems, the proposal is elevated to Internet standard. At any point along the way, if the proposed standard is deemed inappropriate, it becomes an **historic RFC** and is kept

for historical perspective. Internet standards that are replaced or superseded also become historic.

An RFC can also be categorized as experimental or informational. In either case, the RFC is not meant to be an Internet standard, but was created for either experimental reasons or to provide information. Figure 1-12 shows the levels of progression for an RFC.

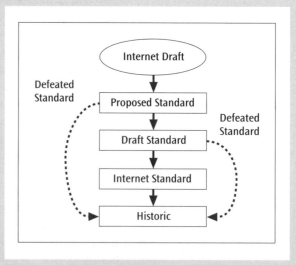

Figure 1-12 *Levels of progression as an RFC moves toward becoming a standard*

The **presentation layer** performs a series of miscellaneous functions necessary for presenting the data package properly to the sender or receiver. For example, the presentation layer can perform ASCII to non-ASCII character conversions, encryption and decryption of secure documents, and the compression of data into smaller units.

The final layer is the **application layer**, which is where the application using the network resides. While many kinds of applications use computer networks, certain ones are in widespread use and thus have been standardized. Applications such as electronic mail, file transfer systems, remote login systems, and directory services have been standardized, or at least attempts have been made to standardize them.

The Internet model

Many experts fully expected the OSI model to be the most commonly used model for designing and creating communications systems and software. However, this has not been the case. The **Internet model**, which incorporates TCP/IP protocols, has surpassed the OSI model in popularity and implementation. The Internet, which incorporates the Internet model, began in the 1960s as a creation of

It is possible to get a printed listing of each RFC. See the author's web page at *http://bach.cs.depaul.edu/cwhite* for the best way to access RFCs.

Internet committees are not the only groups that approve standards for computer networks, data communications, and telecommunications. Another organization that creates and approves network standards is the **International Standards Organization (ISO)**, which is a multinational group composed of volunteers from the standards-making committees of various governments throughout the world. ISO is involved in developing standards in the field of information technology and created the OSI model for a network architecture.

Other standards making organizations include:

▶ **American National Standards Institute (ANSI)** — A private, non-profit organization not associated with the U.S. government, ANSI strives to further the adoption of standards of a variety of items to advance the U.S. economy and protect the interests of the public.

▶ **International Telecommunications Union – Telecommunication Standards Sector (ITU-T)** — Formerly the Consultative Committee on International Telephone and Telegraphy (CCITT), ITU-T is devoted to the research and creation of standards for telecommunications in general and telephone and data systems in particular.

▶ **Institute for Electrical and Electronics Engineers (IEEE)** — The largest professional engineering society in the world, IEEE strives to promote the standardization of the fields of electrical engineering, electronics and radio. Of particular interest to us is the work IEEE has performed on standardizing local area networks.

▶ **Electronic Industries Association (EIA)** — Aligned with ANSI, EIA is a nonprofit organization devoted to the standardization of electronics products. Of particular interest is the work they have performed on standardizing the interfaces between computers and modems.

the U.S. military, but has grown to include a large community of non-military users, including consumers, businesses, academicians, scientists, and other parts of the government. Figure 1-13 shows the layers of the Internet model and how it compares to the OSI model.

Figure 1-13

The layers of the Internet model compared to the layers of the OSI model

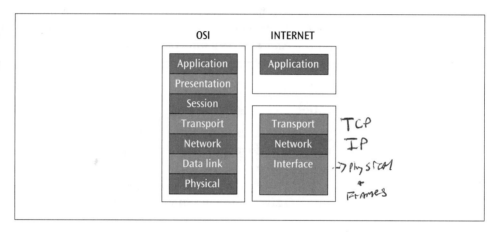

Although the Internet model does not have rigidly defined layers as the OSI model does, several layers have emerged as being the more commonly used ones. These layers include the interface layer, the Internet network layer, the Internet transport layer, and the Internet application layer.

The lowest layer of the Internet model is the **interface layer**, which is roughly equivalent to the physical and data link layers of the OSI model. The Interface layer defines both the physical medium that transmits the signal and the frame that incorporates flow and error control.

Although it may seem strange to roll the services of two layers into one, doing so actually makes good sense. Having distinctly defined layers enables you to "pull" out one layer and insert an equivalent layer without affecting the other layers. For example, let us assume a system was designed for copper-based wire. Later, the system owners decided to replace the copper-based wire with fiber optic cable. Even though a change is being made at the physical layer, it should not be necessary to make any changes at any other layers.

In reality, however, there are relationships between the layers that cannot be ignored. For example, if the physical organization of a local area network is changed, there is a good chance that the frame description at the data link layer will also need to be changed. (We will examine this phenomena in Chapter Seven.) The Internet model recognizes these relationships and merges many of the services of the physical and data link layers into one layer.

The **Internet network layer** is roughly equivalent to OSI's network layer with one major distinction: the Internet model uses the Internet Protocol (IP) to transfer data between networks. The **Internet Protocol** is the software that prepares a packet of data so that it can move from one network to another on the Internet or within a set of networks in a corporation.

The functions of the transport layers of the two models are also fairly equivalent. Once again, however, the Internet model's transport layer represents a variation on the OSI layer. The **Internet transport layer** uses the Transmission Control Protocol (TCP) to maintain an error-free end-to-end connection. To maintain an error-free, end-to-end connection, TCP includes error control information in case one packet from a sequence of packets does not arrive at the final destination; packet sequencing information so that all the packets stay in the proper sequence; and packet age information that keeps a data packet from bouncing around the Internet for too long a period. TCP is not the only possible protocol found at the Internet transport layer. User Datagram Protocol (UDP) is an alternative that is used less frequently in the Internet model.

The final layer of the Internet model, the **Internet application layer,** supports the network applications for which one uses a network and presentation services such as encryption and compression that are necessary to properly present or support the application. Applications at the Internet application layer include many frequently-used programs:

- ▶ **File Transfer Protocol (FTP)** to transfer files from one computer system to another;
- ▶ **Telnet** to allow a remote user to login to another computer system;
- ▶ **Simple Mail Transfer Protocol (SMTP)** to allow users to send and receive electronic mail;
- ▶ **Simple Network Management Protocol (SNMP)** to allow the numerous elements within a computer network to be managed from a single point;
- ▶ **Hyper-Text Transport Protocol (HTTP)** to allow Web browsers and servers to send and receive World Wide Web pages; and
- ▶ **Domain Name System (DNS)** to allow a user to enter a Web address, such as *www.cs.depaul.edu,* and convert it into the proper computer-recognizable binary form.

Although the Internet model is the model of choice for most installed networks, it is important to study both the Internet model and the OSI model. Many books and articles, when describing a product or a protocol, often refer to the OSI model with a statement such as "This product is compliant with OSI layer xxx." If you are not familiar with the various layers of the OSI and Internet models, you would lack some very important basic knowledge, which might impede your understanding of more advanced concepts in the future.

Logical and physical connections

An important concept to understand with regard to the layers of a communication model is the lines of communication between a sender and a receiver. Consider Figure 1-14, which shows a sender using a network application that is designed on the OSI model. The sender is communicating with a receiver who is also using an OSI model system.

Figure 1-14
Sender and receiver communicating using the OSI model

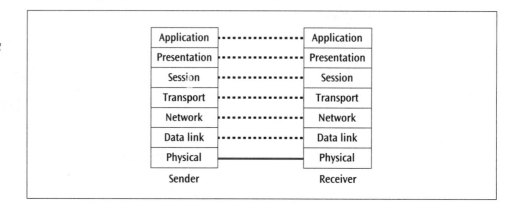

Notice the dashed lines between sender and receiver Application layers, Presentation layers, Session layers, Transport layers, Network layers, and Data link layers. No data flows over these dashed lines. Each dashed line indicates a logical connection—a flow of ideas without a direct physical connection—between sender and receiver at a particular layer. The sender and receiver transport layers, for example, share a set of commands that is used to perform transport-type functions, but the information comes up from the physical layer. Without this logical connection, the two parts of the network would not be able to coordinate their functions. There is not a direct connection between the two transport layers. The only physical connection between sender and receiver is at the physical layer, where actual 1s and 0s—the digital content of the message—are transmitted over wire or airwaves.

As an example of logical and physical connections, consider an imaginary scenario in which the Dean of Arts and Sciences wants to create a new joint degree with the School of Business. In particular, the dean would like to create a degree that is a cross between computer science and marketing. The Dean of Arts and Sciences could call the Dean of Business to create the degree, but the deans are not the experts in creating all the details involved in a new degree. Instead, the Dean of Arts and Sciences starts the process by issuing a request for a new degree from the Dean of Business. Before the request gets to the Dean of Business however, the request must pass through several layers. First the request goes to the chairperson of the computer science department. The chairperson will examine the request for a new degree and add the necessary information to staff the program. The chairperson will then send the request to the Computer Science Curriculum Committee, which will design several new courses. The curriculum committee will send the request to the department secretary who will type up all the memos and create a readable package. This package is then placed in the inter-campus mail and sent to the marketing department in the School of Business.

Once the request arrives at the marketing department, the secretary in the marketing department opens the envelope and gives all the materials to the Marketing Curriculum Committee. The Marketing Curriculum Committee looks at the proposed courses from the Computer Science Curriculum Committee and makes some changes and additions. Once these changes are made, the proposal is given to the chair of the marketing department who looks at the staffing needs suggested by the Chair of Computer Science, checks the request for accuracy, and makes some changes. The Chair of Marketing then hands the request to the Dean of Business who examines the entire document and gives approval with a few small changes. The request then works it way back down to the secretary of the marketing department who sends it back to the secretary of computer science. The computer science secretary then starts the reply to the request up the layers until it reaches the Dean of Arts and Sciences. Figure 1-15 shows how this request for a degree moves up and down through the layers of bureaucracy.

Figure 1-15
Flow of data through the layers of bureaucracy

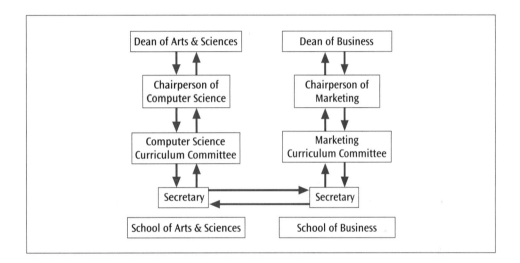

The flow of data was not directly between deans, nor was it directly between department chairpersons or curriculum committees. The data had to flow all the way down to the physical layer (in this case, the secretaries) and then back up the other side. At each layer in the process, information that is used to assist the "peer" layer on the other side was added. This example stretches the truth a little; college curriculums are not designed this way. Therefore, let's examine a more realistic example in which a person using a web browser requests a web page from somewhere on the Internet.

The Internet Model in Action ▶

A more detailed example of a request for a service moving through the layers of a communications model will help make the concepts involved clearer. Consider Figure 1-16 in which a user browsing the Internet requests a web page to be downloaded and then displayed on his or her personal computer.

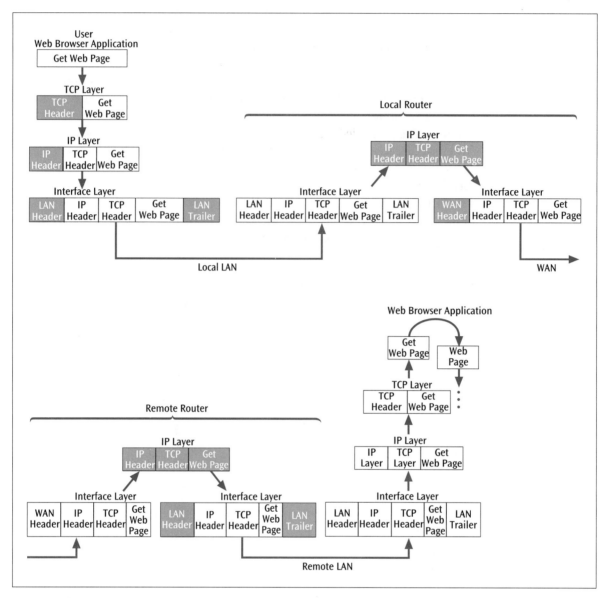

Figure 1-16
Path of a web page request as it flows from browser to Internet server and back

Beginning in the upper left corner of the figure, the user clicks a link on the current web page. In response, the browser software (the application) creates a "Get Web Page" command that is given to the transport layer, TCP. TCP adds various header information, which is used by the TCP layer on the receiving end. This information assists with end-to-end error control and end-to-end flow control and provides the address of the application (the web server).

This enlarged packet is now sent to the Internet layer where IP adds its header. The information contained within the IP header assists the IP layer on the receiving end, as well as assisting the IP layers at each intermediate node during the data's progress through the Internet. This assistance includes the Internet address of the workstation that contains the requested web page.

The packet is now given to the interface layer. Since the user's computer is connected to a local area network (LAN), the appropriate local area network headers are added. Note that sometimes trailers, as well as headers, are added to the data packet. One of the most important pieces of information included in the local area network header is the address of the device (the router) that connects the local area network to the wide area network (the Internet).

Eventually, the binary 1s and 0s of the data packet are transmitted across the user's local area network, where they encounter a router. The router is the gateway to the Internet. The router removes the local area network header and trailer. The information in the IP header is examined, and it is determined that the data packet must go out to the Internet. New wide area network (WAN) header information, which is necessary for the data packet to traverse the wide area network, is applied. Finally, the binary 1s and 0s of the data packet are placed onto the wide area network.

After the data packet moves across the Internet, it will arrive at the router connected to the local area network that contains the desired web server. This remote router removes the wide area network information, sees that the packet must be placed on the local area network, and inserts the local area network header and trailer information. The packet is placed onto the local area network, and using the address information in the LAN header, travels to the computer holding the web server application. As the data packet moves up the layers of the web server's computer, the LAN, IP, and TCP headers are removed. Finally, the web server application receives the Get Web Page command, retrieves the requested web page, and creates a new data packet with the requested information. This new data packet now moves down the layers and back through the routers to the user's network and workstation. Finally, the web page is displayed on the user's monitor.

CBT: To see a visual demonstration of encapsulation, run the first CBT module on the enclosed CD-ROM.

It is interesting to note that as a packet of data flows down through a model and passes through each layer of the system, the data packet grows in size. This growth is attributable to the fact that each layer adds additional information to the original data. This layer-added information is necessary when the data arrives at the destination and aids in providing services such as error detection, error control, flow control, and network addressing. The addition of control information to a packet as it moves through the layers is called **encapsulation**.

◆ ◆

SUMMARY

▶ Computer networks and data communications is a vast and significant field of study. Once considered the province primarily of engineers and technicians, it now involves managers, computer programmers, system designers, and home computer users.

▶ Many services and products that we use every day employ computer networks and data communications in one way or another. Telephones systems, banking systems, cable television, audio and video systems, traffic control systems, and wireless telephones are a few examples.

▶ Wireless networks are networks that use radio waves for communication.

► Local area networks are networks within a small radius, such as a room or building.

► Metropolitan area networks are networks that service a metropolitan area.

► Wide area networks are networks that cover areas such as states and countries.

► Voice networks are networks designed to transmit telephone conversations.

► Data networks are networks designed to transmit computer data.

► The application areas of computer networks and data communications can be understood in terms of general network configurations.

► A computer terminal-to-mainframe computer is one of the oldest network configurations and includes commercial transaction systems.

► A microcomputer-to-mainframe computer configuration is replacing the computer terminal-to-mainframe configuration and includes popular client/server systems in business.

► A microcomputer-to-local area network is found in most businesses and academic settings.

► Microcomputer-to-Internet configurations are systems that dial into the Internet from home or work using a telephone line and modem.

► Sensor-based computer systems are common in production and assembly line systems.

► Local area network-to-local area network connections are for companies that need to support a large number of users.

► Local area network-to-wide area network connections allow business users to access external networks, such as the Internet from the corporate network.

► Wireless telephone systems are for transmitting voice and data.

► Satellite and microwave systems are high speed data transmission systems using line-of-sight microwave transmissions.

► To standardize the design of communication systems, the International Standards Organization (ISO) created the Open Systems Interconnection (OSI) model. The OSI model is based on seven layers.

  ► The application layer is the top layer of the OSI model where the application which is using the network resides.

  ► The presentation layer performs a series of miscellaneous functions necessary for presenting the data package properly to the sender or receiver.

  ► The session layer is responsible for establishing sessions between users.

  ► The transport layer is concerned with an error-free, end-to-end flow of data.

  ► The network layer is responsible for creating, maintaining, and ending network connections.

  ► The data link layer is responsible for taking the raw data and transforming it into a cohesive unit called a frame.

  ► The physical layer handles the transmission of bits over a communications channel.

► The Internet model has surpassed the OSI model in popularity and is composed of four layers.

  ► The application layer contains the network applications for which one uses a network and presentation services that support that application.

  ► The transport layer maintains an error-free end-to-end connection.

  ► The network layer uses the Internet Protocol (IP) to transfer data between networks.

  ► The interface layer defines both the physical medium that transmits the signal and the frame that incorporates flow and error control.

► A logical connection is a flow of ideas, without a direct physical connection, between sender and receiver at a particular layer.

KEY TERMS

bridge
client/server system
codec
computer network
data communications
data link layer
data network
Domain Name System
encapsulation
File Transfer Protocol
frame
hub
Hypertext Transport Protocol
interface layer
Internet application layer
Internet model

Internet Protocol
Internet transport layer
local area network
logical connection
metropolitan area network
modem
multiplexing
network architecture model
network layer
network management
node
Open Systems Interconnection (OSI)
model
physical layer
presentation layer
router

server
session layer
Simple Mail Transport Protocol
Simple Network Management Protocol
subnet
synchronization point
telecommunications
Telnet
token management
transport layer
voice network
wide area network
wireless

REVIEW QUESTIONS

1. Define the following:
   a. a computer network
   b. data communications
   c. telecommunications
   d. a local area network
   e. a wide area network
   f. network management
2. What is the relationship between a subnet and a node?
3. What kind of applications might use a computer terminal-to-mainframe computer connection?
4. What kind of applications might use a microcomputer-to-mainframe computer connection?
5. What language does a microcomputer have to talk to interface to the Internet?
6. What kind of applications might use a sensor-to-local area network connection?
7. Why is a network architecture model useful?
8. List the seven layers of the OSI model.
9. List the four layers of the Internet model.
10. How do the layers of the OSI model compare with the layers of the Internet model?
11. What are some of the more common applications found in the Internet model?
12. What is the difference between a logical connection and a physical connection?

EXERCISES

1. Create a list of all the things you do in an average day that use data communications and computer networks.
2. If you could design your own home, what kinds of computer network or data communications labor-saving devices would you incorporate?

3. Two companies consider pooling resources to perform a joint venture. The CEO of the first company meets with his legal team, and the legal team consults a number of middle managers in the proposed product area. Meanwhile, the CEO of the first company sends an e-mail to the CEO of the second company to offer a couple suggestions concerning the joint venture. Does this scenario follow the OSI model? Explain.

4. Using a laptop computer with a wireless connection into the company's local area network, you download a web page from the Internet. List all the different network configurations involved in this operation.

5. Someone decided that seven layers in the OSI model are too many. You have been asked to reduce the number to four layers. Which layers would you eliminate? Of those layers eliminated, should the functions provided by those layers be placed in some other layers? If yes, which layers?

6. If the data link layer provides error checking, and the transport layer provides error checking, isn't this redundant? Explain.

7. You are watching a television show and somebody is suing somebody else. The lawyers for both parties meet and try to work out a settlement. Is there a logical or physical connection between the lawyers? What about between the two parties?

8. You want to download a file from a remote site using the File Transfer Protocol (FTP). To perform the file transfer, your computer issues a "Get File" command. Show the progression of messages as the Get File command moves from your computer, through routers, into the remote computer, and back.

THINKING OUTSIDE THE BOX

1 You have been asked to create a new network architectural model. Will it be layered or some other form? Show the layers or form and describe the functions performed by each piece.

2 Take an example from your work or school in which a person requests a service and diagram that request. Does the request pass through any layers before it reaches the intended recipient? Are there logical connections as well as physical connections? If so, show them in the diagram.

PROJECTS

1. Since the Internet model is not carved in stone, other books may discuss a slightly different layering. Find two other examples of the Internet model that differ from this text description of layering, and cite the sources. How do the two models compare and differ? How do they compare to the Internet model discussed in this chapter?

2. What is the more precise form of the Get Web Page command shown in Figure 1-16?

3. What types of network applications exist at your place of employment or your college? Are local area networks involved? Wide area networks? List the various network configurations.

4. What other network models exist besides the OSI and Internet models? Describe them.

2

Fundamentals of Data and Signals

◆◆

AM RADIO definitely does not have the allure of FM radio. Its reception is noisy, it is highly susceptible to static discharges such as lightning, and one station can interfere with other stations. Another problem with AM radio is that its signals interfere with the signals of Asymmetric Digital Subscriber Line (ADSL), effectively reducing ADSL's transmission capabilities.

ADSL is one of the newest technologies for transmitting high speed data over existing telephone lines. While ADSL comes in a variety of configurations, most are capable of supporting data transmission rates of millions of bits per second. Unfortunately, ADSL transmits its data in the 138 KHz to 1.1 MHz range, while AM radio broadcasts between 540 KHz and 1.7 MHz. If you compare the numbers, you can see a sizable overlap. If ADSL transmission lines are too close to one or more AM radio transmission towers, or the ADSL signals are weakened from having traveled a long distance, then ADSL data rates can drop by as much

as 40 percent. In some extreme conditions, ADSL will not work at all.

What is being done to lessen this problem? Some telephone companies that provide ADSL service downplay the seriousness of AM radio interference. They argue that this interference happens only in extreme conditions and that the average user of ADSL has little to worry about. Companies that make ADSL modems are considering modifying their products so that ADSL does not use the same frequencies as AM radio. Other experts suggest that better quality wiring will minimize the affect AM radio signals have on ADSL signals. For the present, however, it appears that something as commonplace as AM radio may be hindering the growth of new technology.

Network World, November 22, 1999. Pg. 8.

Is AM radio considered a form of noise?

Will there be other technologies, old and new, that cannot coexist due to signal interference?

Objectives

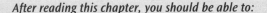

After reading this chapter, you should be able to:

▶ Distinguish between data and signals and cite the advantages of digital data and signals over analog data and signals.

▶ Identify the three basic components of a signal as amplitude, frequency, and phase.

▶ Discuss the bandwidth of a signal and how it relates to data transfer speed.

▶ Identify signal strength and attenuation and how they are related.

▶ Outline the basic characteristics of transmitting digital data with digital signals, analog data with digital signals, digital data with analog signals, and analog data with analog signals.

▶ List and be able to draw diagrams of the basic digital encoding techniques, including the advantages and disadvantages of each.

▶ Identify the different modulation techniques and describe their advantages, disadvantages, and uses.

▶ Identify the two digitization techniques, pulse code modulation and delta modulation, and describe their advantages and disadvantages.

▶ Discuss the characteristics and importance of spread spectrum encoding techniques.

▶ Identify the different data codes and how they are used in communication systems.

Introduction ▶

When the average computer user is asked to list the ingredients of a computer network, most will probably cite computers, wires, modems, telephone lines, and other easily identifiable physical components. This chapter, however, will deal primarily with two ingredients that are more difficult to see physically: data and signals.

Data and signals are two of the basic building blocks of any computer network, but they are not two terms that mean the same thing. Stated simply, a signal is the transmission of data. It is important to understand that a computer system that wishes to transmit data must first convert the data into the appropriate signals. It is equally important to understand *how* a computer converts data into those signals. Both data and signals can be in either analog or digital form, which gives us four possible combinations: transmitting digital data using digital signals; transmitting digital data using analog signals; transmitting analog data using digital signals; and transmitting analog data using analog signals. Each of these four combinations occur quite frequently in computer networks and each has unique applications and properties.

Converting digital data to digital signals is relatively straightforward and involves numerous digital encoding techniques. We will examine a few representative encoding techniques and discuss their basic advantages and disadvantages. Converting digital data to analog signals is the recipe for a modem, a device that allows computers to transmit data over analog transmission systems such as telephone lines. Converting analog data to digital signals is what people generally call digitization. Two basic digitization techniques will be introduced, showing their advantages and disadvantages. Finally, converting analog data to analog signals is fairly common and is found in systems such as cable television.

A big question arises during the study of data and signals: Why should people who are interested in the business aspects of computer networks concern themselves with this level of detail? The answer to that question is that a firm understanding of the fundamentals of communication systems will provide a solid foundation for the further study of the more advanced topics of computer networks.

Data and Signals

Information that is stored within computer systems and transferred over a computer network can be divided into two categories: data and signals. **Data** are entities that convey meaning within a computer or computer system. Common examples of data include:

- ▶ a computer file of names and addresses stored on a hard disk drive;
- ▶ a movie stored on a video tape or video disc;
- ▶ music stored on a compact disc;
- ▶ a collection of samples retrieved from a blood gas analysis machine;
- ▶ a favorite photograph that has been digitized by a flat-bed scanner and stored on a floppy diskette; and
- ▶ all the sales figures for all the salespeople for a company for the last six months.

[handwritten margin notes: Digital DATA in a Digital sense / Digital Data in an Analog sense / Analog DATA in a Digital sense / Analog Data in an Analog sense]

In each of these examples, some kind of information has been electronically captured and stored on some type of computer storage device.

If you want to transfer this data from one point to another, either by using a physical wire or by using radio waves, the data has to be converted into a signal. **Signals** are the electric or electromagnetic encoding of data and are used to transmit data. Common examples of signals include:

▶ a telephone conversation transmitted over a telephone line;

▶ a live television news interview from Europe transmitted over a satellite system;

▶ a term paper transferring over the printer cable between a computer and a printer; and

▶ a web page download as it transfers over the telephone line between your Internet service provider and your home computer.

In each of these examples, data, the static entity, is transmitted over some type of wire or airwave in the form of a signal, the dynamic entity. Some type of device is necessary to convert the static data into a dynamic signal.

However, before examining the basic characteristics of signals and the conversion from data to signal, let's explore some of the more important characteristics of data and signals. Both data and signals share one characteristic: whether the data or signal is analog or digital.

Analog versus digital

*[handwritten margin notes: Force / $E = I * R$ / Force Power = Current * Resistance / (Volts) (AMPS) (ohms) / $1v = 1A * 1\Omega$]*

Although data and signals are two different entities that have little in common, there is one characteristic they do share: whether they are analog or digital. **Analog data** and **analog signals** are represented as continuous waveforms that can be at an infinite number of points between some given minimum and maximum. These minimum and maximum values are presented as voltages by convention. Figure 2-1 shows that between the minimum value X and maximum value Y, the waveform at time *T* can be at an infinite number of places. The most common example of analog data is the human voice. Music and video, when they occur in their natural states, are also analog data. A common example of an analog signal is the telephone system's electronic transmission of a voice conversation.

[handwritten margin notes: Analog Continuous / Digital Discrete / V + 0 / V - / Time]

Figure 2-1
A simple example of an analog waveform

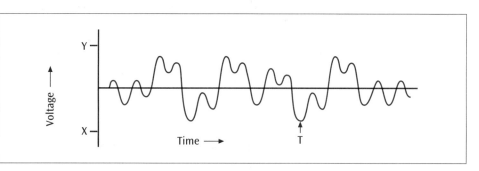

One of the primary shortcomings of analog data and analog signals is how difficult it is to separate noise from the original waveform. **Noise** is unwanted electrical or electromagnetic energy that degrades the quality of signals and data. Since noise is found in every type of data and transmission system, and since its effects range from a slight hiss in the background to a complete loss of data or signal, it is especially important that noise be reduced as much as possible. Unfortunately, noise appears as an analog waveform, which makes it difficult to separate from an analog waveform that represents data.

Consider the waveform in Figure 2-2, which shows the first few notes of a symphonic overture. Noise is intermixed with the music—the data. Can you tell by looking at the figure which is the data and which is the noise? Although this example may border on the extreme, it demonstrates that noise and analog data can appear similar.

Figure 2-2
The waveform of a symphonic overture with noise

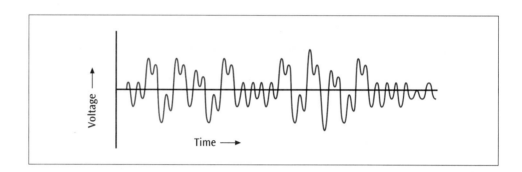

The performance of a record player provides a further example of noise interfering with data. Many people have collections of albums, which when played, produce pops, hisses, and clicks; albums sometimes even skip. Is it possible to create a device that filters out the pops, hisses, and clicks without ruining the original data, the music? Filtering devices were created during the 1960s and 1970s, but only the devices that removed hiss were relatively successful. Devices that removed the pops and clicks sometimes also removed parts of the music. Filters now exist that can fairly effectively remove all forms of noise from analog recordings, but, interestingly, they are expensive digital devices.

An example of noise interfering with an analog signal is the hiss you hear when you are talking on the telephone. Often the background hiss is so slight that most people don't notice it. Occasionally however, the hiss rises to such a level that it interferes with the conversation.

Digital data and **digital signals** are discrete waveforms, rather than continuous waveforms. Between a minimum value X and a maximum value Y, the digital waveform takes on only a finite number of values. In the example shown in Figure 2-3, the digital waveform takes on only two different values.

Figure 2-3
A simple example of a digital waveform

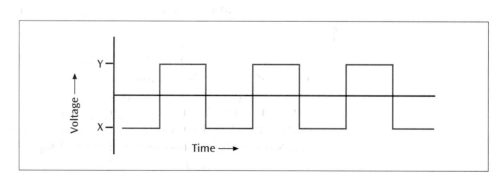

What happens when you introduce noise into digital data and digital signals? Since noise has the properties of an analog waveform, it should be possible to distinguish it from the original digital waveform. Figure 2-4 shows a digital signal with some noise introduced.

Figure 2-4
A digital signal with some noise introduced

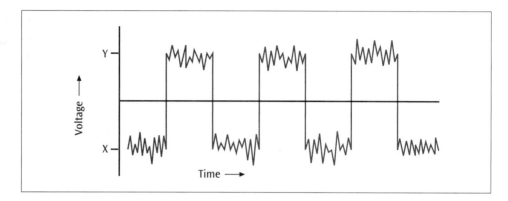

If the amount of noise remains low enough that the original digital waveform can still be interpreted, then the noise can be filtered out leaving the original waveform. In the simple example in Figure 2-4, as long as you can tell a high part of the waveform from a low part, you can still recognize the digital waveform. If the noise becomes so great that it is no longer possible to distinguish a high from a low, as shown in Figure 2-5, then the noise has won and you can no longer understand this portion of the waveform.

Figure 2-5
A digital waveform with noise so great that you can no longer recognize the original waveform

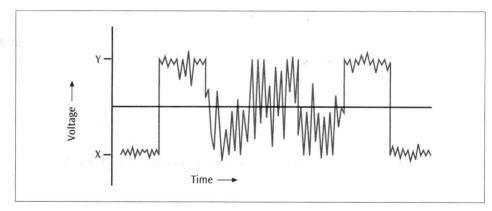

The ability to separate noise from a digital waveform is one of the great strengths of digital systems. When data is transmitted as a digital signal, the signal will always incur some level of noise. It is a relatively simple process to take a noisy digital signal and pass it through a filtering device that removes a significant amount of the noise, leaving the original digital signal.

Despite this strong advantage of digital over analog, not all systems use digital signals to transmit data. The electronic equipment used to transmit a signal through a wire or over the airwaves usually dictates the type of signals the wire can transmit. Certain electronic equipment is capable of supporting only analog signals, while other equipment can support only digital signals.

Fundamentals of signals

CD's Are Analog data
in a digital sense
↑
Carried

Recall that signals are the electric or electromagnetic encoding of data and are used to transmit or carry data. The primary characteristic that separates one form of data from another is the analog versus digital characteristic. Because signals are more complex than data, signals have many more characteristics that are worth examining. Let's begin our study of analog and digital signals by examining their three basic components: amplitude, frequency, and phase.

A sine wave is used to represent an analog signal, as shown in Figure 2-6. The **amplitude** of a signal is the height of the wave above (or below) a given reference point. This height often denotes the voltage level of the signal (measured in volts), but it can also denote the current level of the signal (measured in amps) or the power level of the signal (measured in watts). That is, the amplitude of a signal can be expressed as volts, amps, or watts. Note that a signal can change amplitude as time progresses. In Figure 2-6 you see one signal with two different amplitudes.

Figure 2-6
A signal with two different amplitudes

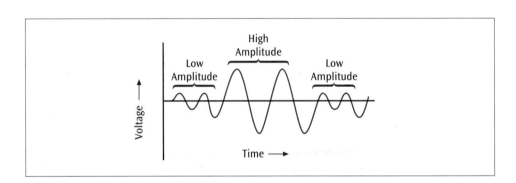

The **frequency** of a signal is the number of times a signal makes a complete cycle within a given time frame. The length, or time interval, of one cycle is called its **period**. Figure 2-7 shows three different analog signals. If the time *T* is one second, the signal in Figure 2-7(a) completes one cycle in one second. The signal in Figure 2-7(b) completes two cycles in one second. The signal in Figure 2-7(c) completes three cycles in one second. Cycles per second, or frequency, is represented by **Hertz (Hz)**. Thus, the signal in Figure 2-7(c) has a frequency of 3 Hz.

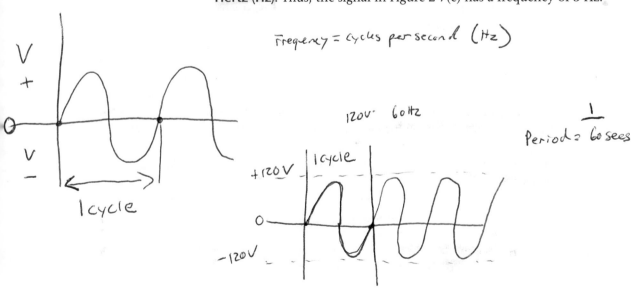

Frequency = cycles per second (Hz)

120v. 60 Hz

$$Period = \frac{1}{60 \, secs}$$

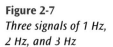

Figure 2-7
Three signals of 1 Hz, 2 Hz, and 3 Hz

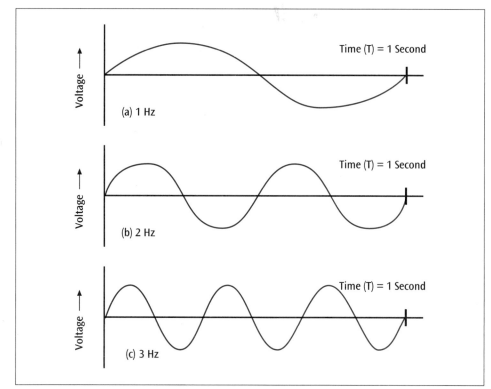

Human voice, audio, and video, as well as most signals, are actually composed of multiple frequencies. These multiple frequencies are what allow us to distinguish one person's voice from another and one musical instrument from another. The average human voice usually goes no lower than 300 Hz and no higher than approximately 3100 Hz. Since a telephone is designed to transmit a human voice, the telephone system transmits signals in the range of 300 Hz to 3100 Hz. The piano has a wider range of frequencies than the human voice. The lowest note possible on the piano is 30 Hz and the highest note possible is 4200 Hz.

The range of frequencies that a signal spans from minimum to maximum is called the **spectrum**. The **bandwidth** of a signal is the absolute value of the difference between the lowest and highest frequencies. The bandwidth of a telephone system that transmits a single voice in the range of 300 Hz to 3100 Hz is 2800 Hz. Because extraneous noise degrades original signals, any electronic device has an **effective bandwidth** that is typically less than its bandwidth. When making communication decisions, many professionals rely more on the effective bandwidth than the bandwidth since most situations deal with real-world problems of noise and interference.

The **phase** of a signal is the position of the waveform relative to a given moment of time or relative to time zero. In the drawing of the simple sine wave in Figure 2-8(a), the waveform oscillates up and down in a repeating fashion. Note that the wave never makes an abrupt change but is a continuous sine wave. A phase change (or phase shift) involves jumping forward (or backward) in the waveform at a given moment of time. Jumping forward one half of the complete cycle of the signal produces a 180 degree phase change (Figure 2-8(b)). Jumping forward one quarter of the cycle produces a 90 degree phase change (Figure 2-8(c)). Some systems, as you will see in this chapter's section on Transmitting Digital Data with Analog Signals, can generate signals that do a phase change of 45, 135, 225, and 315 degrees on demand.

Figure 2-8

A sine wave showing no phase change (a), a 180 degree phase change (b), and a 90 degree phase change (c)

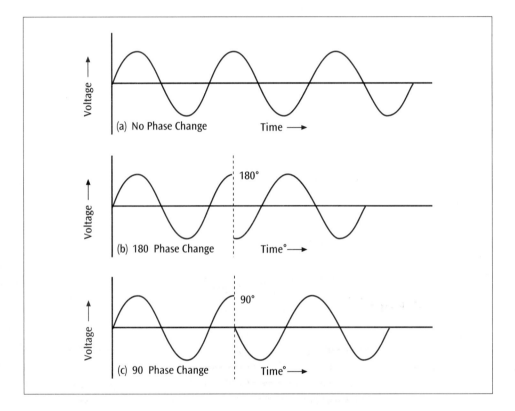

Loss of signal strength

Imagine a scenario in which you are recommending a computer network solution for a business problem. You tell the network specialists that you want to place a computer workstation at the company's reception desk so that the receptionist can handle requests for scheduling meeting rooms. The network specialist says it can't be done because the wire connecting the workstation to the network will be too long and the signal will be too weak. Or worse, the network specialist uses computer jargon: "The signal will have too much attenuation and will drop below an acceptable threshold and noise will take over." Is the network specialist accurate, or is he using computer jargon to dissuade you? A little knowledge of the loss of signal strength will help in such situations.

When traveling through any type of medium, a signal always experiences some loss of its power due to friction. This loss of power, or loss of signal strength, is called **attenuation**. Attenuation in a medium such as copper wire is a logarithmic

[handwritten: Rickter Scale is An example of ~~scale~~ Attenuation and decibel scale]

loss (in which a value increase of 1 represents a ten-fold increase) and is a function of distance and the friction within the wire. Knowing the amount of attenuation in a signal (how much power of the signal was lost) allows you to determine the signal strength. **Decibel (db)** is a relative measure of signal loss or gain and is used to measure the logarithmic loss of a signal.

$$db = 10 \log_{10} \frac{P2}{P1}$$

Details ▶

Composite Signals *[handwritten: ~~on quiz + Test~~ on quiz + Test]*

Almost all of the signals shown in this chapter are simple, periodic sine waves. However, you do not always find simple periodic sine waves in the real world. Instead, you are likely to encounter signals that are complex, periodic waves in which the signal repeats within a particular time span, and not sine waves.

A branch of mathematics called Fourier Analysis shows that any complex periodic waveform is a composite of simpler periodic waveforms. Consider, for example, the first two waveforms shown in Figure 2-9. The equation for the first waveform is 1 sin(2πft), and the equation for the second waveform is 1/3 sin(2π3ft). In the equation, *1* is a value of amplitude, the term *sin* refers to the sine trigonometric function, and *ft* is the frequency over a given period of time. Examining both the waveforms and the equations shows that the amplitude of the second waveform is 1/3 as high as the amplitude of the first waveform. The frequency of the second waveform is 3 times as high as the frequency of the first waveform. The third waveform in Figure 2-29(c) is a composite of the first two waveforms.

If you continue to add more waveforms to this composite signal, the waveform will look more and more like the square wave of a digital signal. You would use this technique of combining waveforms to create a waveform that has the appearance of a digital signal. The additional waveforms would have amplitude values of 1/5, 1/7, 1/9, and so on; the frequency multiplier of the waveforms would have values of 5, 7, 9, and so on. The more waveforms you add to the composite signal, the more the composite signal will look like a square wave. Expressed another way, the higher the frequency of the additional waveforms—the higher the bandwidth of the composite

signal—the more the composite will look like the square wave of a digital signal. Telephone systems use composites of analog signals to create analog signals that look like digital signals. Thus, you have an analog signal that looks like a digital signal from which noise can be removed but can still be transmitted over an analog telephone line.

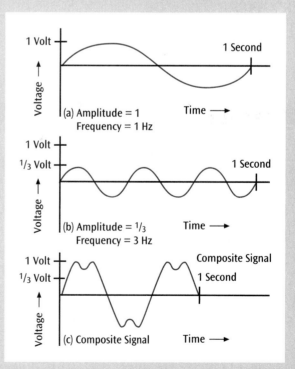

Figure 2-9 *Two simple, periodic sine waves (a) and (b) and their composite (c)*

[handwritten: fourier]

Since attenuation is a logarithmic loss and the decibel is a logarithmic value, calculating the overall loss or gain of a system involves adding all the individual decibel losses and gains. Figure 2-10 shows a communication line running from point A, through point B, and ending at point C. The communication line from A to B experiences a 10 dB loss, point B has a 20 dB amplifier (that is, there is a 20 dB gain at point B), and the communication line from B to C experiences a 15 dB loss. What is the overall gain or loss of the signal between point A and point C? To answer this question, add all dB gains and losses:

-10 dB + 20 dB + -15 dB = -5 db

Figure 2-10
Example demonstrating decibel loss and gain

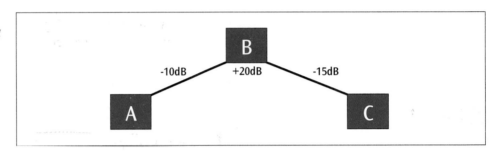

CBT: To see a visual demonstration of decibel loss and gain, run the second CBT module on the enclosed CD-ROM.

Let's return to the earlier example of the network specialist telling you that it may not be possible to install a computer workstation as planned. You now understand that signals lose strength over distance. Although you do not know how much signal would be lost nor at what point the strength of the signal would be weaker than the noise, you can trust part of what the network specialist told you.

Details ▶

Signal Strength   ✳ quiz + Test

The decibel is a relative measure of signal loss or gain and is expressed as

$$dB = 10 \log_{10} (P_2 / P_1)$$

in which P_2 and P_1 are the ending and beginning power levels of the signal expressed in watts. If a signal starts at a transmitter with 10 watts of power and arrives at a receiver with 5 watts of power, the signal loss in dB is calculated as follows:

$$
\begin{aligned}
dB &= 10 \log_{10} (5/10) \\
&= 10 \log_{10} (0.5) \\
&= 10 \, (-0.3) \\
&= -3
\end{aligned}
$$

There is a 3 dB loss between the sender and receiver. Remember that decibel is a relative measure of loss or gain. You cannot take a single power level at time T and compute the decibel value of that signal without having a reference or a beginning power level.

Converting Data into Signals

Like data, signals can be analog or digital. Typically, digital signals convey digital data, and analog signals convey analog data. However, you can use analog signals to convey digital data and digital signals to convey analog data. The choice of using either analog or digital signals often depends on the transmission equipment that is used and the environment in which the signals must travel. Certain electronic equipment is capable of supporting only analog signals, while others support only digital signals. For example, the telephone system was created to transmit human voice, which is analog data. Thus, the telephone system was designed to transmit analog signals. Although telephone wiring is capable of carrying either analog or digital signals, the electronic equipment used to amplify and remove noise from the lines can accept only analog signals. Therefore over telephone lines, it is common to use analog signals to transmit digital data from a computer. Transmitting analog data with digital signals is also fairly common. Some cable television companies transmit 50 or more television channels using digital signals. Digital wireless telephones transmit human voice conversations with digital signals. There are four combinations of data and signals: digital data transmitted using digital signals; digital data transmitted using analog signals; analog data transmitted using analog signals; and analog data transmitted using digital signals.

Transmitting digital data with digital signals: digital encoding schemes

To transmit digital data using digital signals, the 1s and 0s of the digital data must be converted to the proper physical form that can be transmitted over a wire or airwave. Thus, if you wish to transmit a data value of 1, you could transmit a zero voltage on the medium. If you wish to transmit a data value of 0, you could transmit a positive voltage. You could also use the opposite: a data value of 0 is zero voltage and a data value of 1 is a positive voltage. Digital encoding schemes are used to convert the 0s and 1s of digital data into the appropriate transmission form. We will examine five digital encoding schemes that are representative of most digital encoding schemes: NRZ-L, NRZ-I, Manchester, Differential Manchester, and 4B/5B.

Non-Return to Zero Digital Encoding Schemes

The simplest system that transmits 1s as zero voltages and 0s as positive voltages is called the Non-Return to Zero-Level (NRZ-L) digital encoding scheme. The NRZ-L encoding scheme is simple to generate and inexpensive to implement. Unfortunately, it also has a disadvantage, which you will see in a moment. Figure 2-11(a) shows an example of the NRZ-L scheme.

Figure 2-11
Examples of four digital encoding schemes

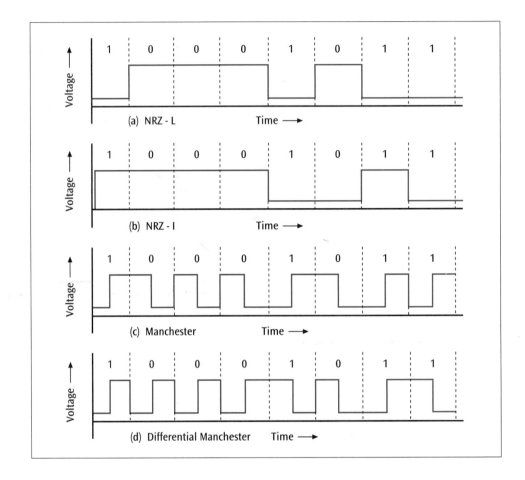

The second digital encoding scheme, shown in Figure 2-11(b), is Non-Return to Zero - Inverted (NRZ-I). This encoding scheme has a voltage change at the beginning of a 1 and no voltage change at the beginning of a 0. There is a fundamental difference between NRZ-L and NRZ-I. With NRZ-L, the receiver has to check the voltage level for each bit to determine whether the bit is a 0 or a 1. With NRZ-I, the receiver has to check whether there is a *change at the beginning* of the bit to determine if it is a 0 or a 1.

An inherent problem with the NRZ-L and NRZ-I digital encoding schemes is that long sequences of 0s in the data produce a signal that never changes. Often the receiver looks for signal changes so that it can synchronize its reading of the data with the actual data pattern. If a long string of 0s is transmitted and the signal does not change, how can the receiver tell when one bit ends and the next bit begins? Many knowledgeable people would answer, "the receiver has an internal clock that knows when to look for each successive bit." Although this statement is true, it contains an inherent problem: the receiver has a different clock than the one the transmitter used to generate the signals. Who is to say that these two clocks keep the same time? A more accurate system would generate a signal that has a change for each and every bit. If the receiver can count on each bit having some form of signal change, then the receiver can stay synchronized with the incoming data stream.

Manchester Digital Encoding Schemes

Manchester class of digital codes ensures that each bit has some type of signal change and thus solves the synchronization problem. The Manchester code shown in Figure 2-11(c) has the following properties: to transmit a 0, the signal changes from high to low in the *middle* of the interval; and to transmit a 1, the signal changes from low to high in the middle of the interval. Note that the transition is always in the middle, a 0 is a high to low transition, and a 1 is a low to high transition. Thus, if the signal is currently low and the next bit to transmit is a 0, the signal has to move from low to high at the beginning of the interval so that it can do the high to low transition in the middle.

The Differential Manchester digital encoding scheme, which is used in most local area networks, is similar to the Manchester scheme in that there is always a transition in the middle of the interval. But unlike the Manchester code, the direction of this transition in the middle does not differentiate between a 0 or a 1. Instead, if there is a transition at the *beginning* of the interval, then you are transmitting a 0. If there is no transition at the beginning of the interval, then you are transmitting a 1. Because the receiver must watch the beginning of the interval to determine the value of the bit, the Differential Manchester is similar to the NRZ-I scheme. Figure 2-11(d) shows an example of differential Manchester encoding.

The Manchester schemes have an advantage over the NRZ schemes: in the Manchester schemes there is always a transition in the middle of a bit. Thus, the receiver can expect a signal change at regular intervals and can synchronize itself with the incoming bit stream. The Manchester encoding schemes are self-clocking, because the occurrence of a regular transition is similar to what a clock does when it ticks off the seconds. As you will see in Chapter Four, it is very important for a receiver to stay synchronized with the incoming bit stream, and the Manchester codes allow a receiver to achieve this synchronization.

The big disadvantage of the Manchester codes is that roughly half the time there will be two transitions during each bit. For example, if the Differential Manchester code is used to transmit a series of 0s, then the signal has to change at the beginning of each bit, as well as change in the middle of each bit. For each data value 0, the signal changes twice. The number of times a signal changes value per second is called the baud rate, or simply **baud**. In Figure 2-12, a series of binary 0s is transmitted using the Differential Manchester encoding scheme. Note that the signal changes twice for each bit. After one second, the signal has changed 10 times. Therefore, the baud rate is 10. During that same time period, only 5 bits were transmitted. The **bits per second**, or **bps**, is 5, which is one half the baud rate. Many individuals mistakenly equate baud rate to bps. Under some circumstances, the baud rate may equal the bps, such as in the NRZ-L or NRZ-I encoding schemes shown in Figure 2-11. There is at most one signal change for each bit transmitted. But with schemes such as the Manchester codes, the baud rate is not equal to the bps.

Figure 2-12
Transmitting five binary 0s using Differential Manchester encoding

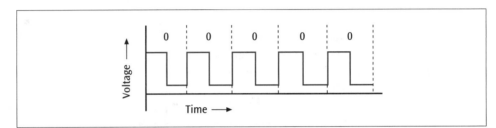

Is there importance to the fact that some encoding schemes have a baud rate twice the bps? Since the Manchester codes have a baud rate that is twice the bps and the NRZ-L and NRZ-I codes have a baud rate that is equal to the bps, hardware that generates a Manchester encoded signal has to work twice as fast as hardware that generates a NRZ encoded signal. If you wish to transmit 100 million 0s per second using Differential Manchester encoding, the signal has to change 200 million times per second. As with most things in life, you don't get something for nothing. Hardware or software that handles the Manchester encoding schemes is more elaborate and probably more costly than the hardware or software that handles the NRZ encoding schemes.

4B/5B Digital Encoding Scheme

The Manchester encoding schemes solve the synchronization problem, but are relatively inefficient because they have a baud rate that is twice the bps. The 4B/5B scheme tries to satisfy the synchronization problem and avoid the "baud equals two times the bps" problem. The **4B/5B** encoding scheme takes four bits of data, converts the four bits into a unique five bit sequence, and encodes the five bits using NRZ-I.

The first step in generating the 4B/5B code is to convert four-bit quantities of the original data into new five-bit quantities. Taking five bits together as one value yields 32 combinations ($2^5 = 32$), only 16 of which are used. The 16 combinations were chosen so that no code has three or more consecutive 0s. If you then transmit the five-bit quantities using NRZ-I encoding, you will never transmit more than two 0s in a row. If you never transmit more than two 0s in a row using NRZ-I encoding, then you will never have a long period of no signal transition. Figure 2-13 shows the 4B/5B code in detail.

How does the 4B/5B code work? Let us say, for example, that the next four bits in a data stream to be transmitted are 0000. Looking at the first column in Figure 2-13, we see that 4B/5B encoding replaces 0000 with 11110. There are no more than two consecutive zeros in 11110. In fact, of all the values in the second column, none of the five-bit codes have more than two consecutive zeros. Having replaced 0000 with 11110, the hardware will now transmit 11110 using NRZ-I encoding. Since the five-bit code is transmitted using NRZ-I, the baud rate equals the bps. Converting a four-bit code to a five-bit code creates a 20 percent overhead. Compare that to a Manchester code in which the baud rate can be twice the bps for a 100 percent overhead.

Figure 2-13
The 4B/5B digital encoding scheme

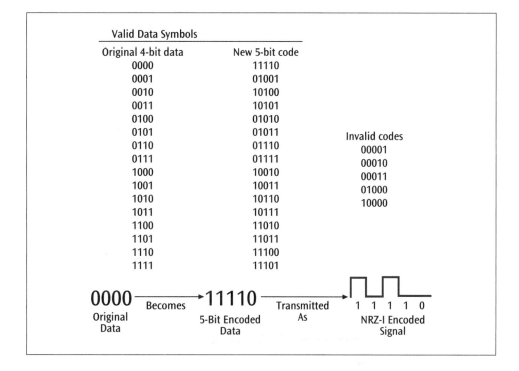

Baud rate

Synchronization

Transmitting digital data with analog signals

Since the U.S. telephone system is still, to a certain degree, an analog system, the transfer of digital data over a telephone line requires conversion of the data to an analog signal. The technique of converting digital data to an analog signal is called **modulation**, or **shift keying**. A device that modulates digital data onto an analog signal and then demodulates the analog signal back to digital data is a (MOdulator/DEModulator). The modem and its characteristics are discussed in detail in Chapter Four. Three currently popular modulation techniques for encoding digital data and transmitting it over analog signals are amplitude modulation, frequency modulation, and phase modulation.

Amplitude Modulation

The simplest modulation technique is **amplitude modulation**, also called amplitude shift keying. As shown in Figure 2-14, a data value of 1 and a data value of 0 are represented by two different amplitudes of a signal. For example, the lower amplitude could represent a 1, while the higher amplitude could represent a 0. *During* each bit period the amplitude of the signal is constant.

Figure 2-14
Example of amplitude modulation

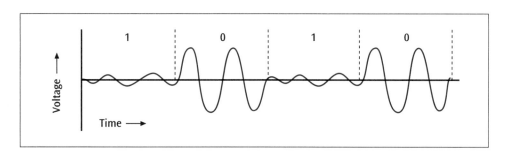

Amplitude modulation is not restricted to two possible amplitude levels. For example, we could create an amplitude modulation technique that incorporates four different amplitude levels, as shown in Figure 2-15. Each of the four different amplitude levels represent two bits. Recall that two bits yields four possible combinations: 00, 01, 10, and 11. Thus, every time the signal changes, two bits are transmitted and the bps is twice the baud rate. This is the opposite of a Manchester code in which bps is one half the baud rate.

Figure 2-15
Amplitude modulation using four different amplitude levels

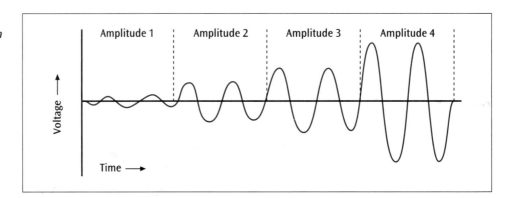

Amplitude modulation has a weakness: it is susceptible to sudden noise impulses such as the static charges created by a lightning storm. When a signal is disrupted by a large static discharge, the signal experiences significant increases in amplitude. For this reason and because it is difficult to accurately distinguish among more than just a few amplitude levels, amplitude modulation is one of the least efficient encoding techniques. Over standard telephone lines, amplitude modulation typically does not exceed 1200 bps.

Frequency Modulation

Frequency modulation (also called **frequency shift keying**) uses two different frequency ranges to represent data values of 0 and 1, as shown in Figure 2-16. For example, the lower frequency signal might represent a 1, while the higher frequency signal might represent a 0. *During* each bit period the frequency of the signal is constant.

Figure 2-16
Simple example of frequency modulation

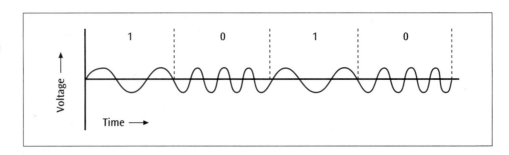

Unlike amplitude modulation, frequency modulation does not have a problem with sudden noise spikes causing loss of data. Thus, frequency modulation is a more robust encoding technique than amplitude modulation. Nonetheless, frequency modulation is not perfect and is subject to intermodulation distortion. **Intermodulation distortion** is caused when the frequencies of two or more signals mix together and create new frequencies.

Phase Modulation

A third modulation technique is phase modulation. **Phase modulation** (also called **phase shift keying**) represents 0s and 1s by different changes in the phase of a waveform. For example, a 0 could be no phase change, while a 1 could be a phase change of 180 degrees, as shown in Figure 2-17.

Figure 2-17
An example of simple phase modulation of a sine wave

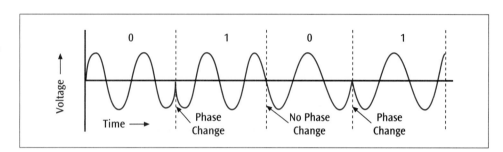

Phase changes are not affected by amplitude changes nor are they affected by intermodulation distortions. Thus, phase modulation is less susceptible to noise and can be used at higher frequencies. Phase modulation is so accurate that the signal transmitter can increase efficiency by introducing multiple phase shift angles. For example, **quadrature phase modulation** incorporates four different phase angles, each of which represents two bits: a 45 degree phase shift represents a data value of 11; a 135 degree phase shift represents 10; a 225 degree phase shift represents 01; and a

315 degree phase shift represents 00. Figure 2-18 shows a simplified drawing of these four different phase shifts. Because each phase shift represents two bits, quadrature phase modulation has double the efficiency of simple phase modulation. With this encoding technique, one signal change equals two bits of information; that is, 1 baud equals 2 bps.

Figure 2-18
Four phase angles of 45, 135, 225, and 315 degrees

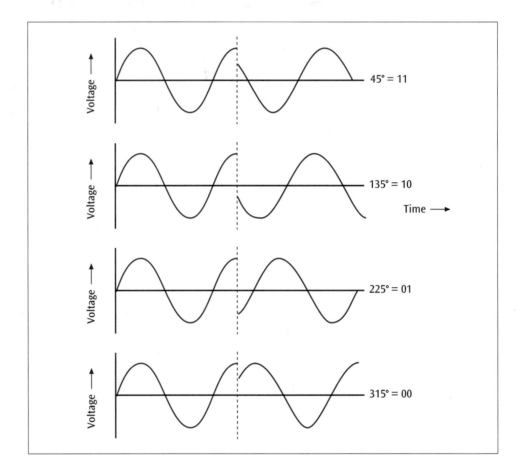

The efficiency of this technique can be increased even further by combining 12 different phase shift angles with two different amplitudes. Figure 2-19(a) shows 12 different phase shift angles with 12 arcs radiating from a central point. Two different amplitudes are applied on each of four angles. Figure 2-19(b) shows a phase shift with two different amplitudes. Thus, eight phase angles have a single amplitude and four phase angles have double amplitudes, resulting in 16 different combinations. This encoding technique, **quadrature amplitude modulation,** uses each signal change to represent four bits (4 bits yield 16 combinations). Therefore, the bps of the data transmitted using quadrature amplitude phase modulation is four times the baud rate. For example, a system using a signal with a baud rate of 2400 achieves a data transfer rate of 9600 bps (4 × 2400). Quadrature amplitude phase modulation is commonly employed in contemporary modems.

Figure 2-19
Figure (a) shows 12 different phases while Figure (b) shows a phase change with two different amplitudes

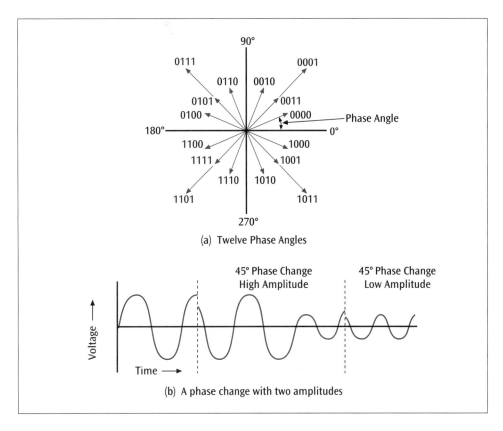

(a) Twelve Phase Angles

(b) A phase change with two amplitudes

Transmitting analog data with digital signals

Many times, it is necessary to transmit analog data over a digital medium. For example, many scientific laboratories contain testing equipment that generate test results as analog data. This analog data is converted to digital signals so that the original data can be transmitted through a computer system and eventually stored in memory or on a magnetic disk. A music recording company that creates a compact disc also converts analog data to digital signals. An artist performs a song that produces music, which is analog data. A device then converts this analog data to digital data so that the binary 1s and 0s of the digitized music can be stored, edited, and eventually recorded onto a compact disc. When the compact disc is used, a person inserts the disc into a compact disc player that converts the binary 1s and 0s back to analog music. Two techniques for converting analog data to digital signals are pulse code modulation and delta modulation.

Pulse Code Modulation

One encoding technique that converts analog data to a digital signal is **pulse code modulation.** Hardware converts the analog data to a digital signal by tracking the analog waveform, taking "snapshots" of the analog data at fixed intervals. A snapshot involves calculating the height, or voltage, of the analog waveform above a given threshold. This height, which is an analog value, is converted to an equivalent fixed-sized binary value. This binary value is then transmitted by means of a digital encoding format. Tracking an analog waveform and converting it to pulses that represent the wave's height above (or below) a threshold is termed **pulse amplitude modulation** (PAM). Conversion of the individual pulses into binary values is **pulse code modulation.** For the sake of brevity, however, we will refer to the entire process as simply pulse code modulation.

Figure 2-20 shows an example of pulse code modulation. At time T (on the X axis), a snapshot of the analog waveform is taken, resulting in the decimal value 12.0 (on the Y axis). The 12.0 is converted to an n-bit binary value (such as 0000 0000 1100) and transmitted to a device for storage. The decimal values can be converted to various sized binary values, such as 8-, 12-, or 16-bit binary values, depending on the desired degree of precision. At time 2t, a second snapshot is taken and the decimal value 4.1 is converted to a second n-bit binary value and stored. This process continues as snapshots are taken, converted to binary form, and stored.

Figure 2-20
Example of taking "snapshots" of an analog waveform for conversion to a digital signal

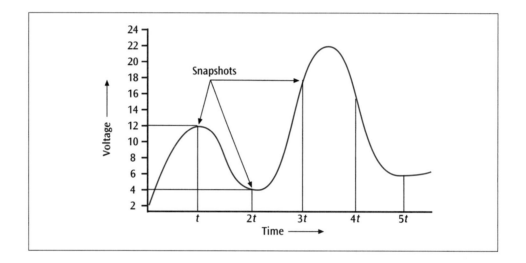

If you want to reconstruct the original analog waveform from the stored digital values, special hardware converts each n-bit binary value back to decimal and generates an electric pulse of appropriate magnitude (height). With a continuous incoming stream of converted values, something close to the original waveform should result, as shown in Figure 2-21.

Figure 2-21
Reconstruction of the analog waveform from the digital "snapshots"

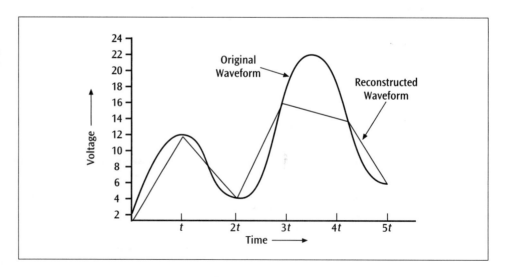

The closer the snapshots are taken to one another (the smaller the time intervals between snapshots, or the finer the resolution), the more accurate will be the reconstructed waveform (Figure 2-22). As always, you don't get something for nothing. To take the snapshots at shorter time intervals, the hardware must be of high enough quality to track the incoming signal quickly and perform the necessary conversions. The frequency at which the snapshots are taken is called the **sampling rate**. If you take samples at an unnecessarily high sampling rate, you will expend much energy for very little gain in the resolution of the reconstruction of the waveform. It is equally possible to have too few samples—a low sampling rate—which would reconstruct a waveform that is not an accurate reproduction of the original.

Figure 2-22
A more accurate reconstruction of the original waveform using a higher sampling rate

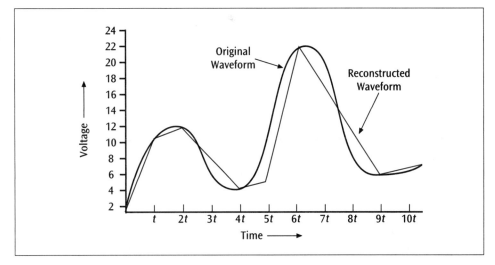

What then is the optimal balance between too high a sampling rate and too low? According to Nyquist's Theorem, to ensure a reasonable reproduction of an analog waveform using pulse code modulation, the sampling rate must be at least twice the highest frequency of the original analog waveform. (You can find more information on Nyquist's Theorem in the Details section.) Using the telephone system as an example, if you assume the highest frequency of a voice is 3100 Hz, you need a sampling rate of at least 6200 samples per second to ensure reasonable reproduction. The telephone system samples at a slightly higher rate of 8000 samples per second to create a slightly better quality reproduction.

A sampling rate of two times the greatest incoming frequency is usually sufficient to reproduce a human voice, but may not be sufficient for analog data with more dynamic properties, such as high quality music. When the reproduced analog waveform is not an accurate representation of the original waveform, **quantizing noise** has been introduced. Some people will tell you that the compact disc offers a poor reproduction of the original music and thus has introduced much quantizing noise. Average listeners, however, cannot hear the difference between a compact disc and the original music, or if they can, don't feel strongly enough to pursue the issue.

Delta Modulation

A second method of analog data to digital signal conversion is delta modulation. Figure 2-25 shows an example. With **delta modulation,** electronic hardware tracks the incoming analog data by assessing up or down "steps." During each time period *T*, the hardware determines whether the waveform has risen one delta step or dropped one delta step. If the waveform rises one delta step, a 1 is transmitted. If the waveform drops one delta step, a 0 is transmitted. With this encoding technique, only one bit per sample is generated. Thus the conversion from analog to digital using delta modulation is quicker than with pulse code modulation, in which each analog value is first converted to a PAM value then the PAM value is converted to binary.

There are two problems with delta modulation. If the analog waveform rises or drops too quickly, the hardware may not be able to keep up with the change, and **slope overload noise** results. Analog waveforms that do not change at all present another problem for delta modulation. Since the hardware outputs a 1 or a 0 only for a rise or a fall, respectively, a non-changing waveform generates alternating 1s and

---

Details ▶

The Relationship Between Frequency and Bits Per Second

When a network application is slow, users often demand that the network specialists transmit the data faster and thus solve the problem. What many network users do not understand is that you cannot keep transmitting data faster and faster without also increasing the frequency of the signal representing the data. And you cannot keep sending the signal faster unless the medium that transmits the signal is capable of supporting the higher frequencies. It is helpful to understand the relationship between bits per second and the frequency of a signal and to be able to use two simple measures to calculate the data transfer rate of a system.

There is an important relationship between the frequency of a signal and the number of bits that a signal can convey per second:

The greater the frequency of a signal, the higher the possible data transfer rate.

The converse is also true:

The higher the desired data transfer rate, the greater the needed signal frequency.

You can see a direct relationship between the frequency of a signal and the data rate (in bits per second, or bps) of the data that a signal can carry. Consider the NRZ-L encoding of the bit string 10101010 shown in Figure 2-23. In the first part of Figure 2-23, the digital signal changes four times during a one second period. The frequency of the digital signal is 2 Hz (2 complete cycles in one second), and the data transfer rate is 4 bps. In the second part of the figure, the digital signal changes eight times during a one second period. The frequency of the digital signal is 4 Hz, and the data transfer rate is 8 bps. As the frequency

of the signal increases, the data transfer rate (in bps) increases. Thus, there is a direct relationship between the frequency of a signal and its data transfer rate.

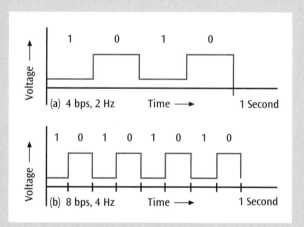

Figure 2-23 *Comparison of signal frequency with bits per second*

This example is simple because there are only two signal levels, one for a binary 0 and one for a binary 1. What if we create an encoding technique with four signal levels, as shown in Figure 2-24? Since there are four signal levels, each signal level can represent two bits. More precisely, the first signal level can represent a binary 00, the second a 01, the third a 10, and the fourth signal level a binary 11. Now when the signal level changes, two bits of data will be transferred.

0s, thus generating quantizing noise. Figure 2-25 demonstrates delta modulation and shows both slope overload noise and quantizing noise.

Figure 2-25
Example of delta modulation that is experiencing slope overload noise and quantizing noise

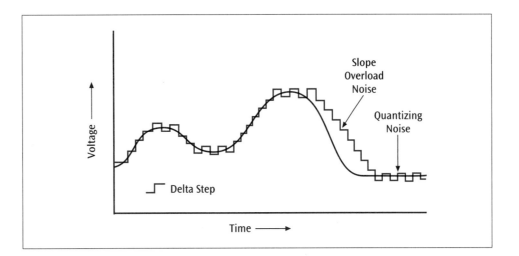

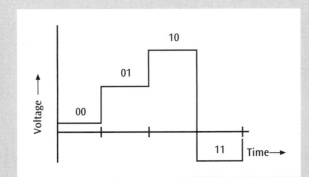

Figure 2-24 *Hypothetical signaling technique with four signal levels*

Two formulas express the direct relationship between the frequency of a signal and its data transfer rate, Nyquist's Theorem and Shannon's formula. **Nyquist's Theorem** calculates the data transfer rate of a signal given its frequency and the number of signaling levels:

$$C = 2 f \log_2 L$$

in which C is how fast the data can transfer over a medium in bits per second (the channel capacity), f is the frequency of the signal, and L is the number of signaling levels. For example, given a

3100 Hz signal and two signaling levels, the resulting channel capacity is 6200 bps ($2 \times 3100 \times \log_2 2 = 2 \times 3100 \times 1$). (Be careful to use $\log_2$ and not $\log_{10}$.) A 3100 Hz signal with four signaling levels yields 12400 bps.

Shannon's formula calculates the maximum data transfer rate of *any* signal (with any number of signal levels) and incorporates *noise*:

$$S(f) = f \log_2 (1 + W/N) \text{ bps}$$

in which S(f) is the data transfer rate in bits per second, f is the frequency of the signal, W is the power of the signal in watts, and N is the power of the noise in watts.

Consider a 3400 Hz signal with a power level of 0.2 watts and a noise level of 0.0002 watts:

$$\begin{aligned} S(f) &= 3400 \times \log_2 (1 + 0.2/0.0002) \\ &= 3400 \times \log_2 (1001) \\ &= 3400 \times 9.97 \\ &= 33898 \text{ bps} \end{aligned}$$

(If your calculator does not have a $\log_2$ key, which many don't, you can always approximate an answer by taking the $\log_{10}$ and then dividing by 0.301.)

Transmitting analog data with analog signals

The combination of analog data and analog signals is similar to the combination of digital data and digital signals: if analog data is transmitted using analog signals, relatively little needs to be done to the original analog data. The most common change that takes place is for analog data to be modulated into a different frequency analog signal. The original analog data can be "moved" to a different frequency range so that it can fit on an appropriate medium or share a medium with other analog signals at other frequencies. When many signals share a medium the effect is called multiplexing. (Multiplexing is discussed in Chapter Five.)

AM radio, FM radio, broadcast television and cable television are the most common examples of analog data to analog signal conversion. Consider Figure 2-26, which shows AM radio, as an example. The audio data that is generated by the radio station might appear like the first sine wave shown in the figure. Then, a carrier wave signal is used to convey the audio data. Essentially, the original audio waveform and the carrier wave are added together, resulting in the third waveform. Note how the dotted lines superimposed over the third waveform follow the same outline as the original audio waveform. The original audio data has been modulated onto a particular carrier frequency (the frequency at which you set the dial to tune in a station) using amplitude modulation. Frequency modulation can be used in similar ways to modulate analog data onto an analog signal.

Figure 2-26
An audio waveform modulated onto a carrier frequency using amplitude modulation

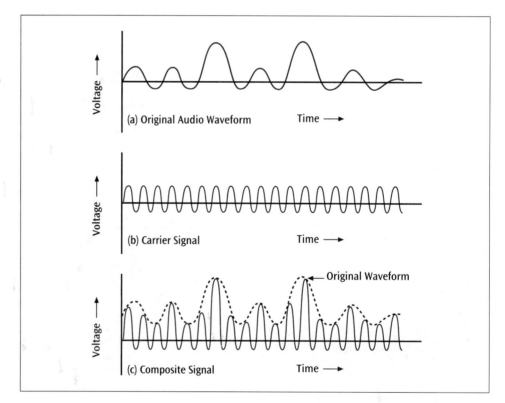

(a) Original Audio Waveform   Time ⟶

(b) Carrier Signal   Time ⟶

(c) Composite Signal   Time ⟶

Original Waveform

Spread Spectrum Technology

Using a spread spectrum transmission system, it is possible to transmit either analog or digital data using an analog signal. However, unlike other encoding and modulation techniques, only an intended receiver with the same type of transmission system can accept and decode the transmissions. The idea behind **spread spectrum** transmission is to bounce the signal around on seemingly random frequencies rather than transmit the signal on one fixed frequency. Anyone trying to eavesdrop will not be able to listen because the transmission frequencies are constantly changing. How does the intended receiver follow this random bouncing around of frequencies? The signal does not actually bounce around on random frequencies; it only *seems* to do so. The transmitter actually follows a *pseudo*random sequence of frequencies. The intended receiver possesses the hardware and software knowledge to follow this pseudorandom sequence of frequencies.

Figure 2-27 demonstrates the basic operation of a spread spectrum receiver and transmitter system. The input data enters a channel encoder, which is a device that produces an analog signal with a narrow bandwidth centered around a particular frequency. This signal is then modulated onto a seemingly random pattern of frequencies using a pseudorandom number sequence as the guide. The pseudorandomly modulated signal is then transmitted to a friendly receiver. The first operation the receiver performs is to "unscramble," or decode, the modulated signal using the same pseudorandom sequence that the transmitter used to encode the signal. The unmodulated signal is then sent to the channel decoder, which performs the opposite operation of the channel encoder. The result is the original data.

Figure 2-27
Basic operation of a spread spectrum receiver and transmitter system

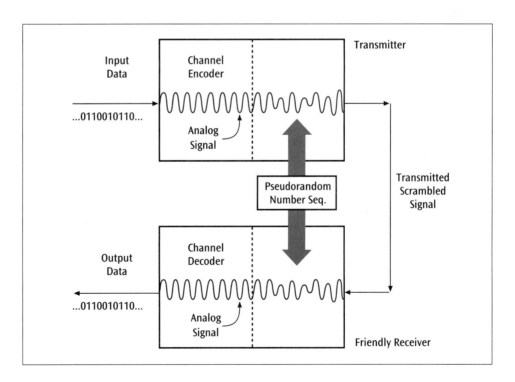

Spread spectrum technology is one of the more recent technologies to filter down from the U.S. military. One of the more common commercial applications of spread spectrum is wireless communications, such as the cordless telephone. Cordless telephones that incorporate spread spectrum technology are impervious to intruder eavesdropping, which is an inherent problem with standard cordless telephones. Spread spectrum technology comes in two flavors: frequency hopping and direct sequence. Both of these techniques are discussed in the Details section.

Data Codes

One of the most common forms of data transmitted between a sender and a receiver is textual data. For example, banking institutions that wish to transfer money often transmit textual information, such as account numbers, names of account owners, bank names, addresses, and the amount of money to be transferred. This textual

Details ▶

Spread Spectrum Frequency Hopping and Direct Sequence

There are two techniques for creating a spread spectrum signal. The first technique, **frequency hopping spread spectrum**, uses a pseudorandom number generator to indirectly control the synthesis of frequencies onto which the original data is modulated. Figure 2-27 shows the basic operation of a spread spectrum receiver and transmitter system. The outgoing spread spectrum signal is not modulated onto a constant set of frequencies, but instead hops around on a seemingly random series of frequencies. Anyone trying to eavesdrop hears only noise, as they do not possess the hardware and software necessary to track the pseudorandom sequence of frequencies. Furthermore, spread spectrum signals are harder to jam than standard radio transmissions should an unfriendly person try to disrupt the transmission of signals.

The second technique for creating a spread spectrum signal is direct sequence spread spectrum. **Direct sequence spread spectrum** spreads the transmission of a signal over a wide range of frequencies using mathematical values. Figure 2-28 shows that as the original data is input into a direct sequence modulator, it is exclusive-ORed with a pseudorandom bit stream. In exclusive-ORing, a zero exclusive-ORed with a zero

equals zero; a one exclusive-ORed with a one also equals zero; and any other combinations equal one. The output is the result of the exclusive-OR between the input data and the pseudorandom bit sequence. When the data arrives at the intended receiver, the spread spectrum signal is again exclusive-ORed with the same pseudorandom bit stream that was used during the transmission of the signal. The result of this exclusive-OR at the receiving end is the original data.

Code division multiple access (CDMA), which is used in digital wireless telephones, is based on direct sequence spread spectrum technology.

Two other spread spectrum techniques exist, but they are not used in commercial applications as are frequency hopping and direct sequence. Time-hopped spread spectrum requires a large amount of random access memory and fast micro-controlled integrated circuits to generate the necessary spread spectrum signals. However, as circuitry becomes smaller, faster, and cheaper, time-hopped systems may become commercially available. Chirp signal spread spectrum is often used in radar systems and is only rarely encountered in commercial spread spectrum systems.

information is transmitted as a sequence of characters. To distinguish one character from another, each character is represented by a unique binary pattern of 1s and 0s. The set of all textual characters or symbols and their corresponding binary patterns is called a **data code**. Two important data codes are EBCDIC and ASCII. A third code, Baudot, is presented for the unique way it handles code assignments.

EBCDIC

The **Extended Binary Coded Decimal Interchange Code**, or EBCDIC code, is an 8-bit code allowing 256 possible combinations of textual symbols ($2^8 = 256$). These 256 combinations of textual symbols include all uppercase and lowercase letters, the 10 digits, a large number of special symbols and punctuation marks, and a number of control characters. The control characters, such as line feed (LF) and carriage return

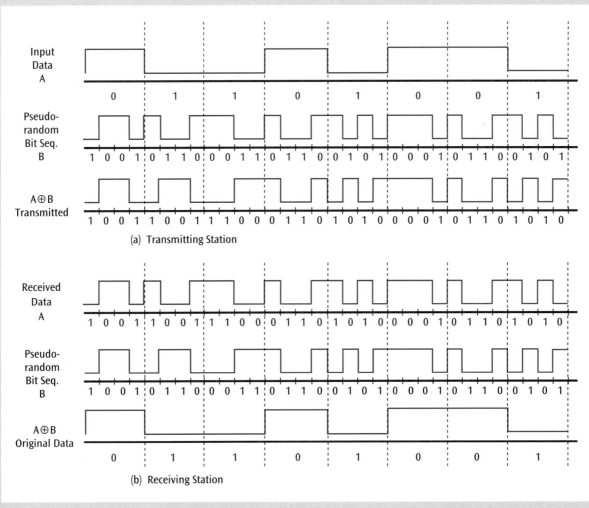

Figure 2-28 *Example of binary data as it is converted into a direct sequence spread spectrum and back*

(CR) provide control between a processor and an input/output device. Certain control characters provide data transfer control between a computer source and computer destination. All the EBCDIC characters are shown in Table 2-1.

Table 2-1
The EBCDIC character code set

Bits	4	0	0	0	0	0	0	0	0	1	1	1	1	1	1	1	1
	3	0	0	0	0	1	1	1	1	0	0	0	0	1	1	1	1
	2	0	0	1	1	0	0	1	1	0	0	1	1	0	0	1	1
	1	0	1	0	1	0	1	0	1	0	1	0	1	0	1	0	1
8 7 6 5																	
0 0 0 0		NUL	SOH	STX	EXT	PF	HT	LC	DEL			SMM	VT	FF	CR	SO	SI
0 0 0 1		DLE	DC_1	DC_2	DC_3	RES	NL	BS	IL	CAN	EM	CC		IFS	IGS	IHS	IUS
0 0 1 0		DS	SOS	FS		BYP	LF	EOB	PRE			SM			ENQ	ACK	BEL
0 0 1 1				SYN		PN	RS	UC	EOT					DC_4	NAK		SUB
0 1 0 0		SP												<	(	+	|
0 1 0 1		&										!	$	.	)	:	¬
0 1 1 0		—												%	-	>	?
0 1 1 1														@		=	"
1 0 0 0			a	b	c	d	e	f	g	h	i						
1 0 0 1			j	k	l	m	n	o	p	q	r						
1 0 1 0				s	t	u	v	w	x	y	z						
1 0 1 1																	
1 1 0 0			A	B	C	D	E	F	G	H	I						
1 1 0 1			J	K	L	M	N	O	P	Q	R						
1 1 1 0				S	T	U	V	W	X	Y	Z						
1 1 1 1		0	1	2	3	4	5	6	7	8	9						

For example, if you want a computer to send the message "Transfer $1200.00" the following EBCDIC characters would be sent:

1110 0011	T
1001 1001	r
1000 0001	a
1001 0101	n
1010 0010	s
1000 0110	f
1000 0101	e
1001 1001	r
0100 0000	space
0101 1011	$
1111 0001	1
1111 0010	2
1111 0000	0
1111 0000	0
0101 1100	.
1111 0000	0
1111 0000	0

IBM mainframe computers are a major user of the EBCDIC character set.

ASCII

The American National Standard Code for Information Interchange (ASCII) is a government standard in the United States and is one of the most widely used data codes in the world. The **ASCII character set** comes in a few different forms including a 7-bit version that allows for 128 possible combinations of textual symbols ($2^7 = 128$), representing upper and lowercase letters, the digits 0 to 9, special symbols, and control characters. Since the byte is a common unit of data and consists of eight bits, the 7-bit version of ASCII characters usually includes an eighth bit. This eighth bit can be used to detect transmission errors (which will be discussed in Chapter Six), can provide for 128 additional characters defined by the application using the ASCII code set, or can be a binary 0. Table 2-2 shows the ASCII character set and the corresponding 7-bit values.

Table 2-2
The ASCII character set

		High-Order Bits (7, 6, 5)							
		000	001	010	011	100	101	110	111
	0000	NUL	DLE	SPACE	0	@	P	`	p
	0001	SOH	DC1	!	1	A	Q	a	q
	0010	STX	DC2	"	2	B	R	b	r
	0011	ETX	DC3	#	3	C	S	c	s
	0100	EOT	DC4	$	4	D	T	d	t
	0101	ENQ	NAK	%	5	E	U	e	u
	0010	ACK	SYN	&	6	F	V	f	v
	0111	BEL	ETB	'	7	G	W	g	w
	1000	BS	CAN	(	8	H	X	h	x
	1001	HT	EM	)	9	I	Y	i	y
	1010	LF	SUB	*	:	J	Z	j	z
	1011	VT	ESC	+	;	K	[	k	{
	1100	FF	FS	,	<	L	\	l		
	1101	CR	GS	-	=	M	]	m	}
	1110	SO	RS	.	>	N	^	n	~
	1111	SI	US	/	?	O	—	o	DEL

Low-Order Bits (4, 3, 2, 1)

Using the example of sending the message "Transfer $1200.00", the corresponding ASCII characters are:

1010100	T
1110010	r
1100001	a
1101110	n
1110011	s
1100110	f
1100101	e
1110010	r
0100000	space
0100100	$
0110001	1
0110010	2

0110000	0
0110000	0
0101110	.
0110000	0
0110000	0

Baudot code

The **Baudot code** was developed by Emile Baudot and uses 5-bit patterns to represent the characters A to Z, the numbers 0 to 9, and several special characters. A 5-bit pattern yields only 32 combinations of textual symbols ($2^5 = 32$). A data code, however, needs to represent 26 letters of the alphabet, 10 digits, and special characters—more than 36 distinct characters.

How can the Baudot code represent more than 32 characters with only a 5-bit code? Table 2-3 shows that each 5-bit pattern in the Baudot code can actually represent two different characters. The Baudot code uses the special bit pattern 11111—called a downshift code—to denote letters. It uses the special bit pattern 11011—called an upshift code—to denote numbers or figures. Every character except for a space is either a downshift character or an upshift character. This process works much like the keyboard on a computer. If the Caps Lock key has been pressed, all characters typed are downshift characters. For example, the bit pattern 10000 represents either the letter T or the number 5. The letter T is a downshift character, and the number 5 is an upshift character. If you wish to transmit a series of downshift characters, then those characters are preceded by the downshift, or letters, bit pattern. If you want to transmit one or more upshift characters, you first transmit the upshift, or figures, bit pattern. If the downshift bit pattern (11111) is first transmitted, then the bit pattern 10000, transmitted immediately following it, represents the letter T. If the figures (or upshift) bit pattern (11011) was most recently transmitted, then the pattern 10000, immediately following it, represents the number 5. By using the special upshift and downshift bit patterns, the 5-bit Baudot code can represent all 36 letters and numbers plus special characters.

Table 2-3
The Baudot code character set

	Letters	Figures		Letters	Figures
Binary	Shift	Shift	Binary	Shift	Shift
00000	blank	blank	10000	T	5
00001	E	3	10001	Z	+
00010	LF	LF	10010	L	)
00011	A	-	10011	W	2
00100	space	space	10100	H	reserved
00101	S	'	10101	Y	6
00110	I	8	10110	P	0
00111	U	7	10111	Q	1
01000	CR	CR	11000	O	9
01001	D	WRU	11001	B	?
01010	R	4	11010	G	reserved
01011	J	BELL	11011	FIGURES	FIGURES
01100	N	,	11100	M	.
01101	F	reserved	11101	X	/
01110	C	:	11110	V	=
01111	K	(	11111	LETTERS	LETTERS

The Baudot code is fairly dated and rarely, if ever, used anymore. However, the Baudot code demonstrates an important concept: A special control character can signal a change in the symbols that follow. The technique of using a special control character to signal something unique appears frequently in computer networks and data communications.

Data and Signal Conversions In Action ▶

Examining two typical business applications in which a variety of data and signal conversions are performed will show how analog and digital data, analog and digital signals, and data codes work together. First, consider a person at work who wants to send an e-mail to a colleague asking about the time for the next meeting. For simplicity, let's assume the message says "Sam, what time is the meeting with accounting? Hannah." and is being sent from a microcomputer that is connected to a local area network. This local area network is connected to the Internet. We'll pretend this is a small business, so the connection to the Internet is over a dial-up modem (Figure 2-29).

Figure 2-29
User sending email from a personal computer over a local area network and the Internet, via a modem

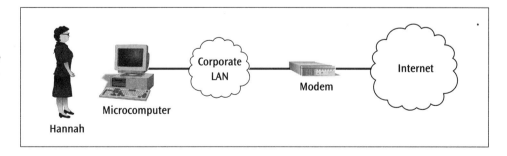

Hannah enters the message into the e-mail program and clicks the Send button. The e-mail program prepares the e-mail message, which contains the data "Sam, what time is the meeting with accounting? Hannah" plus whatever other information is necessary for the e-mail program to properly send it. Since this message contains text, it will be represented in ASCII:

Original message : Sam, what time is the meeting with accounting? Hannah

ASCII string :   1010011 1100001 1101101 ...   (for brevity's sake, only the "Sam" portion   S   a   m   of the message appears in ASCII

Next, the ASCII message is transmitted over a local area network (LAN) within the company. Assume this LAN uses Differential Manchester encoding. The ASCII string now appears as a digital signal as shown in Figure 2-30.

Figure 2-30
The first three letters of the message "Sam, what time is the meeting with accounting? Hannah," using Differential Manchester encoding

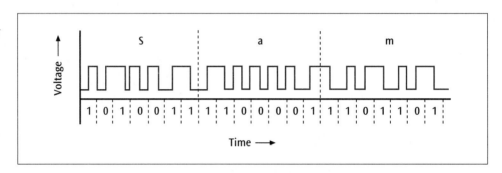

This Differential Manchester encoding of the message travels over the local area network and arrives at another computer, which is connected to a modem. This computer converts the message back to an ASCII string and then transmits the ASCII string to the modem. The modem prepares the message for transmission over the Internet using frequency modulation. For brevity's sake, only the first seven bits of the ASCII string (in this case, the 'S' from Sam) are converted using simple frequency shift keying (Figure 2-31).

Figure 2-31
The frequency modulated signal for the letter 'S'

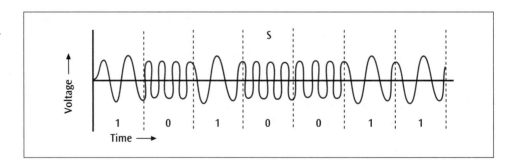

This frequency modulated signal travels over the telephone lines and arrives at the appropriate Internet gateway (the Internet service provider), which demodulates the signal into an ASCII string. From there, the ASCII string representing the original message moves out to the Internet and finally arrives at the intended receiver's computer. The process of transmitting over the Internet and delivery to the intended receiver's computer implies several more code conversions. Since we do not yet have an understanding of what happens over the Internet, nor do we know what kind of connection the receiver has, this portion of the example has been omitted. This relatively simple example demonstrates the number of times a conversion from data to signal to data is performed during a message transfer.

A second example involves the commonplace telephone. The telephone system in the United States is an increasingly complex marriage of traditionally analog telephone lines and modern digital technology. About the only portion of the telephone system that remains analog is the local loop—the wire that leaves your house, apartment, or business and runs to the nearest telephone switching center. Your voice, as you speak into the telephone, is an analog signal that travels over a wire to the local switching center where it is digitized and transmitted to another switching center somewhere in the vast telephone network.

Since the human voice is analog but a good portion of the telephone system is digital, what kind of analog to digital signal conversions are performed? As mentioned earlier in the chapter, the human voice occupies analog frequencies from 300 Hz to 3100 Hz. When transmitted over the telephone system, a bandwidth of 4000 Hz (4 KHz) is used. When this 4000 Hz signal reaches the local telephone office, it is sampled at two times the greatest frequency (Nyquist's Theorem), or 8000 samples per second. Telephone studies have shown that the human voice can be digitized using only 128 different quantization levels. Since 2^7 equals 128, each of the 8000 samples per second can be converted into a 7-bit value, yielding 8000×7, or 56,000 bits per second.

This 56,000 bps digital signal may have to be transmitted again using analog signals, thus there is a need to create an analog signal that looks like the 56,000 bps digital signal. To create an analog signal that closely mimics a digital signal, you need an analog signal that is roughly

Serial port - Serializes the bits
1-Byte -> 8 bits

Serial port -> modem

Sp -> Modem ——— Modem -> SP
↑ serializes byte; ↓ mod-bits; demodulates; ↓ deserializes

half the frequency of the bps signal, or 28,000 Hz. Note in Figure 2-32 that the lower frequency analog signals do a poor job of looking like the digital signal. It is not until an analog signal that is half the frequency of the digital signal is generated that a close representation is created.

Figure 2-32
Analog signal representations of a digital signal

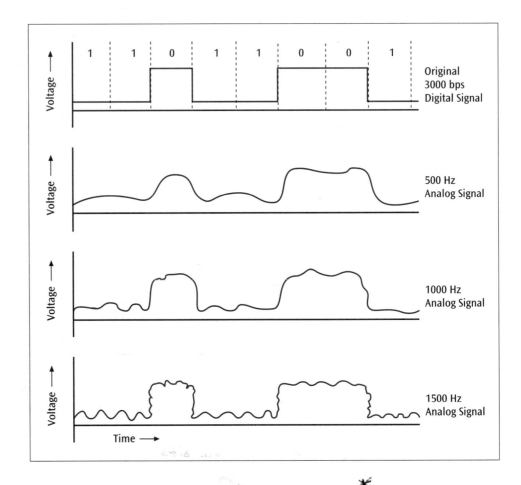

You might notice something interesting about this example. We started with a 4000 Hz voice signal and ended with a 28,000 Hz representation of the original. Is it worthwhile to create a signal that is seven times the frequency of the original? Yes, it is. It is much easier to clean noise out of a digital signal that follows a discrete on and off pattern. In addition, a digital signal is necessary to combine multiple telephone calls over a single wire and to operate the electronic equipment that completes a telephone call, which are both prominent in today's modern telephone systems.

◆ ◆

SUMMARY

▶ Computer networks exist because data and signals need to be transmitted.
▶ All data transmitted over any communications medium is either digital or analog.
▶ Data is transmitted with a signal that, like data, can be either digital or analog.

▸ The most important difference between analog and digital data and signals is that it is easier to remove noise from digital data and signals than from analog data and signals.

▸ All signals consist of three basic components: amplitude, frequency, and phase.

▸ Two important factors affecting the transfer of a signal over a medium are attenuation and noise.

▸ Since both data and signals can be either digital or analog, four combinations of data and signals can be produced: digital data carried by a digital signal, digital data carried by an analog signal; analog data carried by an analog signal, and analog data carried by a digital signal.

▸ Digital data carried by digital signals are represented by digital encoding formats, including the popular Manchester codes.

▸ Manchester codes always have a transition in the middle of the bit, which allows the receiver to synchronize itself with the incoming signal.

▸ For digital data to be transmitted using analog signals, the digital data must first be modulated onto an analog signal.

▸ The three basic techniques of modulation are amplitude modulation, frequency modulation, and phase modulation.

▸ Two common techniques for converting analog data so that it may be carried over digital signals are pulse code modulation and delta modulation

▸ Pulse code modulation converts samples of the analog data to multiple-bit digital values.

▸ Delta modulation tracks analog data and transmits only a 1 or a 0 depending on whether the data rises or falls within the next time period.

▸ To transmit analog data over an analog signal, the data is often modulated to another frequency to achieve efficient transmission.

▸ When a signal is transmitted using spread spectrum techniques, the signal continuously hops from one frequency to another to prevent eavesdropping, disruption of the transmission, or other malicious intervention.

▸ Frequency hopping spread spectrum uses a pseudorandom sequence of frequencies to give the appearance of a signal that randomly jumps from one frequency to another.

▸ Direct sequence spread spectrum involves combining the original data with a pseudorandom bit stream using an exclusive-OR technique, which produces a transmitted signal that is unreadable to anyone not possessing the same technology.

▸ Data codes are necessary to transmit the letters, numbers, symbols, and control characters found in text data. Three important data codes are ASCII, EBCDIC, and Baudot.

▸ The ASCII data code uses a 7-bit code and allows for 128 different letters, digits, and special symbols. ASCII is the most popular data code in the U.S.

▸ The EBCDIC data code uses an 8-bit code and allows for 256 different letters, digits, and special symbols. IBM mainframes use the EBCDIC code.

▸ The Baudot code provides an interesting example of using a shift character to represent a second sequence of characters.

KEY TERMS

amplitude	delta modulation	phase
amplitude modulation	Differential Manchester code	phase modulation
analog data	digital data	pulse amplitude modulation (PAM)
analog signals	digital signals	pulse code modulation
ASCII	EBCDIC	quadrature amplitude modulation
attenuation	effective bandwidth	quadrature phase modulation
bandwidth	frequency	quantizing noise
baud	frequency modulation	self-clocking
baud rate	Hertz (Hz)	shift keying
Baudot Code	intermodulation distortion	signals
bits per second (bps)	Manchester codes	slope overload noise
data	modulation	spectrum
data code	noise	spread spectrum
decibel (db)	period	4B/5B

REVIEW QUESTIONS

1. What is the difference between data and signals?
2. What are the main advantages of digital signals over analog signals?
3. What is the difference between a continuous signal and a discrete signal?
4. What are the three basic components of all signals?
5. What is the spectrum of a signal?
6. What is the bandwidth of a signal?
7. How does a differential code differ from a non-differential code?
8. What does it mean when a signal is self-clocking?
9. What is the definition of baud rate?
10. How does baud rate differ from bits per second?
11. What are the three main types of modulation?
12. What is the difference between pulse code modulation and delta modulation?
13. What is meant by the sampling rate of analog data?
14. Why would analog data have to be modulated onto an analog signal?
15. What are the two basic techniques used to create a spread spectrum signal?
16. What is the primary advantage of spread spectrum?
17. What are the differences between ASCII and EBCDIC?
18. What is the significance of the Shift characters in the Baudot code?

EXERCISES

1. What is the frequency in Hertz of a signal that repeats 80,000 times within one minute? What is its period?
2. What is the bandwidth of a signal composed of frequencies from 50 Hz to 500 Hz?
3. Draw in chart form (as shown in Figure 2-11) the voltage representation of the bit pattern 11010010 for the digital encoding schemes NNZ-L, NRZ-I, Manchester, and Differential Manchester.
4. What is the baud rate of a digital signal that employs Differential Manchester and has a data transfer rate of 2000 bps?

Handwritten notes in left margin:

$2^{numbers} = $ no of phase

$2^{x} = 8$ $x = 3$

$6,000 bps$

$1,536,000 bps$

5. Show the equivalent 4B/5B code of the bit string 1101 1010 0011 0001 1000 1001.

6. What is the data transfer rate in bps of a signal that is encoded using phase modulation with 8 different phase angles and a baud rate of 2000?

7. Show the equivalent analog sine-wave pattern of the bit string 00110101 using amplitude shift keying, frequency shift keying, and phase shift keying.

8. Twenty-four voice signals are to be transmitted over a single high-speed telephone line. What is the bandwidth required (in bps) if the standard analog to digital sampling rate is used and each sample is converted into an 8 bit value?

9. Given the analog signal shown in Figure 2-33, what are the 8-bit pulse code modulated values that will be generated at each time *T*?

Figure 2-33
Analog signal for problem number 9

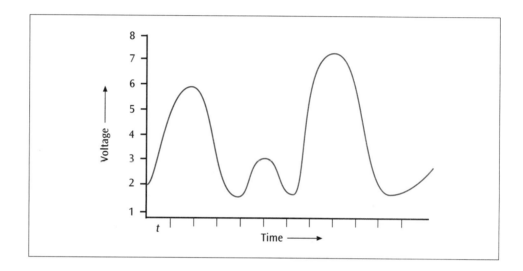

10. Using the analog signal from problem 9 and a delta step that is one-eighth inch long and one-eighth inch tall, what is the delta modulation output? Point out any slope overload noise.

Exercises from the Details sections:

11. Using Nyquist's Theorem, what is the channel capacity C of a signal that has 16 different levels and a frequency of 20,000 Hz?

12. Using Shannon's formula, calculate the data transfer rate given the following information:

    signal frequency = 10,000 Hz

    signal power = 5000 watts

    noise power = 230 watts

13. What is the decibel loss of a signal that starts at point A with a strength of 2000 watts and ends at point B with a strength of 400 watts?

14. What is the decibel loss of signal that starts at 50 watts and experiences a 10-watt loss over a given section of cable?

15. What is the decibel loss of a signal that loses half its power during the course of transmission?

THINKING OUTSIDE THE BOX

1 You are working for a company that has a network application for accessing a dial-up database of corporate profiles. From your computer workstation, a request for a profile travels over the corporate local area network to a modem. The modem, using a conventional telephone line, dials into the database service. The database service is essentially a modem and a mainframe computer. Create a table that shows every time data or signals are converted to a different form. For each entry in the table, show where the conversion is taking place, the form of the incoming information, and the form of the outgoing information.

2 You have been asked to provide input to a trucking problem your company has encountered. While each truck is moving around the metropolitan area, the truck has to communicate its location to a central site every 5 minutes. Should the system use analog signals or digital signals? Explain. What further information do you need to recommend a good solution?

PROJECTS

1. Using sources from the library or the Internet, write a 2-3 page paper that describes how a compact disc player performs the digital to analog conversion of its contents. (*Digital Audio and Compact Disc Technology 2nd ed.*, edited by Luc Baert, Luc Theunissen, and Guido Vergult, B.H. Newnes Publishing Co, 1992 is a very good source.)

2. There are many more digital encoding schemes other than NRZ-I, NRZ-L, Manchester, and Differential Manchester. List three other encoding techniques and show an example of how they encode.

3. Write either an algorithm on paper or a computer program that inputs sample text and outputs the equivalent Baudot code, including the appropriate UPSHIFT and DOWNSHIFT characters.

3

The Media: Conducted and Wireless

◆◆◆

MOST USERS of computer systems spend so much time and energy working with a computer workstation that they give little time or thought to the wiring or media that interconnect the workstations and devices, that is, until something goes wrong. When there is faulty or obsolete wiring, it is customary to pull a new wire or two and hope it rectifies the situation. When something went wrong with the wiring at the McHenry Library at the University of California at Santa Cruz, the staff could no longer tell one wire from another. They couldn't even tell which wires were in use and which were no longer necessary. The wiring conduit that holds the wires had no more room to add additional wires. Likewise, the wiring closets where the runs of wires were interconnected were at capacity. Finally, much of the wiring was so old it could not support the newer technologies and computer applications. It was time to replace the entire wiring system for the library, and it had to be done without disrupting service to library employees and their patrons.

After roughly three months of intense but relatively quiet work, the following had been achieved:

over 20 miles of wire had been pulled;

669 voice and data lines were installed;

60 public terminals and workstations were wired; and

80 staff workstations were wired.

Through proper and thorough planning, continuous consulting with experts, and a high level of communication, the project was completed. More importantly, the appropriate media were chosen, which should support the library well into the 21st century.

Online, January/February 1995, vol.19, pp. 62-68.

Is the choice of media that critical to the success or failure of a system?

How do you choose one type of media over another?

If the McHenry Library were to rewire today (just five years later), would it choose different media?

Objectives ▷

After reading this chapter, you should be able to:

▶ Outline the characteristics of twisted pair wire including the advantages and disadvantages and the differences between Category 1, 2, 3, 4, 5, 5e, 6, and 7 twisted pair wire.

▶ Describe when shielded twisted pair wire works better than unshielded twisted pair wire.

▶ Outline the characteristics of coaxial cable including the advantages and disadvantages.

▶ Outline the characteristics of fiber optic cable including the advantages and disadvantages.

▶ Outline the characteristics of terrestrial microwave systems including the advantages and disadvantages.

▶ Outline the characteristics of satellite microwave systems including the advantages and disadvantages and including the differences between low earth orbit, middle earth orbit, and geosynchronous earth orbit satellites.

▶ Describe the basics of wireless radio, including AMPS, D-AMPS, and PCS systems.

▶ Outline the characteristics of cellular digital packet data including the advantages and disadvantages.

▶ Outline the characteristics of pager systems including the advantages and disadvantages.

▶ Outline the characteristics of broadband wireless systems including the advantages and disadvantages.

▶ Apply the media selection criteria of cost, speed, distance and expandability, environment, and security to various media in a particular application.

Introduction ▶

Physical - Conducted

Radiated - Wireless

The world of computer networks would not exist if there were no medium by which to transfer data. All communications media can be divided into two categories: physical or conducted media, such as wires; and radiated or wireless media, which use radio waves. Conducted media include twisted pair wire, coaxial cable, and fiber optic cable. There have been few recent and unique additions to conducted media. Fiber optic cable, which became widely used by the telephone companies in the 1980s, is the newest member of the family. Nonetheless, improvements in design and technology are continuously increasing the capabilities of the existing media.

In wireless transmission, various types of electromagnetic waves, such as radio waves, are used to transmit signals. Wireless transmission began with AM radio, FM radio, and television in the 1950s, and was followed in 1962 by transmissions through the first orbiting satellite, Telstar. Today, hundreds of applications have emerged over the past 50 years—all having grown from wireless transmission technology. This chapter examines seven basic groups of wireless media used for the transfer of data: terrestrial microwave transmissions; satellite transmissions; cellular radio systems; personal communication systems; pagers; infrared transmissions; and multichannel multipoint distribution service. Microwave systems use line-of-sight transmission towers, which transmit high data rate streams. Satellite systems are microwave transmissions that are directed to a satellite orbiting the earth, are regenerated, and are returned to earth at a different location. Cellular radio systems involve cellular telephones that transmit signals to a nearby cellular transmission tower. Although cellular phone systems are well established and based on older analog transmissions, a new player in the mobile telephone market emerged in November of 1995—personal communication systems (PCS). PCS telephones employ digital transmission techniques and typically offer more services and quieter transmissions. Pagers or beepers are mostly one-way communication devices that transmit short numeric or alphanumeric messages to remote users. Infrared transmission systems, which are similar to those in remote control devices, can transfer data at rates up to 4 Mbps (millions of bits per second) between two devices within a room. Broadband wireless systems, which include multichannel multipoint distribution service (MMDS), is a new technology that uses microwave transmission to send cable television and Internet access signals to an individual user.

Undoubtedly, as you read this paragraph, someone somewhere is designing new materials and building new equipment that is better than what currently exists. The transmission speeds and distances given in this chapter will continue to evolve. Please keep this in mind as you study the media. For the latest information on technological advances in media, consult the author's web page at *http://bach.cs.depaul.edu/cwhite*.

The chapter will conclude with a comparison of all the media types, followed by three examples of selecting the appropriate media for a particular application.

Twisted Pair Wire

The oldest, simplest, and most common type of conducted media is twisted pair wires. Twisted pair is almost a misnomer, as one rarely encounters a single pair of wires. More often, **twisted pair wire** comes as two or more pairs of single conductor wires that have been twisted around each other. Each wire is encased within plastic insulation and cabled within one outer jacket, as shown in Figure 3-1.

Figure 3-1
Example of 4-pair twisted pair wire

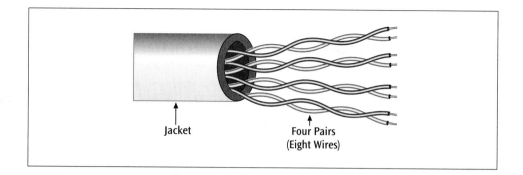

Jacket   Four Pairs
                (Eight Wires)

Unless someone strips back the outer jacket, you will not see the twisting of the wires. The twisting is done to reduce the amount of interference one wire can inflict on the other, one pair of wires can inflict on another pair of wires, and an external electromagnetic source can inflict on one wire in a pair. You might recall two important laws from physics: a current passing through a wire creates a magnetic field around that wire; and a magnetic field passing over a wire induces a current in that wire. Therefore, a current or signal in one wire can produce an unwanted current or signal called **crosstalk** in a second wire. If the two wires run parallel to each other, as shown in Figure 3-2(a), the chance for crosstalk increases. If the two wires cross each other at perpendicular angles, as shown in Figure 3-2(b), the chance for crosstalk decreases. It is the twisting of the two wires around each other, shown in Figure 3-2(c), that keeps the two wires at perpendicular angles as much as possible.

Figure 3-2
(a) Parallel wires – greater chance of crosstalk
(b) Perpendicular wires – lesser chance of crosstalk (c) Twisted wires – note how the wires keep crossing each other at perpendicular angles

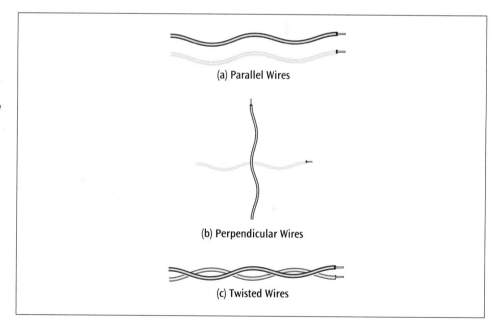

(a) Parallel Wires

(b) Perpendicular Wires

(c) Twisted Wires

You have probably experienced crosstalk many times. Remember when you were talking on the telephone and heard a conversation ever-so-faintly in the background? Your telephone connection, or circuit, was experiencing crosstalk from another telephone circuit.

As simple as twisted pair wire appears to be, it actually comes in many forms and varieties to support a wide number of applications. To help simplify the numerous varieties, twisted pair can be specified as **Category 1-5** and is abbreviated as **CAT 1-5**. **Category 1** twisted pair is standard telephone wire and is designed to carry analog voice or digital data at low speeds (less than or equal to 9600 bps). This wire often works fine for transmitting voice or digital data modulated onto an analog signal for the several miles from your house to the local telephone switch. However, Category 1 twisted pair wire is not recommended for transmitting megabits of computer data. Because of the materials used to make the wire, Category 1 wire is susceptible to noise and attenuation (weakening) of the signal. A few businesses use Category 1 twisted pair to connect computer workstations to local area networks, even though it is not recommended for that use. Often this situation is encountered when a room or building is already wired with Category 1 telephone wire, and it would be too difficult or expensive to install higher quality data cable. Note however, that using Category 1 cable for high speed computer communications can generate an unacceptable number of transmission errors.

Category 2 twisted pair wire is also used for telephone circuits, but is a higher quality wire than Category 1, producing less noise and signal attenuation. It is found on T-1 and ISDN lines and in newer installations of standard telephone circuits. The higher quality is due to the better quality copper used for the wire and better quality insulation surrounding the wire. T-1 is the designation for a digital telephone circuit that transmits voice or data at 1.544 Mbps. ISDN is a newer digital telephone circuit that can transmit voice or data or both from 64 Kbps to 1.544 Mbps. (Chapter Twelve provides more detailed descriptions of T-1 and ISDN.)

Category 3 twisted pair is designed to transmit 10 Mbps data over a local area network for distances up to 100 meters. Although the signal does not magically stop at 100 meters, the signal continues to attenuate and the level of noise continues to grow such that after 100 meters of wire, the likelihood of errors increases. The constraint of no more than 100 meters means 100 meters from the device that generates the signal (the source), to a device that accepts the signal (the destination). This accepting device can be either the final destination or a repeater. A **repeater** is a device that regenerates a new signal by creating an exact replica of the original signal. Thus, category 3 twisted pair can run further than 100 meters from the source to the final destination, as long as the signal is regenerated at least every 100 meters.

Category 4 twisted pair is designed to transmit 20 Mbps for distances up to 100 meters. It was created at a time when a certain type of local area network required a wire that could transmit data faster than the 10 Mbps speed of Category 3. **Category 5** twisted pair is designed to transmit 100 Mbps data for distances up to 100 meters. Both Category 4 and Category 5 twisted pair are manufactured with higher quality materials than the Category 1 to 3 wires and thus introduce less noise.

Therefore, the signals that transmit over Category 4 and 5 wires can run further than 100 meters when the signal is regenerated at least every 100 meters. Category 5 twisted pair costs only pennies per foot more than Categories 1-4 twisted pair. Therefore, it makes sense to install Category 5 wire, regardless of whether you will take immediate advantage of the 100 Mbps transmission speed.

Approved at the end of 1999 is the specification for Category 5e twisted pair. Similar to Category 5, Category 5e will remain at 100 Mbps transmission speed. While the earlier specifications for Categories 1-5 described only the individual wires, the **Category 5e** version includes specifications for exactly four pairs of wires, the connectors on the ends of the wires, patch cords, and other possible components that directly connect with the cable. Thus, Category 5e is a more detailed specification than Category 5 to better support the higher speeds of 100 Mbps local area networks.

Interestingly, work is also progressing on Category 6 and Category 7 twisted pair. Although still a little away from being a published specification, **Category 6 twisted pair** should support data transmission as high as 200 Mbps for 100 meters while **Category 7 twisted pair** will support even higher data rates. Check the author's web page at *http://bach.cs.depaul.edu/cwhite* for the latest updates on Categories 5e, 6 and 7 twisted pair.

All of the above twisted pair wires are unshielded twisted pair. **Unshielded twisted pair (UTP)**, the most common form of twisted pair, does not wrap any of the wires with a metal foil or braid. This shielding is either wrapped around each wire individually, around pairs of wires or around all the wires together and provides an extra layer of isolation from unwanted electromagnetic interference. **Shielded twisted pair (STP)**, which does include this extra level of shielding, is also available in Category 1 through Category 5 and in numerous wire configurations. Figure 3-3 shows an example of shielded twisted pair wire.

Figure 3-3
An example of shielded twisted pair

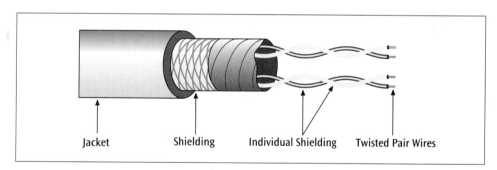

Jacket   Shielding   Individual Shielding   Twisted Pair Wires

If you determine that the twisted pair wire needs to go through walls, rooms, or buildings where there is sufficient electromagnetic interference to cause substantial noise problems, shielded twisted pair can provide a higher level of isolation from that interference than unshielded twisted pair wire, and thus a lower level of errors. Electromagnetic interference is often generated by large motors, such as those found in heating and cooling equipment or manufacturing equipment. Even fluorescent light fixtures generate a noticeable amount of electromagnetic interference. Large sources of power can also generate damaging amounts of electromagnetic interference. For example, it is generally not a good idea to strap twisted pair wiring to a power line that runs through a room or through walls. Furthermore, even though Categories 3-5 shielded twisted pair have improved noise isolation, you cannot expect to push them past the 100 meter limit. Finally, be prepared to pay a premium for shielded twisted pair. It is not uncommon to spend an additional $1 per foot for good quality shielded twisted pair. In contrast, Category 5 UTP often costs between $.10 and $.20 per foot.

Table 3-1 summarizes the basic characteristics of unshielded twisted pair wires. Keep in mind that shielded twisted pair wires have basically the same statistics as unshielded twisted pair wires, but perform better in noisy environments. In addition, shielded twisted pair wires are much more expensive than unshielded twisted pair wires.

Details ▶

Further Characteristics of Twisted Pair

Categories 1 through 7 and shielding make several distinctions about different types of twisted pair cable. In addition, you need to consider where the wire will be placed and the width of the wire. When traveling between the rooms within a building, will the wire be traveling within a plenum or through a riser? A **plenum** is the space within a building that was created by building components and designed for the movement of breathable air, such as the space above a suspended ceiling. A plenum can also be a hidden walkway between rooms that house heating and cooling vents, telephone lines, and other cable services. Plenum cable is designed so that it does not spread flame and noxious fumes in the event of a fire. Thus, the cable's jacket is designed of special materials to meet these standards, which, of course, significantly increases the cost of the cable. Plenum cable can cost as much as double the cost of standard twisted pair cable.

If the cable is going to run through a **riser**, which is a hollow metal tube that runs between walls, floors, and ceilings and encloses the individual wires, then flames and noxious fumes are not as serious an issue. In this case, standard plastic jacketing can be used. This type of twisted pair cable is typically the cable advertised and discussed when dealing with new cable installations since a majority of these installations involve running wires through risers.

In addition to considering the different specifications of twisted pair cable, you need to consider the thickness of the wire. Wire thickness is usually denoted by the American Wiring Gauge (AWG) number. The smaller the AWG number, the thicker the wire and the less resistance to current flowing in the wire. The larger the AWG number, the thinner the wire, and the more resistance to current. If the resistance is too high, the signal transmission will suffer. Standard twisted pair wire is typically 24 AWG. Compare 24 AWG wire to the wire that provides electricity to an outlet in the walls of your house or apartment, which is typically 12 or 14 AWG. 12 or 14 AWG wire is much thicker than the wire used for computer communications. While most twisted pair wire is 24 AWG, you have to be careful that you do not inadvertently purchase and install a thinner wire than 24 AWG since the quality of the signal transmitted on the wire is due in part to the thickness of the wire.

Table 3-1
A summary of the characteristics of twisted pair wires

UTP Category	Typical Use	Signaling Technique	Maximum Data Rate	Maximum Range	Advantages	Disadvantages
Category 1	Telephone wire	Analog and digital	<100 Kbps	3-4 miles	Inexpensive, easy to install and interface, widely used	Insecure, noise
Category	T1, ISDN	Digital	<2 Mbps	3-4 miles	Same as Category 1	Insecure, noise
Category 3	LANs	Digital	10 Mbps	100m (328 feet)	Same as Category 1, with less noise	Insecure, noise
Category 4	LANs	Digital	20 Mbps	100m	Same as Category 1, with less noise	Insecure, noise
Category 5	LANs	Digital	100 Mbps	100m	Same as Category 1, with less noise	Insecure, noise
Category 5e	LANs	Digital	100 Mbps	100m	4-pair specification includes connectors, patch cords, and other components	
Category 6	LANs	Digital	200 Mbps	100m	Draft standard in early stages	
Category 7	LANs	Digital	600 Mbps (?)	100m (?)	Draft standard in very early stages	

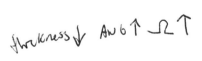

Coaxial Cable

Coaxial cable, in its simplest form, is a single wire wrapped in a foam insulation, surrounded by a braided metal shield, then covered in a plastic jacket. The braided metal shield is very good at blocking electromagnetic signals from entering the cable and producing noise. Figure 3-4 shows an example of coaxial cable and its braided metal shield. Because of its good shielding properties, coaxial cable is very good at carrying analog signals with a wide range of frequencies. Thus, coaxial cable can transmit large numbers of video channels such as those found on the cable television services that are delivered into homes and businesses. Coaxial cable has been used for long-distance telephone transmission, as the cabling within a local area network, and as a connector between a computer terminal and a mainframe computer.

Figure 3-4
Example of coaxial cable showing metal braid

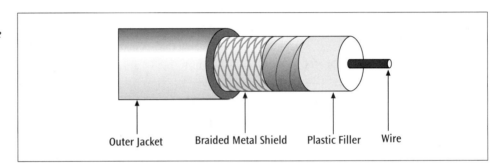

Outer Jacket   Braided Metal Shield   Plastic Filler   Wire

There are two major coaxial cable technologies; each carries a different type of signal—baseband or broadband. **Baseband coaxial** technology uses digital signaling in which the cable carries only one channel of digital data. A fairly common application for baseband coaxial used to be the interconnection of hubs within a local area network. (A hub is the device to which workstations connect. See Chapter Seven for more information on hubs.) The cable would typically carry one 10 Mbps signal and require repeaters at least every kilometer. This use of coaxial cable is being replaced by high quality twisted pair wire and more recently by fiber optic cable.

Broadband coaxial technology transmits analog signals and is capable of supporting multiple channels of data simultaneously. Consider the coaxial cable that transmits cable television. Many cable companies offer up to 50 or more channels. Each channel or signal occupies a bandwidth of approximately 6 MHz. When 50 channels are transmitted together, the coaxial cable is supporting 50 × 6 MHz or a 300 MHz composite signal. This capacity estimate is actually on the low side as many forms of broadband cable can actually support signals as wide as 500 MHz. When comparing the data capacity of broadband cable to twisted pair and baseband cable, each broadband channel can support the equivalent of millions of bits per second. To support such a wide range of frequencies, broadband coaxial cable systems require amplifiers approximately every three to four kilometers. While splitting and joining of broadband signals and cables is possible, it is a rather precise science that is best left to specialists in the field. Thus, many network administrators often hire outside experts to install and maintain broadband systems.

Coaxial cable also comes in two primary physical types: thin coaxial cable and thick coaxial cable, as shown in Figure 3-5. **Thick coaxial cable** ranges in size from approximately 6 to 10 mm in diameter. **Thin coaxial cable** is approximately 4 mm in diameter. Compared to thick coaxial cable, which typically carries broadband signals, thin coaxial cable has limited noise isolation and typically carries baseband signals. Thick coaxial cable has better noise immunity and is generally used for transmission of analog data such as single or multiple video channels. Thick and thin coaxial cable prices vary depending on the quality and construction of the cable, but typically run $0.25 to $1.00 per foot.

Figure 3-5
Examples of thin coaxial cable and thick coaxial cable

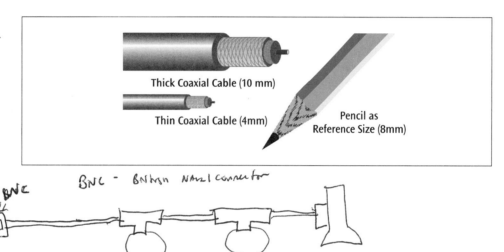

Details ▶

Further Characteristics of Coaxial Cable

An important characteristic of coaxial cable is its ohm rating. **Ohm** is the measure of resistance within a medium. The higher the value, the more resistance in the cable. Although resistance is not a major concern when choosing a particular cable, the ohm value is indirectly important, because coaxial cables with different ohm ratings work better with different kinds of signals and thus with different kinds of applications. Table 3-2 lists some common coaxial cables and their ohm values and applications. The type of coaxial cable is designated by radio guide (RG), which specifies many characteristics including wire thickness, insulation thickness, electrical properties, and more. You can see from the table that it is important to understand what the specific application is before choosing the type of coaxial cable.

Sometimes a second characteristic of coaxial cable is whether the wire that runs down the center of the coaxial cable is braided or single-stranded. **Single-stranded** coaxial cable contains, as the name implies, a single metal con-ductor. **Braided** coaxial cable is composed of many fine small wires twisted around each other, acting as a single con-ductor. If the wire is braided, it is often less expensive and easier to bend than a single, thicker strand, and may require a different kind of connector on the end of the cable.

Type of Cable	Ohm Rating	Application
RG-11	75 Ohm	Used in thick Ethernet applications
RG-58	50 Ohm	Used in thin Ethernet applications
RG-59	75 Ohm	Used in cable television systems
RG-62	93 Ohm	Used to connect IBM 3270 computer terminals

Table 3-2 *Common coaxial cables, ohm values, and applications*

Fiber Optic Cable

All the conducted media discussed so far have one great weakness—electromagnetic interference. Electromagnetic interference is the electronic distortion a metal wire experiences when a stray magnetic field passes over it. This interference can be reduced with proper shielding, but it cannot be completely eliminated unless you use fiber optic cable. **Fiber optic cable** (or optical fiber) is a thin glass cable approximately a little thicker than a human hair surrounded by a plastic coating. When fiber optic cable is packaged into an insulated cable, it is surrounded by a fire resistant yarn and a strong plastic jacket that protects the fiber from bending, heat, and stress. You can see an example of fiber optic cable in Figure 3-6.

Figure 3-6
A person holding plain fiber optic cable and fiber optic cable in an insulated jacket

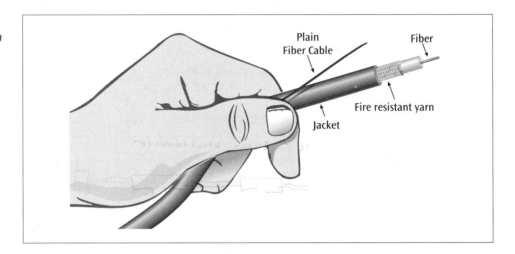

How does a thin glass cable transmit data? A light source, called a **photo diode**, is placed at the transmitting end and quickly switched on and off. The light pulses travel down the glass cable and are detected by an optic sensor called a **photo receptor** on the receiving end. The light source can be either a simple and inexpensive light-emitting diode (LED), such as those found in many pocket calculators, or a more complex laser. Although the laser is much more expensive than the LED, it can produce much higher data transmission rates. Fiber optic cable is capable of transmitting data at over 100 Gbps (that's 100 billion bits per second!) over several kilometers. In 1992, AT&T and a Japanese company tested a 9000 kilometer optically amplified fiber optic system. There were no data transmission errors at 5 billion bps.

In addition to having almost error-free high data transmission rates, fiber optic cable has a number of other advantages over twisted pair and coaxial cable. Since fiber optic cable passes electrically nonconducting photons through a glass medium, it is immune to electromagnetic interference and virtually impossible to wiretap. The only possible way to wiretap a fiber optic line is to physically break into the line, which would be noticed. And, since fiber optic cable cannot generate nor be disrupted by electromagnetic interference, there is no noise generated from extraneous electromagnetic signals. Thus, fiber optic cable has significantly less noise than twisted pair wires or coaxial cable. Fiber optic cable still experiences a small amount of noise as the light pulses bounce around inside the glass cable, but this noise is significantly less than the noise generated in a metallic wire. The lack of significant noise is one of the main reasons it is possible to transmit data for such long distances over fiber optic cable.

Despite these overwhelming advantages, fiber optic cable has two relatively small disadvantages. First, due to the light source and photo receptor, light pulses can travel in one direction only. Thus, to support a two-way transmission of data, two fiber optic cables are necessary. To support two-way transmission, most fiber optic cable is sold with at least two individual strands of fiber bundled into a single package, as shown in Figure 3-7.

Figure 3-7

A fiber optic cable with two strands of fiber, one for each direction of transmission

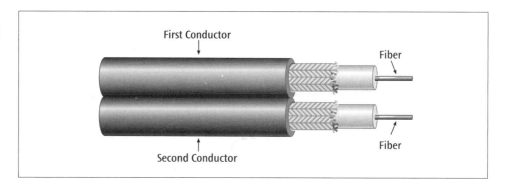

A second disadvantage, which is slowly disappearing, is the higher cost of fiber optic cable. Interestingly, it is not the fiber optic cable itself but the hardware that transmits and receives the light pulses at the ends of the fiber cable that is expensive. For example, it is possible to purchase bulk general purpose duplex (two strands) fiber optic cable for approximately $0.60 per foot, which is close to the price of many types of coaxial cable and shielded twisted pair cable. When you consider the low error rates and high data transmission rates, fiber optic cable is indeed a bargain, even when compared to inexpensive twisted pair wire.

Beginning in 1999, the prices for photo diodes and photo receptors started to drop significantly. Before 1999, it was common to use fiber optic cable only as the **backbone**—the main connecting cable that runs from one end of the installation to another—of a network and use Category 5 twisted pair from the backbone connection up to the back of the workstation. An illustration of a fiber optic backbone is shown in Figure 3-8.

Figure 3-8
A fiber optic backbone with Category 5 twisted pair running to the workstation

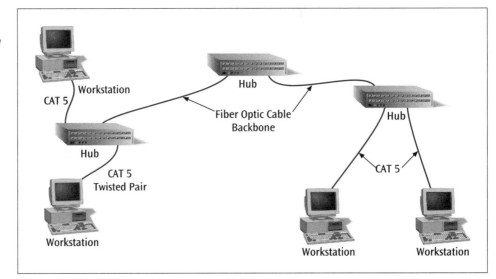

Since the prices for photo diodes and receptors have dropped, many companies are considering using only fiber optic cable. Thus, each workstation would be connected to the network using a high speed fiber optic line. The noise level would be greatly reduced and there would be virtually no limit to how fast or how far the data could be transmitted within the building.

Table 3-3 summarizes the conducted media. Categories 1 and 2 twisted pair have been grouped together since they are usually used for telephone systems, while Categories 3 through 6 have been grouped since they are usually used for

---

Details ▶

Further Characteristics of Fiber Optic Cable

When a light source is passed through a fiber optic cable, there is a tendency for the light wave to bounce around inside the cable or pass through the cable to the outside world. When a light source on the inside of the cable bounces off the cable wall and back into the cable, this is called **reflection**. When a light source passes through a cable wall to the outside, this is called **refraction**. Figure 3-9 demonstrates the difference between reflection and refraction.

There are two basic ways to transmit light through a fiber optic cable. The first technique, called **single-mode transmission**, requires the use of a very thin fiber optic cable and a very focused light source such as a laser. When a laser is fired down a narrow fiber, the light follows a very tight beam and there is less tendency for the light wave to reflect and refract. Thus, this technique allows for a very fast signal with little signal degradation over long distances. Because lasers are used as the light source, single-mode transmission is a more expensive technique then the fiber optic cable signaling technique described below.

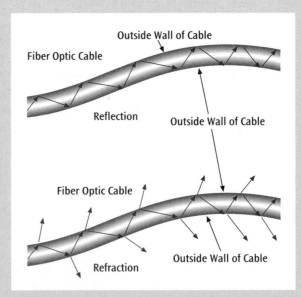

Figure 3-9 *A simple demonstration of reflection and refraction in a fiber optic cable*

local area networks. In almost all cases, maximum data rate and maximum range are typical values and can be less or more depending on environmental factors.

Table 3-3
A summary of the characteristics of conducted media

Type of Conducted Media	Typical Use	Signaling Technique	Maximum Data Rate	Maximum Range	Advantages	Disadvantages
Twisted Pair						
Category 1 – 2	Telephone systems	Analog, digital	<2 Mbps	2 – 3 miles	Inexpensive, common	Noise, security
Category 3 – 6	LANs	Digital	200 Mbps	100m (328 feet)	Inexpensive, versatile	Noise, security
Coaxial Cable						
Thin baseband single channel	LANs	Digital	10 Mbps	100m	Low noise	Security
Thick broadband multi-channel	LANs, cable TV, long distance telephone, short-run computer system links	Analog	10 Mbps	2 – 3 miles	Low noise, multiple channels	Security
Fiber Optic	Data, video, audio, LANs, WANs	Light pulses	10 Gbps	100 miles	Secure, high capacity, very low noise	Interface expensive, but coming down in cost

The second technique, called **multi-mode transmission**, uses a slightly thicker fiber cable and an unfocused light source such as an LED. Since the light source is unfocused, there is more refraction and reflection of the light wave as it propagates through the wire, yielding a signal that cannot travel as far or as fast as the signals generated with the single-mode technique. Multi-mode transmission is correspondingly less expensive than single-mode transmission.

Single-mode and multi-mode transmission techniques use fiber optic cable with different characteristics. The core of single-mode fiber optic cable is 8.3 microns wide and the material surrounding the fiber—the cladding—is 125 microns wide. Single-mode fiber optic cable is labeled **8.3/125 cable**. The core of multi-mode fiber optic cable is most commonly 62.5 microns wide and the cladding is 125 microns. Multi-mode fiber optic cable is labeled **62.5/125 cable**. Other sizes of multi-mode fiber optic cable include 50/125 and 100/140 microns.

Introduction to Wireless Transmissions

Radio, satellite transmissions, visible light, infrared light, X-rays, and gamma rays are all examples of electromagnetic waves or electromagnetic radiation. Electromagnetic radiation is energy propagated through space and, indirectly, through solid objects in the form of an advancing disturbance of electric and magnetic fields.

Radio waves, in particular, are emitted by the accelerations of free electrons, such as occurs in a radio antenna wire. The basic difference between various types of electromagnetic waves is their differing wavelengths, or frequencies, as shown in Figure 3-10.

Figure 3-10
Electromagnetic wave frequencies

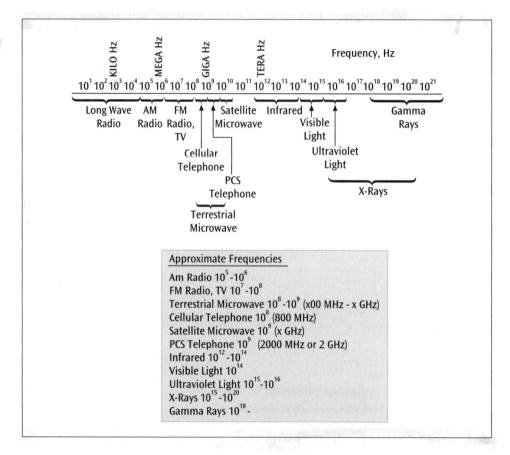

Note how every type of transmission system such as AM radio, FM radio, television, cellular telephones, terrestrial microwaves, and satellite systems are all confined to relatively narrow bands of frequencies. The Federal Communications Commission (FCC) keeps very tight control on what frequencies are used by which application. Occasionally the FCC will assign an unused range of frequencies to a new application. Other times, the FCC will auction off unused frequencies to the highest bidder. The winner of the auction is then allowed to use those frequencies for the introduction of a particular product or service. It is important to note, however, that only so many frequencies are available to be used for applications. Thus, it is very important that each application use its assigned frequencies as best as possible. The cellular telephone system, as we will see, provides a good example of how an application can use its assigned frequencies efficiently.

Terrestrial microwave transmission

Terrestrial microwave transmission systems transmit tightly focused beams of radio signals from one ground-based microwave transmission antenna to another. Figure 3-11 shows a typical microwave antenna installation.

Figure 3-11
A typical microwave tower and antenna

Microwave transmissions do not follow the curvature of the earth, nor do they effectively bounce off objects, which limits transmission distance. Microwave antennas use **line-of-sight transmission**, which means that to receive and transmit a signal, each antenna must be in sight of the next antenna (Figure 3-12). Many microwave antennas are located on top of free-standing towers, and the typical distance between microwave towers is roughly 30 miles. The higher the tower, however, the further the possible transmission distance. Thus, towers located on hills, mountains, or on top of tall buildings can transmit signals further than 30 miles. Another factor that limits transmission distance is the number of objects that might obstruct the path of transmission signals. Buildings, hills, forests, and even heavy rain and snowfall all interfere with the transmission of microwave signals. Assuming there is no interference and that amplifiers are used on the towers to regenerate the signal, terrestrial microwave can run for an unlimited distance.

Figure 3-12
A microwave antenna on top of a free-standing tower transmitting to another antenna on the top of a building

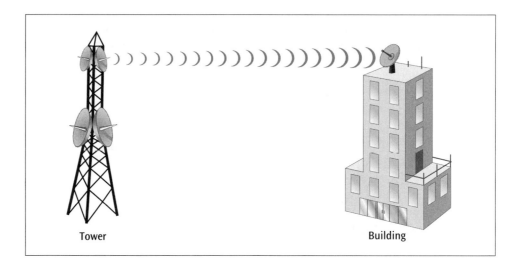

Tower   Building

Possibly the best selling point for terrestrial microwave is its ability to transmit signals up to billions of bits per second without the use of interconnecting wires. The disadvantages of terrestrial microwave include the costs of either leasing the service or installing the antennas, loss of signal strength (or attenuation), and interference from other signals (crosstalk).

The two most common application areas of terrestrial microwave are telephone telecommunications and business intercommunication. Many telephone companies implement a series of antennas placing a combination receiver and transmitter tower every 15 to 20 miles. These systems span metropolitan areas as well as provide intrastate and interstate telephone service. Businesses can also use terrestrial microwave to implement telecommunications systems between corporate buildings. Such an arrangement might be less expensive in the long run than leasing a high speed telephone line from a telephone company, which requires an ongoing monthly payment. With terrestrial microwave, once the system is purchased and installed, no telephone service fees are necessary.

Satellite microwave transmission

Satellite microwave transmission systems are similar to terrestrial microwave systems except that the signal travels from a ground station on earth to a satellite and back to another ground station on earth, thus achieving much greater distances than line-of-sight transmission. In fact, a satellite that is at the furthest possible point from the earth—22,300 miles—can receive and send signals approximately one third the distance around the earth.

One way of identifying a satellite system is by how far the satellite is from the earth. The closer a satellite is to the earth, the shorter the time to send data to the satellite—the **uplink**—and receive data from the satellite—the **downlink.** This delay from ground station to satellite, back to ground station is called **propagation delay.** Although distance from the earth causes a delay, the further a satellite is from earth, the more likely the satellite will remain over the same point on the earth's surface. Satellites that are always over the same point on earth can be used for long periods of high speed data transfers. In contrast, satellites that are close to earth do not stay in the same position over the earth and are used with applications requiring shorter periods of data transfer, such as mobile telephone systems. As Figure 3-13 shows, satellites orbit the earth from three possible ranges: low earth orbit (LEO), middle earth orbit (MEO), and geosynchronous earth orbit (GEO).

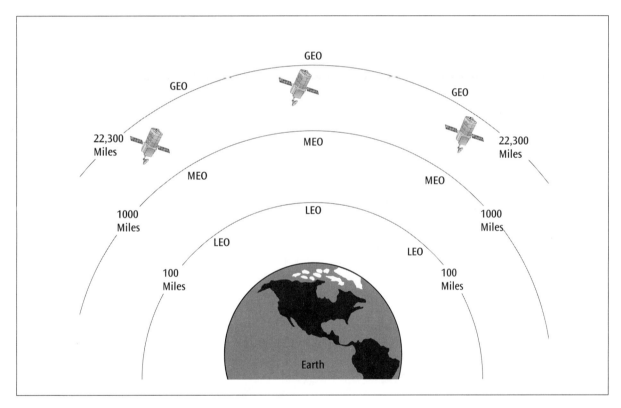

Figure 3-13
The earth and the three earth orbits, LEO, MEO and GEO

Low earth orbit (LEO) satellites are closest to the earth. They can be found as close as 100 miles from earth and as far as 1000 miles from earth. The number of low earth orbit satellites is growing rapidly. At the end of the 20th century, there were approximately 300 LEO satellites. Estimates are that by the year 2003 an additional 700 LEO satellites will be launched. Low earth orbit satellites are used primarily for the wireless transfer of electronic mail, pager systems, worldwide mobile telephone networks, spying, remote sensing, and videoconferencing.

One of the late 20th century's most ambitious projects was Motorola's Iridium handheld global satellite telephone and paging network. The Iridium system was originally designed to use seven layers of satellites with 11 satellites in each layer, or 77 satellites. The network got its name from the element Iridium, which has an atomic weight of 77. After some rethinking it was determined that the system would also work with six layers of 11 satellites and thus was scaled back to 66 satellites (the name however was not changed to Dysprosium; apparently it did not have the same ring). Even with only 66 satellites, from any point on earth, a person can receive or place a telephone call when using an Iridium mobile telephone. Unfortunately, by the summer of 1999, the Iridium system had failed to interest enough subscribers, leaving the future of the service in serious doubt.

Middle earth orbit (MEO) satellites can be found 1000 miles to 22,300 miles from the earth. There are currently approximately 65 MEO satellites orbiting the earth. Although MEO satellite systems are not growing at the same phenomenal rate as LEO systems, industry experts predict that the number of MEO satellites will almost double to approximately 120 by the year 2005.

Details ▶

Satellite Configurations

Satellite systems can be configured into three basic topologies: bulk carrier facilities, multiplexed earth stations, and single-user earth stations. Figure 3-14 illustrates each topology.

Bulk Carrier Facilities

Figure 3-14 shows that in a bulk carrier facility, the satellite system and all its assigned frequencies are devoted to one user. Since a satellite is capable of transmitting large amounts of data in a very short time and the systems are expensive, a very large application would be necessary to economically justify the exclusive use of an entire satellite system by one user. It would make sense for a telephone company, for example, to use a bulk carrier satellite system to transmit thousands of long distance telephone calls. Typical systems operate in the 6/4 GHz bands (6 GHz uplink, 4 GHz downlink) and provide a 500 MHz bandwidth, which can be broken further into multiple 40-50 MHz channels.

Multiplexed Earth Station

In a multiplexed earth station satellite system, the ground station accepts input from multiple sources and in some fashion interleaves the data streams either by assigning different

frequencies to different signals or by allowing different signals to take turns transmitting. Figure 3-14(b) shows a diagram of how a typical multiplexed earth station satellite system operates.

How does this type of satellite system satisfy the requests of users and assign time slots? Each user could be asked in turn if he or she has data to transmit, but so much time would be lost by the asking process that this technique would not be economically feasible. A first-come, first-served scenario, in which each user competed with every other user would also be an extremely inefficient design. The technique that seems to work best for access to multiplexed satellite systems is a reservation system. In a **reservation system**, users place a reservation for future time slots. When the reserved time slot arrives, the user transmits their data on the system. Two types of reservation systems exist: centralized reservation and distributed reservation. In a centralized reservation system, all reservations go to a central location, and that site handles the incoming requests. In a distributed reservation system, no central site handles the reservations, but individual users come to some agreement as to order of transmission.

Middle earth orbit satellites are used primarily for global positioning system-style surface navigation systems. Global positioning systems are complex, but it is worthwhile to take a brief look at how they work. Global positioning system (GPS) is a system of 24 satellites that were launched by the U.S. Department of Defense and are used for identifying locations on earth. By triangulation of signals from at least three GPS satellites, a receiving unit can pinpoint its current location to within a few meters anywhere on earth. The GPS systems that are available for commercial use, however, are not as precise as the government-owned systems (for security reasons) and are accurate only within a few city blocks. One major automobile manufacturer uses a GPS system to locate your automobile anywhere in the U.S. should you become lost while driving, and then sends you directions and the location of the nearest gas station.

Geosynchronous earth orbit (GEO) satellites are found 22,300 miles from the earth and are always over the same point on earth. Thus, two ground stations can conduct continuous transmissions from earth to satellite to earth. Geosynchronous earth orbit satellites are most commonly used for signal relays for broadcast, cable, and direct television; meteorology; government intelligence operations; and mobile maritime telephony. The primary advantage of GEO transmissions is the

Single-user Earth Station

In a single-user earth station satellite system, each user uses their own ground station to transmit data to the satellite. Figure 3-14(c) shows a typical single-user earth station satellite configuration. The **Very Small Aperture Terminal (VSAT)** system is an example of a single-user earth station satellite system with its own ground station and a small antenna (four to six feet across). Among all the user ground stations is one master station that is typically connected to a mainframe computer system. The ground stations communicate with the mainframe computer via the satellite and master station.

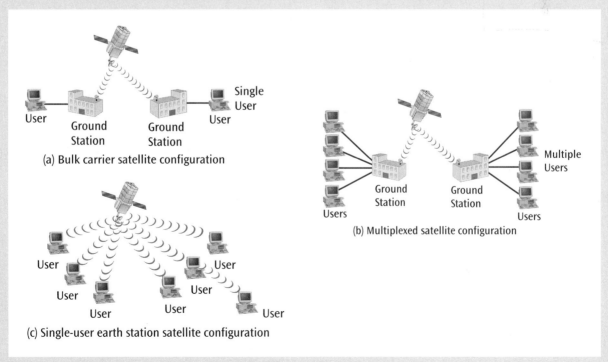

Figure 3-14 *Bulk, multiplexed, and single-user configurations of satellite systems*

capacity for high speed, high quantity bulk transmissions covering up to one third the surface of the earth. Companies that operate GEO satellites can commit all their transmission resources to one client or can share the satellite time with multiple clients. The use of a GEO satellite system by a single client is expensive and usually involves the transfer of great amounts of data.

Mobile telephones

Another wireless transmission technique that uses radio waves instead of wires is the mobile telephone system. Two basic categories of wireless systems currently exist: cellular telephone and personal communication systems (PCS). The older cellular telephone system, which is based on cellular radio technology, has its foundations in analog signaling and is slowly being replaced by the newer, all digital, PCS systems.

The name cellular telephone raises an interesting question: what does the term *cellular* mean? To answer this question, you need to examine the interactive radio network that existed before cellular radio—Improved Mobile Telephone Services (IMTS). IMTS allowed only 12 concurrent users within an entire city. The reason for so few concurrent users was introduced earlier in the chapter—there are only so many radio frequencies available for a particular application. When a user talks to another user, two channels are necessary, one for each direction of transmission. Each channel requires a sufficient range of frequencies to carry a voice signal. To support thousands of simultaneous users within a metropolitan area, an extremely large range of frequencies is required. The FCC could not allocate this many frequencies to a single application, so they divided the country into 728 **Mobile Service Areas (MSAs)**, or markets. Each market, which usually encompasses an entire metropolitan area, is further broken into adjacent cells (Figure 3-15).

Figure 3-15
One mobile telephone market divided into cells

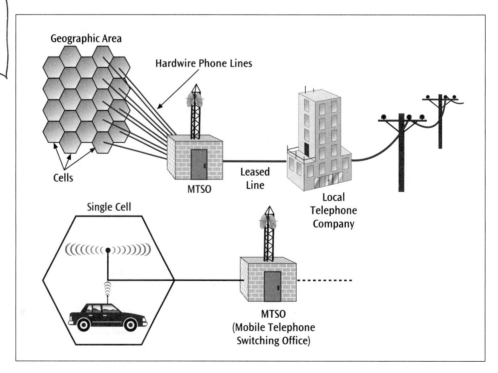

Cells can range in size from one half mile in radius to a 50 mile radius. Located within each cell is a low-power transmitter/receiver, which is often located on a free-standing tower. Lately however, residents have been complaining about the large number of towers in their communities, so mobile telephone companies are getting creative at disguising their antennas.

A mobile telephone within a cell communicates to the cell tower (see Figure 3-16), which in turn is connected to the mobile telephone switching office (MTSO) by a telephone line. The mobile telephone switching office is then connected to the local telephone system. If the mobile telephone moves from one cell to another, the mobile telephone switching office hands off the connection from one cell to another.

Figure 3-16
A mobile telephone tower

Since each cell uses low power transmissions, it is not likely that a transmission within one particular cell will interfere with transmission in another cell that is more than one or two cells away. Only near-adjacent cells need to use different sets of frequencies. Thus, the frequencies that are used within a cell can be reused in other cells, allowing for more simultaneous connections in a market than there are available frequencies.

In each cell, at least one channel, the setup channel, is responsible for the setup and control of calls. A mobile telephone, attempting to place a call, tries to seize a setup channel. Once the mobile telephone has seized a setup channel, it transmits a unique, built-in mobile ID number (MIN). This mobile ID number is transferred to the MTSO, where the user's account is checked for validity. If the telephone bill has been paid, the MTSO assigns a channel to that connection, the mobile telephone releases the setup channel, seizes the assigned channel, and proceeds to place the telephone call.

When someone tries to call a mobile telephone, all cell towers transmit the mobile telephone's ID. When the ID is recognized by the mobile telephone, the mobile telephone tries to seize the local cell's setup channel. When the setup channel is seized, the mobile telephone sends its mobile ID number to the MTSO, the MTSO verifies the mobile ID number, a channel is assigned to the mobile telephone, and the incoming call is connected.

Currently there are three basic mobile telephone technologies in operation in the United States: Advanced Mobile Phone Service, Digital-Advanced Mobile Phone Service, and Personal Communication Systems. **Advanced Mobile Phone Service (AMPS)** is the oldest analog mobile telephone system, and it covers almost all of North America and is found in more than 35 other countries. It uses frequency division multiplexing technology (discussed in detail in Chapter Five) and is the cellular equivalent of the plain old telephone system (POTS). **Digital-Advanced Mobile Phone Service (D-AMPS)** is the newer digital equivalent of analog mobile telephone service. It uses time division multiplexing technology (discussed in Chapter Five) in addition to frequency division multiplexing and provides greater signal clarity and security than AMPS. Because D-AMPS starts with frequency division multiplexing and then adds time division multiplexing techniques, analog mobile telephone systems can be upgraded to D-AMPS. The upgrade increases signal clarity, security features, the number of special services offered, and the number of available channels per cell. Many cellular providers have upgraded their AMPS mobile telephones to D-AMPS service in order to compete with newer all digital services.

The newest category of mobile telephone technology is **Personal Communication Systems (PCS)**, which are all digital and do not rely on more noisy analog techniques. PCS systems were approved by the FCC in 1993, and the first PCS mobile telephone system appeared in Washington, D.C., in November 1995. Since then, three competing (and incompatible) PCS technologies have emerged. The first PCS technology uses a form of time division multiplexing called time division multiple access (TDMA) technology to divide the available user channels by time, giving each transmitting mobile telephone a brief turn to transmit. The second PCS technology uses code division multiple access (CDMA) technology, which spreads the transmission of a mobile telephone signal over a wide range of frequencies using mathematical values. CDMA is based on spread spectrum technology, which was introduced in Chapter Two. The third PCS technology is **Global System for Mobile (GSM)** communication and uses a different form of time division multiple access technology. GSM is currently the least popular mobile telephone technology in the United States and the most popular mobile telephone technology in Europe.

All PCS technologies generally offer a clearer signal than AMPS and D-AMPS, but they are incompatible with each other and incompatible with the older AMPS and D-AMPS technologies. Compared to AMPS and D-AMPS, PCS can be found on a smaller scale throughout the United States. Thus, if you intend to travel and wish to take your PCS mobile telephone with you, be prepared for it not to work in many parts of the country. Many mobile telephone vendors offer **hybrid telephones**, which can send and receive D-AMPS signals as well as PCS signals. With a hybrid telephone, if you are within the market area for PCS signals, your telephone will use the newer all-digital PCS service. When you move out of the PCS transmitting area, your hybrid telephone will revert to the older AMPS service. Also, Motorola announced at the end of 1999 a single integrated circuit which can accept TDMA, CDMA, or GSM signals. Mobile telephones that incorporate this circuit will be able to accept any PCS signal, thus allowing a much wider mobility throughout the United States.

Cellular digital packet data

Cellular digital packet data (CDPD) technology supports a wireless connection for the transfer of computer data from a mobile location to the public telephone network and the Internet. For example, a person using a laptop computer can use CDPD technology to transmit a computer file from the laptop to corporate headquarters or to an Internet service provider. A CDPD connection is performed by "stealing" unused time from mobile telephone frequencies by placing data into packets that are sent during pauses in cellular telephone conversations. Because the time is unused, mobile telephone users are not aware that another wireless service is taking place. CDPD is even capable of moving your data connection from one mobile telephone channel to another to avoid congesting voice communications. Since cellular telephone networks exist in almost every part of the country, and CDPD uses existing frequencies and transmitting towers, it provides a relatively low-cost option for mobile users to transfer computer data. An additional advantage includes the ability to apply digital encryption techniques to the data transferred by CDPD systems, thus creating a secure transmission.

The negative side to CDPD is the fact that many mobile users consider the data transfer speed too slow. A CDPD system can transfer data up to 19,200 bits per second. If you are transferring simple text, this speed is probably sufficient. If you are trying to transfer Internet web pages with graphic images, 19,200 bits per second can be noticeably slow.

CDPD technology is used by emergency vehicles such as police, fire, and paramedics who must transmit and receive information concerning a current event in which they are engaged. As an emergency vehicle is proceeding to the scene, the CDPD receiver housed in the emergency vehicle can receive data such as prior arrests, individual medical information, or hazardous material locations.

Pagers

Another wireless communication technology that has grown immensely in popularity within the last decade is the **pager**, which is also often called a beeper. People who wish to send a message to an individual carrying a pager call the individual's pager telephone number and enter a telephone number for the paged individual to call back. Most systems also allow an individual to leave a short numeric code. Some of the newer systems even allow an individual to leave a short text (alphanumeric) message. The paging system accepts the message and transmits a unique signal either from a tower for a local page or from a satellite for a more distant page. When the pager recognizes its unique signal, the pager displays the message and beeps, producing an audible page, or vibrates, producing a silent page.

Most of the pagers currently in use are one-way pagers; an individual may be paged, but the pager is incapable of returning a message. Some new pagers are capable of two-way conversations and can return a short message. Some pager systems provide users with e-mail notification services. Subscribers of e-mail notification services are able to have an abbreviated e-mail message (up to 208 characters) sent to their pager. The complete message is forwarded to their regular Internet-based e-mail account for retrieval after the message has been sent to their pager. This feature enables users to learn instantly what e-mail they have received and whether it is something that needs immediate attention.

The use of pagers has grown so dramatically that some restaurants now call customers to their tables via pagers. When a customer enters the restaurant and asks for a table, the host or hostess hands the customer a pager, and pages the customer (silently, of course) when his or her table is ready.

Infrared transmissions

Infrared transmission is a special form of radio transmission that uses a focused ray of light in the infrared frequency range ($10^{12} - 10^{14}$ Hz). Much like remote control devices that control television sets work, this focused ray of infrared information is sent from transmitter to receiver over a line-of-sight transmission. Usually these devices are only one to three meters apart, but infrared systems exist that can transmit up to one and one half miles.

Details ▶

AMPS and D-AMPS Cellular Technology

Both AMPS and D-AMPS cellular telephone systems allocate their channels using frequency ranges within the 800 to 900 Megahertz (MHz) spectrum. To be more precise, the 824 - 849 MHz range is used for receiving signals from mobile telephones (the uplink), while the 869 - 894 MHz range is used for transmitting signals to mobile telephones (the downlink). Within a metropolitan area, these two bands of frequencies give approximately 50 MHz in which to transmit. These bands of frequencies are further divided into 30 kHz sub-bands, called channels. This division of the spectrum into sub-band channels is achieved through frequency division multiple access (FDMA), in which each channel is assigned a different set of frequencies on which to transmit (like television and radio).

There are a total of 1666 channels available for signal transmission in a metropolitan area (50 MHz divided by 30 KHz per channel yields 1666 channels). To carry on a two-way conversation on a mobile phone, two channels are required—one for the uplink and one for the downlink. With every conversation requiring two channels, 833 (1666 channels divided in half) two-way connections are available in a metropolitan area. Furthermore, the FCC allows up to two competing carriers to offer AMPS mobile telephone service within a metropolitan area. Thus, 416 connections (833 connections divided by 2 carriers) per carrier per metropolitan area are available for use. Finally, these 416 connections are divided among the number of cells in an area.

Recall that the cells in an area form a honeycomb pattern (on paper it is a nice honeycomb pattern, but in real life it is not such a uniform pattern) and that sets of frequencies can be reused. In fact, seven sets of frequencies are a commonly found occurrence, so 59 two-way connections per cell (416 divided by 7) per carrier are available in a metropolitan area. These 59 connections are further reduced by the fact that a few channels in each cell are used for call setup.

Although 59 available connections per cell does not seem like a large number compared to the total number of cellular telephone users in an area, keep the following facts in mind:

▶ everyone within a cell is not currently using his or her telephone;

▶ cells can be as small as one mile across; and

▶ there are two possible service providers for AMPS systems and up to three service providers for the newer PCS systems.

The number of cellular calls that can be transmitted in a metropolitan area is also influenced by D-AMPS (Digital-Advanced Mobile Phone Service). D-AMPS adds time division multiple access (TDMA) to AMPS to get three channels for each AMPS channel, thus tripling the number of calls that can be handled on a channel.

Infrared transmission systems are growing in popularity and are often associated with laptop computers, handheld computers, peripheral devices such as printers and fax machines, digital cameras, and even children's hand-held electronic games. Infrared transmission works well in the following situations:

► transmitting a document from your laptop computer to a printer or a modem;

► exchanging small files such as business cards between handheld computers;

► synchronizing electronic telephone books and schedulers; and

► retrieving bank records from 24-hour automatic teller machines by walking up to the machine and pointing a hand-held device at the terminal.

In each of these examples, the transmitter and receiver are within the same room or a short distance apart, and data transfer rates are typically no faster than 4 Mbps. Systems exist, however, that can transfer data at speeds up to 16 Mbps, and even higher speeds are in development.

The Infrared Data Association (IrDA) is leading the charge in standardizing infrared technology and in incorporating infrared into many application areas. At the moment, an estimated 70 to 80 percent of notebook computers manufactured have an IrDA port, indicating growing acceptance of infrared technology.

Broadband wireless systems

A broadband wireless system is one of the latest techniques for delivering Internet services into homes and businesses. These systems bypass the telephone company's local loop (the last stretch of telephone line between the telephone central office and the home or business) by transmitting voice, data, and video over very high radio frequencies (2500 MHz). As Figure 3-17 shows, data is sent at fiber optic speeds over limited distances to a base station, then to a switching center, where a switch directs the data traffic to the appropriate voice or data networks.

Figure 3-17
Broadband wireless configuration in a metropolitan area

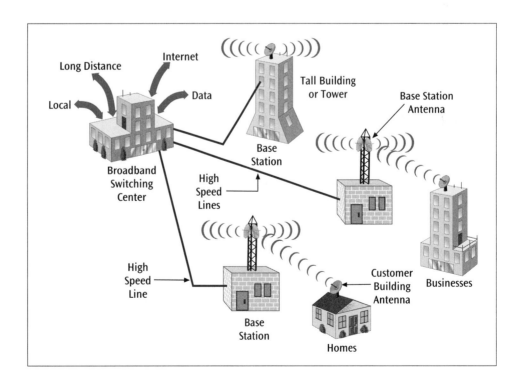

Multichannel Multipoint Distribution Service and Local Multipoint Distribution System are two similar and important methods of broadband wireless transmission in use today. **Multichannel Multipoint Distribution Service (MMDS)** is a new technology that allows one way wireless digital data and cable television transfers from a service provider to a user at speeds of millions of bits per second. Although this service is currently only one way, once the technology is adapted to two way transmissions, it may prove to be an attractive alternative to wired forms of computer data transmissions.

MMDS wireless cable television is now the choice of many cable companies around the world. An MMDS wireless television system can broadcast up to 120 television channels to subscribers with no expensive, time-consuming cable installation. MMDS wireless cable television transmits on the super high frequency (SHF) microwave frequencies (2–3 GHz) and can be digitally encoded for pay-for-view and subscriber services.

The present MMDS systems are download only. If you were to use MMDS for Internet access, you would need a more conventional (and slower) telephone connection for the upload data transfer between user and Internet service provider. The bulk of data transfer when using the Internet, however, is from the Internet service provider to the user, and the MMDS system can handle this download activity very well. Therefore, MMDS may be an attractive alternative to the current telephone-based services that provide users with access to the Internet. A proposal for two-way MMDS transmission was sent to the FCC during the summer of 1999 and is currently awaiting approval.

Local Multipoint Distribution System (LMDS) is also a broadband wireless communication system that can be used to provide digital two-way voice, data, video services, and Internet access. The coverage area of a single cell site in a metropolitan area is in the range of five miles, thus the system is *local*. Unlike MMDS, which was designed for one-way traffic, LMDS was designed for two-way traffic. However, both may be promising technologies for high-speed wireless connections between an Internet service provider and a business or home computer system.

Table 3-4 summarizes the wireless media, including the typical applications, advantages and disadvantages of each. Note that in almost all situations, if you are considering installing a system, costs will be much greater than simply using or leasing a system.

Table 3-4
Summary of wireless media

Type of Wireless Media	Typical Use	Signaling Technique	Maximum Data Transfer Rate	Maximum Range	Advantages	Disadvantages
Microwave	Long-haul telecommunications, building to building	Analog	n-Gbps	20-30 miles	Reliable, high speed and high volume	Long-haul expensive to implement, potential interference
Satellite						
GEO	Signal relays for cable and direct television	Analog	n-Gbps	One third the earth's circumference (8000 miles)	Very long distance, high speed and high volume	Expensive to lease, some interference

Table 3-4
Summary of wireless media (continued)

Type of Wireless Media	Typical Use	Signaling Technique	Maximum Data Transfer Rate	Maximum Range	Advantages	Disadvantages
MEO	GPS-style surface navigation systems	Analog	n-Gbps	Nationwide	High speed transfers, wide distance	Expensive to lease, some interference
LEO	Communications such as email, paging, worldwide mobile phone network, spying, remote sensing, video-conferencing	Analog	n-Gbps	Worldwide	High speed transfers, very wide distance, some applications inexpensive	Some applications expensive, some interference
Cellular (AMPS and D-AMPS)	Wireless telephones	Analog and digital	19.2 Kbps	Each cell: 0.5-50 mile radius, but nationwide coverage	Widespread, inexpensive applications	Noise
CDPD	Mobile data transfer	Digital	19.2 Kbps	Each cell: 0.5-50 mile radius, but nationwide coverage	Uses existing cellular systems	Limited speeds
PCS						
TDMA	Wireless telephones	Digital	56 Kbps	Each cell 0.5-25 mile radius, spotty nationwide coverage	All digital, low noise	Not everywhere
CDMA	Wireless telephones	Digital	56 Kbps	Each cell 0.5-25 mile radius, spotty nationwide coverage	All digital, low noise	Not everywhere
GSM	Wireless telephones	Digital	56 Kbps	Each cell 0.5-25 mile radius, spotty nationwide coverage	All digital, low noise	Not everywhere, least popular PCS technology in U.S.
Pagers	Wireless paging system	Analog		Local and global systems available	Inexpensive use, wide popularity, some systems two-way	Limited text transfer capabilities
Infrared	Short distance data transfer	Infrared light	16 Mbps	1.5 miles	Fast, inexpensive, secure	Short distances
Local Multipoint Distribution Service	Cable television and Internet service	Digital	1 Gbps	Same as microwave	Relatively inexpensive to use	Expensive to implement, not yet widespread

Media Selection Criteria

When designing or updating a computer network, the selection of one type of media over another is an important issue. Computer network-based projects have typically performed poorly and possibly even failed because of a decision to use an inappropriate type of medium. Furthermore, the purchase price and installation costs for a particular medium is often one of the larger costs in a computer network. Once the time and money are spent installing a particular media, a business has to use the chosen media for a number of years to recover the initial costs. The choice of media should not be taken lightly. Assuming you have the option of choosing a medium, you should consider many factors before making that final choice or those choices. The principal factors you should consider in your decision include cost, speed, expandability, distance, environment, and security. The following discussion will only consider twisted pair, coaxial cable, fiber optic cable, terrestrial microwave, satellite microwave, cellular radio, and CDPD. Pagers will not be considered as they are primarily one-way devices, nor will MMDS be considered due to the fact that it is still in an embryonic state.

Cost

Costs are associated with all types of media, and there are different types of cost. For example, twisted pair cable is generally less expensive than both fiber optic cable and coaxial cable. However, it is necessary to consider more than just the cost of the cable—you must also consider the cost of the supporting devices that originate and terminate the cables. For example, twisted pair wire usually ends with a small modular jack, similar to the jack that connects a telephone to a telephone wall jack. These modular jacks are mostly plastic and very inexpensive, costing only pennies each. The connectors that terminate coaxial cable are mostly metal and a little more costly. The connectors that terminate fiber optic cables are more expensive yet. More importantly, if you need to connect a fiber optic cable to a non-fiber optic cable or device, the cost increase is even more dramatic, since you must convert light pulses to electric signals and vice versa.

Terrestrial microwave systems spanning great distances and satellite systems are very expensive media considering the cost of microwave towers and satellites. However, very few businesses install their own towers and launch their own satellites. Most companies lease time from companies that specialize in microwave systems. Considering the lease option, and the fact that microwave systems can send very fast data streams, some applications could find a microwave solution less expensive than a wire-based solution. Privately owned microwave systems, which are mounted on company buildings, are significantly less expensive than wide-area terrestrial microwave and satellite systems. Considering the cost of installing cables, in certain situations short-distance microwave systems may also offer a reasonable alternative to installing cable or leasing time from a service provider.

Each type of media has the additional cost of maintenance. Will a certain type of wire last for X years when subjected to a particular environment? This is a difficult question, which you should ask the company supplying the medium. Whereas it is easy to browse catalogs and learn the initial costs of a particular type of cable, it is more difficult to determine the maintenance costs two, five, or ten years down the road. Too often, initial cost is the main limiting factor to considering better types of media, and long-term maintenance is not taken into account.

Speed

To evaluate media properly you need to consider two types of speed: data transmission speed and propagation speed. **Data transmission speed** is calculated as the number of bits per second that can be transmitted. Maximum bits per second for a particular medium depends proportionally on the effective bandwidth of that medium, the distance over which the data must travel, and the environment through which the medium must pass. If one of the requirements of the network you are designing is a minimally acceptable bits per second speed, then the chosen medium has to support that speed. This issue might sound trivial, but it is complicated by the fact that it is very difficult to predict network growth. Although a chosen medium may support a particular level of traffic at the moment, the medium may not support a future addition of new users or new applications.

Propagation speed is the speed at which a signal moves through a medium. The propagation speed of fiber optic media is very near the speed of light, and wireless media propagates at the speed of light, which is 3×10^8 meters per second. For electrically conducted media (twisted pair and coaxial cable), propagation speed is approximately 2/3 the speed of light, or 2×10^8 meters per second. Although these speeds seem fast enough for most applications, keep in mind that the propagation speed to send a signal to a satellite in an outer earth orbit and back is approximately 0.75 seconds. If you are transferring data from one side of the state to the other, you might want to consider a conducted medium (if possible) rather than a satellite system. Given the relatively short distance, a conducted medium might have a shorter propagation delay.

186,600 MPS

Distance and expandability

Certain media lend themselves more easily to expansion. Twisted pair cable is easier to expand than either coaxial cable or fiber optic cable, and coaxial cable is easier to expand than fiber optic cable. Coaxial cable is more difficult to expand than twisted pair because of the types of connectors on the ends of the cable. Fiber optic connectors are even more elaborate, and joining two pieces of fiber optic cable requires practice and the proper set of tools.

Another consideration for expandability is that most forms of twisted pair can only operate for 100 meters before the signal requires regeneration. Some forms of coaxial cable systems can run for longer lengths (kilometers), and fiber optic cable can extend for many miles before regeneration of the signal is necessary.

Privately-owned terrestrial microwave has a high transmission rate, but if the microwave dishes are mounted on corporate buildings, the buildings can be no further apart than 30 miles. Cellular systems are widespread and expanding continuously. New technologies such as CDMA show great promise for expansion.

If you expect to create a system that may expand in the future, it is worthwhile to consider using a medium that can expand at a reasonable cost. Note, however, that many times the expansion of a system is determined more by the design of the system and the use of the supporting electronic equipment than by the selection of a type of medium. How computer networks are expanded will be examined in Chapter Eight.

Environment

Another factor that must be considered in the media selection process is the environment. Many types of environments are hazardous to certain media. Industrial environments with heavy machinery produce electromagnetic radiation that can interfere with improperly shielded cables. If the cable may be traveling through an electromagnetically noisy environment, shielded cable or fiber optic cable should be considered.

Wireless transmission can also be disrupted by electromagnetic noise and interference from other transmissions. Sunspots, while they do not happen often, can be disruptive to satellite transmissions. Since so many people rely on wireless services for voice, data, and paging, newspapers often warn us when sunspot activity is expected to be high.

Security

If data must be secure during transmission, it is important that the media not be easily tapped. All conductive media, except fiber optic cable, can easily be wiretapped by someone listening to the electromagnetic signal traveling through the wire. Wireless communications can also be intercepted. Data security of conductive media can be improved, however, with the proper use of encryption and decryption software. Spread spectrum technology can be applied to wireless communications making them virtually impervious to interception.

Conducted Media in Action

Let's consider the wiring for a local area network. Figure 3-18 shows a common situation in which a microcomputer workstation that is connected to a local area network must first connect to a device called a hub. A hub is a simple device that connects up to 24 workstations and passes the transmission signal from any workstation on to all other workstations (Chapters Seven and Eight take a detailed look at hubs). In most typical installations, it is quite unlikely that the cable leaving the workstation runs directly to the hub. Instead, the cable that leaves the back of the microcomputer first connects to a wall jack in the room in which the microcomputer is located. This wall jack is a **passive device**, a simple connection point between two runs of cable that does not regenerate the signal on the cable.

Figure 3-18
Example wiring situation involving a microcomputer and a local area network

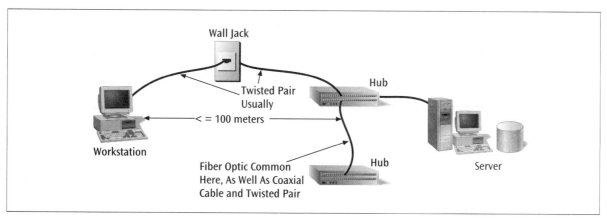

Cable distance is usually an issue that must be taken into account. To account for cable distance in this situation, you must consider the total distance from the back of the microcomputer to the hub. If this distance is less than 100 meters, you can consider using twisted pair to connect the microcomputer to the hub. If the cable does not pass through a noisy environment, then you can consider using unshielded twisted pair, the least expensive and easiest cable to work with. If you can assume that the data rate of the connection will not exceed 100 Mbps, it should be possible to support the connection using category 5 UTP.

If the data rate is higher than 100 Mbps, you will have to consider alternatives to twisted pair. Baseband coaxial cable typically does not support data rates higher than 100 Mbps, thus baseband coaxial cable is probably not worth considering further. If you want to convert the digital signal from the microcomputer to some analog form, then it might be possible to achieve the necessary data rate with broadband coaxial cable. Converting the digital signal to analog at high transmission rates requires additional equipment and is costly, so you might want to consider other alternatives.

Fiber optic cable is a good choice for data transmission rates over 100 Mbps, but will cost a little more than a twisted pair connection. Also, since fiber optic cable is a one-way cable and the data flow between a workstation and a hub is two way, two fiber optic cables are necessary. Since the data rate between the micro-computer workstation and hub is 100 Mbps or less, Category 5 unshielded twisted pair is an acceptable choice. Fiber optic cable would also be a reasonable choice and would be adaptable to higher transmission speeds in the future.

What about cabling the connection between the hub and the next point in the local area network, usually another hub? If the distance between the two hubs exceeds the 100 meter length or if the cable travels through a noisy environment, such as a heating and cooling mechanical room, a cable other than twisted pair should be considered. Under these circumstances, you would consider either a very good coaxial cable or, even better yet, fiber optic cable to connect the two hubs. Why should you consider the more expensive fiber optic cable? The price difference between coaxial cable and fiber optic cable is much closer than the performance differences. Therefore, it makes sense to go with the best cabling and install fiber optic cable. Don't forget that fiber optic is a one-way cable and the traffic that flows between hubs is two way, so there will be a need for two fiber optic cables.

Wireless Media in Action ▶

DataMining Corporation is a large organization that has a main office in Chicago and a second office in Los Angeles. DataMining collects the data from grocery stores from every purchase made by every customer. Using this data, the company extracts spending trends and sells this information to other businesses, which market salable goods. Data is collected at the Chicago office and transmitted to the Los Angeles office where it is stored and later retrieved. Thus, there is a need to transmit large amounts of data between the two sites on a continuous basis.

Currently DataMining is using a leased telephone service between Chicago and Los Angeles, but the telephone bills are very high. The company is trying to reduce costs and is considering alternatives to leased telephone services. (Other forms of telephone service are available, but these will not be introduced until Chapter Twelve.) Let's simply say that DataMining has examined them and found them to be expensive, thus the need to consider other alternatives. DataMining has discovered that a number of companies can provide various levels of satellite communication services. For example, Hughes Network Systems can provide local area network internetworking, multimedia image transfers, interactive voice connections, interactive and batch data transfers, and broadcast video and data communication services. Because DataMining is primarily interested in two-way data transmissions, it is considering a two-way data communications service offered by Hughes through a satellite system called Very Small Aperture Terminal (VSAT).

Very Small Aperture Terminal is a satellite communications system that serves both home and business users. A VSAT end user needs a box that interfaces the user's computer system and an outside antenna with a transceiver. This transceiver, which is small, sends and receives signals to a LEO satellite. VSAT is capable of handling data, voice, and video signals over much of the earth's surface. As shown in Figure 3-19, this two-way data service requires the installation of an individual earth station satellite dish at each of DataMining's corporate locations.

Figure 3-19
VSAT satellite solution for DataMining Corporation

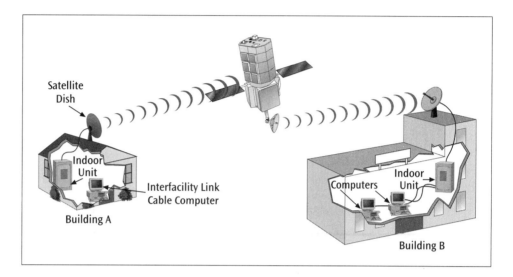

Each individual earth station is composed of two pieces: an indoor unit and an outdoor unit. The outdoor unit consists of the satellite dish and is usually mounted on the roof of the building. The size of the satellite dish depends on the data rates used and the satellite coverage required. The outdoor unit is connected to the indoor unit via a single interfacility link cable. The indoor unit has one or more ports to which the company's data processing equipment can be connected.

Maintenance and support for this VSAT service is provided by the satellite company on a 24-hours-per-day, 7-days-a-week basis, and includes equipment configuration, system status reporting, bandwidth allocation, downloading of any necessary software, and the dispatching of field personnel if necessary.

DataMining Corporation has decided to install the VSAT satellite system. To summarize, let's consider each of the media selection criteria as applied to the VSAT solution:

- ▶ **cost** the VSAT system is relatively expensive, but it delivers a very high data transfer rate with very high reliability;
- ▶ **speed** the VSAT system can support the required data transfer rates of DataMining Corporation;
- ▶ **distance** the satellite system can easily extend from Chicago to Los Angeles;

> ► environment satellite systems can be disrupted by strong electromagnetic forces, which could be a problem. If DataMining cannot tolerate any disruption of service, they might want to consider installing a backup system in case the VSAT system momentarily fails; and

> ► security VSAT satellite systems are difficult to intercept due to the very small transmission beam sent between ground stations and satellite. Additionally, the data stream can be encrypted.

A second company, the American Insurance Company, has two offices, both located in Peoria, Illinois. The first office collects all the premium payments and the second office contains the main data processing equipment. American needs to transfer the collected premium information to the data processing center on a nightly basis. The two offices are approximately two miles apart, and it is not possible to run their own cable over public property and other persons' personal property. Some form of telephone system might provide a reasonable solution, but American Insurance is interested in investing in its own system and would like to avoid the monthly recurring telephone charges. At some point, the monthly recurring telephone charges would exceed the cost of their own system.

ProNet is a company that offers private terrestrial microwave systems that can transfer private voice, local area network data, video conferencing, and high resolution images between remote sites up to 15 miles apart. Figure 3-20 shows a typical microwave communication setup between the two corporate offices of American Insurance.

Let's consider each of the media selection criteria as applied to the ProNet terrestrial microwave solution:

> ► cost the ProNet system is expensive at first when the equipment is purchased, but after that American only has to pay for maintenance;

> ► speed the ProNet system can support the required data transfer rates of American Insurance;

> ► distance the terrestrial microwave system can transmit up to 20 miles. The two corporate buildings are two miles apart;

> ► environment microwave systems can be disrupted by strong electromagnetic forces and inclement weather; ProNet states their system is unaffected by fog or snow and delivers a service reliability and availability of 99.97 percent; and

> ► security terrestrial microwave systems can be intercepted, but the data stream can be encrypted.

American Insurance will seriously consider using ProNet's terrestrial microwave system as it compares favorably to other systems and satisfies their goals of private ownership, high transfer rates, and low recurring costs.

Figure 3-20
Microwave communication between American Insurance's corporate buildings

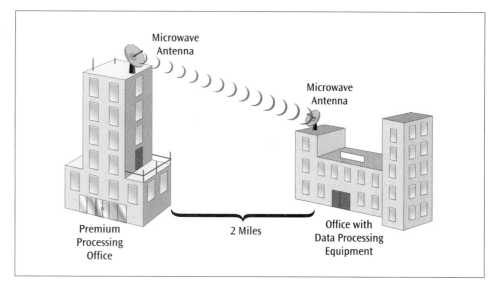

SUMMARY

▶ All data communication media can be divided into two basic categories: the physical or conducted media, such as wires, and the radiated or wireless media, which use radio waves.

▶ There are three types of conducted media: twisted pair, coaxial cable, and fiber optic cable.

▶ Twisted pair and coaxial cable are both metal wires and are subject to electromagnetic interference.

▶ Fiber optic cable is a glass wire and is impervious to electromagnetic interference, which produces a lower noise level than is found with twisted pair and coaxial cable.

▶ Because fiber optic cable has a lower noise level, because light signals do not attenuate as quickly as electric signals, and because light propagates more quickly through glass than electric signals propagate through wire, fiber optic cable has the best transmission speeds and long distance performance of all conducted media.

▶ Satellite, microwave, and radio broadcast networks are an increasingly vital form of communication in the world of data communications and computer networks.

▶ There are seven basic groups of wireless media: terrestrial microwave transmissions, satellite transmissions, cellular radio systems, personal communication systems, pagers, infrared transmissions, and multichannel multipoint distribution service.

▶ When trying to select a particular medium for an application, it helps if you compare the different media using five different criteria: cost, speed, distance and expandability, environment, and security.

KEY TERMS

advanced mobile phone service (AMPS)

backbone

baseband coaxial

broadband coaxial

broadband wireless

Category 1-5 (CAT 1-5)

cellular digital packet data (CDPD)

coaxial cable

crosstalk

data transmission speed

digital advanced mobile phone service (D-AMPS)

downlink

fiber optic cable

geosynchronous earth orbit (GEO)

global system for mobile (GSM)

hybrid telephone

infrared transmission

line of sight transmission

local multipoint distribution system (LMDS)

low earth orbit (LEO)

middle earth orbit (MEO)

mobile service area (MSA)

multichannel multiple distribution service (MMDS)

pager

passive device

personal communication system (PCS)

photo diode

photo receptor

propagation delay

propagation speed

repeater

reservation system

satellite microwave

shielded twisted pair (STP)

terrestrial microwave

thick coaxial cable

thin coaxial cable

twisted pair wire

unshielded twisted pair (UTP)

uplink

very small aperture terminal (VSAT)

REVIEW QUESTIONS

1. Why is twisted pair wire called twisted pair?
2. How does crosstalk occur in twisted pair wire?
3. What are Category 1, 2, 3, 4, and 5 twisted pair wire used for?
4. What are the advantages and disadvantages of shielded twisted pair?
5. What is the primary advantage of coaxial cable compared to twisted pair?
6. What is the difference between baseband coaxial and broadband coaxial cable?
7. Why is fiber optic cable immune to electromagnetic interference?
8. What are the advantages and disadvantages of fiber optic cable?
9. What are the different costs of a conducted media?
10. What is the difference between data transmission speed and propagation speed?
11. What is the difference between terrestrial microwave and satellite microwave?
12. What is an average distance for transmitting terrestrial microwave?
13. What kind of objects can interfere with terrestrial microwave transmissions?
14. List a few common applications for terrestrial microwave.
15. What are the three orbit levels for satellite systems?
16. List a few common application areas for each orbit level satellite system.
17. What is the sequence of events when placing a call from a cellular telephone?
18. What is the function of a mobile telephone switching office?
19. What is the primary difference between AMPS and D-AMPS cellular systems?
20. What is the primary difference between AMPS (or D-AMPS) cellular systems and the newer PCS mobile telephones?
21. What are the three competing technologies used in PCS mobile telephones?
22. What is a hybrid mobile telephone?
23. With what system does CDPD share its frequencies?
24. What is a common application area of CDPD?

25. Which of the following are true concerning pagers:
    a. they transmit in one direction only
    b. they transmit in two directions
26. Infrared transmission can be used for which type of applications?
27. Broadband wireless service supports what kind of applications?

EXERCISES

1. Table 3-2 shows Category 1 wire transmitting a signal for 2-3 miles but, Category 5 for only 100 meters (328 feet). Is Category 1 the best wire for long distance transmissions? Explain.

2. List three different examples of crosstalk that don't involve wires and electric signals. (*Hint: look around you.*)

3. Is there an official Category 6 twisted pair yet? Use either paper sources or the Internet to find the answer.

4. Can you transmit a video signal over twisted pair wire? Explain. Be sure to consider multiple scenarios.

5. The local cable TV company is considering removing all the coaxial cable and replacing it with fiber optic cable. List the advantages and disadvantages of this plan.

6. The local cable TV company has changed its mind. It is now going to replace all the existing coaxial cable with unshielded twisted pair. List the advantages and disadvantages of this plan.

7. Terrestrial microwave is a line-of-sight transmission. What sort of objects are tall enough to interfere with terrestrial microwave?

8. Using an outside source, such as the Internet or the library, what is the typical height of a terrestrial microwave tower? If the tower's height is raised by 10 meters, how much farther will the tower be able to transmit?

9. Given that a satellite signal travels at the speed of light, exactly how long does it take for a signal to go from the earth to a satellite in geosynchronous orbit?

10. How long does it take a signal to reach a satellite in low earth orbit?

11. You are walking down the street and your cell phone rings. What was the sequence of events that allowed a person with a conventional telephone to call you on your cellular telephone?

THINKING OUTSIDE THE BOX

1 You have been asked to recommend the type of wiring for a manufacturing building. The size of the building is 200 meters by 600 meters and houses large, heavy machinery. Over 200 devices in this building must be connected to a computer system. Each device transmits data at 2 Mbps and sends a small packet of data every 2 to 3 seconds. Any cable recommended has to be suspended high overhead in a hard-to-reach conduit system. What type of cable would you recommend? Use the media selection criteria introduced in this chapter to arrive at your answer.

2 GJ Enterprises has hired you as a productivity consultant. Currently, they employ six people who routinely exchange information via "sneakernet" (floppy disk transfer). They want the least expensive network wiring solution and only minimal required hardware support. All employees are in the same brick building but not in the same room. They will be sending word processing documents and small spreadsheets as well as e-mail. What media would you recommend and why?

3 You are the technology guru for an interstate trucking company. You need to maintain constant contact with your fleet of trucks. Which wireless technologies will enable you to do this?

4 You need to connect two buildings across a public road, and it is not feasible to use a direct cable connection. When would you use a microwave link? When would you use a radio link? When would you use an infrared laser link?

5 For many years you have had a computer at home for all members of the family to use. Recently however, you added a second computer. You want to connect the two computers so they can share data and a good quality printer. Unfortunately, one computer is on the main floor and the second one is upstairs. How are you going to connect the two computers? List as many possibilities as you can with the advantages and disadvantages of each.

PROJECTS

1. Collect and label samples of as many kinds of conducted media as possible and nicely display them on cardboard.

2. Using the Internet, locate the technical specifications of two different types of cables.

3. How many different types of conducted media are in place in your business or school? How are they used? Draw a rough diagram that shows the approximate locations and types of wire.

4. Using any sources possible, investigate a company that can offer a microwave service in your area. Report on what kinds of applications can be supported, what equipment will be necessary, where the equipment will be located, and what services this company offers.

5. Using any sources possible, investigate a company that can offer VSAT satellite service in your area. Report on what kinds of applications can be supported, what equipment will be necessary, where the equipment will be located, and what services this company offers.

6. Does anyone in your area offer an MMDS or LMDS satellite service? Is yes, write a one-page summary that includes the service's main features.

7. How many different cellular telephone companies offer services within your area? Are the services AMPS or PCS? If PCS, are they CDMA, TDMA, or GSM? Do they have an estimate of the number of current subscribers? Do they know how many cells are in your market?

8. Ask your local cable television company how they receive their television signals. If it is a satellite service, is the service LEO, MEO, or GEO? What is the frequency range of the signals they receive? Do they receive signals from multiple sources? Do they have a backup plan if one of their services is disabled?

9. Visit the FCC's web site (*http://www.fcc.gov*) and report what frequencies are currently being auctioned.

4

Making Connections

◆◆

CONNECTING PERIPHERAL DEVICES to a computer has never been an easy task. The interface between a computer and a peripheral is complex and contains many layers of hardware and software. Many experts in the field have been working for years on simplifying the interconnection process, and the Universal Serial Bus appears to be one of the best contenders for a new interface standard.

Unfortunately, new interface standards rarely work as intended on the first try. Consider the Comdex Spring '98 computer show in which Microsoft Corp. CEO Bill Gates and an associate, in an attempt to demonstrate the soon to be released Windows 98 and its Universal Serial Bus interface, tried to connect a page scanner to a computer while the

computer was turned on. Despite the fact that the Universal Serial Bus and Windows 98 were designed to automatically accept peripherals whenever they are plugged in, Gates' computer did not accept the device and crashed. To the delight of the audience, Gates quickly responded that "It's a good thing that (Windows 98) is still in beta."

Computing Canada, vol.24, Issue 17, p.17.

What is involved with connecting a computer to other devices?

What is the status of the Universal Serial Bus today?

Objectives ▶

After reading this chapter, you should be able to:

▶ Identify a standard modem and cite its basic operating characteristics.

▶ Discuss the advantages of the newer digital modems and recognize why they do not achieve the high transfer speeds as advertised.

▶ List the alternatives to traditional modems, including T1 modem, cable modems, IDSN modems, and DSL modems.

▶ Recognize the uses of a modem pool and its advantages and disadvantages.

▶ List the four components of all interface standards.

▶ Discuss the basic operations of the EIA-232E interface standard.

▶ Cite the advantages of FireWire and Universal Serial Bus interface standards.

▶ Outline the characteristics of asynchronous and synchronous data link interfaces.

▶ Recognize the difference between half duplex, full duplex, and simplex connections.

▶ Identify the operating characteristics of terminal to mainframe connections and why they are unique from other types of computer connections.

Introduction ▶

A computer would be of no use if we could not connect it to anything. Imagine, if you will, a computer with no monitor to view output and no keyboard to enter data. Many people also feel that their computers would not be worth much if there was no way to connect the computer to a printer or if they could not connect it to a modem to dial into the Internet or a remote computer system. Many people in the corporate world depend almost exclusively on an interconnection between their computer and a company local area network. Through this connection they are able to access corporate databases, e-mail, the Internet, and other software applications. In fact, computer and computer-related manufacturers are constantly creating new devices to interconnect computers and other devices. These devices include document scanners, digital cameras, video cameras, and music systems, among others.

Connecting a peripheral device to a computer can be a challenging task. Various levels of hardware and software have to agree completely before the computer can talk to the device or vice versa. Questions arise such as: "Is the connector on the end of the cable coming from the device compatible with the plug on the back of the computer?" and "will the electrical properties of the two devices be compatible?" Even if the answers to these questions are "Yes," will the computer and device "speak the same language"? Connecting computers to other devices has many pitfalls and obstacles.

One of the more common connections found in computer systems is the connection of a computer to a modem. Using a modem and a standard telephone line, a computer can connect to remote computers, remote computer networks, and the Internet. Although the basic operating principles of a modem have not changed much over the years, advances in technology have increased a modem's data transfer speed. Even with these advances in technology, it appears the data transfer speed of modems in conjunction with standard telephone lines will not get much higher than they are now. Thus, computer users are turning their attention to faster connections such as cable modems and digital subscriber line. Both of these technologies, as well as others, will be introduced in this chapter.

To better understand the interconnection between a computer and a device such as a modem requires you to understand the concept of interfacing. Interfacing a device to a computer is considered a physical layer activity since it deals directly with analog signals, digital signals, and hardware components. We will examine the four basic components of an interface—electrical, mechanical, functional and procedural—and then introduce several of the more common interface standards. Although the older interface standards were designed for computer to modem connections, the newer interface standards are designed to support a wider range of devices.

Connecting a computer to a device requires more than just physical layer connections. It is also necessary to define the packaging of the data as it transfers between computer and device. More generally speaking, connections at the data link layer define the basic configuration of the data that is passed between sender and receiver. We will examine two popular data link layer configurations: asynchronous and synchronous connections.

Finally, we will examine the connection between a terminal and a mainframe computer. Since a terminal is a relatively unintelligent device, a mainframe computer creates a unique dialog called polling, which prompts the terminals if they have data to submit to the mainframe.

Modems

Today's modems are complex and offer so many functions and features that the user manuals accompanying them are sometimes hundreds of pages long. For example, most contemporary modems include functions such as support for multiple transmission rates, standard telephone operations, connection negotiation, compression, error correction, facsimile transmission, security, and loop-back testing. Modems can also be physically characterized by whether they are internal or external models and whether they are suitable for use with a laptop computer. How does a modem work? What are all these functions, and what are they doing in your modem?

Basic modem operating principles

Since portions of the U.S. telephone system transmit analog signals, the transfer of a computer's digital data over a telephone line requires conversion of the data to an analog signal. The technique of converting digital data to an analog signal is called modulation. The device that modulates digital data onto an analog signal and then demodulates the analog signal back to digital data is the **modem**. Recall from Chapter Two that three currently popular modulation techniques for encoding digital data and transmitting it over analog signals are amplitude modulation, frequency modulation, and phase modulation. Modern modems use various combinations of these modulation techniques to convert digital data into an analog signal. The newer modems that transmit data at speeds of 33,600 bits per second and higher use modulation techniques that are very advanced and well beyond the scope of this textbook. One of these techniques, reviewed in Chapter Two, is quadrature amplitude modulation, which uses 12 different phase angles in combination with two different amplitudes. The combination of phase changes and amplitudes results in 16 different signal changes and is used in modems that transmit data at 9600 bits per second. The faster speed modems use higher levels of phase changes in combination with more advanced technologies to achieve their transfer rates.

Data transmission rate

One of the most obvious characteristics of a modem is the maximum rate at which it can transfer data. Most modern modems are capable of supporting a number of data transmission rates. For example, a modem that is commonly used to connect a personal computer to the Internet and transmit data at a rate of 56,000 bits per second can support a wide range of data rates from 0 bps up to 56,000 bps.

Why does a 56,000 bps modem accommodate so many different data rates? There are two very good answers to this question. First, not everyone has the same speed modem. So that a newer modem can talk to an older and possibly slower modem, the newer modem is designed to accommodate the slower speed. Second, telephone lines and their connections experience varying levels of noise at differing times. When two modems establish a connection, they agree on an acceptable transmission rate. That acceptable transmission rate depends on the level of noise present during establishment of the connection.

Standard telephone operations

Most modern modems can perform a set of standard telephone functions that support basic modem operations. This set of functions includes, among others, auto answer, auto dial, auto disconnect, and auto redial. Auto answer means that the modem automatically answers "the telephone" when it detects the ringing signal on the line. This operation makes it possible for a computer and its modem to be left in unattended mode. Auto dial means that the modem will automatically place a call when given the appropriate signal followed by the telephone number. This function can be extended to dial any special codes before dialing the actual telephone number. For example, many businesses require a number such as 9 to be dialed before an outside line can be accessed. Auto disconnect means that the modem will terminate a call and reset itself when it detects that the line signal has been dropped. Auto redial means that the modem will try to redial the last number dialed if it receives a busy signal when attempting to place a call.

Connection negotiation

Most modern modems can dynamically perform fallback or fall forward negotiation. When two modems first establish a connection, they agree on an acceptable rate of data transfer based on the maximum capabilities of the modems and the level of noise on the connection. If two modems are not capable of supporting the highest possible rate of traffic, they **fallback** to a slower speed. This fallback continues until the two modems can agree on an acceptable rate. If a telephone is so noisy that it cannot support a connection at all, the two modems will fallback all the way to zero and disconnect. Modems can also **fall forward**, or dynamically negotiate a higher rate of data transfer. If the noise level on the connection drops, the two modems may renegotiate and agree on a faster transmission speed.

Compression and error correction

Back in the 1980's, Microcom Corporation created a series of modems that could perform automatic error correction and data compression. For example, if a Microcom modem was transmitting data to another Microcom modem and the data arrived garbled, the receiving modem would return a message to the transmitting modem stating that the last data transmitted contained an error. These messages would occur without the knowledge of the upper layers of software and without the knowledge of the user. Compression worked in a similar fashion in that compression and decompression would be handled by the modems and no further software was needed. If the two modems are engaged in data transfer and the data lends itself nicely to being compressed (such as long strings of zeros), the MNP 5 compression standard can deliver the equivalent of 115,200 bits per second over a 33,600 bits per second connection.

The two features of error correction and compression became so popular that the two Microcom features became standards found in virtually all modern modems. Microcom named the standards **MNP 1-5** depending on the function and version of the software.

Facsimile

Many modems are capable of sending facsimile, or fax, communications. Since fax transmissions require a different protocol, a modem that transmits faxes must be able to support the fax standards, such as V.17, V.27ter, and V.29.

Security

Modems can perform a number of security functions that can prove to be quite valuable when it is desirable to limit who has access to a computer network or system. Standard security features include blacklisting, callback security, and backdoor entry with password protection. Blacklisting is the ability of a modem to block out certain callers trying to dial in from a remote location. If a remote modem tries to dial into a local modem with callback security, the local modem will hang up and call back the remote modem, but only at a predetermined telephone number. Callback security prevents unauthorized dialing in from unscrupulous parties. Backdoor entry with password protection allows a system administrator to dial into a modem from a remote location, enter a password or series of passwords, and check or set the configuration of the modem.

Self-testing (loop back)

Most modern modems are capable of performing two testing functions: local loop-back testing and remote loop-back testing. These two features are useful when trying to debug a computer to modem connection. Questions like "Is the computer correctly connected to the modem?" and "Is the telephone circuit between the local modem and remote modem operating properly?" can be answered with loop-back testing. In **local loop-back testing**, computer or terminal A sends data to its local modem, which immediately returns the data to computer A. This type of testing can determine if the local computer is correctly connected to the local modem. In **remote loop-back testing** computer A transmits data to its local modem, which then transmits the signal over the interconnecting medium to the remote modem B. The remote modem B immediately returns the signal to modem A, which returns the data to computer A, as shown in Figure 4-1. Remote loop-back testing not only tests the local computer to modem connection but also tests the interconnection medium and remote modem connection.

Figure 4-1
Local loop-back testing and remote loop-back testing

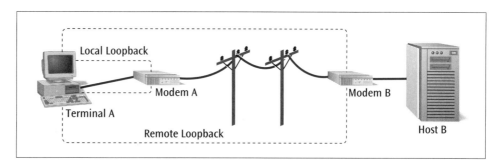

Internal versus external models

Many modems, particularly those purchased for use with personal computers, can be either internal models or external models, as shown in Figure 4-2.

Figure 4-2

Example of an external modem and an internal modem side by side

An **internal modem** is a printed circuit board that is designed to plug into an available printed circuit card slot within a personal computer. Internal modems have the advantage of being enclosed within the computer, and do not require extra cabling, a separate power supply, or space to sit on the desktop or elsewhere. The chief disadvantages of internal modems are that they require an empty slot and an interrupt request (IRQ) assignment that cannot conflict with any other IRQ assignments within the computer. An IRQ assignment is an assigned location where the computer can expect a particular device to send the computer signals about its operation. These signals momentarily interrupt the computer so that it can decide what processing to perform next.

An **external modem** is a modem with its own power supply and is located outside the computer's cabinet. External modems are advantageous if there is no additional room within the computer (that is, there is no available slot) and if there is an IRQ conflict that cannot be resolved. External modems also require a special cable and a port to plug into on the back of the computer. Usually this port is a serial I/O port, but on the newest computers this port could be a USB or a FireWire port.

Modems for laptops

Modems designed to operate in laptop computers can also be classified as internal or external. Many laptop computers, as well as desktop computers, come from the manufacturer with an internal modem installed. If a laptop computer is purchased without an internal modem, and you decide later on to add a modem, you should select an external modem specifically designed for a laptop. These laptop external modems are a cross between internal modems and external modems. They are partially external in that they plug into a connector on the back or side of the laptop and extend outside the laptop's case, as you can see in Figure 4-3.

Figure 4-3
Example of a laptop computer with an external modem sticking out and a phone line connected to the modem

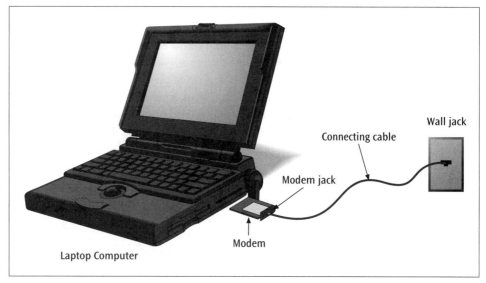

They are partially internal in that they do not have their own power supply, but draw their power from the connection instead. To achieve some level of standardization, most laptop modems conform to the **PCMCIA** (Personal Computer Memory Card International Association) standard. PCMCIA is a non-profit trade association and standards body that promotes a variety of technologies, including:

► **PC card** technology—Credit card-sized, rugged peripherals that add memory, mass storage, and I/O capabilities to laptop computers;

► miniature card technology—Memory cards that support DRAM, flash memory, and ROM and are smaller than PC Cards;

► smart media card technology—Memory cards that support flash memory and ROM and are roughly the same size as miniature cards, but much thinner; and

► **CardBus**—A newer 32-bit version of PC card technology; adds memory, mass storage, and I/O capabilities.

In the past, these cards were known as *PCMCIA Cards*, but the industry now refers to products based on the technology as *PC Cards*, *PC Card Hosts* and *PC Card Software*, and refers to the association as PCMCIA.

Breaking Bandwidth Limitations

When the 33,600 bps modem became available, many industry experts believed that this was the fastest speed a modem would ever achieve using the standard telephone lines that connect our homes and businesses to the telephone network. This belief was based on the fact that the telephone connection into a home or business (the local loop) is an analog connection, and that the telephone signal is transmitted with a certain signal level and background noise level. (If you are interested in the mathematical calculation, read the Details section accompanying this section of the chapter.) Approximately two years after the 33,600 bps modem was announced, the 56,000 bps modem was announced. Did something change to allow the faster transmission speed, or were the industry experts wrong? The experts were correct—two important facts changed with the 56,000 bps modems: digital signaling was introduced, and the signal power level was increased.

The newer 56K modems are a hybrid design combining both analog signaling and digital signaling. Figure 4-4 shows that the only analog portion of most telephone networks is the local loop that runs from the local user to the telephone central office. When data leaves the local user's computer and enters the 56,000 bps modem, the digital data is modulated onto an analog signal. This analog signal travels across the local user's local loop until it arrives at the central office, where it is converted from analog back to digital. A problem occurs when the analog data is converted to digital. Recall that one of the biggest problems with analog to digital and digital to analog conversion is quantization noise—not recreating the original signal precisely. Every time a signal is converted from analog to digital or digital to analog, some quantization noise is introduced. It is this quantization noise that causes the data rate to slow down and limits transmission from local user to remote system (upstream transmission) to 33,600 bps.

Figure 4-4
The analog portion or local loop of the local telephone network

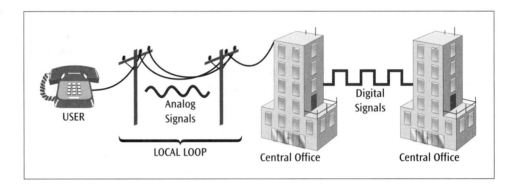

As long as there are no analog to digital conversions, quantization noise is not as serious a problem on the link from the remote system back to the local user (downstream transmission). No quantization noise should be the case if the remote system has a digital connection to the telephone company, as shown in Figure 4-5.

Figure 4-5
No analog to digital conversions on the downstream link

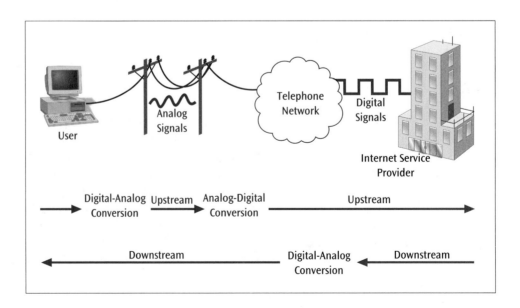

If the remote system has an analog connection to its telephone company, as shown in Figure 4-6, then an analog to digital conversion occurs, more noise will be introduced, and the data rate will have to drop back to 33,600 bps.

Figure 4-6

Remote system with an analog connection to the telephone system

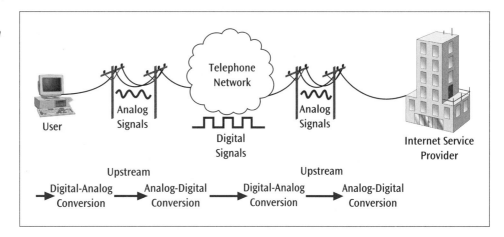

If the remote system has a digital connection to its telephone company and there are no other analog to digital conversions along the way, 56,000 bps modems can create a digital connection that transfers digital signals over the digital portion of the telephone network at speeds up to 64,000 bps. These digital values arrive at the local user's central telephone office where they are converted to corresponding discrete analog signals and sent to the 56,000 bps modem over the analog local loop. If there was no noise on the local loop going into the local user's residence, then the 56,000 bps modem could actually receive the full 64,000 bps signal. However, there is always noise. This noise reduces the 64,000 bps digital signal to 56,000 bps. As it turns out, receiving a 56,000 bps signal is not possible either. So that the digital signals on one telephone line do not interfere with the digital signals on an adjacent telephone line, the FCC requires that the power level of the signal be lowered by a small amount, making noise a bigger factor. Because noise is a bigger factor, the signal has to slow down even more to approximately 53,000 bps.

The picture gets even worse. Some telecommunications specialists estimate that 50 to 60 percent of the telephone lines in the United States cannot support 33,600 bps connections, let alone 56 Kbps connections. These telephone lines cannot support higher speed connections because they use old telephone system hardware, which introduces too much noise. If those estimates are accurate, then half the telephone lines in the United States can support only 28,800 bps, if that.

To further complicate the issues, when 56,000 bps modems first arrived on the market, there were two competing formats for establishing a 56,000 bps connection: x2 and K56flex. Each format was created by a modem manufacturing company and was incompatible with the other. Eventually, the International Telephone Union (ITU) created a real standard named V.90. The **V.90 standard**, which is a 56,000 bps dial-up modem standard approved by a standards making organization and not a single company, is slightly incompatible with both x2 and K56flex. Most x2 and K56flex modems, however, will accept upgrades to make them compatible with the V.90 standard.

Alternatives to Traditional Modems

According to our definition, a modem modulates the digital data of a computer onto a signal appropriate for a standard telephone line and then demodulates the signal back to digital data form. All of the modems that have been discussed so far qualify as modems by this definition. There are, however, four alternative transmission technologies available, other than the traditional telephone line, that can be used to connect a computer into a remote network system: T-1 digital telephone lines, cable television networks, Integrated Services Digital Network (ISDN), and Digital Subscriber Line (DSL). Each of these transmission technologies requires a particular kind of device that converts the digital data of a computer to the proper form for transmission. Although these technologies will be discussed in more detail in Chapter 12, let's spend a few minutes introducing their modem-like devices: channel service unit/data service unit, cable modems, ISDN modems, and DSL modems.

Channel Service Unit/Data Service Unit (*CSU/DSU*)

A **Channel Service Unit/Data Service Unit (CSU/DSU)** is a hardware device about the size of an external modem that converts the digital data of a computer into the appropriate form for transfer over a 1.544 Mbps T1 digital telephone line or over a 56 Kbps to 64 Kbps leased telephone line. If your company has a local area network that connects to a leased digital telephone line, such as a T1 line, you will need a CSU/DSU at your end and the telephone company will have a CSU/DSU at its end, as shown in Figure 4-7.

Details ▶

Calculating the Maximum Data Transfer Rate

Claude Shannon created a relatively simple and elegant formula for determining the maximum data transfer rate of *any* signal given the frequency of the signal and given the strengths of the signal and the always present background noise:

$$S(f) = \text{signal frequency} \times \log_2 (1 + \text{signal level in watts} / \text{noise level in watts})$$

For example, consider a typical telephone system that transmits a 3200 Hz signal with a power level of 15 watts and a noise level of 0.01 watts:

$$S(f) = 3200 \times \log_2 (1 + 15 / 0.01)$$
$$= 3200 \times \log_2 (1501)$$
$$= 3200 \times 10.55$$
$$= 33760 \text{ bps}$$

Given that the standard telephone line that runs from the central office into a home or business is transmitting a 3200 Hz analog signal and, assuming the power levels of the signal and the noise are as given, the maximum number of bits per second that can be transmitted over a telephone line lies somewhere in the low- to mid-30,000 bits per second.

It appears Shannon's value is accurate. Experts have been agreeing that the modem which delivers 33,600 bits per second over a standard telephone line using analog signaling and the given power levels will not get any faster.

Figure 4-7
CSU/DSU connection between a LAN and a T1 line

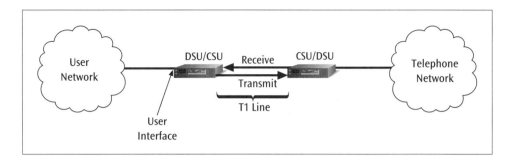

The Channel Service Unit (CSU) transmits signals to and receives signals from the telephone company's T1 line and provides an isolation barrier for electrical interference from either side of the unit. The CSU can also echo loopback signals from the telephone company for testing purposes and has storage for keeping track of line usage statistics. The Data Service Unit (DSU) manages line control for each T1 connection and converts the data to be transmitted into the proper frames for transmission on the T1 line and converts the incoming T1 data frames into a non-T1 format. CSU/DSUs can be purchased as separate products or are sometimes integrated with a router.

Cable modems

A **cable modem** is a communications device that allows high-speed access to wide area networks such as the Internet via a cable television connection. A cable modem is the interface between a personal computer and the cable connection that enters the room, as Figure 4-8 shows.

Figure 4-8
Cable modem connecting a personal computer to the Internet via a cable television connection

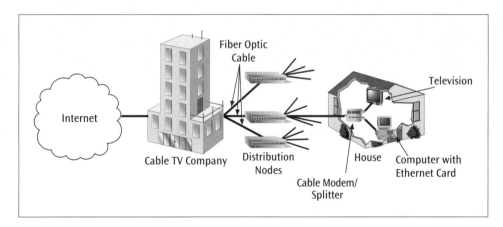

Most cable modems are external devices that connect to the personal computer through a common Ethernet card which is either provided by the cable company or is purchased at most stores that sell computer equipment. As such, the connection is capable of transmitting megabits of data, with transmission rates ranging anywhere from 500 Kbps to 2.5 Mbps. These connections, however, are usually asymmetric—the downstream data transmission speed is faster than the upstream transmission speed. This asymmetry is reasonable since most cable modems are used for connection to the Internet and the World Wide Web, where downstream transmissions consist of web pages and graphic images and upstream transmissions usually consist of much smaller get-web-page commands.

Although most cable modems use the cable television connection for both the upstream and downstream transmissions, a small percentage of systems use a telephone line for the upstream transmissions, as shown in Figure 4-9. Since the upstream connection is over a telephone line, data transmission is restricted to the slower 33,600 bits per second. This slower transmission rate is acceptable because high data transfer rates are not as necessary in the upstream connection as they are in the downstream connection. Using a telephone line for the upstream connection has a major disadvantage for the customer: the customer has to install an additional telephone line just to support the upstream cable connection. Fortunately, most cable modem systems do not require the use of a telephone connection.

Figure 4-9
Cable modem with a telephone line for the upstream connection

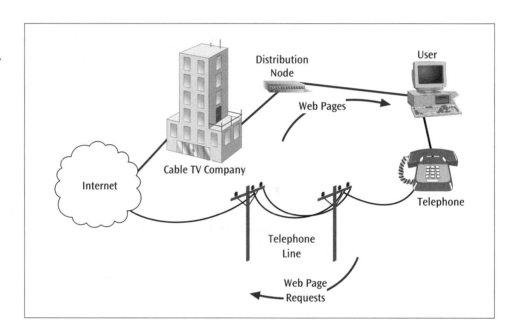

Details ▶

Cable Modem Operation

Cable companies that offer a cable modem service use a standard called Data Over Cable Service Interface Specification (DOCSIS). DOCSIS was designed to include all the operational elements used in delivering a data service over a cable television system, including service provisioning, security, data interfaces, and radio frequency interfaces. The basic architecture of a cable modem system is shown in Figure 4-10.

The system consists of six major components. The first component, the Cable Modem Termination System (CMTS), is located at the main facility of the cable operator and translates the incoming web page data packets into radio frequencies. The frequencies used for the transfer of the web page data packets are unused frequencies in the 6 MHz range. The second component is the Combiner, which combines the frequencies of the web page data packets with the frequencies of

the common cable television channels . These signals are then sent over a full-duplex fiber optic line, which is the third component in the system, to a fiber distribution node, the fourth component in the system. The fiber distribution node resides in a user's neighborhood and can distribute the signals to 500 to 2000 homes. From the fiber distribution node, coaxial cables carry the signals into homes where a splitter, the fifth component, separates the web page data packet frequencies from the other cable channel frequencies. The other cable channel frequencies go to the television set, and the web page data packet frequencies that have been split off go to a cable modem. The cable modem (the sixth component of the system) converts the radio signals back to digital data packets for delivery to your computer.

Although cable modems provide high speed connection to the World Wide Web and other Internet services, their deployment across the United States is quite scattered. This scattered deployment is due to the large investments cable companies must make to upgrade their systems with the appropriate technology. One of the more expensive improvements required to offer cable modem service and high-speed Internet access is replacing a large segment of existing coaxial cable with fiber optic cable. The fiber optic cable is necessary to support the very high bandwidths needed to provide multiple users with high speed Internet access. Similar to the local telephone system local loop, the cable "local loop" that runs into a home or business will remain coaxial, while main distribution runs will be replaced with fiber optic cable and repeaters.

Wide-spread deployment of cable modems may also be hindered by a characteristic we will study in Chapter Seven on local area networks: as traffic on Ethernet-based local area networks increases, overall throughput—the ability to send or receive a complete message—decreases. As more customers within a local geographic area, such as a small number of neighborhood blocks, subscribe to cable modem service, traffic will increase to the point where throughput may suffer noticeably.

ISDN modems

Integrated Services Digital Network (ISDN) is an all-digital telephone service provided by most local telephone companies that can provide voice and data transmissions at speeds up to 128 Kbps. Since ISDN is an all-digital service, data that originates at the computer in digital form does not have to be modulated onto an analog signal for transfer over a telephone line. Nonetheless, an **ISDN modem** is necessary to convert the digital data from the computer into the proper form necessary for transmission over an ISDN network. ISDN networks, which will be covered in more detail in Chapter Twelve, are a reasonably attractive alternative to a conventional telephone line and modem. An even more attractive alternative is the digital subscriber line.

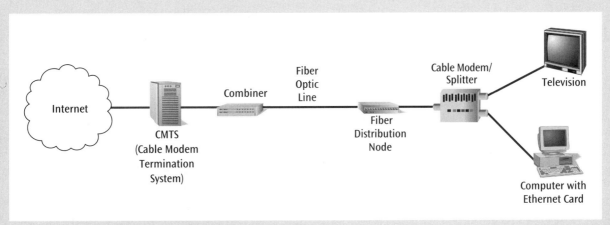

Figure 4-10 *Basic architecture and components of a cable modem system*

DSL modems

Digital Subscriber Line (DSL) is an all-digital phone service that can provide data transmission at speeds of 100s of thousands up to millions of bits per second. For a user to connect to a DSL service, a **DSL modem** is necessary. DSL is a fairly new and very advanced technology that is still experiencing many growing pains. There is a wide variety of DSL formats, and each requires its own type of DSL modem. Anyone interested in acquiring a DSL service should first check with their local service providers to see if the service is available to their home or place of business and then obtain the necessary hardware directly from that service provider. The complexity of DSL is well beyond the level of this textbook.

Modem Pools

A **modem pool** is a relatively inexpensive technique that allows multiple workstations to access a modem without placing a separate modem on each workstation. Modem pools are also used to allow multiple outside users to dial into a computer system. Consider a situation in which 100 workstations are connected to a local area network. You would like to provide each workstation with access to a modem, but doing so would require purchasing 100 modems and installing 100 telephone lines. Even though the price of modems continues to drop, the combined cost of modems and leasing 100 telephone lines on a month-to-month basis makes this setup an expensive proposition, especially if all 100 workstations are not using their modems at the same time. If approximately one third of the workstations would be using a modem at one time, you could install a modem pool, as shown in Figure 4-11, that would provide 30 modems in a single unit. A setup of only 30 modems would require only 30 outgoing telephone lines.

Details ▶

ISDN Modem Types

All ISDN modems should have two features: they should support three different ISDN protocols, and they should support Multi-Link Point to Point Protocol. First, there are three different ISDN switch type protocols used by different telephone companies in the United States: AT&T 5ESS, NI1, and DMS100. So that an ISDN modem will work anywhere in the United States where ISDN service exists, the modem should be able to support all of these protocols. Second, when a person subscribes to an ISDN service and requests the more common Basic Rate Interface, the subscriber actually receives two 64 Kbps channels on which they can transfer either voice or data or both. Most ISDN users wish to tie the two 64 Kbps channels together to achieve one 128 Kbps data channel. The two channels can be tied together if the ISDN modem supports Multi-Link Point to Point Protocol (ML-PPP).

Figure 4-11
*Modem pool providing
30 modems to a local
area network*

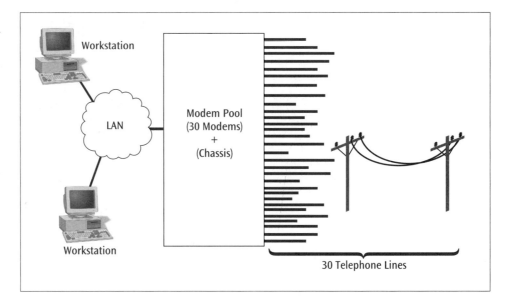

Modem pools have some financial advantages and some technical drawbacks. The device that houses and controls access to the 30 modems—the modem pool chassis—typically costs $2000 to $3000. When you add in the cost of the 30 modems, the modem pool is roughly the same cost as installing a modem into each of the 100 workstations. The fact that you need only 30 telephone lines with the modem pool, instead of 100 with individual modems, clearly reduces the monthly recurring costs. In addition, modem pools come in a variety of configurations and support a range of modem types, such as 56K and ISDN modems, in the same chassis. The major drawback to using a modem pool is that only 30 workstations can access a modem at the same time. If a 31st workstation attempts to access a modem, the system will return a "No Modem Available" message, and the user of the workstation will have to wait until a modem becomes available.

Interfacing a Computer to Modems and Other Devices

Now that we have discussed modems in detail, let's turn our attention to the interconnection between a computer and other devices, including modems. This connection is often called the interface, and the process of providing all the proper interconnections between a computer and a modem (or other devices) is called interfacing. Interfacing is a complex area of study. It is a relatively technical process and varies greatly depending upon the type of device, the computer, and the desired connection between the device and computer. Let's begin our discussion of interfacing by exploring the basic concepts of data terminal equipment (DTE) and data circuit terminating equipment (DCE), followed by the basics of interfacing. We will conclude with several examples of the more commonly used interface standards.

Data terminal equipment and data circuit-terminating equipment

One of the first items standardized for the interface between a computer and a modem was the terminology for talking about the devices commonly found on either end of the interface. **Data terminal equipment** (DTE) identifies terminating devices such as terminals or computers. **Data circuit-terminating equipment** (DCE) identifies the device that terminates the end of the circuit and is attached to the DTE. The most common DCE is the modem. An interface between the DTE and the DCE is called an interchange circuit. Figure 4-12 shows a typical interface with DTEs, DCEs, and interchange circuits.

Figure 4-12
Typical interface showing DTEs and DCEs

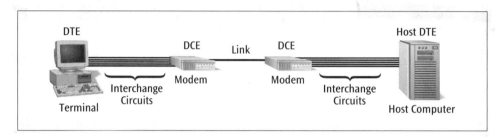

Interface standards

Many years ago, computer and peripheral manufacturers and users realized that if one company made a computer or computer terminal and another company made a modem, the odds of the two "talking" to one another were slim. Thus, various organizations set about creating a standard interface between devices such as computers and modems. Because there are so many different transmission and interface environments, one standard alone will not suffice, and multiple standards have been created. Nonetheless, all interface standards have two basic characteristics: they have been created and approved by an acceptable standards-making organization, and they consist of four components.

The primary organizations involved in making standards are:

- International Telecommunications Union (ITU), formerly Comite Consultatif International Telephonique et Telegraphique (CCITT);
- Electronics Industries Association (EIA);
- Institute for Electrical and Electronics Engineers (IEEE);
- International Standards Organization (ISO); and
- American National Standards Institute (ANSI).

Often individual companies, in a hurry to rush a product to market, will create a new product that incorporates a non-standard protocol. While there is a definite marketing advantage to being the first to offer a new technology, there is the disadvantage that the non-standard protocol has not yet been approved by one of the standards-making organizations. If a new protocol is created by a standards-making organization and it is not the same as the one created by the company, the new protocol may make the company's product obsolete. In the quickly changing world of computer technology, creating a product that conforms to an approved interface standard is difficult but highly recommended.

The second basic characteristic of an interface standard is its composition. An interface standard consists of four parts or components:

- ▸ the electrical component;
- ▸ the mechanical component;
- ▸ the functional component; and
- ▸ the procedural component.

All of the standards that exist address one or more of these components. The **electrical component** deals with voltages, line capacitance, and other electrical components. The electrical components are more the study of a technician, and we will not discuss them in great detail. The **mechanical component** deals with items such as the connector or plug description. Questions typically addressed by the mechanical component include "What is the size and shape of a connector, how many pins are found on the connector, and what is the pin arrangement?" The **functional component** describes the function of each pin or circuit that is used in a particular interface. The **procedural component** describes how the particular circuits are used to perform an operation. For example, the functional component may describe two circuits, Request to Send and Clear to Send. The procedural component describes how those two circuits are used so that the DTE can transfer data to the DCE and vice versa. We will examine four common interface standards for connecting computers to modems: EIA-232E, EIA RS-449, X.21, and the Hayes standards.

RS-232 and EIA-232E

EIA-232E is an interface standard for connecting a DTE to a voice-grade modem (DCE) for use on analog public telecommunications systems. The EIA-232E interface standard is the current incarnation of the RS-232 standard, which was created in 1962. **RS-232** was designed for the interface between a terminal or computer (the DTE) and its modem (the DCE). Although the RS-232 and EIA-232E standards are a little bit different, they are similar enough that in describing EIA-232E, you can grasp the most important elements of RS-232. Note that the EIA-232E interface standard uses the ITU v.28 standard to define the electrical component, the ISO 2110 standard to define the mechanical interface, and the ITU V.24 standard to define functional and procedural components.

The electrical component of EIA-232E incorporates ITU's V.28 standard, which describes the electrical characteristics for an interchange circuit. With the V.28 standard, voltage levels are detected at the receiving modem by the relative voltage difference between two different circuits. A voltage difference of more than +3 volts is considered to be a value of 1, and a voltage difference of more than -3 volts is considered to be a value of 0.

EIA-232E incorporates the ISO 2110 standard to define the mechanical interface. ISO 2110 precisely defines the size and configuration of a 25 pin connector (DB-25). Figure 4-13 shows the connector's configuration.

Figure 4-13
The ISO 2110 connector

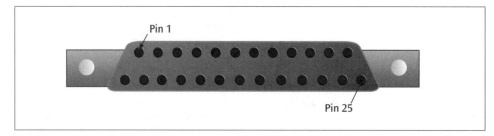

To define functional and procedural components, the EIA-232 incorporates the ITU's V.24 standard. V.24 defines a list of 43 interchange circuits that can be used by other standards for defining an interface. Each of these interchange circuits defines a particular function. By using the appropriate functions in the proper sequence, a computer (DTE) and modem (DCE) can create a connection between themselves and between the local computer and the remote computer. EIA-232E uses 22 of those 43 interchange circuits, which are shown in Table 4-1. All 22 circuits are rarely used in a single interface. In fact, it is common to find a DTE/DCE interface that consists of only three wires: signal ground, transmitted data, and received data.

Table 4-1
EIA-232E interchange circuits

Pin Number	Circuit Originates From:	Circuit Names
1	—	Shield
2	DTE	Transmitted Data
3	DCE	Received Data
4	DTE	Request to Send
5	DCE	Clear to Send
6	DCE	DCE Ready
7	—	Signal Ground
8	DCE	Received Line Signal Detector
12	DCE	Secondary Received Line Signal Detector
13	DCE	Secondary Clear to Send
14	DTE	Secondary Transmitted Data
15	DCE	Transmitter Signal Element Timing
16	DCE	Secondary Received Data
17	DCE	Receiver Signal Element Timing
18	—	Local Loopback
19	DTE	Secondary Request to Send
20	DTE	Data Terminal Ready

Table 4-1
EIA-232E interchange circuits (continued)

Pin Number	Circuit Originates From:	Circuit Names
21	DCE	Remote Loopback
22	DCE	Ring Indicator
23	DTE	Data Signal Rate Selector
24	DTE	Transmitter Signal Element Timing
25	—	Test Mode

Transmitted Data (pin 2) is the line on which the DTE sends data to the DCE. Received Data (pin 3) is the line on which the DCE sends data to the DTE. Request to Send (pin 4) is sent from the DTE to the DCE when the DTE wishes to transmit data. Clear to Send (pin 5) is sent in response to Request to Send when the DCE is ready to accept data from the DTE. DCE Ready (pin 6) is a signal sent to the DTE indicating the status of the local DCE. Received Line Signal Detector (pin 8) indicates to the DTE that the DCE is receiving a suitable carrier signal. Transmission Signal Element Timing (pin 15) and Receiver Signal Element Timing (pin 17) provide timing information between the computer and the modem. The secondary data and control lines are equivalent in function to the primary data and control lines, except they are for use on a reverse or backward channel. A reverse channel is used for transmission of supervisory or error control signals and flow in the opposite direction of the data being transferred. Local Loopback (not available on the early versions of RS-232) permits a test on the loop between the DTE and the local DCE. Remote Loopback (also not available on the early versions of RS-232) permits a test on the loop from the local DTE, through the local DCE, over the transmission lines to a remote DCE, and then back. Ring Indicator indicates to the DTE that a ringing signal is being received at the local DCE. Test Mode indicates whether the local DCE is in a test condition.

you only need pins 2, 3, 7 to send + receive data

CBT: To see a visual demonstration of two modems establishing a connection, run the third CBT module on the enclosed CD-ROM.

Table 4-2 examines the sequence of events that allows a DTE, such as a microcomputer, to instruct a DCE, such as its modem, to place a call to a remote modem (DCE) and computer (DTE). Walking through this example carefully will enable you to better understand the dialogue represented by the procedural component of the EIA-232E standard. The text on the left side of Table 4-2 shows the operation of the local DTE and DCE, and the text on the right side of the page shows the operation of the remote DTE and DCE.

Table 4-2

Sequence of commands sent between local DTE and DCE and between remote DCE and DTE to establish a connection

Local DTE and DCE	Remote DCE and DTE
1. Local DTE turns on Data Terminal Ready (pin 20) to tell its DCE (modem) it wants to establish a circuit.	
2. Local DTE sends phone number to its modem over Transmitted Data (pin 2).	
	3. Remote DCE (modem) alerts its DTE to incoming call via Ring Indicator (pin 22).
	4. Remote DTE turns on its Data Terminal Ready (pin 20).
	5. Remote modem sends a carrier signal back to Local modem.
	6. Remote modem sends a DCE Ready signal (pin 6) to its DTE.
7. Local modem detects carrier and alerts its DTE via Received Line Signal Detector (pin 8).	
8. Local modem sends a DCE Ready signal (pin 6) to its DTE.	
9. Local modem sends a carrier signal to remote modem.	
	10. Remote modem detects carrier and alerts its DTE via Received Line Signal Detector (pin 8).
11. Local DTE wishes to send data, so asserts Request to Send (pin 4).	
12. Local modem responds with Clear to Send (pin 5).	
13. Local DTE sends digital data over Transmitted Data (pin 2) to its modem, which converts it to analog and sends over carrier.	
	14. Signal arrives at remote DCE, which converts the analog signal to digital and sends it to its DTE via Received Data (pin 3).

EIA RS-449

After several years of use, many computer users realized the short-comings of the original RS-232 interface standard. In particular the data transmission rate between the terminal/computer and modem was too slow, and there were no built-in functions to perform testing of the connection (available in later versions of RS-232). In 1975, EIA introduced a new **RS-449** interface standard to replace RS-232. The RS-449 interface standard is based on the original RS-232 standard and includes several major improvements. RS-449 was designed to:

- ► allow for two possible standards (RS-422A and RS-423A) that can be used as the electrical component—both of these newer electrical standards provide for faster transmission speeds between computer and modem;
- ► add 10 additional circuits, including loop-back testing (the original RS-232 did not contain loop-back testing); and
- ► change the connector from a 25-pin connector to a 37-pin connector.

Due to widespread acceptance of the older RS-232 standard, RS-449 never replaced RS-232, even though it was superior due to the additional features.

X.21

The **X.21** interface was another standard that was designed to replace the aging RS-232 standard. X.21 did not completely replace RS-232, but it has become a popular standard as a digital interface found between a computer and an ISDN connection. In the EIA-232E and RS-449 standards, each wire (or circuit) has a single function. Thus to support 20 or 30 functions, the interface needs 20 or 30 separate wires. In contrast, the X.21 standard uses a 15-pin connector, from which only four circuits carry the bulk of the interface information. Each circuit in the X.21 standard can contain many different signals. Since each circuit can transmit different signals, the combination of signals on the four circuits is much larger than if each circuit performed only a single function (as in EIA-232E).

▌ Details ▌ ▶

The Electrical Component of RS-449

The electrical component of RS-449 can be defined by selecting one of two different standards: RS-423A (X.26) and RS-422A (X.27). The RS-423A (X.26) standard consists of a signal wire plus a ground wire and provides an unbalanced transmission. The unbalanced transmission of RS-423A provides for data transmission between DTE and DCE of 3 Kbps at 1000 meters and 300 Kbps at 10 meters. EIA-232E also operates using an unbalanced transmission. The RS-422A (X.27) standard consists of three wires—one for the signal, one for a voltage reference, and one for the ground—and provides a balanced transmission. The balanced transmission of RS 422A provides for superior data transmission between DTE and DCE at 100 Kbps at 1200 meters and 10 Mbps at 12 meters.

Why is there such a big difference in the data transmission speeds between an unbalanced transmission and a balanced one? All electrical circuits need two paths: one path for providing current to a device and a second as a return path back to the source of the current. An **unbalanced circuit**, or unbalanced transmission, relies on a *ground line* for the return path. When people think of a ground line, they often think of something with zero voltage. The grounds from one location to another, however, can be different and actually produce a voltage difference. Additionally, a ground line can have noise on it, which will interfere with a signal's voltages. **Balanced transmission** does not rely on the ground line as a reference point, but provides a third wire for the reference signal instead. This third wire is isolated from the ground line and thus contains much less noise and spurious voltages. Although this concept may seem highly technical, or electrical, it is important to understand. Balanced transmission is a common way to fight noise—the never ending foe—in communications systems.

Consider the simple dialog shown in Table 4-3, which shows a sequence of events for creating a connection and then transmitting data.

Table 4-3
Sequence of events using X.21 standard

DTE		DCE		
Transport (T)	Control (C)	Receive (R)	Indication (I)	Comment
1	OFF	1	OFF	Ready
0	ON	1	OFF	Call request from DTE
0	ON	+	OFF	DCE says go ahead
ASCII address	ON	+	OFF	DTE transmits address
1	ON	c.p.s.	OFF	Connection in progress
1	ON	1	ON	DCE says ready for data
Data	ON	Data	ON	DTE transfers data

c.p.s. = call progress signals

When the DTE and DCE are both ready, the DTE transmits a 1 on the T line and an OFF on the C line, while the DCE transmits a 1 on the R line and an OFF on the I line. When the DTE wishes to place a call, it changes the 1 on the T line to a 0 and the OFF on the C line to an ON. When the DCE wants to tell the DTE to go ahead and send the number of who to call, the DCE transmits a + on the R line and maintains an OFF on the I line. As the call progresses, the appropriate signals are generated, as further shown in Table 4-3.

Hayes Interface Standards

In the 1980s when the microcomputer industry started to explode, Hayes modems developed by Hayes Microcomputer Products Inc. became the industry standard for microcomputer systems. The most important feature of the Hayes modems was the **AT command set**. Their use of AT commands simplified microcomputer-to-modem connections and became a worldwide standard. Prior to the 1980s, different modems and microcomputers needed different interface settings to work together. Cumbersome switches and hardware jumpers were needed to allow people to customize the interface settings on modems so they could work with specific microcomputers. Hayes simplified and standardized modem commands, thereby replacing the cumbersome switches and hardware jumpers with simple, powerful, and flexible commands sent from the microcomputer to the modem. For example, using EIA-232E, the instruction sent to a modem to dial a particular telephone number required a series of complex EIA-232E signals. Using the AT command set, a computer simply had to send "AT D 555-1212" to the modem and the modem would dial (D) the supplied telephone number.

The first modem to feature this AT command set was the 1981 Hayes Smartmodem, which entered the market at the same time that the IBM Personal Computer was introduced. The Hayes AT Command Set quickly became the industry standard. Because of the widespread success of the AT command set, all modern modems, regardless of the manufacturer, are compatible with the **Hayes standard.**

Interfacing a Computer and a Peripheral

Interface standards such as EIA-232E, X.21, and Hayes have existed for many years and, by current standards, are relatively complex to create and difficult to support. They were designed primarily to support modems, or in the case of the Hayes standards, only to support modems. Computer designers have been working for many years trying to create a new interface that is flexible and fast and supports not only modems but the growing array of peripheral devices such as document scanners and video cameras. It would be even better if the interface was so simple to use that when the peripheral was plugged into the computer, the interface between the computer and peripheral configured itself. Two new interface standards that have great potential are FireWire and Universal Serial Bus.

FireWire

Used for multimedia Applications

FireWire is the name of a bus that connects peripheral devices such as wireless modems and high speed digital video cameras to a microcomputer and is currently IEEE standard 1394. Conceived by Apple Computer, FireWire is an easy to use, flexible, and low cost digital interface that is capable of supporting transfer speeds of up to 400 Mbits/second. Since FireWire provides a digital interface, there is no need to convert the digital signals of the microcomputer to analog signals for transfer over a connection. Furthermore, FireWire is a relatively thin, space-saving cable to which devices can be added and removed while the bus is active (called hot pluggable).

FireWire supports two types of data connections: an asynchronous connection and an isochronous connection. The asynchronous connection supports the more traditional peripheral devices such as modems and printers. An **isochronous** connection provides guaranteed data transport at a pre-determined rate, which is essential for multimedia applications. Multimedia applications are relatively unique in that it is important to provide an uninterrupted transport of time-critical data and just-in-time delivery. This continuous high-speed delivery of data reduces the need for difficult and costly buffering. FireWire is a very good choice for interfacing digital consumer electronics and audio and visual peripherals, such as digital video cameras and digital cameras.

Details ▶

Bell Interface Standards

Before the early 1980s, the Bell telephone system would not allow any devices made by another company to interconnect to their phone system. As a consequence, there were few players in the modem market, and the Bell System of modems and their interfaces became the industry standard. Since then, Bell modems have provided a set of standards on which other modem companies have modeled their modems. The more common Bell modems include:

103/133 Series — 300 bps, frequency shift keying, full duplex on two wire systems;

202 Series — 1800 bps on conditioned leased lines, 1200 bps on dial-up lines, frequency shift keying, half duplex on two wire systems;

201 Series — 2400 bps, phase shift keying, half duplex on two wire systems, full duplex on four wire systems;

208 Series — 4800 bps, phase shift keying, half duplex on two wire systems, full duplex on four wire systems;

212A Series — Asynchronous 1200 bps with phase shift keying or 300 bps with frequency shift keying, synchronous 1200 bps, full duplex on two wire systems; and

209A Series — Synchronous 9600 bps on leased lines.

All the modems produced today support all of the Bell standards. Even though the Bell standards support slow data transfer speeds, recall that a modem can fall back to a slower transmission speed if the connection encounters noise.

Universal Serial Bus (USB)

The **Universal Serial Bus** is also a modern standard for interconnecting modems and other peripheral devices to microcomputers. Similar to FireWire, USB is also a digital interface with a single standardized connector for all serial and parallel type devices. The idea behind USB is to simply plug in the peripheral and turn it on—the computer will dynamically recognize the device and perform the interface. The case of the computer does not have to be opened nor do any software or

Details ▶

Other V-series Standards

Since modern modems can support a wide range of transmission speeds, they incorporate a wide range of interface standards. Most of these standards were created by ITU and have the V-designation. The more commonly found V-series recommendations are listed in Table 4-4. You can see from the list that many interface standards are available.

Series	Purpose
V.1	Brief description of terminology pertaining to binary symbols and signals.
V.2	Provides direction on permissible power levels to be used on equipment.
V.5	Describes signaling rates (bps) over half duplex or duplex dial-up links.
V.6	Describes signaling rates (bps) over dedicated links.
V.7	Definition of terms concerning data communications over the telephone network.
V.10	Defines the electrical characteristics of an unbalanced interchange circuit. Equivalent to RS-423A.
V.11	Defines the electrical characteristics of a balanced interchange circuit. Equivalent to RS-422A.
V.17	Standard for transmitting facsimile (FAX) at 14,400 bps.
V.22	Defines 1200 bps over telephone lines.
V.22bis	Defines 2400 bps over telephone lines.
V.24	Provides a list of definitions for the interchange circuits.
V.25	Describes the conventions for automatic dial-and-answer interfaces.
V.28	Defines the electrical characteristics for an unbalanced interchange circuit.
V.32	Defines 9600 bps over telephone lines.
V.32bis	Defines 14,400 bps over telephone lines.
V.34	Defines 28,800 bps over telephone lines.
V.34+	Defines 33,600 bps over telephone lines.
V.42	Defines a standard for error correction, based on Microcom techniques.
V.42bis	Defines a standard for data compression, based on Microcom techniques.
V.50	Standard limits for transmission quality for data transmission.
V.54	Loop test devices for modems.
V.57	Comprehensive data test set for high data signaling rates.
V.90	Defines 56,000 bps digital transmission of telephone lines.

Table 4-4   *More common V-series standards*

hardware switches have to be set. Using peripherals that are designed with a USB connector, it is possible to connect one USB peripheral onto another, or "daisy-chain" multiple peripherals together. Another unique feature about USB is that it is possible for the USB cable to provide the electrical power necessary to operate the peripheral. With this option, it is not necessary to find multiple electrical outlets, one for each peripheral. Finally, data transfer over a USB cable is bi-directional. With bi-directional data transfer one USB cable can both send and receive data from one or more peripherals.

Clearly, interfaces such as FireWire and USB have come a long way from the early RS-232 days. If this trend continues (which it undoubtedly will), the topic of interfacing may one day become a history lesson.

Data Link Connections

As we have seen, interface standards such as EIA-232E and RS-449 consist of four components: the electrical, mechanical, functional and procedural components. Since these four components define the physical connection between a DTE and a DCE, these four components reside at the physical layer of the OSI model. But there is more to a connection than just a definition of the physical components. When data is transmitted between two points on a network, such as between a DTE and a DCE, or between a network sender and a network receiver, we need to define the data link connections. If we once again compare this to the OSI model, the definition of the data link connection is performed at the data link layer.

For example, assuming that the physical layer connections are already defined by some protocol such as EIA-232E, what is the basic form of the data that is passed between sender and receiver? Is the data transmitted in single-byte blocks, or does the connection create a larger, multiple-byte block? The former connection is an example of an asynchronous connection, while the latter is a synchronous connection. Can the connection transmit data in both directions at the same time, or only in one direction at a time? As we will shortly see, this is the difference between a full duplex connection and a half duplex connection. Examples such as these define the type of connection at the data link layer.

While examining the data link connection, recall the duties of the data link layer from the OSI model. Two important tasks the data link layer should perform is to create a frame of data for transmission between sender and receiver, and to provide some way for checking for errors during transmission. Keep these tasks in mind as we examine several different types of data link connections.

Asynchronous connections

An asynchronous connection is one of the simplest examples of a data link protocol and is found primarily in microcomputer to modem connections. In an **asynchronous connection**, a single character, or byte of data is the unit of transfer between the sender and receiver. The sender prepares a data character for transmission, transmits that character, then begins preparing the next data character for transmission. An indefinite amount of time may elapse between the transmission of one data character and the transmission of the next character.

To prepare a data character for transmission, a few extra bits of information are added to the data bits of the character creating a **frame** or small packet of data. A

start bit, which is always a 0, is added to the beginning of the character and informs the receiver that an incoming data character (frame) is arriving. The start bit allows the receiver to synchronize itself to the character. At the end of the data character, one or two **stop bits**, which are always 1s, are added to signal the end of the frame. Although there is usually only one stop bit, some systems still allow a choice of one or two. The start and stop bits have, in essence, provided a beginning and ending frame around the data. Finally, a single parity bit, which is inserted between the data bits and the stop bit, may be added to the data. This **parity bit** may be either even parity or odd parity and performs an error check on only the data bits. This error check is achieved by adding a 0 or 1 such that an even or odd, respectively, number of 1s is maintained. Figure 4-14 shows an example of the character A (in ASCII) with one start bit, one stop bit, and an even parity bit added.

Figure 4-14
Example of the character A with start bit, one stop bit, and even parity

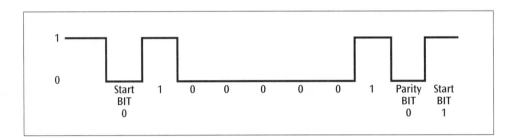

Because each character has its own start, stop, and parity bits, the transmission of multiple characters, such as HELLO, is possible. Figure 4-15 demonstrates the transmission of HELLO.

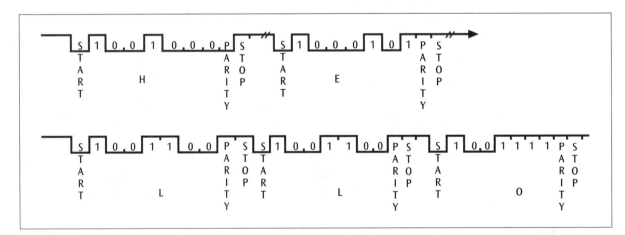

Figure 4-15 *Example of the character string HELLO with included start, stop and parity bits*

An asynchronous connection has advantages and disadvantages. On the positive side, generation of the start, stop, and parity bits is simple and requires little hardware or software. During the early years of the microcomputer industry, the simplicity of an asynchronous connection lent itself nicely to the hobby segment of microcomputer-to-host computer links. This association still exists today within the microcomputer industry.

An asynchronous connection has one disadvantage in particular that cannot be overlooked. Given that seven data bits (ASCII character code set) are often combined with one start bit, one stop bit, and one parity bit, the resulting transmitted character contains three check bits and seven data bits, for a 3:7 ratio. Given a total of 10 check and data bits, 3 of the 10 bits, or 30 percent of the bits, are used as check bits. This ratio of check bits to data bits is not very efficient for high amounts of data transfer.

Recall from Chapter Two the importance of a receiver staying synchronized with the incoming data stream, especially if the data stream contains a long sequence of non-changing values. The Manchester codes were designed to help with this problem, but they were not yet created when asynchronous connections were developed. Thus, an asynchronous connection incorporates its own methods for keeping the receiver synchronized with the incoming data stream. How does an asynchronous connection keep the receiver synchronized? Two key features of asynchronous connections help to maintain synchronization:

▶ the frame size — since each frame in an asynchronous connection is one character plus a few check bits, the receiver will only receive a small amount of information at one time. It should not be difficult for the receiver to stay synchronized for that short of a period; and

▶ the start bit — when the receiver recognizes the start bit, the synchronization begins. Since there are only approximately eight or nine bits following, there will not be a long sequence of non-changing values.

Interestingly, the term asynchronous connection is misleading because, despite the word asynchronous, the protocol does maintain synchronization with the incoming data stream. *Even though it is called Async, the receiver must be in sync with the sender*

Synchronous connections

The second technique for maintaining synchronization between a receiver and the incoming data stream is a synchronous connection. With **synchronous connections** the unit of transmission is a sequence of characters. This sequence of characters may be thousands of characters in size. Much as start, stop, and parity bits frame the data bits in an asynchronous connection, a start sequence, a control byte, a checksum, and an end sequence frame the data bits in a synchronous connection, as shown in Figure 4-16.

Figure 4-16

Block diagram of the parts of a generic synchronous connection

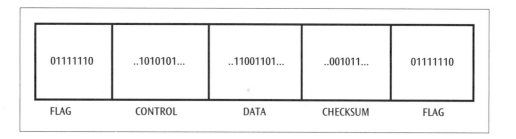

01111110	..1010101...	..11001101...	..001011...	01111110
FLAG	CONTROL	DATA	CHECKSUM	FLAG

The starting and ending sequences, called flags, are typically eight bits (a byte) each in length. Following the start sequence flag is usually one or more bytes of control information. This control information provides information about the enclosed data or provides status information pertaining to the sender or receiver or both. For example, a unique bit in the control information byte may be set to one, indicating the enclosed data is high priority. Often the control byte contains addressing information that indicates where the data is coming from or for whom it is intended. Following the data is almost always some form of error checking sequence, such as the cyclic checksum. Cyclic checksum is a more advanced error checking technique than parity checking and is used in many modern implementations of computer networks. After the error-checking sequence is the end sequence flag.

How does a synchronous connection keep the sender and receiver synchronized? This question is especially important since it is possible to send thousands of characters in a single package. There are three ways to maintain synchronization in synchronous connections:

1. Send a synchronizing clock signal over a separate line that runs parallel with the data stream. As the data arrives on one line, a clock signal arrives on a second line. The receiver can use this clock signal to stay synchronized with the incoming data.

2. If transmitting a digital signal, use a Manchester code. Since a Manchester code always has a signal transition in the middle of each bit, the receiver can anticipate this signal transition and read the incoming data stream with no error. A Manchester encoded digital signal is an example of a **self-clocking** signal.

3. If transmitting an analog signal, use the properties of the analog signal itself for self-clocking. For example, an analog signal with a periodic phase change can provide the necessary synchronization.

Asynchronous and synchronous connections are the two most common techniques for creating a connection at the data link level.

Half Duplex, Full Duplex, and Simplex Connections

A connection between a sender and receiver can also be classified by whether both sides can talk at the same time or whether only one side at a time can talk. In a **half duplex** connection, data is transmitted from the sender to the receiver and from the receiver to the sender in only one direction at a time. Computer networks with half duplex connections were more prevalent in the 1960s and 1970s in systems that were called stop-and-wait. A sender would transmit data, then wait for a response from the receiver before sending more data. A data transmission system that employs a half duplex connection is slower than a system that allows a simultaneous two-way connection. Nonetheless, half duplex connections still exist today. A pager that allows the paged individual to return a message and walkie-talkie radios are two modern examples of half duplex connections.

In a **full duplex** connection, data can be transmitted from sender to receiver and from receiver to sender in both directions at the same time. To allow this two-direction, simultaneous flow of data, something has to be done to enable the two signals traveling in opposite directions to coexist. One possible method is to use two sets of cables; one cable transfers signals in one direction, and the other cable transfers a second set of signals in the other direction. For example, in fiber optic systems the hardware that creates and detects a light source works in only one direction. Thus, it is necessary to use two fiber optic cables, one for each direction. A second common possible solution is to transmit the two signals with two different sets of frequencies so that the signals coming from multiple devices do not interfere with each other. Most modern computer networks as well as telephone networks support full duplex connections. A third type of connection is the simplex connection. A **simplex** connection transmits data in one direction only. Although most computer communications systems transmit in more than one direction, many other forms of communication systems are simplex. Television and radio are both examples of simplex connections. Television and radio signals travel only from the transmitting tower to the receiving antenna.

Terminal-to-Mainframe Computer Connections

One of the common configurations introduced in the first chapter was the terminal-to-mainframe computer configuration. Since terminals possess little processing power compared to a microcomputer workstation, the mainframe computer has to take control and perform all the data transfer operations. The operations performed by the mainframe computer depend upon the type of physical connection between a terminal and mainframe. A direct connection between a terminal and a mainframe computer, as shown in Figure 4-17(a), is a **point-to-point** connection. There is a single wire between the two devices and no other terminals or computers share this connection. When multiple terminals share one direct connection to a mainframe, as shown in Figure 4-17(b), the connection is called **multipoint**. A multipoint connection is a single wire with the mainframe connected on one end and multiple terminals connected on the other end.

Figure 4-17
Point-to-point and multipoint connections of terminal and main-frame computer

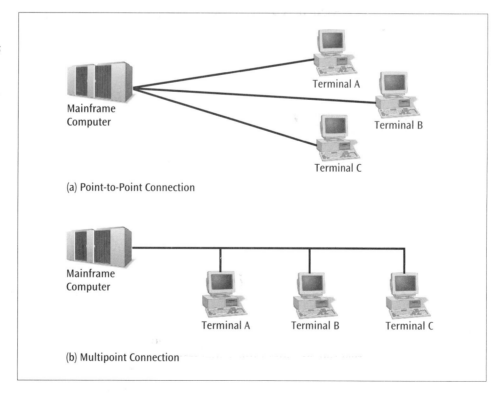

(a) Point-to-Point Connection

(b) Multipoint Connection

Once again, if multiple devices are to share a single line, something special must be done so that more than one device does not try to transmit at the same time. A technique called **polling**, which allows only one terminal to transmit at one time, is successful in controlling multiple terminals sharing a connection to a mainframe computer. Polling originated in the early years of computing when terminals were relatively dumb devices incapable of performing many operations beyond data entry and display. During this period, a mainframe computer was called the **primary** and terminals were called the **secondaries**. When using the polling technique, a terminal or secondary would only transmit data when prompted. **Roll-call polling** is the polling method in which the mainframe computer (primary) polls each terminal (secondary), one at a time, in round-robin fashion. If three terminals A, B, and C shared one connection, as in Figure 4-18, the primary would begin by polling terminal A. If A had data to send to the host, it would do so. When A was done transmitting, the primary would poll terminal B. If B had nothing to send, it would inform the primary accordingly, and the primary would poll terminal C. When terminal C was finished, the primary would return to terminal A and continue the polling process.

Figure 4-18
Terminals A, B, and C being polled by a primary

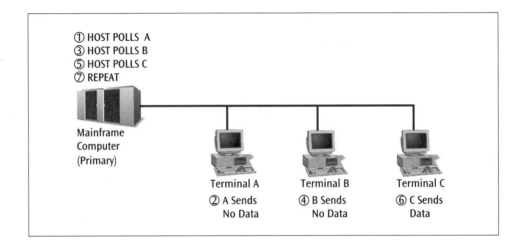

An alternative to roll-call polling is hub polling. A primary that incorporates **hub polling** polls only the first terminal, which then passes the poll to the second terminal, and each successive terminal passes along the poll. For example, after being polled by the primary when terminal A is finished responding, terminal A passes the poll to terminal B. When terminal B is finished transmitting, it passes the poll to terminal C. In this fashion, the primary does not need to poll each terminal separately. The process of the primary sending a poll to a terminal and waiting for a response takes time. When large amounts of data are being transmitted this time may be significant.

When the primary wishes to send data to a terminal, it uses a process called selection. In **selection**, the primary creates a packet of data with the address of the intended terminal and transmits the packet. Only that terminal recognizes the address and accepts the incoming data. A primary can also use selection to broadcast data to all terminals.

If control simplicity is your primary goal, point-to-point connection of terminals is clearly superior to multipoint connections. With point-to-point connections, polling is not necessary because there is only one terminal per line. In multipoint connections, the terminal must possess software necessary to support polling. Another disadvantage of multipoint connections in which several terminals share one connection is that each terminal has to wait while another terminal transmits. Although point-to-point connections make more efficient use of transmission time, they also require more expensive hardware. When each terminal has a direct connection to the primary, more cabling is necessary.

Making Computer Connections In Action ▶

The personal computer that your company ordered for your office has just arrived. You called technical support to ask when someone would be available to come and set up your new machine. Unfortunately, they are swamped and won't be able to get to your computer until early next week. Since you are really itching to test drive the new computer, you decide to open the boxes and set it up yourself. You've hooked up your stereo system at home—how much different can a computer be?

After unwrapping the main system unit, you notice the back panel of the computer has a large number of different sized connectors. The layout of the back panel looks something like Figure 4-19. After looking at 10 different connectors, you begin to realize that maybe you are in over your head and perhaps should wait for a technical support person to visit next week. What are all these connectors for? Do they have anything to do with interfacing?

Figure 4-19
Layout of the back panel of a typical personal computer

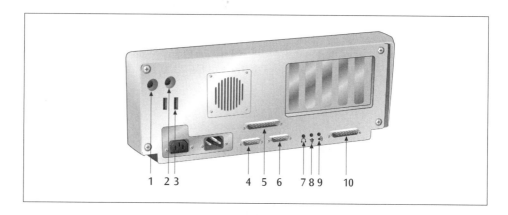

At the left side of the panel are two round connectors. The first connector (1) is for attaching a mouse, and the second connector (2) is for attaching the keyboard. These mouse and keyboard ports are standard on most microcomputers and are usually round, indicating either a 5-pin DIN connector or a 6-pin Mini-DIN connector. A **DIN** (Deutsches Institut fur Normung) **connector** is a family of plugs and sockets used to connect a variety of devices, such as audio and video equipment. Although it is not terribly important to know the procedural specification, it is interesting to list the mechanical, electrical, and functional specifications on the newer 6-pin Mini-DIN; Table 4-5 shows a number of interesting features. There is only one data line, which more than likely indicates a simplex connection. Pins 3 and 4 provide the voltage to operate the device, while pin 5 provides some form of clock signal for synchronization.

Table 4-5
Mechanical, Electrical, and Functional specifications of 6-pin Mini-DIN

Pin Number	Signal Name
1	Data
2	Reserved
3	Ground
4	+5 V DC
5	Clock
6	Reserved

The next connector shown on the back panel is a small rectangular Universal Serial Bus connector (3). As noted earlier in the chapter, the Universal Serial Bus is slowly becoming a popular means for interconnecting peripheral devices to a personal computer. One peripheral device that can use the Universal Serial Bus is an external modem. If our computer did not come with a modem and one is desired in the future, this would be a good connector to use due to its speed and simplicity of use.

The fourth and sixth connectors are DB-9 9-pin serial port connectors. Two serial ports are provided so that two devices requiring a serial connection may be connected to the computer at the same time. **Serial ports** can be used to connect devices such as modems and mice to personal computers. It is a serial connection because there is one wire for the data, and all the data transmits over this wire in a serial fashion, or one bit after another. The DB-9 serial port connector uses a subset of the signals from the EIA-232E standard, thus its nine pins have the same names and functions as those of EIA-232E, as shown in Table 4-6.

Table 4-6
Pin Assignments of the DB-9 Serial Port Connector

Pin Number	Signal Name
1	Received Line Signal Detector
2	Received Data
3	Transmitted Data
4	Data Terminal Ready
5	Signal Ground
6	Data Set Ready
7	Request to Send
8	Clear to Send
9	Ring Indicator

The fifth connector is a parallel port connector. A **parallel port** is a connection in which there are eight data lines transmitting an entire byte of data at one moment in time. Parallel ports are most often used to connect printers to personal computers. The most commonly found parallel port is the Centronics parallel interface, which is an older and still very widely used interface for connecting printers and occasionally other devices to computers. This interface begins with a 25-pin DB-25 connector on the back of the computer and ends with a rather cumbersome 36-pin connector at the printer. Even though it is a 36-pin connector, all 36 pins are not used. If we were to interconnect a printer to our computer, the parallel port would be the most likely place to make our connection.

The seventh, eighth, and ninth connectors are simple phono jacks for connecting speakers for audio output and a microphone for audio input.

The final connector (10) is a game port, which uses a DB-15 connector to create a MIDI interface. MIDI (Musical Instrument Digital Interface) is a standard designed for the recording and playback of music using digital synthesizers. Rather than represent musical sound directly, the MIDI interface transmits information about how music is produced. The command set for this interface includes note-ons, note-offs, key velocity, pitch bend and other methods of controlling a synthesizer. Since we will not be interfacing any MIDI devices in our office, we will not use this port at this time.

Now we should have a basic understanding of the different types of connectors and connections available on a typical microcomputer. Although we did not examine each type of connection in detail, you should understand that many different types of connections are possible and each connection has many possible layers of interface standards.

◆ ◆

SUMMARY

▶ A standard modem has several basic characteristics, including varying data transmission rates, internal versus external operation, support for standard telephone operations, laptop capabilities, connection negotiation, data compression, error control, facsimile capabilities, and security features.

▶ The data rate of standard modems using voice-grade telephone lines has peaked at 33,600 bits per second. The newer digital modems are capable of speeds near 56,000 bits per second, depending on line conditions.

▶ New technologies such as cable modems and digital subscriber line have increased the data transfer into homes and businesses to rates up to millions of bits per second.

▶ A modem pool can be an attractive alternative to installing a modem into each computer workstation and attaching a separate telephone line to each modem. Modem pools are also attractive for systems that allow remote users to dial into the local computer system.

▶ A DTE is a data terminating device such as a computer, and a DCE is a data circuit-terminating device such as a modem.

▶ To support the interface between a computer (or terminal) and a modem many standards have emerged over the years, with the RS232 standard, now in the EIA-232E form, being one of the most well-known standards.

▶ The EIA-232E standard provides mechanical, electrical, functional, and procedural specifications for an interface between a computer and a modem.

> Older interface standards that have played pivotal roles in establishing current modem standards include the Bell standards, the Hayes Smartmodem standards, and Microcom Networking Protocols.

> New peripheral interfacing standards that promise to provide power, flexibility, and ease of installation include FireWire and the Universal Serial Bus.

> Besides the physical layer interface standards such as EIA-232E, when data is transmitted between two points on a network a data link connection is also required. Two common data link connections include asynchronous connections and synchronous connections.

> Asynchronous connections use single character frames and start and stop bits to establish the beginning and ending points of the frame.

> Synchronous connections use multiple-character frames sometimes consisting of thousands of characters. A synchronizing clock signal or some kind of self-clocking must be used with synchronous transmission to enable the receiver to remain synchronized during this large frame.

> Transmission systems that are half duplex can transmit data in both directions, but in only one direction at a time.

> Full duplex systems can transmit data in both directions at the same time. To achieve full duplex communications, something has to be done to allow the two signals traveling in opposite directions to coexist. One possible solution is to use two sets of cable. One cable transfers signals in one direction, and the other cable transfers a second set of signals in the other direction.

> A connection between a computer terminal and a mainframe computer that is dedicated to one terminal is called a point-to-point connection.

> A shared connection between more than one computer terminal and a mainframe computer is called a multipoint connection.

> Mainframe computers use polling techniques such as roll call polling and hub polling to support multipoint connections.

KEY TERMS

asynchronous connection	functional component	primary
AT command set	Hayes standards	procedural component
cable modem	hub polling	remote loop-back testing
CardBus	internal modem	roll-call polling
CSU/DSU	ISDN modem	RS-232
DIN connector	isochronous connection	RS-449
DSL modem	local loop-back testing	secondary
DCE	mechanical component	selection
DTE	MNP 1-5	self-clocking
EIA-232E	modem	serial port
electrical component	modem pool	simplex
external modem	multipoint connection	start bit
half duplex	parallel port	stop bit
fall forward	parity bit	synchronous connection
fallback	PC Card	Universal Serial Bus (USB)
FireWire	PCMCIA	V.90 standard
frame	point-to-point connection	X.21
full duplex	polling	

REVIEW QUESTIONS

1. What are the basic characteristics of modern modems?
2. What is meant by fall forward and fallback negotiation?
3. Why is the modem for a laptop different than a modem for another type of computer?
4. Why are the 56 Kbps modems faster than the older 33,600 bps modems?
5. Why don't the 56 Kbps modems transmit at 56 Kbps?
6. What are the alternatives to traditional modems?
7. What is the primary advantage of a modem pool?
8. What is a DTE, and what is a DCE?
9. What are the four components of an interface?
10. What three circuits are needed, at minimum, to create an EIA-232E connection?
11. What are the primary differences between X.21 and EIA-232E?
12. What is the significance of the Bell standards and the Hayes standards?
13. FireWire and USB are standards to interconnect what to what?
14. What are the advantages of FireWire and Universal Serial Bus?
15. What are the primary differences between asynchronous connections and synchronous connections?
16. In asynchronous connections, what additional bits are added to a character to prepare it for transfer?
17. In asynchronous connections, how many characters are placed into one frame?
18. What are the advantages and disadvantages of asynchronous communication?
19. What is the basic block diagram of a synchronous frame?
20. What are the advantages and disadvantages of synchronous communication?
21. What is the difference between half duplex and full duplex communications?
22. What is the difference between a point-to-point connection and a multipoint connection?
23. How does a mainframe computer ask a terminal to send it data?

EXERCISES

1. If you install a 56 Kbps modem into your computer and dial into a remote network that only has 33,600 bps modems, is your modem useless?
2. You are in charge of a highly sensitive application which accepts requests and supplies confidential data. You don't want unauthorized individuals to access the data. But you do want to allow remote dial in using a modem. What security feature will allow an authorized remote user to access the database?
3. How many different "things" can prevent a 56 Kbps modem from transmitting at 56 Kbps?
4. List as many reasons as possible for a company using a modem pool.

5. List which of the EIA-232E interface signals are used only between a DTE and its DCE and list which signals travel over the phone line to the remote side.

6. Create a table that compares the advantages and disadvantages of the Universal Serial Bus to the RS-232 interface.

7. Show the sequence of start, data, and stop bits that are generated during asynchronous transmission of the character string "LUNCH."

8. List two examples each of simplex, half duplex, and full duplex connections not mentioned in the book.

9. Terminals A, B, and C are connected to a mainframe computer. Only terminal C has data to transmit. Show the sequence of messages sent between the mainframe and the three terminals using roll call polling.

10. Suppose you want to send 1000 characters of data. How many check bits will you need using asynchronous transmission? How many check bits will you need using synchronous transmission? Assume that all 1000 characters will fit within one synchronous transmission frame.

THINKING OUTSIDE THE BOX

1 You have a company with 200 computer workstations. You would like to provide as many as possible with access to a dial-out modem. Assume you can purchase a 56 Kbps modem for $60, a modem pool chassis costs $2500, and a telephone line costs $25. What is the point at which a modem pool is less expensive than individual workstation modems?

2 You are designing an application at work that transmits data records to another building within the same city. The data records are 500 bytes in length and your application will send one record every 0.5 seconds. Is it more efficient to use a synchronous connection or an asynchronous connection? What speed transmission line is necessary to support either type of connection? Show all your work.

3 You are using a computer system that has 20 terminals connected to a mainframe computer using roll-call polling. Each terminal is polled once a second by the mainframe and sends a 200 byte record every 10 seconds. Your boss says you should remove the terminals, install computer workstations, and replace the polling with asynchronous connections. Which is more efficient: keeping the terminals and polling or using workstations and asynchronous connections?

PROJECTS

1. What signals in addition to T, C, R and I are used in the X.21 interface standard?

2. The asynchronous protocol, as it first appeared, described an option for 1, 1.5, or 2 stop bits. Why would anyone need to use or want to use 1.5 or 2 stop bits?

3. Are there any external modems that have a Universal Serial Bus interface? If so, are those modems more or less expensive than a modem without a USB interface?

4. Does the Hayes company still exist? If so, is it still making modems? If not, what exactly does it produce now?

5. AT&T did not make its own modems, but bought another company to make modems. What was the name of this company? How much did AT&T pay for it? In what year did this happen?

6. Modern modems can also perform *spoofing*. What is meant by the term spoofing as applied to modems, and what is being spoofed?

5

Multiplexing: Sharing a Medium

◆◆◆

"IF I WERE TO PICK a product category of the year, it would have to be DWDM. There would be no contest." Thus spoke Scott Bradner of *Net Insider*. Dense wavelength division multiplexing (DWDM) is the technology that combines multiple data streams onto a single fiber optic cable using different colors of laser light. Each color of laser light is termed a *lambda*. Current DWDM technologies provide approximately 64 lambdas per strand of fiber, with possibilities of up to 1000 or more lambdas per strand within a few years. If each lambda is capable of supporting 96 gigabits of information, multiple lambdas will support terabits (trillions of bits) of information per second.

As exciting as DWDM sounds, it is not without a price tag. The cost of a single DWDM installation can run tens or hundreds of thousands of dollars. Where it is possible,

installing multiple fiber optic cables is much less expensive than DWDM. But where it is not possible to install multiple fibers, DWDM looks extremely promising. Like many new communication technologies, DWDM has a bright future.

Network World Fusion, http://www.nwfusion.com/bes99/ wares-bradner.html, 11/15/99.

How can multiplexing increase the capacity of a particular medium?

Are there less expensive multiplexing techniques than DWDM?

What types of applications can benefit from multiplexing?

Objectives ▶

After reading this chapter, you should be able to:

▶ Describe frequency division multiplexing and list its applications, advantages, and disadvantages.

▶ Describe synchronous time division multiplexing and list its applications, advantages, and disadvantages.

▶ Outline the basic multiplexing characteristics of both T1 and ISDN telephone systems.

▶ Describe statistical time division multiplexing and list its applications, advantages, and disadvantages.

▶ Cite the main characteristics of dense wavelength division multiplexing and its advantages and disadvantages.

▶ Apply a multiplexing technique to an example business situation.

Under the simplest conditions, a medium can carry only one signal at any moment in time. For example, the twisted pair cable that connects a keyboard to a microcomputer carries a single digital signal. The category 5 twisted pair wire that connects a microcomputer to a local area network carries only one digital signal at a time. The fiber optic cable that connects the major components of a local area network carries only one signal at a time. Many times, however, we want a medium to carry multiple signals at the same time. When watching television, we want to receive multiple television channels in case we don't like the program on the channel we are currently watching. We have the same expectations of broadcast radio. When you walk or drive around town and see many people talking on cellular telephones, something allows this simultaneous transmission of multiple signals to happen. This technique of transmitting multiple signals over a single medium is **multiplexing**. Multiplexing is a technique performed at the physical layer of the OSI model or the interface layer of the Internet model.

For multiple signals to share one medium, the medium must somehow be divided, giving each signal a portion of the total bandwidth. Presently there are three basic ways to divide a medium: frequency division multiplexing, time division multiplexing, and dense wavelength division multiplexing.

Frequency division multiplexing involves assigning non-overlapping frequency ranges to different signals. For example, cellular telephone systems that use AMPS technology divide the available bandwidth into multiple channels. Thus, the telephone connection of one user is assigned one set of frequencies for transmission, while the telephone connection of a second user is assigned a second set of frequencies.

In **time division multiplexing**, the available transmission time on a medium is divided among users. As a simple example, suppose two users A and B wish to transmit data over a shared medium to a distant computer. We can create a rather simple time division multiplexing scheme by allowing User A to transmit during the first second, then User B during the following second, followed again by User A during the third second, and so on. **Statistical time division multiplexing** is a form of time division multiplexing in which the multiplexor creates a data packet of only those devices that have something to transmit.

Although frequency division and time division are the two most common multiplexing techniques, a third technique, dense wavelength division multiplexing, has emerged in the last couple years and promises to be a very powerful multiplexing technique. **Dense wavelength division multiplexing** is used with fiber optic systems and involves the transfer of multiple streams of data over a single optical fiber using multiple-colored laser transmitters. Let's examine each of these multiplexing techniques in detail.

Frequency Division Multiplexing

Frequency division multiplexing, as stated earlier, is the assignment of non-overlapping frequency ranges to each "user" of a medium. A user might be the cellular telephone you are talking on, or it could be a computer terminal sending data to a mainframe computer. A user can also be a television station that has been assigned a set of frequencies on which to transmit its television channel into homes and businesses. So that multiple users can share a single medium, each user is assigned a channel. A channel is an assigned set of frequencies that is used to transmit the user's signal. There are many examples of frequency division multiplexing in business and everyday life. One of the first computer-based examples of frequency division multiplexing is the connection of multiple computer terminals to a mainframe computer using a pair of multiplexors. The **multiplexor** is the device that accepts input from multiple users, converts the digital data streams to analog signals using assigned frequencies, and transmits the combined analog signals over a medium that is fast enough to support the total of all the assigned frequencies. A second multiplexor, or demultiplexor, is attached to the receiving end of the medium and splits off each signal, delivering it to the appropriate receiver. Figure 5-1 shows a simplified diagram of frequency division multiplexing.

Figure 5-1
Simplified example of frequency division multiplexing

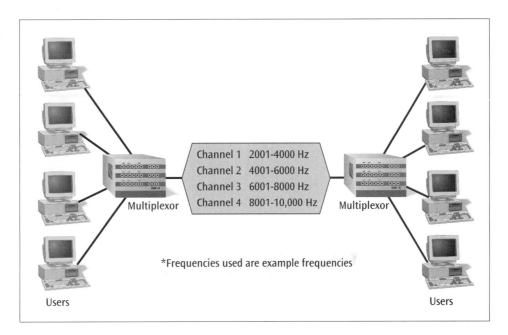

Frequency division multiplexing is used with analog signaling. The multiplexor inputs multiple data sources, assigns a different range of frequencies to each source, and transmits the multiple sources on analog signals using the assigned frequency ranges. To keep one signal from interfering with another signal, a set of unused frequencies called a guard band is usually inserted between the two signals to provide a form of insulation. The demultiplexor on the receiving end receives the combined set of signals and, using electronic filtering, separates the signals back into the individual sources. The medium that transfers the combined set of signals from sender to receiver must be capable of carrying a range of frequencies that can support the sum of all the individual frequency ranges. Most frequency division multiplexors that are used to connect computer terminals to a mainframe computer use a static assignment of frequencies. If a computer terminal has no data to transmit, a range of frequencies is still assigned to that input source. If multiple devices are not transmitting, the need to assign a range of frequencies to all input sources, can lead to a waste of frequencies. This form of static frequency assignment for terminal connections is decreasing in use and is being replaced with time division multiplexing techniques.

A still widely-found application of frequency division multiplexing is cable television. Each cable television channel is assigned a unique range of frequencies, as shown in Table 5-1. The television set, cable television box, or a video cassette recorder contains a tuner, or channel selector, which is the demultiplexor. The tuner separates one channel from the next and presents each as an individual data stream to you, the viewer. Note that each channel is assigned a range of frequencies by the Federal Communications Commission (FCC) and that the frequencies from one channel to another do not overlap. For example, Channel 2's frequencies might end at approximately 60.2 MHz, while Channel 3's frequencies might begin around 60.4 MHz.

Table 5-1 *Assignment of frequencies for cable television channels*

	Channel	Frequency in MHz
Low-Band VHF and Cable	2	54-60
	3	60-66
	4	66-72
	5	76-82
	6	82-88
Mid-Band Cable	95	90-96
	96	96-102
	97	102-108
	98	108-114
	99	114-120
	14	120-126
	15	126-132
	16	132-138
	17	138-144
	18	144-150
	19	150-156
	20	156-162
	21	162-168
	22	168-174
High-Band VHF and Cable	7	174-180
	8	180-186
	9	186-192
	10	192-198
	11	198-204
	12	204-210
	13	210-216

Some companies use frequency division multiplexing with broadband coaxial cable to deliver multiple audio and video channels to computer workstations. Videoconferencing is a common application in which two or more users transmit frequency multiplexed signals, often over long distances.

The cellular telephone system that employs AMPS technology is another very common example of frequency division multiplexing. As explained in Chapter Three, cellular telephone systems allocate channels using frequency ranges within the 800 to 900 Megahertz (MHz) spectrum. To be more precise, the 824-849 MHz range is used for receiving signals from cellular telephones (the uplink), while the 869-894 MHz range is used for transmitting to cellular telephones (the downlink). To carry on a two-way conversation, two channels must be assigned to each telephone connection. The signals coming into the mobile telephone come in on one 30 kHz band, while the signals leaving the mobile telephone go out on a different 30 kHz band. Cellular telephones are an example of dynamically assigned channels. When a user enters a telephone number and presses the Send button, the cellular network assigns a range of frequencies based on current network availability. Dynamic assignment of frequencies is less wasteful than the static assigned frequencies found in terminal-to-computer multiplexed systems.

Frequency division multiplexing is the oldest multiplexing technique and is used in many fields of communication, including cable television, broadcast television and radio, cellular telephones, and pagers. It is also one of the simplest multiplexing techniques. It is not difficult to construct an electronic device that separates one range of frequencies from another. Separating frequencies is performed with a series of electronic filters that allow a given range of frequencies to pass through while blocking all other frequencies.

Frequency division multiplexing suffers from two major disadvantages. The first disadvantage is found in computer-based systems that multiplex multiple channels over a single medium. Since the frequencies are statically assigned, devices that do not have anything to transmit are still assigned frequencies and thus bandwidth is wasted.

The second disadvantage of frequency division multiplexing occurs because the technique uses analog signals exclusively, and analog signals are more susceptible to noise than digital signals. Nonetheless, many different types of applications use frequency division multiplexing, and it will more than likely be with us for a long period of time.

Time Division Multiplexing

Signal division using frequency assignments is not terribly efficient, and frequency division multiplexing cannot be used with digital signaling techniques unless the digital signals are first converted to analog signals. In contrast, time division multiplexing (TDM) directly supports digital signals. In time division multiplexing, sharing of the signal is accomplished by dividing available transmission time on a medium among users.

How does time division multiplexing work? A simple example illustrates the technique. Suppose an instructor in a classroom poses a controversial question to students. In response, a number of hands shoot up, and the instructor calls on each student, one at a time. It is the instructor's responsibility to maintain that only one student talks at any given moment so that each individual's conversation is understandable. In a relatively crude way the instructor is a time division multiplexor, giving each user (student) a moment in time to transmit data (express an opinion to the rest of the class). In the same way, a time division multiplexor calls on one input device after another, giving each device a turn at transmitting its data over a high speed line. Since time division multiplexing was introduced in the 1960s, it has split into two roughly parallel but separate technologies: *synchronous* time division multiplexing and *statistical* time division multiplexing.

Synchronous time division multiplexing

Synchronous time division multiplexing, the original technology, gives each incoming source a turn to transmit, proceeding through the sources in round-robin fashion. Given *n* inputs, a synchronous time division multiplexor accepts one piece of data, such as a bit, from the first device, transmits it over a high speed link, accepts one bit from the second device, transmits it over the high speed link, and continues this process until a bit is accepted from the *n*th device. After the *n*th device's first bit is transmitted, the multiplexor returns to the first device and continues in round-robin fashion. Rather than accepting a single bit at a time from each source, the multiplexor may accept bytes as the unit input from each device. The resulting output stream from the multiplexor is demonstrated in Figure 5-2.

Figure 5-2
Sample output stream generated by a synchronous time division multiplexor

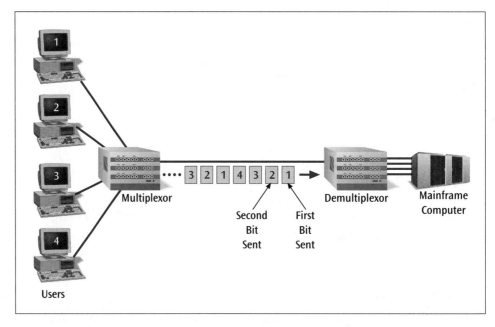

Note that the demultiplexor on the receiving end of the high speed link must disassemble the incoming bit stream and deliver each bit (or byte) to the appropriate destination. Since the high speed output data stream generated by the multiplexor does not contain addressing information for individual bits, a precise order must be maintained so that the demultiplexor can disassemble and deliver the bits to the respective owners in the same sequence as what was input.

The synchronous time division multiplexor, under normal circumstances, maintains a simple round-robin sampling order of the input devices, as shown in Figure 5-2. What would happen if one input device is sending data at a much faster rate than any of the others? You could provide an extensive buffer to hold the data from the faster device, but this buffer would provide only a temporary solution to the problem. A better solution is to sample the faster source multiple times during one round-robin pass. Figure 5-3 demonstrates how the input from device A is sampled twice for every one sample of the other input devices. As long as the demultiplexor understands this arrangement and this arrangement doesn't change dynamically, there should be no problems. This technique will only work, however, if the faster device is two, three, or four times (an integer multiple) faster than the other devices. If device A is two and a half times faster than the other devices, this technique will not work. Device A's input stream will have to be padded with additional "unuseable" bits to make its input stream three times faster than the other devices.

Figure 5-3
A synchronous time division multiplexor system which samples device A twice as fast as the other devices

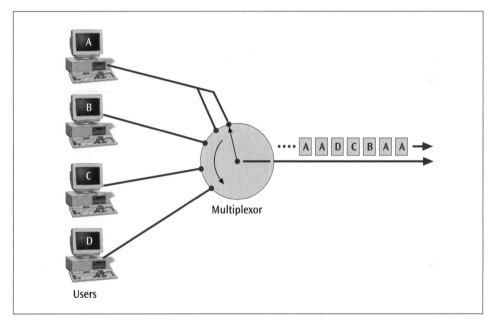

CBT: To see a visual demonstration of synchronous time division multiplexing, see the enclosed CD-ROM.

What happens if a device has nothing to transmit? If a device has nothing to transmit, the multiplexor must still allocate a slot for that device in the high speed output stream. In essence, that time slot is empty. Since each time slot is statically fixed, the multiplexor cannot take advantage of the empty slot and reassign busy devices to it. If, for example, only one device is transmitting, the multiplexor still transmits a bit from each input device (Figure 5-4). In addition, the high speed link that connects the two multiplexors must always be capable of carrying the total of all possible incoming signals, even if all input sources are not transmitting data.

Figure 5-4
Multiplexor transmission stream with only one input device transmitting data

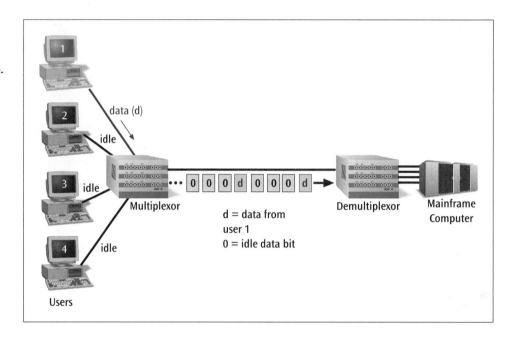

multiplexors are DCE

Terminals & Sensors are DTE

As with a simple connection between one sending device and one receiving device, maintaining synchronization across a multiplexed link is important. To maintain synchronization between sending multiplexor and receiving demultiplexor, the data from the input sources are often packed into a simple frame with synchronization bits added somewhere within the frame (Figure 5-5). Depending on the TDM technology used, anywhere from one bit to several bits can be added to a frame to provide synchronization. The synchronization bits act in a fashion similar to Differential Manchester's constantly changing signal—they provide a constantly reappearing bit sequence that the receiver can anticipate and lock onto.

Figure 5-5
Transmitted frame with added synchronization bits

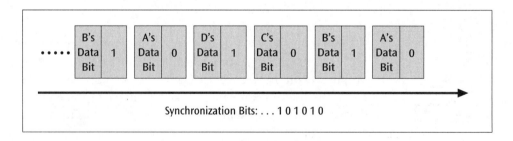

Two types of synchronous time division multiplexing that are popular today are T1 multiplexing and ISDN multiplexing. Although the details of both T1 and ISDN are very technical, a brief examination will show how multiple channels of information are multiplexed together into a single stream of data.

T1 multiplexing

In 1984, AT&T presented a service that multiplexed digital data and digitized voice onto a high-speed T1 line with a data rate of 1.544 megabits per second. The T1 line's original purpose was to provide a high speed connection between AT&T's

switching centers. When businesses learned of this high speed service, they began to request the service to connect their computer and voice communication systems into the telephone network.

The **T1 multiplexor's** output stream is divided into 24 separate digitized voice/data channels of 64 Kbps each (Figure 5-6). Users who wish to use all 24 channels are using a full-T1, while other users who need to use only part of the 24 channels are using a fractional-T1. The T1 multiplexed stream is a continuous repetition of frames. Each frame consists of one byte from each of the 24 channels (users) plus one synchronization bit. Thus, data from the first user is followed by the data from the second user, and so on, until data from the 24th user is once again followed by data from the first user. If one of the 24 input sources has no data to transmit, the space within the frame is still allocated to that input source.

Figure 5-6
T1 multiplexed data stream

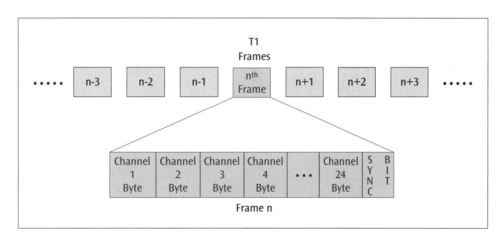

Details ▶

T1 Multiplexing

T1 communication lines are a popular technology for connecting businesses to high speed sources such as Internet service providers and other wide area networks. Because T1 multiplexing is a classic example of synchronous time division multiplexing, it merits further examination.

A T1 telecommunications line uses DS-1 signaling, which provides for the multiplexing of 24 separate channels at a total speed of 1.544 Mbps. How does the T1 line achieve a transmission of 1.544 Mbps? To answer this question, let's consider an example in which the T1 line supports 24 voice channels. Recall that Figure 5-6 shows the T1 frame sequence.

Since the average human voice occupies a relatively narrow range of frequencies (approximately 3000 - 4000 Hz), it is fairly simple to digitize voice. In fact, an analog-to-digital converter needs only 128 different quantization levels to achieve a fair digital representation of the human voice. Since 128 equals 2^7, each pulse code modulated voice sample can fit into a 7-bit value. 256 quantization levels would allow for an even more precise representation of the human voice. Since

$256 = 2^8$, and eight bits is one byte, the telephone system uses 256 quantization levels to digitize the human voice.

Recall that to create an accurate digital representation of an analog signal, you need to sample the analog signal at twice the highest frequency. When digitizing voice, you need to sample the analog voice signal 8000 times per second, given that the highest voice frequency is 4000 Hz. Since each T1 frame contains one byte of voice data for 24 different channels, the system needs 8000 frames per second to maintain 24 simultaneous voice channels. Since each frame is 193 bits in length (24 channels * 8 bits per channel + 1 control bit = 193 bits), 8000 frames per second * 193 bits per frame = 1.544 Mbps.

T1 can be used to transfer data as well as voice. If data is being transmitted, the 8-bit byte for each channel is broken into 7-bits of data and 1 bit of control information. Seven data bits per frame * 8000 frames per second = 56,000 bits per second per channel. Thus, each of the 24 T1 channels that is used for data is capable of supporting a 56 Kbps connection.

Although newer technologies such as frame relay and asynchronous transfer mode have grown in popularity, the T1 system remains a classic application of synchronous time division multiplexing. The input data from a maximum of 24 devices are assigned to fixed intervals. Each device can only transmit during that fixed interval. If a device has no significant data to transmit, the time slot is still assigned to that device and data such as blanks or zeros are transmitted.

ISDN multiplexing

Integrated Services Digital Network (ISDN) is a digital telephone service that provides voice and data transfer services over standard twisted pair into a home or small business. Although ISDN was designed to provide a number of services in addition to voice and data, data is the more popular use of most ISDN installations.

Details ▶

SONET/SDH

Synchronous Optical Network (SONET) and synchronous digital hierarchy (SDH) are very powerful standards for multiplexing data streams over a single medium. **SONET**, developed in the United States by ANSI, and **SDH**, developed in Europe by ITU-T, are two almost identical standards for the high bandwidth transmission of a wide range of data types over fiber optic cable. SONET and SDH have two features that are of particular interest in the context of multiplexing. First, they are both synchronous multiplexing techniques . A single clock controls the timing of all transmission and equipment across an entire SONET/SDH network. Using only a single clock to time all data transmissions yields a higher level of synchronization, since the system does not have to deal with two or more clocks having slightly different times. This high level of synchronization is necessary to achieve the high level of precision required when transmitting data at hundreds and thousands of megabits per second.

Second, SONET and SDH are able to multiplex varying speed streams of data onto one fiber connection. SONET defines a hierarchy of signaling levels, or data transmission rates, called synchronous transport signals (STS). Each STS level supports a particular data rate, as shown in Table 5-2. Each of the STS signaling levels is supported by a physical specification called the optical carriers (OC). Note that the data rate of OC-3 is exactly 3 times the rate of OC-1; this relationship carries throughout the entire table of values. SONET is designed with

this data rate relationship so that it is relatively straight forward to multiplex signals. For example, it is relatively simple to multiplex three STS-1 signals into one STS-3 signal. Likewise, four STS-12 signals can be multiplexed into one STS-48 signal. The STS multiplexor in a SONET network can accept electrical signals from copper-based media, convert those electrical signals into light pulses, and then multiplex the various sources onto one high speed stream.

STS Level	OC Specification	Data Rate (in Mbps)
STS-1	OC-1	51.84
STS-3	OC-3	155.52
STS-9	OC-9	466.56
STS-12	OC-12	622.08
STS-18	OC-18	933.12
STS-24	OC-24	1244.16
STS-36	OC-36	1866.23
STS-48	OC-48	2488.32
STS-96	OC-96	4976.64
STS-192	OC-192	9953.28

Table 5-2 *STS signaling levels, corresponding OC levels, and data rates*

ISDN multiplexing comes in two basic forms: primary rate interface (PRI) and basic rate interface (BRI). PRI was designed for business applications and, similar to T1, multiplexes 24 input channels together onto one high speed telephone line. BRI is the interface more often used by consumers to connect their home and small business computers to the Internet. BRI multiplexes only three separate channels onto a single medium speed telephone line. Two of the three channels—the B channels—carry either data or voice, while the third channel—the D channel—carries the signaling information that controls the two data/voice channels. Since most consumers already have a standard telephone line, it is very common to use both of the B channels for data.

Each SONET frame contains the data that is being transmitted plus a number of control bits, which are scattered throughout the frame. Figure 5-7 shows the frame layout for the STS-1 signaling level. The STS-1 signaling level supports 8000 frames per second, and each frame contains 810 bytes (6480 bits). 8000 frames per second times 6480 bits per frame yields 51,840,000 bits per second, which is the OC-1 data rate. The other STS signaling levels are similar except for the layout of data and the placement and quantity of control bits.

SONET and SDH are used in numerous applications in which very high transfer rates of data over fiber optic lines are necessary. For example, two common users of SONET are the telephone company and companies that provide an Internet backbone service. Both telephone companies and Internet backbone providers have very high speed transmission lines, which span parts of the country. Since these transmission lines must transmit hundreds and thousands of millions of bits per second over long distances, fiber optic lines supporting SONET transmission technology is the best solution.

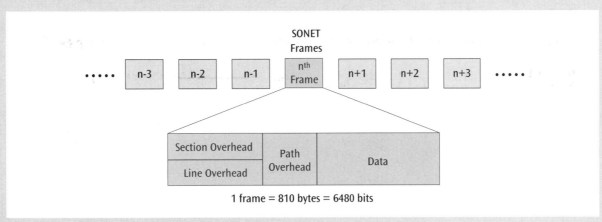

Figure 5-7 *SONET STS-1 frame layout*

Figure 5-8 shows how the data from the two B channels (B1 and B2) plus the signaling information from the D channel are multiplexed together into a single frame. Note that 8 bits of data from the first B channel are followed by signaling control information, which is then followed by 8 bits of data from the second B channel and more signaling control information. These four groups of information repeat again to form a single frame.

Figure 5-8
ISDN frame layout showing B channel bits and signaling control information bits

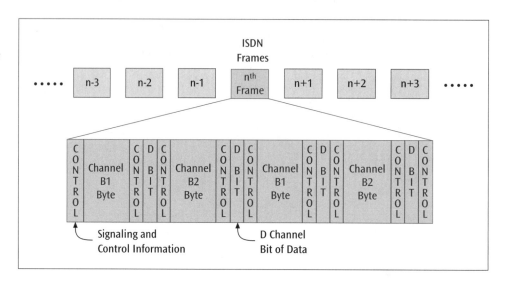

Statistical time division multiplexing

Both frequency division multiplexing and synchronous time division multiplexing can waste unused transmission space. One solution to this problem is statistical time division multiplexing. **Statistical time division multiplexing** (sometimes called asynchronous time division multiplexing) transmits data only from active users and does not transmit empty time slots. To transmit data only from active users, a more complex frame is created that contains data only from those input sources that have something to send. For example, if four stations A, B, C, and D are connected to a statistical multiplexor but only stations A and C are currently transmitting, the statistical multiplexor transmits only the data from stations A and C, as shown in Figure 5-9. Note that at any moment, the number of stations transmitting can change from two to zero, one, three or four. If that happens, the statistical multiplexor creates a new frame containing data from the currently transmitting stations.

Test

*

Statistical is the only multiplexing compression that the sum of current
Doesn't have to equal the sum of the ~~components~~ components

Figure 5-9
Two stations out of four transmitting via a statistical multiplexor

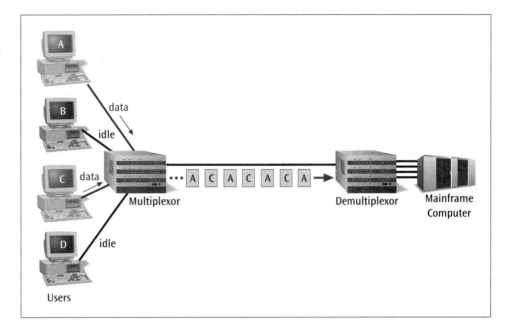

Since only two of the four stations are transmitting, how does the demultiplexor on the receiving end recognize the correct recipients of the data? Some type of address must be included with each byte of data to identify who sent the data and for whom it is intended (Figure 5-10). The address can be as simple as a binary number that uniquely identifies the station that is transmitting. For example, if the multiplexor is connected to four stations, then the addresses can simply be 0, 1, 2, and 3 for stations A, B, C, and D. In binary, the values would be 00, 01, 10, and 11.

Figure 5-10
Sample address and data in a statistical multiplexor output stream

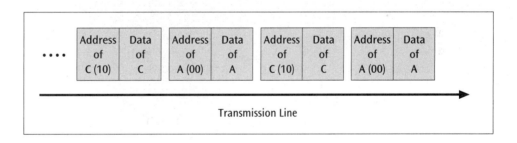

If the multiplexor transmits more than one byte of data at a time from each source, then an alternative form of address and data are required (Figure 5-11). To transmit a variable sized piece of data, a length field defining the length of the data block is included along with the address and data. This packet of *address/length/data/address/length/data...* is then packaged into a larger unit by the statistical multiplexor.

Figure 5-11

Packets of address and data fields in a statistical multiplexor output stream

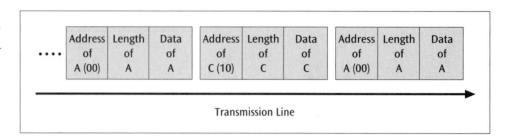

This larger unit, shown in Figure 5-12, looks much like the frame that is transmitted using a synchronous connection. The FLAG at the beginning and end delimits the beginning and end of the frame. The CONTROL field provides information that is used by the pair of multiplexors to control the flow of data between sending multiplexor and receiving multiplexor. Finally, the Frame Check Sequence (FCS) provides information that the receiving multiplexor can use to detect transmission errors within the frame.

Figure 5-12

Frame layout for the information packet transferred between statistical multiplexors

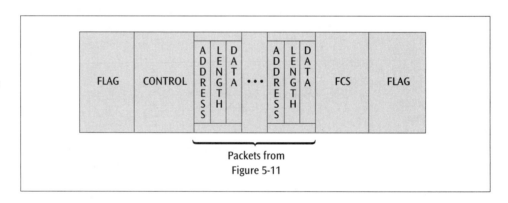

Statistical multiplexors have one very good advantage over synchronous time division multiplexors. Although both synchronous time division multiplexing and statistical time division multiplexing can transmit data over a high speed link, statistical time division multiplexing does not require a line as high speed as synchronous time division multiplexing does. Statistical time division multiplexing assumes that all devices do not transmit at the same time; therefore, it does not require a high speed link that is the total of *all* the incoming data streams. Since it is assumed that only a percentage of input sources transmit at one time, the output line capacity coming from the statistical multiplexor should be less than the output line capacity from the synchronous multiplexor, thus allowing for a slower speed link between multiplexors. This slower speed link usually translates into less cost.

One small disadvantage of statistical multiplexors is their increased level of complexity. Synchronous TDM simply accepts the data from each attached device and transmits that data in a non-ending cycle. The statistical multiplexor must collect and buffer data from active attached devices and, after creating a frame with necessary control information, transmit that frame to the receiving multiplexor. Although this slightly higher level of complexity translates into higher initial costs, those costs are usually offset by the ability to use a smaller capacity interconnecting line.

CBT: To see a visual demonstration of statistical time division multiplexing, see the enclosed CD-ROM.

Statistical time division multiplexing is a good choice for connecting a number of lower speed devices that do not transmit data on a continuous basis to a remote computer system. Examples of these systems include data entry systems, point of sale systems, and many other commercial applications in which users enter data at computer terminals.

Dense Wavelength Division Multiplexing

When transmission systems employing fiber optic cable were first installed, the explosive growth of the Internet and other data transmission networks had not even been imagined. Now that the 21st century has begun, it is painfully obvious that early growth forecasts were gross underestimates. With Internet access growing by more than 100 percent per year, for individuals requesting multiple telephone lines for faxes and modems, video transmissions, and teleconferencing, a single fiber optic line transmitting billions of bits per second is simply no longer sufficient. This inability of a single fiber optic line to meet users needs is called fiber exhaust. Technology specialists saw few ways to resolve fiber exhaust other than by installing additional fiber lines, often at great expense. Now there appears to be an attractive solution that takes advantage of currently installed fiber optic lines. **Dense wavelength division multiplexing,** sometimes called wave division multiplexing, is a fairly new technology that multiplexes multiple data streams onto a single fiber optic line. Unlike frequency division multiplexing, which assigns input sources to separate sets of frequencies, and time division multiplexing, which divides input sources by time, wave division multiplexing uses different wavelength lasers to transmit multiple signals. The wavelength of each different colored laser is called the **lambda**. Thus, DWDM supports multiple lambdas.

Figure 5-13 illustrates how dense wavelength division multiplexing works. The technique takes each input source, assigns a uniquely colored laser to that source, and combines the multiple optical signals of the input sources so that they can be amplified as a group and transported over a single fiber. Interestingly, because of the properties of the signals, light, and glass fiber, each signal carried on the fiber can be transmitted at a different rate from the other signals. This means that a single fiber optic line can support simultaneous transmission speeds such as 51.84 Mbps, 155.52 Mbps, 622.08 Mbps, and 2.488 Gbps. These speeds are the speeds defined by OC-1, OC-3, OC-12, and OC-48—the optical carrier specifications for high speed fiber optic lines. In addition, a single fiber optic line can support different transmission formats such as synchronous optical network (SONET), synchronous digital heirachary (SDH), asynchronous transfer mode (ATM), and numerous other data transmission formats, in various combinations (Figure 5-14).

Figure 5-13
Multiple lasers transmitting data signals down a single fiber optic line

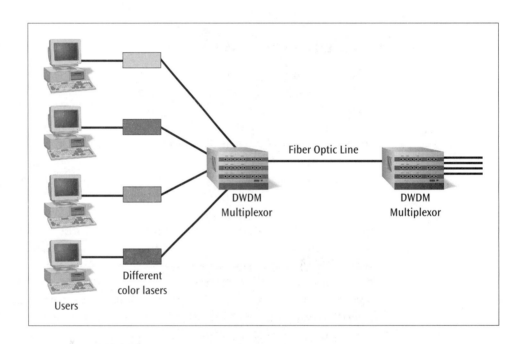

Blue has a shorter wave length than RED

Blue = DVD   Blue can hold more bits, and they are close together
Red = CD

Figure 5-14
Fiber optic line using dense wavelength division multiplexing and supporting multiple-speed transmissions

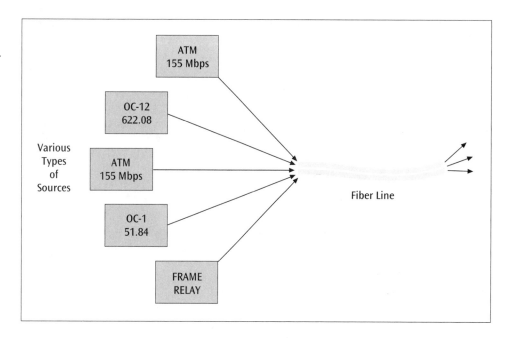

Dense wavelength division multiplexing is scalable. As the demands on a system and its applications grow, it is possible to add additional wavelengths or lambdas onto the fiber, thus further multiplying the overall capacity of the original fiber optic system. Current technology allows for as many as 64 lambdas per fiber, with the possibility for 1000 lambdas within a few years. This power does not come without a price tag, however. Dense wavelength division multiplexing is an expensive way to transmit signals from multiple devices. Many technology experts predict, despite its cost, dense wavelength division multiplexing will continue to grow in popularity and may someday even replace technologies such as SONET and SDH.

Comparison of Multiplexing Techniques

The advantages and disadvantages of each multiplexing technique are summarized in Table 5-3. Frequency division multiplexing relies on analog signaling and is the simplest and most noisy of all the multiplexing techniques. Synchronous time division multiplexing is also relatively straight forward, and like frequency division multiplexing, input devices that have nothing to transmit can waste transmission space. The big advantage of synchronous TDM is the lower noise during transmission. Statistical TDM is one step above synchronous TDM because it transmits data only from those input devices that have data to transmit. Thus, statistical TDM wastes less bandwidth on the transmission link. Finally, dense wavelength division multiplexing is a very good, albeit expensive, technique for transmitting multiple concurrent signals over a fiber optic line.

Table 5-3   *Advantages and disadvantages of multiplexing techniques*

Multiplexing Technique	Advantages	Disadvantages
Frequency Division Multiplexing	Simple	Analog signals only
	Popular with radio, TV, cable TV	Limited by frequency ranges
	Relatively inexpensive	
	All the receivers, such as cellular telephones, do not need to be at the same location.	
Synchronous Time Division Multiplexing	Digital signals	Wastes bandwidth
	Relatively simple	
	Commonly used with T1 and ISDN	
Statistical Time Division Multiplexing	More efficient use of bandwidth	More complex than synchronous time division multiplexing
	Packets can be variable sized	
	Frame can contain control and error information	
Dense Wavelength Division Multiplexing	Very high capacities over fiber	Cost
	Scalable	Complexity
	Signals can have varying speeds	

Business Multiplexing In Action   ▶

XYZ Corporation has two buildings, A and B, separated by a distance of 300 meters (see Figure 5-15). A 3-inch diameter tunnel runs underground between the two buildings. Building B contains 66 text-based terminals that need to be connected to a mainframe computer in Building A. The text-based terminals transmit data at 9600 bits per second. What are some good ways to connect the terminals in Building B to the mainframe computer in Building A?

Figure 5-15
Buildings A and B and the 3-inch tunnel connecting the buildings

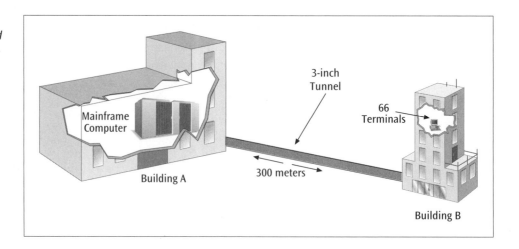

Considering the technologies that have been introduced thus far, there are four possible scenarios for connecting the terminals and mainframe computer:

1. Connect each terminal to the mainframe computer using separate point-to-point lines. Each line will be some form of conducted media.

2. Connect all the terminals to the mainframe computer using one multipoint line. This one line will be some form of conducted media.

3. Collect all the terminal outputs and use microwave transmissions to send the data to the mainframe computer.

4. Collect all the terminal outputs using multiplexing and send the data to the mainframe computer over a conducted media line.

Let's examine the pros and cons of each solution. The first solution of connecting each terminal to the mainframe computer using point-to-point conducted media has some advantages but also some serious drawbacks. The distance of 300 meters poses an immediate problem. When transmitting data in millions of bits per second, twisted pair typically has a maximum distance of 100 meters. XYZ Corporation's data is not being transmitted at millions of bits per second, but at 9600 bits per second instead. At the slower rate, we may be able to extend the distance further than 100 meters, but doing so may not be a good idea. Electromagnetic noise is potentially always a problem, and we may discover after the installation of the wires that there is too much noise. A more noise resistant medium, such as coaxial cable or fiber optic cable, might be a reasonable option. But 66 coaxial cables probably will not fit in a 3-inch diameter tunnel. Fiber optic cable has roughly the same dimensions as coaxial cable and a much higher expense, considering the 66 pairs of optical devices that would be needed at the ends of the cables. What if we could run 66 wires of some media through the tunnel, and a month after the installation management decides to add 10 more terminals to Building B? Ten additional cables are not likely to fit through the tunnel, and if they could, it would be time consuming and expensive to pull ten more cables through a tunnel.

The main advantages of the first solution include the lower cost of using a relatively inexpensive conducted media, and the fact that multiple point-to-point lines eliminate the need for additional services such as polling or multiplexing.

The chief advantage of the second solution—using one multipoint line—is that it would require only a single wire running through the tunnel. There are two major problems, however. First, it may not be possible to create or purchase a system that polls 66 devices fast enough to maintain 9600 bits per second transmission on each terminal line. Second, even if you could poll 66 devices at 9600 bits per second, the solution is not scalable. We could not add more terminals in the future and still have the system work at an acceptable level.

The third solution of transmitting the data using microwave signals is interesting. Microwave is very fast, and private ownership between buildings can be attractive. The following concerns, however, are worth investigating:

▶ Is the line-of-sight between Building A and Building B unobstructed by trees or other buildings? If there is an obstruction, microwave will not work.

▶ What is the cost of installing a microwave system between the two buildings? If the cost is high, there may be a more reasonable alternative.

▶ The system would still need some kind of device that collects data from the 66 terminals and prepares a single data stream for microwave transmission. Will the microwave system handle this collection, or will we need something like a multiplexor?

The fourth solution—to install multiplexors at each end of the tunnel and connect the multiplexors with some type of high speed medium—also requires some forethought and investigation. The following issues are worth considering:

► Can one pair of multiplexors handle 66 terminals? If not, we may have to install two pairs of multiplexors.

► What does a pair of multiplexors cost? Will this cost be so high that we are forced to consider other alternatives?

► What kind of medium could we use to connect the multiplexors? How many wires would we need to run? Fiber optic cable or even coaxial cable would be a good choice. Even if the system requires multiple strands of fiber or coaxial cable, they should fit within the 3-inch diameter tunnel.

► Is the multiplexor solution scalable? Can the system expand to include additional terminals in the future? In the worst case scenario, we would have to add an additional pair of multiplexors and another cable in the tunnel. We could plan ahead and pull several strands of fiber or coaxial cable in order to be prepared for future expansion.

In conclusion, it appears a multiplexing scheme provides the most efficient use of a small number of cables running through the small tunnel. If a high quality cable such as fiber optic wire is used, it will minimize noise intrusion, and allow for the greatest amount of future growth. The microwave solution is also attractive, but will undoubtedly cost much more than a pair of multiplexors and connecting cables.

◆ ◆

SUMMARY

► For multiple signals to share a single medium, the medium must be divided into multiple channels.

► There are three techniques for dividing a medium into multiple channels: frequency division multiplexing, time division multiplexing, and dense wavelength division multiplexing.

► Frequency division multiplexing involves assigning non-overlapping frequency ranges to different signals.

► Frequency division multiplexing uses analog signals while time division multiplexing uses digital signals.

► Time division multiplexing of a medium involves dividing the available transmission time on a medium among the users.

► Time division multiplexing has two basic forms: synchronous time division multiplexing and statistical time division multiplexing.

► Synchronous time division multiplexing accepts input from a fixed number of devices and transmits their data in a non-ending repetitious pattern.

► T1 and ISDN telephone systems are common examples of systems that use synchronous time division multiplexing.

► The static assignment of input devices to particular frequencies or time slots can be wasteful if the input devices are not transmitting data.

► Statistical time division multiplexing accepts input from a set of devices that have data to transmit, creates a frame with data and control information, and transmits that frame. Input devices that do not have data to send are not included in the frame.

► Dense wavelength division multiplexing involves fiber optic systems and the transfer of multiple streams of data over a single fiber using multiple colored laser transmitters.

KEY TERMS

dense wavelength division multiplexing

frequency division multiplexing

ISDN multiplexing

lambda

multiplexing

multiplexor

statistical time division multiplexing

synchronous time division multiplexing

time division multiplexing (TDM)

T1 multiplexor

REVIEW QUESTIONS

1. List three common examples of frequency division multiplexing.

2. Is frequency division multiplexing associated with analog signals or digital signals?

3. In what order does synchronous time division multiplexing sample each of the incoming signals?

4. What would happen if a synchronous time division multiplexor sampled the incoming signals out of order?

5. How does a synchronous time division multiplexor stay synchronized with the demultiplexor on the receiving end?

6. How many separate channels does a T-1 multiplexor multiplex into one stream?

7. How many channels does Basic Rate ISDN support over a single connection?

8. What are the main differences between statistical time division multiplexing and synchronous time division multiplexing?

9. If a statistical multiplexor is connected to 20 devices, does it require a high speed output line that is equivalent to the sum of the 20 transmission streams?

10. Why is addressing of the individual data streams necessary for statistical multiplexing?

11. What type of medium is required to support dense wavelength division multiplexing?

12. How many different wavelengths can dense wavelength division multiplexing place onto one connection?

EXERCISES

1. State two advantages and two disadvantages of:
   a. frequency division multiplexing compared to the other multiplexing techniques
   b. synchronous time division multiplexing compared to the other multiplexing techniques
   c. statistical time division multiplexing compared to the other multiplexing techniques
   d. dense wavelength division multiplexing compared to the other multiplexing techniques

2. When data is transmitted using a statistical multiplexor, the individual units of data must have some form of address that tells the receiver the intended recipient of each piece of data. Instead of assigning absolute addresses to each piece of data, is it possible to incorporate relative addressing? If so, explain its benefits.

3. A benefit of frequency division multiplexing is that all the receivers do not have to be at the same location. Explain what this benefit means and give an example.

4. Using statistical time division multiplexing with 24 stations inputting data at 9600 bps, and assuming that only 60 percent of those stations transmit at one time, what is the minimal speed necessary for the high speed link connecting multiplexor and demultiplexor?

5. Twenty-four *voice signals* are to be multiplexed and transmitted over twisted pair. What is the total bandwidth required if frequency division multiplexing is used?

6. Twenty-four voice signals are to be multiplexed and transmitted over twisted pair. What is the bandwidth required (in bps) if synchronous time division multiplexing is used, if we use the standard analog to digital sampling rate, and if each sample is converted into an 8-bit value?

THINKING OUTSIDE THE BOX

1 A company has two buildings 50 meters apart. In between the buildings is private land owned by the company. There is a large walk-through tunnel connecting the two buildings. In one building is a collection of 30 computer workstations, and in the other building is a mainframe computer. What is the best way to connect the workstations to the mainframe computer? Show your reasoning and all the possible solutions you considered.

2 A company has two buildings 100 meters apart. In between the buildings is public land with no access tunnel. In one building is a collection of 30 computer workstations, and in the other building is a mainframe computer. What is the best way to connect the workstations to the mainframe computer? Show your reasoning and all the possible solutions you considered.

3 What is the relationship, if any, between synchronous time division multiplexing and synchronous and asynchronous connections from Chapter 4.

PROJECTS

1. Locate advertising material that lists the maximum number of devices a frequency or time division multiplexor can handle.

2. Digital broadcast television will someday replace conventional analog television. What form of multiplexing is used to broadcast digital television signals?

3. Broadcast radio is one of the last forms of entertainment to go digital. Find the latest material describing the current state of digital broadcast radio and write a two- or three-page report that includes the type of multiplexing envisioned and the impact digital radio will have on the current radio market.

4. There is a new set of frequencies created by the FCC for walkie-talkie radios. This set is called the family radio service and allows two radios to transmit up to a two mile distance. What kind of multiplexing is used with these radios? How many concurrent channels are allowed?

5. The local loop of the telephone circuit that enters your house uses multiplexing so that the people on the two ends of the connection can talk at the same time (if they wish). What kind of multiplexing is used? State the details of the multiplexing technique.

6

Errors, Error Detection, and Error Control

◆ ◆

DURING THE 1970s, an unmanned rocket that was supposed to launch into outer space never made it. Moments after the rocket took off, it strayed off course and the rocket was destroyed. Afterwards, all hardware and software systems were examined in an attempt to determine what caused the rocket to fail. Legend has it that after much work, a problem was discovered with the software program, which was written in Fortran. The program contained a DO loop that did not loop. The code might have looked something like the following:

DO 200 I=1,20

This statement is supposed to execute the statements down to statement 200 twenty times. Unfortunately, this statement was not what was coded. Instead, the following statement was coded:

DO 200 I=1.20

The only difference between the two statements is a period in place of a comma. Because of that one difference, the statement is no longer a DO loop, but a simple assignment statement. Because of this type of error, a multi-million dollar rocket was lost.

Why didn't the computer catch this typographic error?

How can a simple error like this ruin an entire project?

Could some type of built-in error detection have been used to avoid this problem?

Objectives ▶

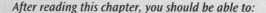

After reading this chapter, you should be able to:

▶ Identify the different types of noise commonly found in computer networks.

▶ Specify the different error prevention techniques and be able to apply an error prevention technique to a type of noise.

▶ Compare the different error detection techniques in terms of efficiency and efficacy.

▶ Perform simple parity and longitudinal parity calculations and enumerate their strengths and weaknesses.

▶ Cite the advantages of cyclic redundancy checksum and specify what types of errors cyclic redundancy checksum will detect.

▶ Differentiate the three basic forms of error control and describe under what circumstances each may be used.

▶ Follow an example of stop-and-wait ARQ, go-back-N ARQ, and selective reject ARQ.

Introduction ▶

The process of transmitting data over a medium often works according to Murphy's Law: If something can go wrong, it probably will. Even if all possible error-reducing conditions are applied before and during data transmission, something invariably alters the form of the original data. If this alteration is serious enough, the original data becomes corrupt, and the receiver does not receive the data that was originally transmitted. Even using the highest quality fiber optic cable, noise eventually creeps in and begins to disrupt the data transmission.

Given that noise is inevitable and errors happen, something needs to be done to detect error conditions. This chapter examines some of the more common error detection methods and compares them in terms of efficiency and efficacy.

Before beginning to learn about error detection techniques, it is vital to understand the different forms of noise that commonly occur during data transmission. Having a better understanding of the different types of noise and what causes them to occur will enable you to apply noise reduction techniques to communication systems to limit the amount of noise before it reaches the threshold at which errors occur. Despite the best efforts to control noise, some noise is inevitable. When the ratio of noise power to signal power becomes such that the noise overwhelms the signal, errors occur. It is at this point that error detection techniques become valuable tools.

Once an error has been detected, what action should a receiver take? There are three options: ignore the error, return an error message to the transmitter, or correct the error without help from the transmitter. Although ignoring the error seems like an irresponsible position, it has merit and is worth examining. The second option—return an error message to the transmitter so that the transmitter can re-send the errant data—is the most common. The third option—correcting the error without asking for additional help from the transmitter—may sound like the ideal situation, but it is difficult to support and requires a significant amount of overhead.

How does error detection and control fit into the OSI model introduced in Chapter One? Error detection and error control are details covered in the data link layer of the OSI model. All the topics discussed so far, with the possible exception of asynchronous and synchronous connections, have been details of the physical layer. To understand how error detection and control fits in, it is important to understand the basic concept of the OSI model: the details of one layer should not be affected by the details of another layer. Thus, the selection of an error detection scheme or error control technique is an issue separate from the type of media selected or the choice of multiplexing technique. Any of the error detection and control schemes presented in this chapter can be applied to any type of communication system.

Noise and Errors

As you might expect, many things can go wrong during data transmission. From a simple blip to a massive outage, transmitted data—both analog and digital—are susceptible to many types of noise and errors. Copper-based media have traditionally been plagued with many types of interference and noise. Satellite, microwave, and radio networks are also prone to interference and crosstalk. Even near-perfect fiber optic cables can introduce errors into a transmission system, though the probability of this happening is less than with the other types of media. Let's examine several of the major types of noise that occur in transmission systems.

White noise

White noise, which is also called thermal noise or Gaussian noise, is a relatively constant type of noise and is much like the static you hear when a radio is tuned between two stations. It is always present to some degree in transmission media and electronic devices and depends on the temperature. As the temperature increases, the level of noise increases. Since white noise is relatively constant, it can be reduced significantly, but never completely eliminated. White noise is what makes an analog or digital signal become fuzzy (Figure 6-1).

Figure 6-1
White noise as it interferes with a digital signal

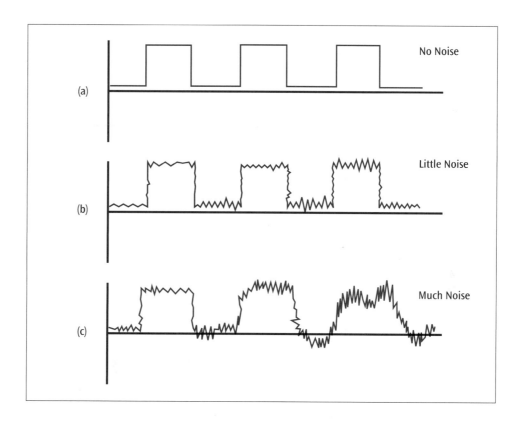

Removing white noise from a digital signal is relatively straight forward if the signal passes through a signal regenerator before the noise completely overwhelms the original signal. Removing white noise from an analog signal is also possible by passing the noisy analog signal through an appropriate set of filters, which (hopefully) leaves nothing but the original signal.

Impulse noise

Impulse noise, or spike, is a non-constant noise and one of the most difficult errors to detect because it can occur randomly. The difficulty comes in separating the noise from the signal. Typically the noise is an analog burst of energy. If the impulse spike interferes with an analog signal, removing it without affecting the original signal is extremely difficult. Consider an old LP album (the thing we listened to before compact discs). As the album aged, scratches on the surface created clicks and pops that interfered, in varying degrees, with enjoyment of the music. Why didn't someone invent a device to remove these clicks and pops? Actually, analog devices existed in the 1960s and 1970s, but they required some adjustment so they could distinguish noise from music. These devices used fast-acting partial mute or clipping circuitry to remove the pops and clicks. Even with these adjustments, the devices would occasionally remove some of the music, thinking it was noise when it wasn't. Thus, the devices never became popular. Interestingly, a device now exists to remove clicks and pops from albums. This device performs digital signal processing and allows workstation-style editing operations that actually splice and rebuild damaged portions of the music. These operations are not performed in real time and come with much higher time and cost price tags.

If impulse noise interferes with a digital signal, often the original digital signal can be recognized and recovered. When the noise completely obliterates the digital signal, the original signal cannot be recovered (Figure 6-2).

Figure 6-2
The affect of impulse noise on a digital signal

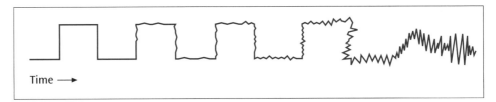

With digital signals, you can see quite dramatically how transmission speed can affect whether or not noise is significant. Figure 6-3 shows a digital signal transmission at relatively slow speed and at relatively high speed. Notice in the figure that when transmission speed is slower, you can still determine the value of a signal. But when transmission speed increases, you can no longer determine whether the signal is a 0 or a 1.

Figure 6-3
Transmission speed and its relationship to noise

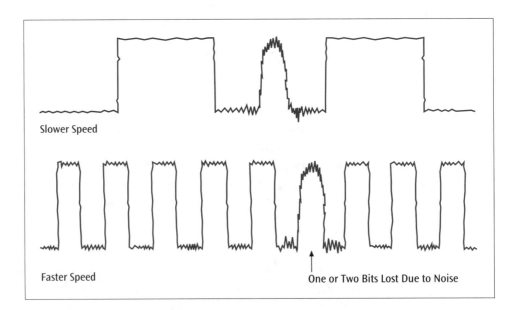

Crosstalk

Crosstalk is an unwanted coupling between two different signal paths. This unwanted coupling can be electrical, as might occur between two sets of twisted pair wire (as in a phone line), or it can be unwanted signals picked up by microwave antennas. Telephone signal crosstalk was more common 20 to 30 years ago before telephone companies used fiber optic cables and other well-shielded wires. Today, when crosstalk appears, you can hear another telephone conversation in the background (Figure 6-4). High humidity and wet weather can cause electrical crosstalk over a telephone system to increase. Crosstalk is relatively constant and can be reduced with proper precautions and hardware, as you will see shortly.

Figure 6-4
Two telephone circuits experiencing crosstalk

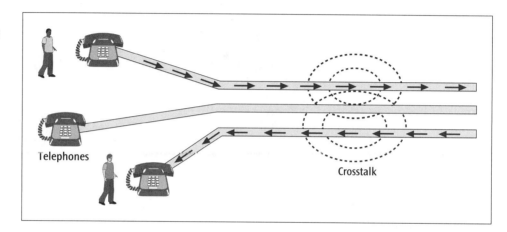

Echo

Echo is the reflective feedback of a transmitted signal as the signal moves through a medium. Much like a voice will echo in an empty room, a signal can hit the end of a cable, bounce back through the wire, and interfere with the original signal. This error occurs most often at junctions where wires are connected or at the open end of a coaxial cable. Figure 6-5 demonstrates a signal bouncing back from the end of a cable and creating an echo. To minimize echo, a device called an echo suppressor can be attached to a line. An echo suppressor is essentially a filter that allows the signal to pass in one direction only. For local area networks that use coaxial cable, a small filter is placed on the open end of a wire to absorb any incoming signals.

Figure 6-5
A signal bouncing back at the end of a cable and causing echo

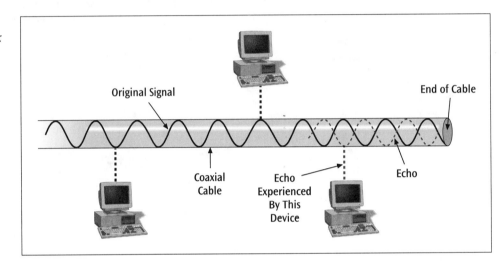

Jitter

Jitter is the result of small timing irregularities during the transmission of digital signals that become magnified as the signals are passed from one device to another. If unchecked, jitter can cause video devices to flicker, audio transmissions to click and breakup, and transmitted computer data to arrive erroneously. If jitter becomes too great, it can limit overall system performance and require the transmitting devices to slow down their transmission rates. Figure 6-6 shows a simplified example of a digital signal experiencing jitter.

Figure 6-6
Original digital signal and digital signal with jitter

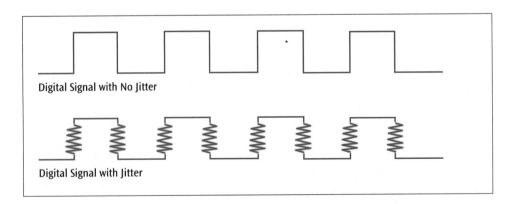

Causes of jitter can include electromagnetic interference, crosstalk, passing the signal through too many repeaters, and the use of lower quality equipment. Possible solutions to the jitter problem involve the reversal of the above conditions.

Proper shielding can reduce or eliminate electromagnetic interference and crosstalk, while limiting the number of times a signal is repeated may also help reduce the occurrence of jitter.

Delay distortion

Delay distortion can occur because the velocity of propagation of a signal through a wire varies with the frequency of the signal. Since a signal may be composed of multiple frequencies, some of those frequencies may arrive at the destination sooner than others. Delay distortion can be particularly harmful to digital signals. Proper electronic techniques that equalize transfer speeds of faster and slower signals can significantly reduce this type of error.

Attenuation

Attenuation is the continuous loss of a signal's strength as it travels though a medium. It is not necessarily a form of error. As you learned in Chapter Two, attenuation can be eliminated with the use of amplifiers for analog systems or repeaters for digital systems.

Error Prevention

Since there are so many forms of noise and errors, and since one form of noise or another is virtually a given, every data transmission system must include precautions to reduce noise and the possibility of errors. An unfortunate side effect of noise during a transmission is that the transmitting station has to slow down its transmission. Recall from Chapter Four the concept of fallback negotiation. If a modem sends data and the data arrives garbled, the receiving modem may ask the transmitting modem to step back to a slower transmission speed. This slow down creates a signal in which the bit duration of each 0 and 1 is longer, thus giving the receiver a better chance of distinguishing one value from the next despite the introduction of noise. However, if you can reduce the possibility of noise before it happens, the transmitting station may not have to slow down its transmission stream. With proper error prevention techniques, many types of errors can be reduced.

The following techniques can be applied to reduce the possibility of transmission errors:

- ► Proper shielding of cables to reduce electromagnetic interference and crosstalk is an essential error prevention technique.
- ► Telephone line conditioning or equalization—in which filters are used to help reduce signal irregularities—is provided by the telephone company. For an additional charge, the telephone company will provide various levels of conditioning to leased lines. This conditioning provides a quieter line, which minimizes data transmission errors.
- ► Replacing older equipment with more modern, digital equipment, although initially expensive, is often the most cost effective way to minimize transmission errors in the long run.

> ► Proper use of digital repeaters and analog amplifiers can increase signal strength, thus decreasing the probability of errors.
>
> ► Observing the stated capacities of a medium and not pushing transmission speeds beyond their recommended limits can reduce the possibility of errors. For example, recall from Chapter Two that twisted pair category 5 cable should not be longer than the recommended 100 meter distance when transmitting at 100 Mbps.

Reducing the number of devices, decreasing the length of cable runs, or reducing the transmission speed of the data may be effective ways to reduce the possibility of errors. Although choices like these are not always desirable, sometimes they are the most reasonable choices available.

Table 6-1 summarizes the different types of errors and includes one or more possible error prevention techniques for each.

Table 6-1   *Summary of errors and error prevention techniques*

Type of Error	Error Prevention Technique
White noise	Filters for analog signals, signal regeneration for digital signals
Impulse noise	Special filters for analog signals, digital signal processing for digital signals
Crosstalk	Proper shielding of cables
Echo	Proper termination of cables
Jitter	Better quality electronic circuitry, slowing the transmission
Delay distortion	Added circuitry that equalizes the transmission speeds of the different frequencies
Attenuation	Amplification of analog signals, regeneration of digital signals

Do not be lured into thinking that by simply applying various error prevention techniques, errors will not happen. Appropriate error detection methods still need to be implemented. Let's examine the main error detection techniques next.

Error Detection Techniques

Despite the best attempts at prevention, errors still occur. Since most data transferred over a communication line are important, it is necessary to apply an error detection technique to the received data to ensure that no errors were introduced into the data during transmission. If an error is detected, it is typical to perform some type of request for retransmission.

There are several places within a communications model in which to check for errors. One of the most common places to perform error detection is the data link layer. At the data link layer, when a device creates a frame of data, it inserts some type of error detection code. When the frame arrives at the next device in the transmission sequence, the receiver extracts the error detection code and applies it to the data frame. Then a data frame is reconstructed and sent to the next device in the transmission sequence.

New technologies are beginning to abandon error detection at the data link layer. Instead of performing error detection at each device in the transmission sequence, error detection is applied only once—at the final destination. When error detection is applied at the final destination only, a higher layer such as the transport layer or the application layer is then responsible for performing error detection. With error detection applied only once, data transmission across a network of computers and devices is faster. And with an increase in the use of fiber optic cable, errors occur less frequently, making error detection at every device in the transmission sequence unnecessary.

Regardless of where the error detection is applied, all systems still recognize the importance of checking for transmission errors. The error detection techniques themselves can be relatively simple or relatively elaborate. Generally, simple techniques do not provide the same degree of error checking as the more elaborate schemes. For example, the simplest error detection technique is simple parity. It adds a single bit to a character of data, but catches the fewest number of errors. At the other end of the spectrum is the most elaborate and most effective technique available today—cyclic redundancy checksum. Although it is more complex and typically adds 16 bits of error detection code to a block of data, it is the most effective error detection technique ever devised. Let's examine both of these error detection techniques and evaluate the strengths and weaknesses of each.

Parity checks

Parity checks are the most basic error detection technique. In the simplest form, parity involves adding a single bit to a single character of data. This error detection scheme is used with asynchronous connections. Although there are more elaborate forms of single-character parity checking, one fact remains constant—they let too many errors slip through undetected. For this reason alone, parity checks are rarely, if ever, used on any kind of serious data transmissions. There are, however, two forms of parity checks that are worthwhile examining: simple parity and longitudinal parity.

Simple parity

Simple parity (occasionally known as vertical redundancy check) is the easiest error detection method to incorporate into a transmission system and comes in two basic forms: even parity and odd parity. The basic concept of parity checking is the addition of another bit to a string of bits. With **even parity**, another bit is added to produce an even number of binary 1s. With **odd parity**, another bit is added to produce an odd number of binary 1s. If you use the seven-bit ASCII character set, you can add a parity bit as the eighth bit. For example, if you assume even parity and wish to transmit the character "k," which is 1101011 in binary, you would add a parity bit of 1 to the end of the bit stream, as follows: 11010111. (If you are using odd parity, you would add a 0 at the end resulting in 11010110.)

Now the eight-bit stream contains an even number of 1s. If a transmission error causes one of the bits to be flipped (the value is erroneously interpreted as a 0 instead of a 1, or vice versa), the error can be detected, assuming the receiver understands it needs to check for even parity. For example, if you send 11010111, and 01010111 is received, the receiver counts the 1s and sees that there is an odd number, which indicates an error.

What happens if we send 11010111 with even parity and two bits are corrupted? For example, 00010111 is received. Will an error be detected? No, an error will not be detected since the number of 1s is still even. Simple parity can detect only an odd number of bits in error per character. Is it possible for more than one bit in a character to be altered due to transmission error? Yes. Isolated single-bit errors occur 50 to 60 percent of the time. Error bursts, with two erroneous bits separated by less than 10 non-erroneous bits, occur 10 to 20 percent of the time.

Note that for every seven bits of data a parity bit is added. Thus, the ratio of parity bits to data bits is 1:7. Simple parity produces high ratios of check bits to data bits, while achieving only mediocre error detection results.

Longitudinal parity

Longitudinal parity, sometimes called longitudinal redundancy check or horizontal parity, tries to solve the main weakness of simple parity in which all even numbers of errors are not detected. To provide this extra level of protection, additional parity check bits are necessary. In this scheme, as you can see in Table 6-2, individual characters are grouped together in a block. Each character (also called a row) in the block has its own parity bit. In addition, after a certain number of characters are sent, a row of parity bits, or a block character check, is included. Each parity bit in this last row is a parity check for all the bits in the column above. If one bit is altered in Row 1, the parity bit at the end of Row 1 signals an error. In addition, the parity bit for the corresponding column signals an error. If two bits in Row 1 are flipped, the Row 1 parity check will not signal an error, but two column parity checks will signal an error. If two bits are flipped in Row 1 and two bits are flipped in Row 2, and the errors occur in the same column, as shown in Table 6-3, no errors will be detected.

Table 6-2 *Simple example of longitudinal parity*

	Data							Parity
Row 1	1	1	0	1	0	1	1	1
Row 2	1	1	1	1	1	1	1	1
Row 3	0	1	0	1	0	1	0	1
Row 4	0	0	1	1	0	0	1	1
Parity Row	0	1	0	0	1	1	1	0

Table 6-3 *The second and third bits in Rows 1 and 2 have errors, but longitudinal parity does not detect the error*

	Data							Parity
Row 1	1	ꓕ0	ꓳ1	1	0	1	1	1
Row 2	1	ꓕ0	ꓕ0	1	1	1	1	1
Row 3	0	1	0	1	0	1	0	1
Row 4	0	0	1	1	0	0	1	1
Parity Row	0	1	0	0	1	1	1	0

Although longitudinal parity provides an extra level of protection by using a double parity check, this method also introduces a higher level of check bits to data bits. If N characters in a block are transmitted, the ratio of check bits to data bits is $N + 8 : 7N$. In other words, to transmit a 20-character block of data, you need to add a simple parity bit to each of the 20 characters, plus a full 8-bit block character check at the end. As with simple parity, longitudinal parity produces a high ratio of check bits to data bits, while achieving only mediocre error detection results.

Cyclic redundancy checksum

The cyclic redundancy checksum (CRC), or cyclic checksum, is one of the few cases in the field of computer science in which you get *more* than you paid for. The simple parity and longitudinal parity techniques of error detection produce high ratios of check bits to data bits, while achieving only mediocre error detection results. The **cyclic redundancy checksum (CRC)** method typically adds either 16 or 32 check bits to potentially large data packets and approaches 100 percent error detection.

The CRC error detection method treats the packet of data to be transmitted (the message) as a large polynomial. The right-most bit of the data becomes the x^0 term, the next data bit is the x^1 term, and so on. If a bit in the message is 1, the corresponding polynomial term is included. Thus, the data 101001101 is equivalent to the polynomial

$$x^8 + x^6 + x^3 + x^2 + x^0$$

(Since any value raised to the 0th power is 1, you could replace the x^0 term with 1.) The transmitter takes this message polynomial and, using polynomial arithmetic, divides it by a given generating polynomial. A **generating polynomial** is an industry-approved bit string that is used to create the cyclic checksum remainder. This remainder is attached to the end of the transmitted data. When the data plus remainder arrive at the destination, the same generating polynomial is used to detect an error. Some common generating polynomials that are in wide-spread use include:

- ▶ CRC-12: $x^{12} + x^{11} + x^3 + x^2 + x + 1$
- ▶ CRC-16: $x^{16} + x^{15} + x^2 + 1$
- ▶ CRC-CCITT: $x^{16} + x^{15} + x^5 + 1$
- ▶ CRC-32: $x^{32} + x^{26} + x^{23} + x^{22} + x^{16} + x^{12} + x^{11} + x^{10} + x^8 + x^7 + x^5 + x^4 + x^2 + x + 1$
- ▶ Asynchronous Transfer Mode CRC: $x^8 + x^2 + x + 1$

Dividing the message polynomial by the generating polynomial results in a quotient and a remainder. The quotient is discarded but the remainder (in bit form) is appended to the end of the original message polynomial, and this combined unit is transmitted over the medium.

The receiver divides the incoming data (the original message polynomial plus the remainder) by the exact same generating polynomial that was used by the transmitter. If no errors were introduced during data transmission, the division should produce a remainder of zero. If an error was introduced during transmission, the altered original message polynomial plus the remainder will not divide evenly by the generating polynomial and will produce a non-zero remainder, signaling an error condition.

In real life, the transmitter and receiver do not perform polynomial division with a software program. Instead, hardware designed into an integrated circuit can perform the process much more quickly.

The CRC method is almost goof proof. Table 6-4 summarizes the performance of the CRC technique. In cases where the size of the error burst is less than r+1, where r is the degree of the generating polynomial, error detection is 100 percent. For example, if CRC-CCITT is used, the highest power of the polynomial, or the degree, is 16. If the error burst is less than r+1 or 17 bits in length, CRC will detect it. Only in cases where the error burst is greater than or equal to r+1 bits in length is there a chance that CRC may not detect the error. Nonetheless, the chance or probability of an error burst of size r+1 being caught is $1-(\frac{1}{2})^{(r-1)}$. Assuming again that r = 16, $1-(\frac{1}{2})^{(16-1)}$ = 1- 0.0000305, which equals 0.999969. Thus, the probability of a large error being detected is very close to 1.0 (100 percent).

Table 6-4 *Error detection performance of cyclic redundancy checksum*

Type of Error	Error Detection Performance
Single bit errors	100 percent
Double bit errors	100 percent, as long as the generating polynomial has at least three 1s (they all do)
Odd number of bits in error	100 percent, as long as the generating polynomial ends with a Factor of + 1) (they all do)
An error burst of length < r+1	100 percent
An error burst of length = r+1	probability = $1 - (\frac{1}{2})^{(r-1)}$
An error burst of length > r+1	probability = $1 - (\frac{1}{2})^{r}$

Details ▶

Cyclic Redundancy Checksum Calculation

Hardware performs the process of polynomial division very quickly. The basic hardware used to perform the calculation of CRC is a simple register that shifts all data bits to the left every time a new bit is entered. The unique feature of this shift register is that the left most bit in the register feeds back around at select points. At these points, the value of this fed-back bit is exclusive-ORed with the bits shifting left in the register. Figure 6-7 shows a schematic of a shift register used for CRC. You can see from the figure that where there is a term in the generating polynomial, there is an exclusive-OR (indicated by a plus sign in a circle ⊕) between two successive shift boxes. As a data bit enters the shift register (is shifted in on the right), all bits shift one position to the left. Before a bit shifts left, if there is an exclusive-OR to shift through, the left-most bit currently stored in the shift register wraps around and exlusive-ORs with the shifting bit. Table 6-5 shows an example of CRC generation using the message 1010011010 and the generating polynomial $x^5 + x^4 + x^2 + 1$—not a standard generating polynomial, but one created for this simplified example.

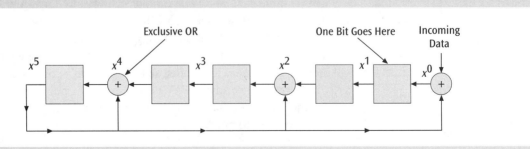

Figure 6-7 *Hardware shift register used for CRC generation*

Compared to parity checking, cyclic redundancy checksum detects almost 100 percent of all errors. Recall that parity checking, depending on whether it is simple parity or longitudinal parity, can only detect between 50 percent and approximately 80 percent of errors. You can perform manual parity calculations quite quickly, but the hardware methods of cyclic redundancy checksum are also very quick. As you saw earlier, parity requires a high number of check bits per data bits. In contrast, cyclic redundancy checksum requires that a remainder-sized number of check bits (either 8, 16, or 32, as you can see from the list of generating polynomials) be added to a message. The message itself can be hundreds to thousands of bits in length. Therefore, in cyclic redundancy, a small number of check bits can perform error detection on a large number of data bits.

Cyclic redundancy checksum is a very powerful error detection technique and should be seriously considered for all data transmission systems. Indeed, all local area networks use CRC techniques (CRC-32 is found in Ethernet LANs), the Internet uses a 16-bit CRC, and most other wide area network protocols incorporate a cyclic checksum.

0 ⊕ 0		0 ⊕ 0		0 ⊕	Incoming data ↓
0	0	0	0	1	1
0	0	0	1	0	0
0	0	1	0	1	1
0	1	0	1	0	0
1	0	1	0	0	0
1	1	1	0	0	1
0	1	1	0	0	1
1	1	0	0	0	0
0	0	1	0	0	1
0	1	0	0	0	0
1	0	0	0	0	0
1	0	1	0	1	0
1	1	1	1	1	0
0	1	0	1	1	0
Remainder 1	0	1	1	0	0

Table 6-5 *Example of CRC generation using shift register method*

Error Control

Once an error is detected in the received data transmission stream, what is the receiver going to do about it? The receiver can do one of three things:

- ▶ nothing;
- ▶ return a message to the transmitter asking it to resend the data packet that was in error;

  or
- ▶ correct the error.

Let's examine each of these options in more detail.

Do nothing

The first error control option—doing nothing—hardly seems like an option at all. Yet, doing nothing for error control is becoming a mode of operation for some newer wide area network transmission techniques. For example, frame relay, which has only been in existence since 1994 offered by telephone companies to transfer data over wide areas, supports the do nothing approach to error control. If a data frame arrives at a destination and after performing the cyclic checksum, an error is detected, the frame is simply discarded. The rationale behind this action is twofold. Frame relay networks are created primarily of fiber optic cable. Since fiber optic cable is the least prone to generating errors, it is assumed that the rate of errors is low and that error control is unnecessary. If a frame is in error and is discarded, frame relay assumes that the higher layer application that is using frame relay to transmit data will keep track of the frames and will notice that a frame has been discarded. It would then be the responsibility of the application to request that the dropped frame be retransmitted.

Return a message

The second option—sending a message back to the transmitter—is probably the most common form of error control. Returning a message was also one of the first error control techniques. Over the years, three basic versions of this technique have emerged: **stop-and-wait ARQ**, **go-back-N ARQ**, and **selective-reject ARQ**. The name of each technique includes the acronym ARQ, which stands for Automatic Repeat reQuest. Go-back-N ARQ and selective-reject ARQ are part of another protocol called the **sliding window protocol**. To better understand Go-back-N ARQ and Selective-reject ARQ, it will be necessary to take a look at the Sliding window protocol. Let's look at the stop-and-wait ARQ protocol first.

Stop-and-wait ARQ

Stop-and-Wait ARQ is an error control technique usually associated with a class of protocols also called stop-and-wait. This protocol and its error control technique are the simplest and thus most restrictive. Station A transmits one packet of data to Station B, then stops and waits for a reply from Station B. If the packet of data arrives without error, Station B responds with a positive acknowledgment, such as ACK. If the data arrives with error, Station B responds with a negative acknowledgment, such as NAK or REJ (for reject). When Station A receives an ACK, it transmits the next data packet. If Station A receives a NAK, it resends the previous data packet. Figure 6-8 demonstrates a sample dialog between Station A and Station B using stop-and-wait ARQ.

Figure 6-8

Sample dialog using stop-and-wait ARQ

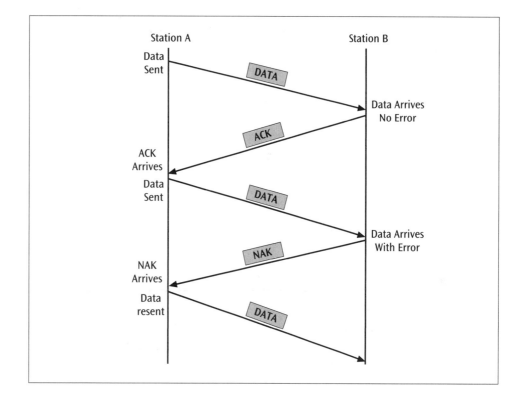

Four things can happen when a packet is transferred between the two stations. First, if a packet is transmitted and it arrives uncorrupted, the receiving station returns an ACK. Second, if the packet arrives corrupted, the receiving station returns a NAK.

Third, a packet arrives at Station B uncorrupted, Station B transmits an ACK, but the ACK is lost or corrupted. Since Station A must wait for some form of acknowledgment, it will not be able to transmit any more packets. After a certain amount of time (called a **timeout**), Station A resends the last packet. If this packet arrives uncorrupted at Station B, Station B will not know that it is the same packet as the last one received. To avoid this confusion, the packets are numbered 0, 1, 0, 1, and so on. If Station A sends packet 0, and the ACK for packet 0 is lost, Station A will resend packet 0. Station B will notice two packet 0s in a row (the original and a duplicate) and deduce that the ACK from the first packet 0 was lost.

Fourth, Station A sends a packet, but the packet is lost. Since the packet did not arrive at Station B, Station B will not return an ACK. Since Station A does not receive an ACK, it will timeout and resend the previous packet. For example, Station A sends packet 1, times out, and resends packet 1. If this packet 1 arrives at Station B, Station B responds with an ACK. How does Station A know whether the ACK is acknowledging the first packet or the second? To avoid confusion, the ACKs are numbered, just like the packets are numbered. In contrast to packets (which are numbered 0, 1, 0, 1, and so on), ACKs are numbered 1, 0, 1, 0, and so on. If Station B receives packet 0, it responds with ACK1. The ACK1 tells Station A the *next* packet is expected. Since packet 0 just arrived, packet 1 is expected next.

One of the most serious drawbacks to the simple stop-and-wait error control technique is its high degree of inefficiency. Stop-and-wait error control is a half duplex protocol. Only one station can transmit at one time. The time the transmitting station wastes waiting for an acknowledgment could be better spent transmitting additional packets. There are more efficient protocols than stop-and-wait. One of these protocols is the sliding window protocol, which can incorporate either go-back-N ARQ or selective-reject ARQ.

Sliding window protocol

A **sliding window** protocol allows a station to transmit a number of data packets at one time before receiving an acknowledgment (a Receive Ready, or RR command). When the acknowledgment arrives, it can acknowledge multiple packets.

Sliding window protocols have been around since the 1970s. In the 1970s, computer networks had two important limitations. First, line speeds and processing power were much less than they are today. It was important that a station transmitting data did not send data too quickly and overwhelm the receiving station. Second, memory was more expensive. Network devices had limited buffer space in which to store incoming and outgoing data packets. Because of these limitations, standard sliding window protocols set their maximum window size to 7 packets. A station that had a maximum window size of 7 could transmit only 7 data packets at one time before it had to stop and wait for an acknowledgment. Extended sliding window systems could support 127 packets. In our examples, let's consider the standard system with a maximum window size of 7.

To follow the flow of data in a sliding window system, packets are assigned numbers 0, 1, 2, 3, 4, 5, 6, and 7. Even though the packets are numbered 0 through 7, which is eight different packets, only seven data packets can be outstanding at one time. (The reasoning for this will be seen shortly.) Since a maximum of 7 data packets can be outstanding at one time, it should never be the case that two data packets both numbered 4, for example, will be transmitting at the same time. If a sender has a maximum window size of 7 and transmits 4 packets, it can still transmit 3 more packets before it has to wait for an acknowledgment. If the receiver acknowledges all 4 packets before the sender transmits any more data, the sender's window once again returns to 7.

Consider a scenario in which the sender transmits all 7 packets and stops. The receiver, for reasons unknown, acknowledges only the first 5 packets. On receipt of the acknowledgment, the sender can send up to 5 more packets before it has to stop again.

The acknowledgment that a receiver transmits to the sender is numbered. In the sliding window scheme, acknowledgments always contain a value equal to the number of the *next expected* packet. For example, if the sender transmits 3 packets numbered 0, 1, and 2, and the receiver wishes to acknowledge all of them, the receiver will return an acknowledgment with value 3, since packet 3 is the next packet the receiver expects. Figure 6-9 demonstrates this scenario.

Figure 6-9
Example of sliding window

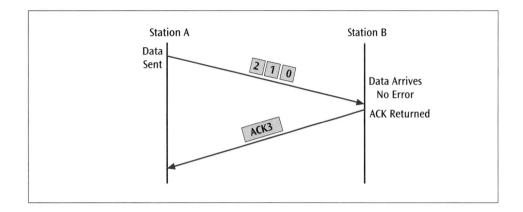

Let's go back now and consider what would happen if the protocol had allowed 8 packets to be sent at one time. Assume the sender sends packets numbered 0 through 7. The receiver receives all of them and acknowledges all by sending an acknowledgement numbered 0 (the next packet expected). What if none of the packets had arrived? If the sender asks the receiver what is the next packet expected, the receiver would answer with 0. The sender would never know if that meant all the packets were received or none of the packets were received.

Go-back-N ARQ

Now, let's add error control to the sliding window protocol. Two basic forms of error control are found with sliding window protocols: Go-back-N ARQ and Selective-reject ARQ. **Go-back-N ARQ** simply informs the transmitting station to go back to the Nth numbered data packet and resend it along with *all* of the data packets that were sent after it. For example, if a station transmits packets 4, 5, 6, and 7, and packet 5 arrives corrupted, the receiving station will return a go-back-N ARQ message (REJect) with the value 5. Upon receipt, the transmitting station will resend packets 5, 6, and 7.

Selective-reject ARQ

Selective-reject ARQ is a more efficient technique. If packet N arrives corrupted, the receiving station will return a selective-reject ARQ message asking for *only* packet N to be resent. If a receiving station receives a number of packets and one of those packets is in error, the receiving station needs sufficient memory space to hold the uncorrupted packets and insert the resent packet into the proper order.

To more closely examine Go-back-N ARQ and selective-reject ARQ, let's try to consider all the possible things that can go wrong. Packets can be garbled and lost. Acknowledgments can be garbled and lost. Go-back-N or selective-reject commands can be garbled or lost. Let's list all the possible things that can go wrong followed by the corrective action taken.

What happens when nothing goes wrong? Figure 6-10 shows the scenario. Station A starts by transmitting 4 packets numbered 0, 1, 2, and 3. Station B receives them and sends an RR command acknowledging all 4. Station A responds by sending 5 more packets numbered 4, 5, 6, 7, and 0. Station B acknowledges the first 3 (packets 4, 5, and 6). After a timeout, Station A wonders what has happened to the acknowledgment for the remaining 2 packets (packets 7 and 0) and sends a Receive Ready command. Station B then responds with an acknowledgment for the remaining 2 packets.

Figure 6-10
Normal transfer of information between stations

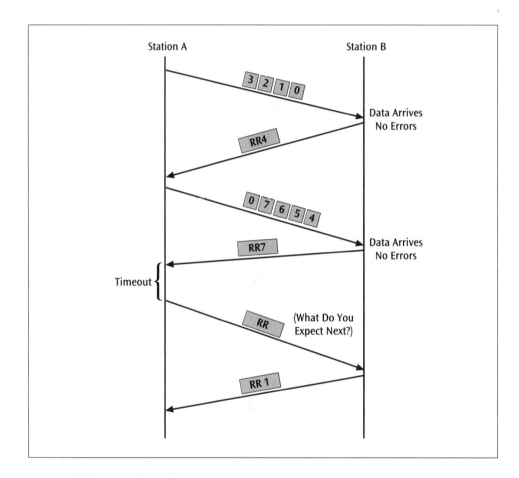

What happens when packets are lost? Figure 6-11 illustrates the situation in which Station A transmits 5 packets but the last two are lost. Station B acknowledges only the first three packets. After a timeout, Station A sends an RR command to inquire about the remaining packets. Station B sends an acknowledgment saying it still expects packet 3 next (RR3). Station A resends packets 3 and 4.

Figure 6-11
Two packets are lost

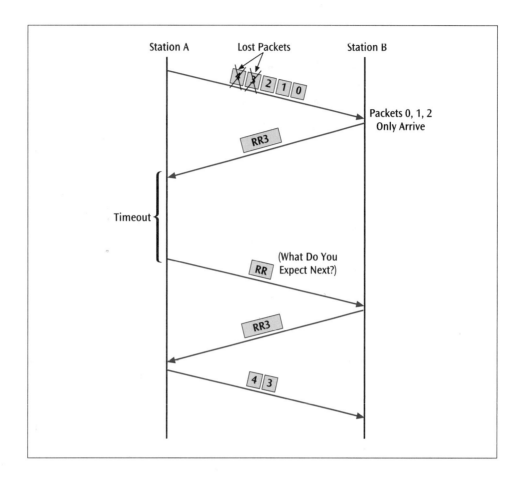

In the second scenario, packets were lost from the end of the transmission. What happens if packets are lost from the middle of a transmission? Figure 6-12 illustrates. Station A sends packets 4, 5, and 6, but packet 5 is lost. Since Station B received packet 6 but not packet 5, Station B sends a REJ command saying it is still expecting packet 5.

Figure 6-12
Station B responds to a lost packet

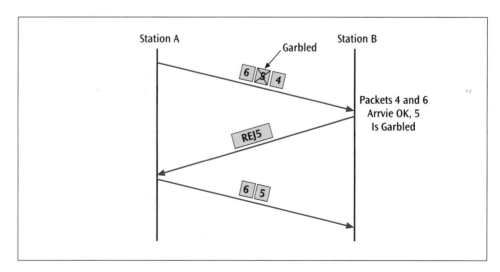

CBT: For a visual example of sliding window ARQ error control, see the enclosed CD-ROM.

What happens if an acknowledgment (RR) is lost? There are two possible scenarios. If a lost RR command is followed shortly by another RR command that does not get lost, no problem should occur, since the acknowledgments are cumulative (the second acknowledgment will have an equal or later packet number). If an RR command is lost and is not followed by any subsequent RR commands, the transmitting station will eventually timeout and ask the receiving station for the number of the most recent packet that was received. The receiving station will then respond with the appropriate RR command.

Correct the error

The beginning of this section on error control listed three things a receiver can do if an error packet is deemed corrupted: do nothing, return an error message, or correct the error. Correcting the error seems like a reasonable solution. The data packet has been received, and error detection logic has determined that an error has occurred. Why not simply correct the error and continue processing? Unfortunately, correcting an error is not that simple. For a receiver to be able to fix an error, called **forward error correction**, additional information must be present so that the receiver knows which bit or bits are in error and what their original values were. For example, if you were given the data 0110110 and informed that a parity check had detected an error, can you determine which bit or bits are corrupted? No, you do not have enough information.

Detect + Correct on the Fly

Details ▶

Forward Error Correction and Hamming Distance

For a data code to perform forward error correction, redundant bits must be added to the original data bits. These redundant bits allow a receiver to look at the received data and, if there is an error, recover the original data using a consensus of received bits. As a simple example, if three bits are received with the values 101, the 0 bit is assumed to be a 1, since the majority of bits are 1. To understand how these redundant bits are created, you need to examine the **Hamming distance** of a code, which is the smallest number of bits by which character codes differ. The Hamming distance is a characteristic of a code. To create a self-correcting, or forward error correcting code, you must create a code that has the appropriate Hamming distance.

Using the ASCII character set as an example, the letter B in binary is 1000010 and the letter C is 1000011. The difference between B and C is one bit—the rightmost bit. If you compare all the ASCII characters, you will find that some pairs of characters differ by one bit, and some differ by two or more bits. Since the Hamming distance of a code is based on the *smallest* number of bits by which character codes differ, the ASCII character set has a Hamming distance of 1. If a character set has a Hamming distance of 1, it is not possible to detect errors nor correct them. Ask yourself the following question: If a receiver accepts the character 1000010, how does it know for sure it is the letter B and not the letter C with a 1 bit error?

When a parity bit is assigned to ASCII, the Hamming distance becomes 2. Since the rightmost bit is the parity bit, and assuming even parity, the character B becomes 10000100, and the character C becomes 10000111. Now, the last two bits of B have to change from 00 to 11 for the character B to become the character C, and the difference between the two characters is two bits. When the Hamming distance is 2, it is possible to detect single bit errors, but you still cannot correct errors. If the character B is transmitted, but one bit is flipped by error, a parity check error occurs. However, you still cannot tell what the intended character was supposed to be. For example, the character B with even parity added is 10000100. If one bit is altered, such as the second bit, the binary value is now 11000100. This character would cause a parity check, but what was the original character? Any one of the bits could have changed, thus allowing for many possible original characters.

You can correct single bit errors only when the Hamming distance of a character set is 3. Achieving a Hamming distance of 3 requires an even higher level of redundancy. A very simple example of how redundant bits can be used to provide error correction is shown in Figure 6-13. The figure shows a 4-bit code with the data bits labeled D_3, D_5, D_6, and D_7. Added to these four data bits are three parity check bits: C_1, C_2, and C_4. C_1 provides even parity for D_3, D_5, and D_7. C_2 provides even

To see the full extent of the problem, consider what would happen if you transmitted three identical copies of each bit. For example, you would transmit 0110110 as 000 111 111 000 111 111 000. Now, let's corrupt one bit: 000 111 111 001 111 111 000. Can you determine which bit was corrupted? If you assume that only one bit has been corrupted, you can apply the majority rules principle and determine the error bit. In this example, forward error correction entails, transmitting *three times* the original amount of data, and it provides only a small level of error correction. This level of overhead limits the application of forward error correction.

Despite the additional costs of using forward error correction, are there applications that would benefit from the use of this technology? Two major groups of applications can reap a benefit: applications that send sensitive data and applications that send data over very long distances. Any organization that sends highly sensitive information such as financial data, health records, or government data does not want to transmit the same record multiple times. Despite the fact that anyone sending sensitive information uses encryption software, every time a record is sent it is vulnerable to interception. Thus, if a record is transmitted, arrives in error, and is corrected without retransmission, it will decrease the likelihood of interception.

Likewise, if data has to be sent over a long distance, it costs time and money to retransmit a packet if it arrives in error. For example, the time required for NASA to send a message to a Martian probe is several minutes. If the data arrives garbled, it

parity for D_3, D_6, and D_7. C_4 provides even parity for D_5, D_6, and D_7. The seven bits are then transmitted in the order C_1, C_2, D_3, C_4, D_5, D_6, and D_7. If you wish to transmit the data 0101, the seven bits actually transmitted would be 0100101. If data bit D_3 is corrupted during transmission, C_1 an C_2 will detect parity errors. If data bit D_5 is corrupted, C_1 and C_4 will detect parity errors. If data bit D_6 is corrupted, C_2 and C_4 will detect parity errors. If data bit D_7 is corrupted, C_1, C_2, and C_4 will detect parity errors. If only C_1 detects a parity error, then C_1 itself must be in error, since there are always at least two C bits that denote an error on a D bit. This situation also applies to C_2 and C_4. The various combinations of C parity bits will point directly to the D bit that is in error.

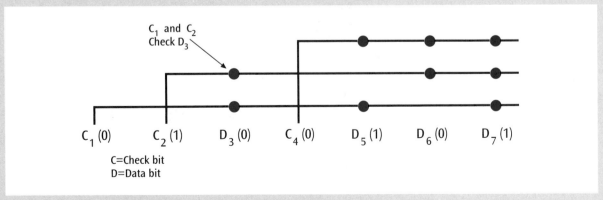

Figure 6-13 *With a Hamming code, redundant bits (C_1, C_2, and C_4) provide error correction*

will be another several minutes before the negative acknowledgment is received, and another several minutes before the data can be retransmitted. If a large number of data packets arrive garbled, transmitting data to Mars could be a very long, tedious process.

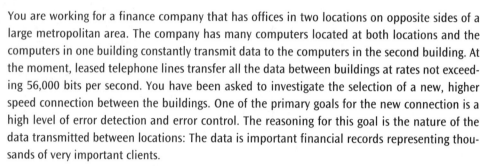

Error Detection and Error Control in Action

You are working for a finance company that has offices in two locations on opposite sides of a large metropolitan area. The company has many computers located at both locations and the computers in one building constantly transmit data to the computers in the second building. At the moment, leased telephone lines transfer all the data between buildings at rates not exceeding 56,000 bits per second. You have been asked to investigate the selection of a new, higher speed connection between the buildings. One of the primary goals for the new connection is a high level of error detection and error control. The reasoning for this goal is the nature of the data transmitted between locations: The data is important financial records representing thousands of very important clients.

You have been handed a very short list of possible technologies that can be used for this high speed connection: asynchronous transfer mode (ATM) or frame relay. You know relatively little about these technologies other than they both are services offered by information service providers, such as telephone companies, and they can transfer data at millions of bits per second. ATM is unique in that it was designed to carry most types of traffic—real-time data, computer data, music, and video. Frame relay is best suited for computer data and voice, but can also carry other types of traffic. A primary question remains: How do these two technologies support error detection and control?

Without going into heavy technical details of each technology, you begin by examining ATM. ATM, like any complex communications protocol, is composed of layers. Each layer is designed to support one or more functions that another layer can use. The lower ATM layers insert a cyclic redundancy checksum into a cell (the name for a packet of ATM data at this layer). ATM performs a cyclic redundancy checksum, which is the best error detection method, on the data within a cell. However, upon closer examination, you see that the checksum remainder comes after the cell's control fields and before the cell's data and is only used to check for errors in the control fields. Thus, this checksum cannot be used to detect errors in the data. Interestingly, this checksum can perform forward error correction as well as error detection, eliminating the need for error control messages, but once again only on data within the control fields.

After further examining ATM, you see there is a second layer to the protocol. This next higher layer has control information that is used to support the handling of lost or incorrectly directed cells (data packets) and a cyclic checksum that is applied to the data portion of the transmission. If an error is detected by this second cyclic checksum, is there an error control procedure? The error control procedure depends on the type of data being transmitted. For example, if the data is real-time data, asking a receiver to retransmit garbled data is useless. The whole idea of real-time data is that it arrives in real-time, such as live video or live audio. Thus, ATM has two levels of error detection: one on the control information and one on the overall data packet. The error detection performed on the control information includes forward error correction. The error detection on the data packet includes error control but only on certain types of data.

Upon examination of the layers of frame relay, you see that a cyclic redundancy checksum is applied to the data portion of a frame relay frame (or packet). However, there is no error control in frame relay. Perhaps it would be better to say that frame relay's error control technique is Do Nothing. If a frame arrives at a destination and the cyclic checksum detects an error, the frame is simply discarded. No information is sent to anyone regarding the deleted frame. It is completely up to a higher layer protocol or application to keep track of all the frames transmitted. The easiest way to keep track of all the frames transmitted is to sequentially number each transmitted frame. As the frames arrive, the application follows the sequential numbers. If there is a number missing, it is the responsibility of the application to ask the transmitter to retransmit a particular frame. Thus, frame relay has good error detection, but it has no error control.

In conclusion, both ATM and frame relay provide error detection and the error detection is performed by cyclic redundancy checksum. ATM has two levels of error detection and can provide forward error correction on control information and error control on certain types of data. Frame relay has one level of error detection and no error control. Error control is the responsibility of a higher layer protocol or application. This is acceptable, as long as someone selects a higher layer protocol or application that can work in conjunction with frame relay to provide error control. Otherwise, frame relay may not be an acceptable choice for your high risk data.

◆ ◆

SUMMARY

▸ It is a fact of computer network life that noise is always present, and if the noise level is too high, errors will be introduced during the transmission of data.

▸ The types of noise include white noise, impulse noise, crosstalk, echo, phase jitter or timing jitter, delay distortion, and attenuation.

▸ Only impulse noise is considered a non-continuous noise, while the other forms of noise are continuous.

▸ Among the techniques for reducing noise are proper shielding of cables, telephone line conditioning or equalization, using modern digital equipment, using digital repeaters and analog amplifiers, and observing the stated capacities of media.

▸ Other reasonable options for reducing the possibility of errors include reducing the number of devices in a transmission stream, decreasing the length of cable runs, or reducing data transmission speed.

▸ Two basic forms of error detection are parity and cyclic redundancy checksum.

▸ Simple parity adds one additional bit to every character and is a very simple error detection scheme that suffers from low error detection rates and high numbers of check bits to data bits.

▸ Longitudinal parity adds an entire character of check bits to a block of data and improves error detection, but still suffers from inadequate error detection rates and high numbers of check bits to data bits.

▶ Cyclic redundancy checksum is a superior error detection scheme with almost 100 percent ability to recognize corrupted data packets. Calculation of the checksum remainder is fairly quick when performed by hardware, and it adds relatively few check bits to potentially large data packets.

▶ Once an error has been detected, there are three possible options: do nothing, return an error message, and correct the error.

▶ Doing nothing is used by some of the newer transmission technologies, such as frame relay. Frame relay assumes that fiber optic lines will be used, which significantly reduces the chance for errors. If an error does occur, a higher layer protocol will note the frame error and will perform some type of error control.

▶ The option of returning an error message to the transmitter is the most common response to an error and uses stop and wait protocols and sliding window protocols.

▶ A stop and wait protocol allows only one packet to be sent at a time. Before another packet can be sent, the sender has to receive a positive acknowledgment.

▶ A sliding window protocol allows multiple packets to be sent at one time. A receiver can acknowledge multiple packets with a single acknowledgment.

▶ Go-back-N ARQ is used in conjunction with sliding window and informs the sender to go back and resend the Nth packet along with all the following packets.

▶ Selective-reject ARQ is used in conjunction with sliding window and informs the sender to go back and resend only the Nth packet.

▶ Error correction is a possibility if the transmitted data contains enough redundant information so that the receiver can properly correct the error without asking the transmitter for additional information. This form of error control requires a high amount of overhead and is only used in special applications in which the retransmission of data is not desirable.

KEY TERMS

attenuation	jitter
crosstalk	longitudinal parity
cyclic redundancy checksum (CRC)	odd parity
delay distortion	selective-reject ARQ
echo	simple parity
even parity	sliding window protocol
forward error correction	stop-and-wait ARQ
generating polynomial	timeout
go-back-N ARQ	white noise
impulse noise	

REVIEW QUESTIONS

1. What is white noise?

2. What is impulse noise?

3. What is crosstalk?

4. What is echo?

5. What is jitter?

6. Which of the noises introduced in this chapter are continuous and which are non-continuous?

7. Will proper shielding of a media increase or decrease the chance of errors?

8. What is the difference between even parity and odd parity?

9. What is the ratio of check bits to data bits for simple parity?

10. What is the ratio of check bits to data bits for longitudinal parity?

11. What types of errors will simple parity not detect?

12. What types of error will longitudinal parity not detect?

13. What is a generating polynomial?

14. What types of errors will cyclic checksum not detect?

15. Frame relay practices which form of error control?

16. How many packets can be sent at one time using stop-and-wait ARQ?

17. What is the function of an ACK in the stop-and-wait ARQ?

18. What is the function of a NAK in the stop-and-wait ARQ?

19. In communication systems what does timeout mean?

20. What are the two window sizes found in most sliding window formats?

21. What is the difference between go-back-N ARQ and selective-reject ARQ?

22. What is necessary for error correction to be performed?

EXERCISES

1. Which is the most difficult type of noise to remove from an analog signal? Why?

2. Which is the most difficult type of noise to remove from a digital signal? Why?

3. Explain the relationship between twisted pair wires and crosstalk.

4. Given the character 0110101, what bit will be added to support even parity?

5. Generate the parity bits and longitudinal parity bits for even parity for the characters 0101010, 0011010, 0011110, 1111110, and 0000110.

6. In a stop-and-wait ARQ system, Station A sends packet 0, and it is lost. What happens next?

7. In a sliding window system with go-back-N ARQ, Station A sends packets 4, 5, 6, and 7. Station B receives them and wants to acknowledge all of them. What does Station B send back to Station A?

8. In a sliding window system with go-back-N ARQ, Station A sends packets 0, 1, 2, 3, 4, 5, and 6. Packet 0 arrives corrupted. What does Station B send back to Station A? What is Station A's response?

9. In a sliding window system with selective-reject ARQ, Station A sends packets 0, 1, 2, 3, 4, 5, and 6. Packet 0 arrives corrupted. What does Station B send back to Station A? What is Station A's response?

10. With a sliding window protocol and a maximum window size of 7, a three-bit field is used to hold the window size. A three-bit value gives 8 combinations, from 0 to 7. Why then can a transmitter send out only a maximum of 7 packets at one time and not 8?

11. Is stop-and-wait ARQ a half duplex protocol or a full duplex protocol?

12. Devise a code set for the digits 0 to 9 that has a Hamming distance of 2.

THINKING OUTSIDE THE BOX

1 Your company is transmitting 500-character (byte) records at a rate of 400,000 bits per second. You have been asked to determine the most efficient way to transmit these records. If a stop-and-wait protocol is used, how long will it take to transmit 1000 records? If a sliding window protocol is used with a window size of 7, how long will it take to transmit 1000 records? If on the average one record in every 200 records is garbled, how long will it take to transmit 1000 records using selective-reject ARQ? What are your conclusions?

2 Your boss has heard a lot about cyclic redundancy checksum but is not convinced that it is much better than simple parity. You quote some of the numbers and probabilities of error detection but they mean little. If you could give a more concrete example of how good cyclic redundancy checksum is, you might win the boss over. Recalling the probabilities given in Table 6-4, if your company transmits a continuous stream of data at 128,000 bits per second, how much time will pass before cyclic redundancy checksum lets an error slip through?

PROJECTS

1. What other error detection techniques are available? How do they compare to parity and cyclic redundancy checksum?

2. What is intermodulation distortion? What sort of signals are susceptible to this form of distortion?

3. Find an example of a real Hamming code and describe how it works.

4. Describe two situations in which error free transmission is crucial to communications.

7

Local Area Networks: The Basics

◆ ◆

If you drop something very small on the floor in the front seat of your car and get down on your hands and knees to look for it, you might look up under the dashboard. And if you look up, you will see a mass of wires and connectors. This jungle of wires makes it very difficult to perform any repairs under the dashboard. As automobiles become more and more complex, including services such as satellite navigation systems, advanced audio and video components, and an increasing array of other electric and electronic convenience features, this wiring situation will continue to worsen. Wouldn't it be interesting if all the wires under the dashboard could be replaced with a single set of wires, or a bus? Then, like adding a workstation to a local area network bus, you could add a new electronic option to your automobile simply by connecting it to the bus. One company, OKI Electric Industry Co., Ltd., is helping the industry move in just this direction.

OKI Electric Industry Co., Ltd. recently unveiled a new high speed integrated circuit (chip) that speeds communications in internal automobile networks. Called the OKI MSM9225, this new chip controls the signals that are passed between the electronic devices within an automobile—anti-lock brakes, the engine, air bags, and the suspension system. "As automobiles become increasingly intelligent, high speed data transmission on in-vehicle networks will become increasingly vital." Thus, the automobile is slowly being transformed the way business offices, school computer labs, and home business offices are being transformed with the introduction of local area networks of computers and supporting devices.

OKI Electric Industry Company press release, March 17, 1999.

http://www.oki.co.jp/OKI/Home/English/New/OKI-News/1000/z9886e.html

If your office, your home, and your automobile all include local area networks, what will be next?

Are local area networks really that vital to everyday computing?

Objectives ▶

After reading this chapter, you should be able to:

▶ State the definition of a local area network.

▶ List the functions and application areas of a local area network.

▶ Cite the advantages and disadvantages of local area networks.

▶ Identify the physical and logical local area network topologies.

▶ Specify the different medium access control techniques.

▶ Describe the common local area network systems, such as Ethernet, token ring, and FDDI.

Introduction

Microsoft Didn't invent the Local Area Network

August 81 IBM PC.

A **local area network (LAN)** is a communication network that interconnects a variety of data communicating devices within a small geographic area and broadcasts data at high data transfer rates with very low error rates. There are several points in this definition that merit a closer look. The phrase *data communicating devices* can include computers such as personal computers, workstations, and mainframe computers. The phrase can also include peripheral devices such as disk drives, printers, and modems. Data communicating devices could also include items such as motion and heat sensors, as well as motors and speed controls. These latter devices are often found in manufacturing environments where assembly lines and robots are commonly found.

Within a *small geographic area* usually implies that a local area network can be as small as one room, or can extend over multiple rooms, multiple floors within a building, and even multiple buildings within a single campus area. The most common geographic areas, however, are within a room or multiple rooms within a single building.

Finally, local area networks differ from many other types of networks in that most *broadcast* their data to all the workstations connected to the network. Thus, if one workstation has data it wishes to send to a second workstation, that data is transmitted to all workstations on the network. When data arrives at a workstation that is not the requested recipient, the data is simply discarded. As you will see later in this chapter, it is possible to design a local area network that does not transmit data to all workstations, thus reducing the amount of traffic on the network.

Two additional phrases in the definition of a local area network—*high data transfer rates* and *very low error rates*—used to separate local area networks from other forms of networks but are no longer as significant as the other phrases. These two phrases are no longer as significant because many wide area networks are now transmitting data at rates equal to or higher than many local area networks with equally low error rates. The primary reason for this increased performance is fiber optic cable. Many wide area network technologies, such as frame relay and asynchronous transfer mode, can transfer data over fiber optic cable for long distances at millions and hundreds of millions of bits per second.

Perhaps the strongest characteristic of a local area network is its ability to share hardware and software resources. For example, a special version of a popular word processing program can be purchased and installed in one location on a local area network. When any user of the local area network wishes to use that word processor, a copy of the software is transmitted over the local area network to the desired workstation where it is loaded and executed. Similarly, a high-quality printer can be installed on the network, and all users can share access to this relatively expensive peripheral.

Since the local area network first appeared in the 1970s, its use has become widespread in commercial and academic environments. It would be very difficult to imagine a collection of personal computers within a computing environment that does not employ some form of local area network. Even computer users at home are beginning to install simple local area networks to interconnect two or more computers. Again, one of the driving forces for a home local area network is to share peripherals such as printers and modem connections to the Internet. To better understand this phenomenon, it is necessary to examine several "layers" of local area network technology. This chapter begins by discussing the basic layouts or topologies of the most commonly found local area networks. We then discuss the medium access control protocols that allow a workstation to transmit data on the network. An examination of most of the common local area network products such as Ethernet and token ring follows. Chapter Eight will introduce the common techniques for interconnecting two or more local area networks so that data may be shared between multiple local area networks and between a local area network and a wide area network. Finally, Chapter Nine will introduce the software that operates on local area networks, including the ever-so-important network operating system. Before we begin examining the basic topologies, let's discuss the common functions and the advantages and disadvantages of local area networks.

Functions of a Local Area Network

File server

Novell

To better understand the capabilities of a local area network, let's examine their typical functions and applications. The majority of users expect a local area network to perform the following functions and provide the following applications: file serving, database and application serving, print serving, electronic mail, video transfers, process control and monitoring, and distributed processing.

A local area network, with the aid of a workstation that has a large storage disk drive, can act as a central storage repository, or **file server**. For example, if the local area network offers access to a commercial spreadsheet application, the network stores the spreadsheet software on the file server and transfers a copy of it to the appropriate workstation on demand. As a second example, suppose two or more users wish to share a data set. Once again the network stores this data set on the file server, providing access to those users who have the appropriate permissions.

Similar to file serving, a local area network can act as a repository for one or more databases and multiple applications. A user sitting at a workstation connected to a local area network can perform a database query or request an application from a server located on the local area network.

The second most common function of a local area network is to provide access to one or more high quality printers. The local area network software, called a **print server**, can provide the authorization to access a particular printer, accept and queue print jobs, print the jobs with the appropriate cover sheets, and provide users access to the job queue in order to provide routine administrative functions.

Most local area networks can provide the service of sending and receiving electronic mail. This electronic mail service can occur both within the local area network and between the local area network and other networks.

A local area network can have the capability to interface with other local area networks, wide area networks (such as the Internet), and mainframe computers. Thus, a local area network can be the glue that holds together many different types of computer systems and networks. For example, it is this glue that allows employees at a company to interact with people external to the company, such as customers and suppliers. If employees wish to send purchase orders to vendors, they enter transactions on their workstations. The transactions travel across the local area network, which is connected to a wide area network. The suppliers are connected to the wide area network through their own local area network, and eventually receive the purchase orders.

Figure 7-1 shows typical interconnections between a local area network and other entities. For example, it is common to interconnect one local area network to another local area network via a device such as a bridge. Equally common is the interconnection of a local area network to a wide area network via a router. Finally, a local area network can connect to a mainframe computer, and the two entities can share each other's resources. (More precise details on local area network interconnections are presented in Chapter Eight.)

Figure 7-1
A local area network interconnecting another local area network, the Internet, and a mainframe computer

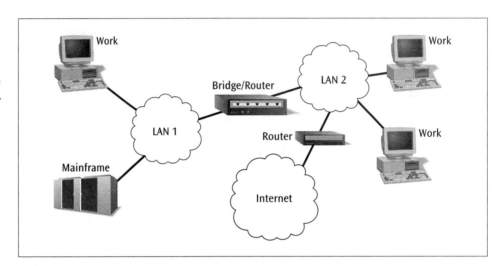

Many higher speed local area networks provide the capabilities for transferring video images and video streams. For example, a local area network could allow a user to transfer high resolution graphic images, transfer video streams, and perform teleconferencing between two or more users.

In manufacturing and industrial environments, local area networks are often used to monitor manufacturing events and report and control their occurrence. The local area network provides process control and monitoring. An automobile assembly line, which uses sensors to monitor partially completed automobiles and control robots for assembly, is an excellent example of local area networks performing process control functions.

Depending on the type of network and the choice of network operating system, a local area network may support **distributed processing**, in which a task is subdivided and sent to remote workstations on the network for execution. The results of these remote executions are then returned to the originating workstation for dissemination or further processing. The distribution of tasks or parts of tasks can lead to an increase in execution speed and can delegate tasks to those computers that are most capable of handling specific chores.

These functions demonstrate that a local area network can be an effective tool in many application areas. One of the most common application areas is the electronic office. A local area network in an electronic office can provide word processing, spreadsheet operations, database functions, electronic mail access, Internet access, electronic appointment scheduling, and graphic image creation capabilities over a wide variety of platforms and at a large number of workstations. Completed documents can be routed to high quality printers, which can produce letterheads, graphically designed newsletters, and formal documents.

A second common application area for a local area network is an academic environment. In a laboratory setting, a local area network can provide students with access to the tools necessary to complete homework assignments, send electronic mail, and interact with the Internet. In a classroom setting, a local area network can transfer to students tutorials and lessons using high-quality graphics and sound. Multiple workstations can provide student-paced instruction as the instructor monitors and records each student's progress at every workstation.

A third common application area for a local area network is manufacturing. Modern assembly lines operate exclusively under the control of local area networks. As products move down the assembly line, sensors control position, robots perform mundane, exacting, or dangerous operations, and product subassemblies are inventoried and ordered. The modern automobile assembly line is a technological tour de force incorporating numerous local area networks and mainframe computers.

Advantages and Disadvantages of Local Area Networks

One of the strongest advantages of a local area network is its ability to share hardware and software resources. Local area networks can share high-quality printers, modems, tape-backup systems, plotters, mass storage systems, and other hardware devices in an economical and efficient manner. On the software end, local area networks can share commercial applications, in-house applications, and data sets over one or all user workstations. Another advantage of local area networks is that an individual workstation can survive a network failure if the workstation does not rely on software or hardware found elsewhere on the network. A further advantage is the fact that component or system evolution, independent of each other, is often possible. For example, if workstations become old and new workstations are desired, it is possible to replace older workstations with the newer ones with few, if any, changes to the network itself. Likewise, if one or more network components become obsolete, it is possible to upgrade the network component without replacing or radically altering individual workstations.

Under some conditions, equipment from different manufacturers can be mixed on the same network. For example, it is possible to create a local area network that incorporates both IBM-type personal computers with Sun workstations and Apple microcomputers. An additional advantage that you have already seen is that local area networks provide a means to access other local area networks, wide area networks, such as the Internet, and mainframe computer systems.

Two former advantages of local area networks in comparison to wide area networks are transfer rates and error rates. Local area networks typically have high data transfer rates and low error rates. Because of these rates, documents can be transferred quickly and confidently across a local area network. Finally, since local area networks can be purchased outright, the entire network and all workstations and devices can be privately owned and maintained. Thus, a company can offer the services it wishes to offer using the hardware and software it deems best for its employees. Interestingly, some companies are beginning to view equipment purchases as a disadvantage. Supporting an entire corporation with the proper computing resources is expensive. It does not help that as a computer reaches its first birthday, there is a newer, faster, and less expensive computer waiting to be purchased.

There are also disadvantages to local area networks. Local area network hardware, the operating systems, and the software that runs on the network can be expensive. Items that require significant funding include the network server, the network operating system, the network cabling system including hubs and switches, the network-based applications, and support and maintenance. Despite the fact that a local area network can support many types of hardware and software, the different types of hardware and software may not inter-operate with one another. If a local area network supports two different types of database systems, users may not be able to share data between the two database systems. Another disadvantage is the potential for purchasing software with the incorrect user license. For example, it is almost always illegal to purchase a single-user copy of software and then install it on a local area network for multiple use. Special licensing agreements must be maintained to protect the software from illegal use.

An important disadvantage that has often been overlooked in the past is that management and control of the local area network requires many hours of dedication and service. A manager, or system administrator, of a local area network should be properly trained and should not assume that the network can support itself with only a few hours of attention per week. Therefore, a local area network requires specialized staff, the right diagnostic hardware and software, and specialized knowledge.

Finally, a local area network is only as strong as its weakest link. For example, a network may suffer terribly if the file server cannot adequately serve all the requests from users of the network. Upon upgrading a server, a company may discover that the cabling is no longer capable of supporting the higher traffic. Upon upgrading the cabling, it may become apparent that the network operating system is no longer capable of performing the necessary functions. Upgrades to part of a network can cause ripple effects throughout the network. Usually the cycle of upgrades continues until it is once again time to upgrade the server.

When you understand the advantages and disadvantages of local area networks, you can see that the decision to incorporate a local area network into an existing environment requires much planning, training, support, and money.

Basic Network Topologies

Local area networks are interconnected using one of four basic configurations, or **topologies:** bus/tree, star, ring, and wireless. The choice of topology is occasionally dictated by the physical environment in which the local area network is to be placed. More than likely, the choice of a topology is determined by other factors such as a preferred access method, data transfer speeds, and brand loyalty. Let's examine each of the four topologies, paying special attention to advantages and disadvantages. It is important to note that the following discussion is about topologies only and not about the local area network access protocol that operates over a topology. For example, a bus topology can support multiple types of access protocols such as Ethernet and token bus. An in-depth introduction to the access protocols follows in a later section.

Bus/tree topology

The **bus topology** was the first topology used when local area networks became commercially available in the late 1970s. As is shown in Figure 7-2, the bus is simply a linear coaxial cable that multiple devices or workstations tap into.

Figure 7-2
Simple diagram of a local area network bus topology

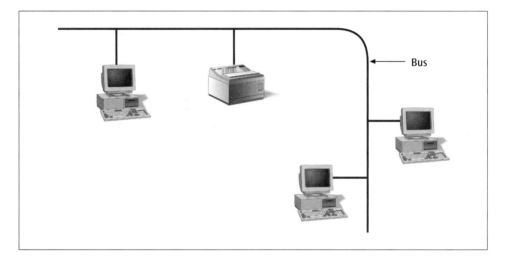

Connecting to the cable requires a simple device called a **tap** (Figure 7-3). This tap is a **passive device** because it does not alter the signal and does not require electricity to operate. On the workstation end of the cable is a network interface card. The **network interface card (NIC)** is an electronic device, typically in the form of a computer circuit board, that performs the necessary signal conversions and protocol operations so that the workstation can send and receive data on the network.

Figure 7-3
Tap used to interconnect a workstation to a LAN cable

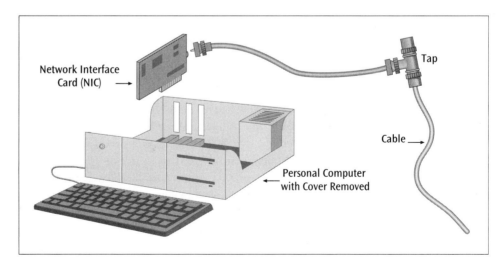

Two different signaling technologies exist when using a bus network: baseband signaling and broadband signaling. (Recall that baseband signaling and broadband signaling were introduced in Chapter Three during the discussion of coaxial cable.) **Baseband** signaling uses a single digital signal (such as a Manchester encoding) to transmit data over the bus. This one digital signal uses the entire spectrum of the cable, allowing only one signal at a time on the cable. All workstations must be aware that someone else is transmitting so they do not attempt to transmit and destroy the signal of the first transmitter. Allowing only one workstation access to the medium at one time is the responsibility of the medium access control protocol, which will be discussed in detail a little later in the chapter.

Another characteristic of baseband technology worth noting is that baseband transmission is **bidirectional**; when the signal is transmitted from a given workstation, the signal propagates in both directions on the cable away from the source (Figure 7-4).

Figure 7-4
Bidirectional propagation of a baseband signal

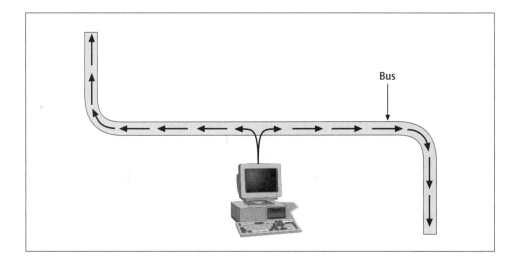

Baseband local area networks are moderately easy to install and maintain, requiring minimal physical maintenance and service. Although many factors influence the maximum number of workstations connected to a local area network, baseband bus local area networks typically have fewer than 100 attached workstations and transmit data at a rate of 10 Mbps (10 million bits per second).

The second type of signaling technology used on the bus local area network is broadband technology. **Broadband** technology uses analog signaling in the form of frequency division multiplexing to divide the available medium into multiple channels. Each channel is capable of carrying a single conversation between two workstations. Since the medium can be divided into multiple channels, broadband signaling allows multiple concurrent conversations.

Broadband signals incorporate analog signals and frequency division multiplexing to separate one channel from another. Since analog signals need to be amplified to maintain an acceptable signal strength over a long distance, broadband transmissions are unidirectional because the amplifiers are unidirectional. (Bidirectional amplifiers do exist but they are more expensive.) Small systems that do not require amplifiers or systems that use bidirectional amplifiers can use frequency modulation to transmit multiple signals in both directions over a single cable. Normally, however, to establish a two-way conversation (either half duplex or full duplex) between sender and receiver, two sets of cables, or their equivalent, are required. Figure 7-5 shows three basic techniques that allow signals to transfer in both directions. In the first example (Figure 7-5(a)), two separate cables are used with one cable transmitting in one direction and the second cable transmitting a signal in the opposite direction. With this technique, each workstation requires two NICs and two taps, one for each cable. In the second example (Figure 7-5(b)), a single cable leaves the workstation and eventually forms a loop that returns to the workstation. Each workstation still requires two NICs and two taps, one for each transmission direction. In the final example (Figure 7-5(c)), a single cable transmits multiplexed signals. Two different channels, each with its own set of frequencies, are used for the two-way transmission of signals. Note that no amplifiers are used in this third system.

Figure 7-5
Three different techniques to allow a two-way conversation using broadband signaling

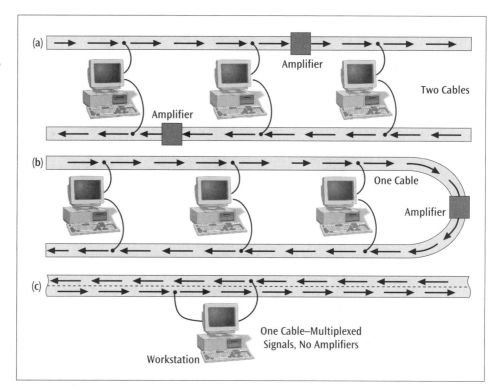

Several characteristics of broadband systems make them an attractive alternative in the proper situation. Broadband signals are relatively easy to amplify, a trait that allows a broadband network to extend for 100s to 1000s of meters supporting 100s to 1000s of users. And because broadband signals have a wide bandwidth from which multiple channels can be created, broadband can support data, video, and radio signal transmissions. In fact, a broadband bus system is the only local area network topology that allows multiple concurrent channels.

It is also possible to split and join broadband cables and signals to create configurations more complex than a single linear bus. These more complex bus topologies that consist of multiple cable segments that are all interconnected are termed **trees.**

Figure 7-6 shows an example of a tree local area network. Attached to Bus 1 is a second cable, Bus 2. The two cables are attached via some form of bridging device that allows only the appropriate traffic to pass from one bus to another. The third bus is attached to the second bus in a similar fashion. Under normal circumstances, traffic that originates on Bus 2 remains on Bus 2. However, if a workstation attached to Bus 2 requests information from a workstation on another bus, the appropriate bridging device will forward the message.

Figure 7-6
Simple example of a broadband tree topology

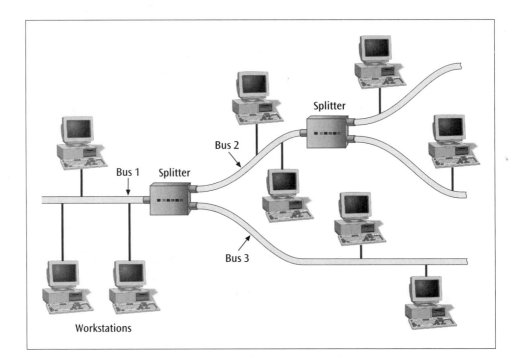

A popular example of a broadband transmission system is cable television. Coaxial cables carry analog signals with a wide range of frequencies supporting many channels. As the signals travel through the neighborhoods, the cables split into smaller segments until they arrive at their final destination.

As you might expect, broadband networks have disadvantages. Broadband networks are more susceptible to noise and are more expensive than baseband networks to maintain because broadband networks are composed of analog frequencies. A specialist is often needed to monitor a broadband network periodically to ensure that the analog frequencies of newly added channels or modifications to the network do not affect the already existing channels.

Because broadband transmissions are unidirectional, the average propagation delay is twice that of baseband technology. Consider a city that is composed of only one-way streets. While occasionally a one-way street is going in the direction you desire, many times you have to drive out of your way to reach your destination. Returning to your original location requires you to find yet another route, different from the original route.

All bus networks—whether broadband or baseband—share a major disadvantage: in general, it is difficult to add a new workstation if no tap currently exists. Since there is no tap, the cable has to be cut and a tap has to be inserted. Cutting the cable and inserting a tap disrupts the traffic on the network and is a somewhat messy job. If possible, it is best to anticipate where workstations will be and have the installation team install all the necessary taps in advance. As you might expect, trying to predict the exact number and location of taps is virtually impossible. Bus-based networks have lost popularity with the introduction of other topologies. The star-wired bus, or simply star topology, is one of those other topologies.

Star-wired bus topology

Test question

The most popular configuration for a local area network is the **star-wired bus topology,** or simply, star topology. This topology should not be confused with an older topology, also called the star topology. The older star topology supported a local area network called the StarLAN in which one computer at the center of the star controlled the transmissions of all the other workstations. Today's modern star topology acts like a bus but looks like a star. To be a little more precise, it *logically* acts as a bus but it *physically* looks like a star. The **logical design** of a network is how the data moves around the network from workstation to workstation. The **physical design** is how the network would appear if drawn on a large sheet of paper.

Logical – look like a bus

Physical – looks like a star

In a star-wired bus topology all workstations connect to a central device called a hub, as you can see in Figure 7-7. The **hub** is an unintelligent device that simply and immediately retransmits received data out all connections. When any workstation transmits data to the hub, the hub sends a copy of the data to *all* other workstations (or devices) connected to the hub. All workstations hear the transmitted data because there is only a single transmission channel, and all workstations are using this one channel to send and receive. Sending data to all workstations and devices generates a lot of traffic but keeps the operation very simple because there is no routing to a particular workstation. Thus, when viewing its logical design, the star-wired bus is acting as a bus; when a workstation transmits, everyone immediately receives the data. The physical design however, is a star because the devices are connected to the hub and radiate outward in a star-like pattern.

Figure 7-7
Simple example of a star-wired bus topology for a local area network

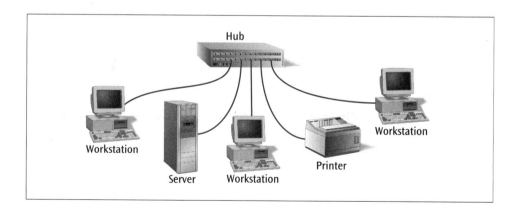

Twisted pair cabling has become the preferred medium, although it is possible to use coaxial cable and fiber optic cable, especially as a connector between multiple hubs. The connectors on the ends of the cables are simple-to-use modular RJ-45 connectors. The RJ-45 connector is very similar to the modular connector that connects a telephone to the wall jack. Because of the star-wired bus' twisted pair cable and modular connectors, it is much simpler to connect workstations to a star-wired bus than to a coaxial-cabled bus.

The hub comes in a variety of designs. One of the most common is the 24-port hub, which will interconnect 24 workstations or devices. If more than 24 workstations are desired, it is fairly simple to interconnect two or more hubs. Figure 7-8 demonstrates that to interconnect two hubs, you simply run a cable from a special connector on the front or rear of the first hub to a special connector on the front or rear of the second hub. Many hubs support multiple types of media for this inter-hub connection—twisted pair, coaxial cable, and fiber optic cable.

Figure 7-8

Interconnection of two hubs in a star-wired bus local area network

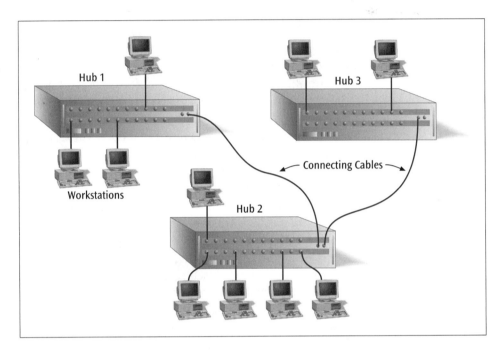

When two or more hubs are interconnected and a workstation transmits data, *all* workstations connected to *both* hubs receive the data. As you can see, the hub is a relatively non-intelligent device. It does not filter out any data frames and it does not perform any routing. A local area network in which all workstations immediately hear a transmission is an example of a **shared network**. In Chapter Eight, we will examine the opposite of a shared network—a switched network.

Hubs Are Not for segmenting networks.
Hubs Are used to inter connect PC in A specefic segment

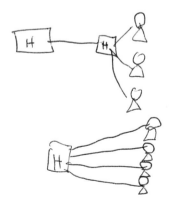

The many advantages of a star-wired bus topology include simple installation and maintenance, low cost components such as hubs and twisted pair wiring, and high volume of compatible products due to major market share. Perhaps the only disadvantage of a star design is the hub and the amount of traffic it must handle. Later you will see that the hub can be replaced with a more advanced device, called a switch, that can reduce the amount of traffic on the network.

Ring topology

The **ring topology**, when viewed logically, is a circular connection of workstations, as Figure 7-9 demonstrates. Since ring topologies support baseband signals, the ring is capable of supporting only one channel of information. This channel of information flows in one direction around the ring, moving from workstation to workstation. Since the ring is a closed loop of wire, it is important that some device removes a circling piece of data from the ring, otherwise, the piece of data will keep circling. The device that removes the data is the workstation that originally transmitted the data.

Figure 7-9
Ring topology viewed logically

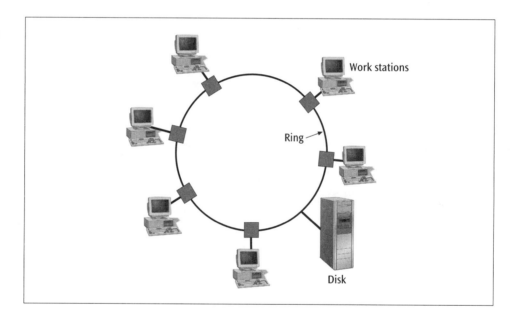

Similar to bus and star-wired bus LANs, each workstation is attached to the ring via a network interface card (NIC). One of the key components of a ring NIC is a repeater. A **repeater,** as shown in Figure 7-10, performs three basic functions:

> ► Bypass, in which data does not copy to the workstation. The bypass function is used for inactive nodes, which are workstations that are turned off or malfunctioning.

> ► Copy, in which data on the ring is copied to the workstation. This function allows a workstation to listen to the data on the ring and accept the data that is addressed to this particular workstation.

> ► Write, in which data from a workstation is copied onto the ring. The write function allows a workstation to place data onto the ring. This data is intended for the next workstation downstream and will circle around the ring to the receiving workstation and eventually back to this originating workstation.

In addition, the repeater regenerates the incoming digital signal into a clean digital signal as it leaves the repeater.

Figure 7-10
Three possible operations of the workstation repeater on a ring topology

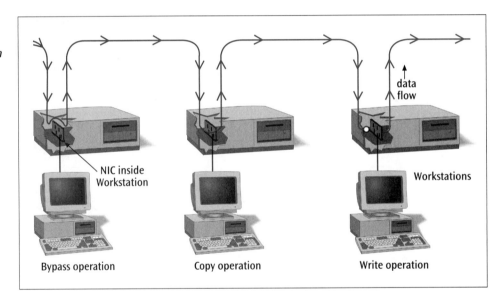

The repeater is very fast; there is only a 1-bit delay from when a bit enters the repeater until it leaves the repeater. That is, as the third bit is entering the repeater, the first bit is leaving. Since the data is leaving the repeater almost as fast as it enters, the flow of data experiences almost no delay. Despite a potentially large number of active workstations, the transfer speed of data is not significantly slowed by passing through each repeater in a ring topology.

MAU

Although logically a ring is organized as a circular connection of workstations, the physical organization of a ring is not circular. Physically, a ring looks much like a star design with all workstations connected to a central device (Figure 7-11). This central device is not a hub but a multi-station access unit. A **multi-station access unit (MAU)** accepts data from a workstation and transmits this data to the next workstation *downstream* in a ring fashion (Figure 7-12). Thus, it operates a bit differently from a hub in that every connection does not immediately receive a copy of the incoming data. If a workstation is not connected to a particular port on the MAU, that port simply closes itself so that a continuous ring is maintained. Some people call a ring topology based on MAUs a star-wired ring topology.

Figure 7-11
Physical organization of the ring topology

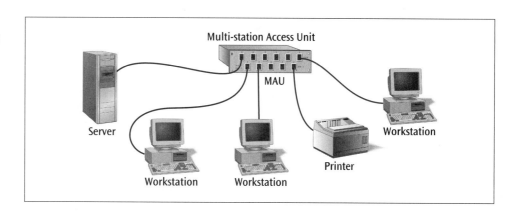

Figure 7-12
Multi-station access unit on a ring topology

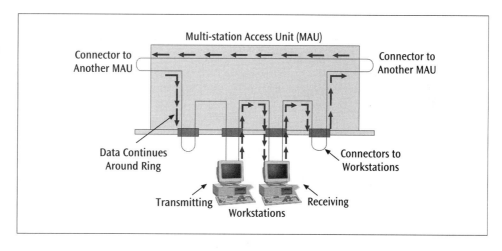

Like hubs in the star design, it is possible to interconnect multiple MAUs to extend the size of a ring local area network (Figure 7-13). As the data passes around the ring in the first MAU, it encounters the connector to the second MAU. The signal then passes over the cable to the second MAU and begins its journey around the ring in the second MAU. When all workstations have been accessed on the second MAU, the signal passes again over the cable and returns to the first MAU.

Figure 7-13
Interconnection of multiple MAUs on a ring topology

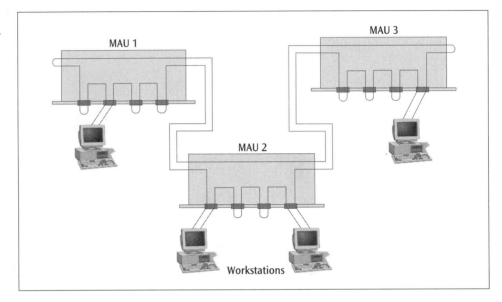

The ring topology has many of the same advantages as the star-wired bus topology because it is easy to install new workstations and easy to maintain. Ring networks contain mostly inexpensive twisted pair wire, especially between the workstation and the MAU. It is not unusual, however, to find coaxial cable or fiber optic cable interconnecting multiple MAUs.

Rings have very few disadvantages. This lack of disadvantages might make you wonder why the ring local area network is not the most common topology. As will be shown shortly, other factors have had a major impact on the acceptance of ring local area networks.

Wireless LANs

Wireless topology, as you might suspect, does not follow any set physical pattern. (You could question that if it doesn't follow a set physical pattern, is it even a topology?) By attaching a transmitter/receiver to a special network interface card on a workstation or laptop and the similar hardware on a device called an access point, it is possible to transmit data between a workstation and network server at speeds up to 20 Mbps. A workstation can be located anywhere within the acceptable transmission range. This acceptable range varies with the wireless technology used, but typically varies between 50 and 800 feet.

The acceptable transmission range of wireless local area networks is broken into two areas. The first area, called the **basic service set,** resembles a cell in a cellular telephone network and is the transmission area surrounding a single access point. A collection of basic service sets forms the second area, the **extended service set**. The extended service set appears as one logical local area network. Note how all of the basic service sets attach to the local area network through an access point and together create the extended service set. Figure 7-14 shows the differences between the basic service set and the extended service set.

Figure 7-14
Basic service set and extended service set of a wireless local area network

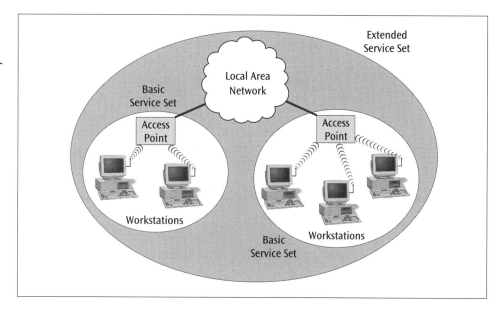

Details ▶

IEEE 802.11 Wireless LANs

IEEE approved the standard for wireless local area networks in June 1997 and labeled it 802.11. The specification defined three different types of physical layer connections to match price, performance, and operations to a particular application. The first type of physical layer defines infrared transmissions. Infrared, as you might recall from Chapter Three, is line of sight and cannot pass through solid walls. Transmission rates for infrared wireless range from 1 to 2 Mbps. The second type of physical layer defines a direct sequence spread spectrum technology and can transmit data at rates up to 2 Mbps in quiet environments and 1 Mbps in noisy environments for distances up to 800 feet. The third type of physical layer is also spread spectrum but uses frequency hopping and can transmit data at 2 Mbps for 300 feet.

There are two interesting and useful characteristics of the 802.11 standard that are worth considering. The first is that the OSI layers of the 802.11 standard work seamlessly with existing Ethernet. Thus, it is relatively straight-forward to connect a wireless workstation to an existing wired network. The second characteristic is that the 802.11 standard defines three different types of wireless workstations, depending on the degree of mobility. A no-transition workstation is either stationary or moves, but remains within the transmission range of the basic service set. A BSS-transition workstation can move from one basic service set to another basic service set. An ESS-transition workstation can move completely out of one extended service set into another extended service set.

Most wireless LAN systems transmit at approximately 2 to 11 Mbps. Although a transmission of 2 Mbps is not fast compared to wire-based local area networks (10 Mbps to 100 Mbps is fairly common), the convenience of not having to run wires can be extremely attractive in certain environments. Consider installing a local area network workstation in the middle of a large warehouse, in a historical building that would lose authenticity if holes were drilled and wires pulled, or in a building with marble walls that would make drilling holes very difficult. In all of these examples, a wireless local area network would be preferable to a wired network. Another good environment for a wireless network is a **nomadic** application in which a workstation is moving around. Clearly, a wire would not be an acceptable solution if a workstation needs to move. In addition, since there is no cabling between a workstation and a hub, installation of new workstations is very simple, as long as the workstation is located within the acceptable transmission range of the wireless signals.

The number of concurrent channels available with wireless local area networks depends on the technology incorporated. The simpler wireless schemes, which have all wireless workstations within the same room and use infrared transmission signals, all share the same set of frequencies and thus have only two channels—one for transmission to the access point and one for transmission from the access point. The more advanced wireless schemes, which can cover larger geographic areas, may incorporate multiple sets of frequencies, thus creating a system with multiple channels.

When wireless LANs first appeared, organizations were slow to accept them. Approval of the IEEE 802.11 wireless standard (described in the Details section) has greatly helped with standardization of wireless networks and has sped their growth and acceptance. Today, there are many businesses that could not survive without wireless LAN technology. Businesses with large rooms such as stock exchanges and warehouses, and employees who have to move around inside a building have embraced wireless LANs. As the technology improves and speeds and distances increase, the demand for wireless LANs will also increase.

Comparison of bus, star-wired bus, ring and wireless topologies

Table 7-1 summarizes the characteristics of the bus, star-wired bus, ring, and wireless topologies. Note that the topology of a local area network is only one part of the whole picture. To properly compare local area networks, you must also consider the medium access control protocols that operate over the topologies, as well as other factors such as transmission speeds, network operating systems, and equipment costs. In spite of working with an incomplete picture, you can still see some interesting profiles emerging from the three topologies. The broadband bus and wireless LAN technologies are the only ones that use analog signals. All the others use digital or baseband signals. The original bus topology is a linear wire, while star-wired buses and rings have a central hub-like device. Finally, broadband bus is the only technology that allows multiple concurrent channels.

Table 7-1
Comparison of the bus, star, ring, and wireless topologies

Characteristic	Baseband Bus	Broadband Bus	Star-Wired Bus	Ring	Wireless
Signaling technique	Digital	Analog	Digital	Digital	Analog
Physical layout	Linear	Linear	Central	Central	None
Usual media type	Coaxial cable	Coaxial cable	Twisted pair	Twisted pair	None
Installation ease	Moderate	Moderate	Easy	Easy	Easy
New workstation installation	Hard (if no tap available)	Hard (if no tap available)	Easy (if port available)	Easy (if port available)	Easy
Concurrent channels	No	Yes	No	No	No

Medium Access Control Protocols

A medium access control protocol is part of the software that allows a workstation to place data onto a local area network. Depending on the network's topology, several types of protocols may be applicable. The bottom line with all medium access control protocols is this: since a local area network is a broadcast network, it is imperative that only one workstation at a time be allowed to transmit its data onto the network. In the case of a broadband local area network, which can support multiple channels at the same time, it is imperative that only one workstation at a time be allowed to transmit its data onto *a channel* on the network.

There are three basic categories of medium access control protocols for local area networks:

> ▶ contention-based protocols, such as carrier sense multiple access with collision detection;
> ▶ round robin protocols, such as token passing; and
> ▶ reservation protocols, such as demand priority.

Let's examine each of these protocols.

Contention-based protocols

A **contention-based** protocol is a first-come first-served protocol—the first station to recognize that no one is transmitting data is the first station to transmit. The most popular contention-based protocol is carrier sense multiple access with collision detection (CSMA/CD). The CSMA/CD medium access control protocol is found almost exclusively on bus and star-wired bus local area networks.

The name of this protocol is so long that it almost explains itself. With the **carrier sense multiple access with collision detection (CSMA/CD)** protocol, only one workstation at a time can transmit. Because only one workstation can transmit at a time, the CSMA/CD protocol is a half-duplex protocol. A workstation listens to the medium (senses for a carrier on the medium) to learn whether any other workstation is transmitting. If another workstation is transmitting, the workstation wanting to transmit will wait and try again to transmit. The amount of time the workstation waits depends on the particular type of CSMA/CD proto-col used (see Details accompanying this section). If no other workstation is cur-rently transmitting, the workstation transmits its data onto the medium. In most situations, the data is intended for one other workstation. Nonetheless, all work-stations on the network receive the data. Only the intended workstation (the workstation with the intended address) will do something with the data. All the other workstations will discard the frame of data.

As the data is being transmitted, the workstation continues to listen to the medium, listening to its own transmission. Under normal conditions the worksta-tion should just hear its own data being transmitted. If the workstation hears garbage, it assumes a collision has occurred. A collision results when two or more workstations listen to the medium at the same moment, hear nothing, and then transmit their data at the same moment. In the case of a broadband network, a colli-sion results when two or more workstations transmit data on a particular channel.

The two workstations do not need to begin transmission at exactly the same moment for a collision to occur. Consider a situation in which two workstations are at opposite ends of a bus. A signal propagates from one end of the bus to the other end in time *n*. A workstation will not hear a collision until its data has, on average, traveled halfway down the bus, collided with the other workstation's sig-nal, and then propagated back down the bus to the first station (Figure 7-15). This interval during which the signals propagate down the bus and back is the **collision window**. During this collision window, a workstation might falsely hear no one transmitting and then transmit its data.

Details ▶

Persistence Algorithms

Suppose a workstation wishes to transmit data and listens to the medium. What happens if the medium is busy? The workstation does not transmit, but waits. How long does the workstation wait? What degree of listening persistence does the workstation exhibit? Three different persistence algorithms have been created: non-persistent, 1-persistent, and p-persistent.

With the **non-persistent algorithm**, if the workstation finds the medium is busy, it waits for a random amount of time (*t*), then listens again. What if the medium had become free immediately after the workstation listened and had found it busy? That would be too bad; if the workstation had been more persistent, it would have learned sooner that the medium had become free and it would not have wasted time waiting.

The **1-persistent algorithm** takes this condition into account and listens continuously until the medium is free, then transmits immediately. What happens if two worksta-tions following the 1-persistent algorithm are both listening and waiting? They will probably try to transmit at the same moment and cause a collision.

With the **p-persistent algorithm**, if the medium is busy, the workstation continues listening. When the medium becomes idle, the workstation transmits with probability *p* or delays the standard random amount of time with probability 1-*p*. The p-persistent algorithm is a compromise between non-persistent and the 1-persistent algorithms. The workstation continuously listens until the medium becomes free, but then does not transmit immediately. It transmits only with proba-bility *p*. If *p* = 0.1, then nine times out of ten the workstation waits, and one time out of ten it immediately transmits. The addition of probability to the algorithm makes collisions less likely to occur. The odds are more in our favor that if two or more workstations are waiting for a free medium, both of them will *not* begin transmitting at the exact moment the medium is idle. The selection of the value of p is often deter-mined by the number of workstations on the network. The larger the number, the smaller the p value.

Figure 7-15
Two workstations at opposite ends of a bus experiencing a collision

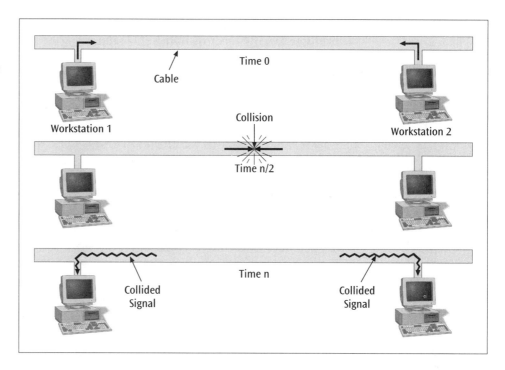

CBT: For a visual example of a frame moving through a CSMA/CD network, see the CD-ROM that accompanies this text.

If the network is experiencing a small amount of traffic, the chances for collision are small. The chance for a collision increases dramatically when the network is under a heavy load and many workstations are trying to access it simultaneously. Studies have shown that as the traffic on a CSMA/CD network increases, the rate of collisions increases, which further degrades the service of the network. If a workstation detects a collision, it will immediately stop its transmission, wait some random amount of time, and try again. If another collision occurs, the workstation will wait once more.

The CSMA/CD access protocol is analogous to humans carrying on a conversation. If no one is talking, someone can speak. If someone is talking, everyone else hears this and waits. If two people start talking at the same time, they both stop immediately (or at least polite humans do) and wait a certain amount of time before trying again.

Since the number of times a workstation will have to wait is unknown, it is not possible to determine exactly when a workstation will be allowed to transmit its data without collision. Thus, CSMA/CD is a non-deterministic protocol. A **non-deterministic protocol** is one in which you cannot calculate the time at which a workstation will transmit. If your application must have its workstations transmit data at known times, you might want to consider a medium access control protocol other than CSMA/CD.

Round robin protocols

The second basic category of medium access control protocols is the round robin protocol. A **round robin protocol** is one in which each workstation takes a turn at transmission and the turns are uniformly distributed over all workstations. Unlike the first-come first-served contention-based protocols, if multiple workstations are waiting to transmit, each workstation will have to wait until its turn comes around. The advantage is that each workstation will eventually get a turn and cannot be forced out if another workstation seizes the communications channel first.

The most popular example of round robin protocol is the token passing protocol. Before a workstation can transmit, it must possess the one and only token. Once the transmission is complete, the workstation releases the token to the next workstation in round robin order. Eventually, the token is passed around to all workstations and returns to the first workstation to begin another cycle. Two types of token passing protocols exist: token ring and token bus. While they both use a token passing algorithm, they have different underlying topologies. Let's introduce the more popular of the two algorithms first—token ring.

Token ring

The **token ring** local area network uses the ring topology for the hardware and a round robin protocol for the software. It operates on the principle that to transmit data onto the ring, your workstation must be currently in possession of a software token. There is only one token in the entire network, so only one workstation may transmit at a time. When a workstation has completed its transmission, it passes the token on to the downstream neighboring workstation. Only the workstation holding the token can transmit, so there is no need for any workstation to listen for a collision while transmitting, because collisions cannot occur.

Collisions are one of the main problems of CSMA/CD. As the number of concurrent users rises, the number of collisions rises. As the collisions rise, more workstations are forced to retransmit their messages and overall throughput declines. Since the token ring does not experience any collisions, overall throughput remains high even under heavy loads. This ability of token ring to give every workstation a turn is attractive and is valuable for applications that require uniform response times. Since the order of transmission by each workstation is known, the wait time to transmit is deterministic rather than non-deterministic, thus token ring is a **deterministic protocol.**

Let's take a look at how the token ring protocol works. Consider Figure 7-16 in which Station A has just released the token. Since Station B is the next downstream neighbor from Station A, and Station B has data to transmit on the ring, Station B seizes the token. After seizing the token, Station B transmits its data, which is destined for Station M. As Station M copies in the data frame, the data continues around the ring until it returns to Station B, which removes the data from the ring. After Station B has removed its data from the ring, it passes the token to Station C.

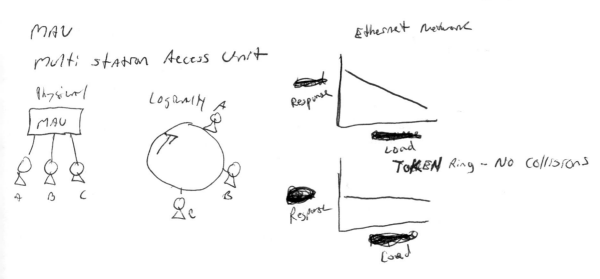

Figure 7-16
Data transmission on a token ring local area network

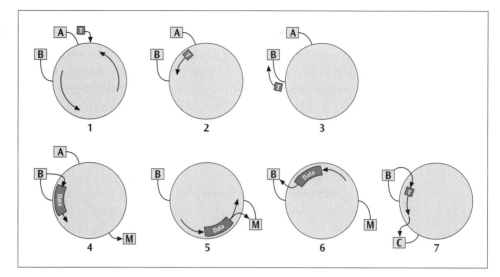

In the token ring protocol, a workstation also has the option of reserving the token. If a workstation does not wish to wait for the token to be passed all the way around the ring and if it has the proper priority, the workstation can reserve the token and seize it before its normal turn. It is debatable whether this feature is an advantage or a disadvantage of token ring.

A major disadvantage of the token ring access protocol is the complexity of the software needed to maintain the token. This software has to address important questions such as:

► What happens if the token disappears? (A workstation does not forward it.)

► If the token disappears, who generates a new token?

► Is it possible for two stations to generate a token, thus resulting in two tokens on the ring?

Although token ring has the definite advantage of being a deterministic protocol and performing quite well under heavy loads, it has had a difficult time competing with CSMA/CD protocol networks. A clear majority of local area networks use CSMA/CD as the medium access control protocol of choice. Some reasons that CSMA/CD is more popular than token ring are as follows.

► CSMA/CD was the first local area network medium access control method, and thus got a good jump on installations and support.

► Token ring local area networks have almost always lagged behind CSMA/CD networks with regard to transmission speed. When CSMA/CD first became popular, the typical transmission speed was 10 Mbps. Token ring, when it first appeared, had a transmission speed of only 4 Mbps. For a while, token ring jumped ahead with a 16 Mbps version, but CSMA/CD caught up with a 100 Mbps version, and now a 1000 Mbps version. Token ring finally announced a 100 Mbps version, but it appears it was too late to save the protocol in the marketplace. Many people feel it will just be a matter of time before token ring fades into the history books.

► CSMA/CD is less expensive to implement, due in part to its widespread marketing and acceptance.

► CSMA/CD is a simpler protocol.

Token bus

The **token bus** local area network was designed to be a deterministic protocol like token ring but to operate on a bus topology, not a ring. For a workstation to transmit data onto the bus, the workstation must possess a software token. Since there is only one token in the network, only one workstation may transmit at a time, thus eliminating collisions. When the workstation has completed its transmission, the token is passed to the neighbor workstation. Each workstation maintains a list of neighbors, thus creating a defined order of token passing. A neighbor need not be a physical neighbor, only a logical neighbor. Thus, the token bus protocol creates a token-passing logical ring on top of a physical bus.

The token bus protocol was designed primarily for manufacturing plants in which a non-deterministic protocol is unacceptable. A manufacturing plant's assembly line dictates a linear bus topology, not a ring or a star. By layering a logical token-passing ring on a physical bus, the token bus protocol is able to provide a deterministic medium access control protocol for use with a linear topology. One rarely, if ever, sees a token bus in any application outside of manufacturing.

Reservation protocols

The third major category of medium access protocols is the reservation protocol, in which a workstation must submit a request for transmission. Reservation protocols do not operate on a first-come first-served basis as CSMA/CD does, nor does it depend on the passing of a token. The most common example of a reservation protocol for local area networks is the **demand priority protocol,** which is based on the IEEE 802.12 standard.

To transmit data using a reservation protocol such as the demand priority system a workstation must ask the central hub for a turn at transmission. When a workstation wants to transmit a frame, the workstation sends a request to a central hub, and then waits for the hub to return permission to transmit. A workstation, during the request, can specify whether the request is a high priority request or a normal priority request. The central hub scans all incoming lines in a round robin fashion. For example, if there are three workstations attached to the central hub and they are numbered 1, 2, and 3, the central hub will first look at the line from workstation 1, then workstation 2, and finally workstation 3. Once all requests have been received, the central hub looks at all the requests and separates the high priority requests from the normal priority requests. The hub services all high priority requests before any normal priority requests. If there are multiple high priority requests, each is handled in similar round robin fashion. Once the high priority requests are serviced, the normal priority requests are handled in round robin fashion. If a high priority request arrives during the time that normal priority requests are being serviced, the normal priority requests are set aside and the high priority request is serviced.

The demand priority algorithm is effective. When multiple workstations are trying to transmit their frames, demand priority acts like a token passing algorithm and services each workstation request in a round robin manner. When few workstations are trying to transmit, demand priority acts more like a CSMA/CD protocol and services each request similar to a first-come first-served manner.

Medium Access Control Sublayer

Although the seven-layer OSI model was designed to support most types of communication systems, it fell short in several areas. For example, broadcast networks did not fit well into the seven-layer model. Broadcast networks, such as local area networks, as well as some radio and satellite systems, have three characteristics that make them significantly different from other communication systems.

First, one of the main concerns of the network layer of the OSI model is routing a message through the network. This routing often involves deciding which station should receive the message next in order to follow some optimal path. Broadcast networks, as the name implies, broadcast a message from a given point to all other stations on the network. Thus, there is generally no need to make routing decisions. Some very large broadcast networks do make routing decisions, but this is the exception, not the rule.

Second, there is a close link between the data link layer and the physical layer of most broadcast networks. For example, if someone wishes to create a local area network with all fiber connections, the frame created in the data link layer will be different than if the medium had remained copper based. Because of this relationship, it is difficult to discuss the data link layer without specifying a particular type of hardware, and vice versa. Recall that the original OSI model was designed to keep each layer separate from the others. A change in the data link layer that precipitates a change in the physical layer violates the original intent of OSI.

Third, broadcast networks differ from other communication systems in that some type of medium access control procedure is necessary to control who talks when on the one and only medium that interconnects the network.

Test For these three reasons, the OSI model was modified, as shown in Figure 7-17. The data link layer has been split into two sublayers: the medium access control sublayer and the logical link control sublayer. The **medium access control (MAC) sublayer** works more closely with the physical layer and contains a header, computer (physical) addresses, error detection codes, and control information. Because of this closeness with the physical layer, there is not a strictly defined division between the MAC sublayer and the physical layer. The **logical link control (LLC) sublayer** is primarily responsible for logical addressing and providing error control and flow control information. Since many of the responsibilities of the network layer do not exist in broadcast networks, the network layer is (figuratively) smaller.

Figure 7-17
Modification of OSI model to split data link layer into two sublayers

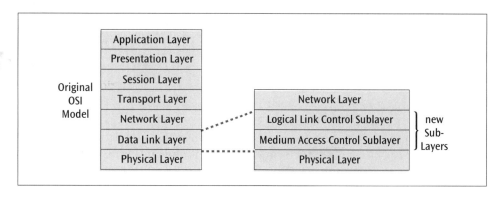

The medium access control sublayer defines the layout of the data frame, called the frame. As you will see in the next section, there are a number of different frame formats depending on the type of local area network. For example, CSMA/CD LANs have one frame format while token ring LANs have another format. Within this frame format are the fields for error detection, workstation addressing, and various control information. Thus, the MAC sublayer is a very important layer when describing a local area network. For this reason, many people refer to one LAN or another by its MAC sublayer.

IEEE 802 Frame Formats

To standardize many of the local area network protocols, IEEE created a series of protocols called the **IEEE 802 suite of protocols**. With the IEEE 802 standards, the frame formats for data at the medium access control (MAC) sublayer were also created. Thus, as data comes down from the application layer through the lower layers of the communications model and arrives at the MAC sublayer, MAC software places the data into a unique frame format, ready for transmission across the medium (the physical layer).

The IEEE 802.3 standard for CSMA/CD uses the frame format shown in Figure 7-18. The preamble and start of frame byte combine together to form an 8-byte flag that the receiver locks onto for proper synchronization. The destination address and source address are the 2- or 6-byte addresses of the receiving computer and sending computer. More precisely, each network interface card in the world has a unique 6-byte (48-bit) address. When CSMA/CD sends data to a particular computer, it creates a frame with the appropriate NIC address of the intended computer. The data length is simply the length in bytes of the following data field. The PAD field adds characters to the frame (pads the frame). The minimum size frame that any station can transmit is 64 bytes long. Frames shorter than 64 bytes are considered **runts**, or frame fragments, that result from a collision and are automatically discarded. Thus, if a workstation wishes to transmit a frame in which the data field is very short, PAD characters are added to ensure that the overall frame length equals at least 64 bytes. Finally, the checksum field is a 4-byte cyclic redundancy checksum.

Figure 7-18
Frame format for IEEE 802.3 CSMA/CD

Preamble	Start of Frame Byte	Destination Address	Source Address	Data Length	Data	Pad	Checksum
7 bytes of 10101010	10101011	2 or 6 bytes	2 or 6 bytes	2 bytes	0–1500 bytes	0–46 bytes	4 bytes

The IEEE 802.5 standard for token ring is shown in Figure 7-19. The first field is the starting delimiter (SD) and is VV0VV000 in binary, where V is a code violation bit. An example of a code violation bit is a Manchester code that does not change in the middle of the bit. Since all bits transmitted using a Manchester code are supposed to change values in the middle, a bit that does not change is a clear violation of the code and can be recognized as something unusual. The second field is access control (AC), and the third field is frame control (FC). Together, these two fields provide priority information along with other control information. The destination and sources address fields are 2- or 4-byte addresses of a unique token ring NIC. The eighth field is the ending delimiter (ED). The value of the ED field is VV1VV1IE in binary, where V is a code violation bit, I is a special control bit called an intermediate frame bit, and E is an error-detected bit that can be set to 1 by a bridge if a checksum error is detected.

Figure 7-19
Frame format for IEEE 802.5 token ring protocol

Starting Delimeter (SD)	Access Control (AC)	Frame Control (FC)	Destination Address (DA)	Source Address (SA)	Data	Checksum	Ending Delimeter (ED)	Frame Status (FS)
1 byte	1 byte	1 byte	2 or 4 bytes	2 or 4 bytes	n bytes	4 bytes	1 byte	1 byte

The final field, the frame status (FS) field, contains the bit sequence ACRRACRR. The A bit is set to 1 by the receiving station if the address for which the frame is intended is recognized. The C bit is set to 1 if the receiving station accepts (copies) the frame. The R bits are reserved for future use. Since a frame is removed from the ring by the sending workstation, the A and C bits can be set by the receiving workstation to inform the sender quickly if the frame was recognized and accepted. Note that there are two A bits and two C bits. The second bit of each is simply a copy of the first to provide a small degree of error protection. For example, if a receiver wants to acknowledge the incoming frame, it will set the two A bits to 1. If one of the A bits is corrupted on the frame's trip back to the source workstation, the source will still see one of the A bits set to 1 and assume that the frame was acknowledged. Why does the FS byte come after (outside of) the checksum field? If the FS byte came before the checksum, and a receiving station changed either the A or C bits, a checksum error would occur. Recall that any data that changes *before* the checksum field will cause a checksum error to result. The receiving station wants to acknowledge receipt of the frame. If the receiving station changes a bit before the checksum field, a checksum error will result. By placing the A (and C) bits after the checksum field, the bits will not affect the outcome of the error detection code.

Even though the details of the IEEE 802 frame formats are fairly technical, a number of interesting observations can be made. First, all frame formats have a NIC address that identifies the sending and receiving workstations. All frame formats also use the powerful cyclic redundancy checksum to check for errors. And all frame formats begin with some type of header to prepare the receiver for incoming data. Unfortunately, there is one big difference between the frame formats—their overall layout. CSMA/CD frames have certain fields that token ring frames do not, and vice versa. Because of these differences, it is not possible to connect a CSMA/CD workstation to a token ring network, or a token ring workstation to a CSMA/CD network. Each type of workstation has to operate on its own type of network. It is possible to connect a CSMA/CD network to a token ring network, but this requires a device that can convert between the two frame formats. This interconnection of networks is discussed in more detail in Chapter Eight.

Local Area Network Systems

The discussion of local area networks started by examining the main types of network topologies: bus, star-wired bus, ring, and wireless. Then the three major categories of medium access control protocols that can operate on these different topologies were introduced: CSMA/CD, token passing, and demand priority. Let's now turn our attention to actual products or local area network systems that are found in a typical computer environment. Four of the most popular local area network systems are Ethernet, IBM token ring, fiber data distributed interface, and 100VG-AnyLAN.

Ethernet

Ethernet was the first commercially available local area network system and remains the most popular system today. It is based on the bus and star-wired bus topologies and uses the CSMA/CD medium access protocol. Since Ethernet is so popular and has been around the longest, it has evolved into a number of different forms. To avoid mass mayhem, the IEEE created a set of individual standards for Ethernet or CSMA/CD local area networks called IEEE 802.3. All the 802.3 standards are summarized in Table 7-2.

Table 7-2
Summary of Ethernet standards

Ethernet Standard	Basic Features
10Base5	10 Mbps baseband signals over coaxial cable, maximum segment length 500 meters
10Base2	10 Mbps baseband signals over coaxial cable, maximum segment length 185 meters
1Base5	1 Mbps baseband signals over unshielded twisted pair, maximum segment length 250 meters
10BaseT	10 Mbps baseband signals over unshielded twisted pair, maximum segment length 100 meters
10Broad36	10 Mbps broadband signals over coaxial cable, maximum segment length 3600 meters
100BaseTX	100 Mbps baseband signals over 2-pair Category 5 or higher unshielded twisted pair, maximum segment length 100 meters
100BaseT4	100 Mbps baseband signals over 4-pair Category 3 or higher unshielded twisted pair, maximum segment length 100 meters
100BaseFX	100 Mbps baseband signals over fiber optic cable, maximum segment length 1000 meters
1000BaseSX	1000 Mbps baseband signals over fiber optic cable, maximum 100 meters
1000BaseLX	1000 Mbps baseband signals over fiber optic cable, maximum 100 meters
1000BaseCX	1000 Mbps baseband signals over fiber optic cable, maximum 100 meters
1000BaseT	1000 Mbps baseband signals over twisted pair cable, maximum 100 meters

The more common 802.3 standards include 10Base5, 10Base2, 1Base5, and 10BaseT. The **10Base5** standard was one of the first Ethernet standards approved. The term Base is an abbreviation for baseband signals using Manchester encoding. Recall that baseband signals are digital signals. Since there is no multiplexing of digital signals on any baseband LANs, there is only one channel of information on the network. The 10 represents a 10 Mbps transmission speed and the 5 represents a 500 meter maximum cable segment length. **10Base2** (nicknamed Cheapernet) was designed to allow for a less expensive network by using less expensive components. The network can transmit 10 Mbps digital signals over coaxial cable but only for a maximum of 185 meters (the value 2 in 10 Base2). **1Base5** was a system designed for twisted pair wiring, but with only a 1 Mbps data transfer rate for 250 meters. Perhaps the most popular Ethernet system (and local area network system overall) is 10BaseT. A **10BaseT** system transmits 10 Mbps baseband (digital) signals over twisted pair for a maximum of 100 meters per segment length. Many businesses, schools, and, now, homes are using 10BaseT as their local area network.

One of the most common standards for broadband (analog) Ethernet is the **10Broad36** specification. Using coaxial cable to transmit analog signals, 10Broad36 can transmit data at 10 Mbps for a maximum segment distance of 3600 meters. Note the much longer distance due to the use of coaxial cable and analog signals. And since broadband signals can support multiple channels, 10Broad36 can deliver many concurrent channels of data, each supporting a 10 Mbps transmission stream.

When 10 Mbps Ethernet was first available, it was a fast protocol for many types of applications. As in most computer-based technologies, it was not fast enough for very long. In response to the demand for faster Ethernet systems, researchers created the 100 Mbps Ethernet. The following 100 Mbps Ethernet standards are called **Fast Ethernet** to distinguish them from the 10 Mbps standards. **100BaseTX** was designed to support 100 Mbps baseband signals using two pairs of Category 5 unshielded twisted pair. Like its 10BaseT counterpart, 100BaseTX was designed for 100 meter segments. It is similar to 10BaseT systems that use twisted pair wiring and hubs with up to 24 workstation connections. **100BaseT4** was created to support older category wire. Thus it could operate over Category 3 or 4 twisted pair wire, as well as the newer Category 5 unshielded twisted pair. It too would transmit 100 Mbps baseband signals for a maximum of 100 meters. Finally, **100BaseFX** was the standard created for fiber optic systems. It would support 100 Mbps baseband signals using two strands of fiber but for much greater distances—1000 meters.

CBT: For the latest in CSMA/CD and Ethernet standards, visit the author's web page at http://bach.cs.depaul.edu/ cwhite

The latest set of Ethernet standards to be developed are based on 1000 Mbps transmission speeds, or 1 gigabit (1 billion bits) per second. These standards define the new **gigabit Ethernet**, which is quickly becoming the newest darling of local area networks. The gigabit standards all use fiber optic cable to achieve the 1000 Mbps speeds. IEEE 802.3z includes three substandards: **1000BaseSX**, which supports the interconnection of relatively close clusters of workstations and other devices, **1000BaseLX**, which is designed for longer distance cabling within a building, and **1000BaseCX**, which is designed for backbone cabling within a building or across a campus. All of these standards transmit 1000 Mbps (1 Gbps) baseband signals. While all of these standards were designed to employ fiber optic cable, a more recent standard termed simply **1000BaseT** is capable of using the new Category 5e cable specification.

IBM token ring

Token ring was popularized by IBM, which remains one the few manufacturers of token ring products. Thus, token ring and IBM token ring are usually two names for the same thing. As its name implies, IBM token ring uses the ring topology and token ring access method described earlier. IEEE 802.5 defines a token ring standard for 4 Mbps, 16 Mbps, and the newer 100 Mbps data transmission rate. The 100 Mbps token ring (IEEE 802.5t) is designed for workstation-to-MAU connections using either Category 5 twisted pair wire or fiber optic cable. Work is currently in progress for a gigabit token ring protocol (IEEE 802.3v), but the standards making committee that is supporting any new versions of token ring has been having difficulty generating interest in the business sector due to the intense competition from CSMA/CD. Because of the decreased demand for new token ring products, it appears likely that the standards making committee will no longer recommend new versions of token ring.

Fiber data distributed interface (FDDI)

Although many people like the token ring local area network for its deterministic protocol and its high throughput under heavy loads, many people were not happy with its slow 4 Mbps and 16 Mbps transmission speeds. In an attempt to marry a deterministic access protocol with a high transmission rate, the **fiber data distributed interface (FDDI) ring** was created. The FDDI protocol resembles a token ring network that has been lifting weights. With a data transmission speed of 100 Mbps, network distances up to 200 km, and a possible interconnection of 500 stations, an FDDI network is a vastly updated token ring network. To achieve these impressive figures several modifications were made to the original token ring design. The first and major modification is the use of fiber optic cable to connect each workstation to the central MAU.

The second modification was to create a topology of a ring within a ring. The FDDI dual ring topology, shown in Figure 7-20, uses an inner ring and an outer ring. The inner ring transmits data in the opposite direction of the outer ring and acts as a backup to the outer ring. A workstation is either a dual-attached workstation and connects to both rings, or is a single-attached station and connects to only one ring.

Figure 7-20
FDDI dual ring topology

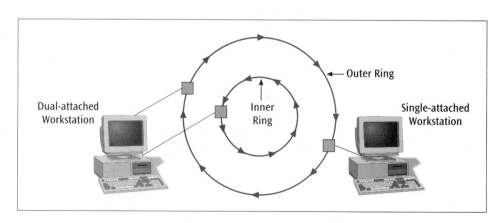

The third modification was to use the newer 4B/5B encoding instead of Differential Manchester encoding. Recall that 4B/5B encoding has the advantage of not having a signal transition in the middle of each bit frame; thus its baud rate is *not* twice the bps. At the higher transmission rate of 100 Mbps, a baud rate double the bits per second would generate a signal that changes 200 million times per second. When transmission speeds reach into the millions of bits per second, a jump from 100 million changes per second to 200 million changes per second is significant. The hardware necessary to support this jump is going to be more expensive.

Finally, the token passing protocol itself was modified. In standard token ring, a workstation seizes the token, transmits its data, removes its data from the ring, and passes the token on to the next workstation. In FDDI, a workstation seizes the token, transmits its data, which may consist of multiple frames, and appends the token to the end of the last data frame. As the data frame travels to the next workstation, the neighbor workstation can strip the token off the end of the neighbor's frame and transmit its own frame. This modification was made because data transmission is so fast and distances are possibly so long that too much time would be wasted using the older method.

During the period when CSMA/CD networks transmitted at 10 Mbps and token ring networks transmitted at 16 Mbps, FDDI networks transmitting at 100 Mbps were clearly the network of choice for high power systems such as real-time applications and bandwidth-intensive graphics applications. High-end applications with powerful workstations and more powerful network servers relied heavily on FDDI for delivering the data in the least amount of time.

Despite the promise of FDDI, it appears once again that the faster versions of Ethernet are providing fierce competition and are driving out the installation of new FDDI systems. Industry experts indicate that FDDI may not be around much longer. Although this could be premature speculation, it is worth noting, particularly if your company decides to invest a large amount of money in a technology that may not be supported in the near future. Once again, the popularity of Ethernet and all its various forms may have forced another communications protocol into near oblivion.

100VG-AnyLAN

100VG-AnyLAN is based on the demand priority access method, which was approved as a local area network standard (IEEE 802.12). 100VG-AnyLAN transmits data at 100 Mbps and borrows elements from both Ethernet and token ring local area networks and thus can support both Ethernet and token ring formats (hence the "Any" designation in 100VG-AnyLAN). It was designed to run on Category 3 twisted pair, or voice grade wire (thus, the VG in 100VG-AnyLAN), but like many twisted pair installations, Category 5 is now the preferred cable.

Similar to the star-wired bus topology, 100VG-AnyLAN's topology is based on a hub topology arranged in a star formation. Similar to round robin protocols such as token ring, requests from workstations to transmit are handled in a round-robin fashion by the root hub using a demand priority access method. Unlike a token ring protocol, 100VG-AnyLan can have two levels of priorities—normal priority and high priority. All high priority requests are serviced before any normal priority requests are serviced.

One of the more interesting topological features of 100VG-AnyLAN is that hubs can be cascaded up to three levels (Figure 7-21). This cascading allows for a high number of workstations—theoretically 1024. Actual implementations have shown that 1024 workstations is not realistic and a maximum of 250 is more reasonable. A special **cascade port** is used to connect a hub to its upstream connection. In Figure 7-21, the cascade port is shown figuratively as the port on the top of the Level 2 and Level 3 hubs. Thus, the Level 1 hub is upstream from the Level 2 hubs, and the cascade port makes that upstream connection.

Figure 7-21
Three levels of the 100VG-AnyLAN topology

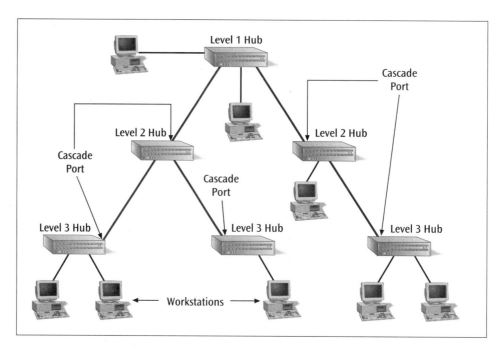

Several other interesting characteristics of 100VG-AnyLAN include the following:

▶ Because 100VG-AnyLAN uses demand priority access, frames with a higher priority are serviced before lower priority frames. High priority frames could include audio and video streams, which operate in real time.

▶ The round robin algorithm of demand priority means that 100VG-AnyLAN uses a deterministic protocol.

100VG-AnyLAN can operate over a wide variety of media types, including 4-pair Category 3 and 4 unshielded twisted pair, 2- or 4-pair Category 5, and 2-strand fiber optic cable.

Although 100VG-AnyLAN is a viable, sound local area network system, it too is having difficulties competing with the latest versions of Ethernet. Ethernet's high transmission speeds, low cost of ownership, and high market share make it difficult to beat.

LANs in Action: A Small Office Solution

Hannah is the computer specialist for a small business on the west side of Chicago. Her company currently has approximately 20 workers, each with his or her own workstation. They use the computers mainly for word processing and spreadsheets. The owner of the company would like to update their computer services by offering the following applications to employees:

- ► electronic mail;
- ► a database that contains information on all their past and present customers;
- ► limited access to the Internet; and
- ► one or two high quality color laser printers.

After hearing these requests from the owner, Hannah concludes that the best way to offer these services is to install a local area network. Since the 20 workstations are only one to two years old, they will need only the appropriate network interface cards and software to connect them to a local area network. The problem is deciding which network to install. After reading the literature and talking to some fellow computer specialists, Hannah creates the following list of possible local area network systems:

- ► 10 Mbps or 100 Mbps CSMA/CD using baseband signaling;
- ► 10Broad36 CSMA/CD using broadband signaling;
- ► Token ring 16 or 100 Mbps;
- ► 100VG-AnyLAN; or
- ► FDDI.

The employees are not running real-time applications, nor will they be very heavy users placing a high demand on the network. If either situation were true, Hannah would give serious consideration to a fast round robin or reservation medium control access protocol that gives equal access to all workstations. These protocols are found in 16 Mbps and 100 Mbps token ring, FDDI, and 100VG-AnyLAN. If she considered one of these protocols, however, her decision would have to be tempered by the fact that these systems are experiencing serious sales growth problems and support for new hardware and software may be difficult to find in the future. None of the employees will be using video applications, which would make systems like broadband CSMA/CD look attractive.

Since her employees are not running real-time or video applications and since baseband CSMA/CD is the most popular and least expensive form of local area network, Hannah decides to install that type of system. She has the choice of installing either a 10BaseT system or the faster 100BaseTX system. 1000 Mbps systems are available but are relatively new, high priced, and more powerful than she needs for the given user applications. Since prices of 100BaseTX equipment are dropping, and 100BaseTX systems are one of the simplest systems to install and maintain, it seems like a reasonable decision to plan for the near future and install a 100BaseT network. Since Hannah is learning about local area networks (the same as you), she figures she better keep things as simple as possible.

Next she has to decide how to configure the system. She knows that the workstations plug into a hub, and that the typical hub supports 24 workstations. Even though she would only need one hub for 20 workstations, the distance between the workstations furthest apart might exceed 100 meters. She also wants to leave some room for future growth. Therefore, she plans for two hubs, a network server, 20 NICs, and the appropriate wiring. The network server is a substantially-powered workstation with 128 Mb RAM, multiple gigabyte hard drives, a tape backup, and a backup power supply should the power go out. Hannah has also decided to install 100 Mbps NICs in each workstation since the price difference between 10 Mbps and 100 Mbps NICs is growing smaller every day.

The office space has the configuration shown in Figure 7-22 and dictates some of Hannah's decisions. Category 5 unshielded twisted pair should work fine as the wiring between workstations and the hubs. To interconnect the two hubs, Hannah decides to install fiber optic cable, because the distance between hubs may be close to the 100 meter twisted pair limit and the cables may have to run past electromagnetically noisy equipment.

Figure 7-22
Office layout for Hannah's company

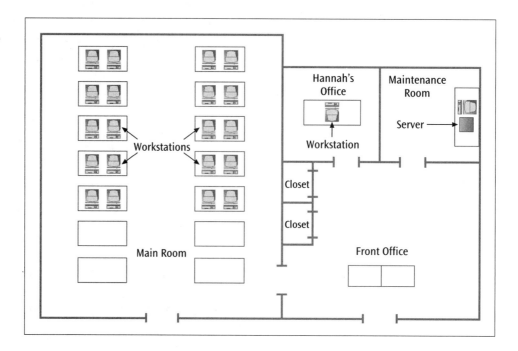

Hannah decides to place one hub in a closet just off the main room where the 20 employees work. This closet can be locked for security and has simple ventilation to keep the equipment cool. She decides to place the second hub and the network server in a small maintenance room adjacent to her office. The maintenance room can also be locked and has ample ventilation to keep the electronic equipment and the room cool. All of the ceilings are false, so the wiring can run up the walls and through conduits between workstations and hubs. The rough wiring diagram should look something like Figure 7-23. In this configuration, all twisted pair wires run for less than 100 meters.

Figure 7-23
Wiring diagram of Hannah's office space showing hub and server placement

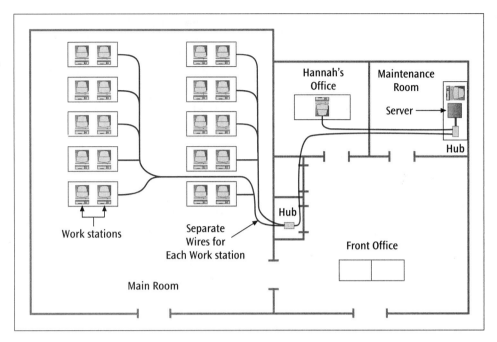

Hannah is off to a good start with setting up a new LAN in her company's small office. In a later chapter you will learn that she also needs to consider routers and similar equipment, as well as the choice of a network operating system.

LANs in Action: A Home Office Local Area Network Solution

Sam has a nice computer system set up at home, but is getting tired of sharing it with his wife and children. Since he is not willing to give up time on the computer, his only other option is to purchase a second computer. Having purchased a second computer, Sam realizes that he does not want to purchase another printer, most certainly does not want to install another telephone line, and does not want to establish another Internet connection with an Internet service provider. Yet both computers have a need for printing and accessing the Internet. The solution for Sam is what some people call a "Network in a Box." A Network in a Box is one of the fastest growing segments of the networking market and is geared towards the small office/home office (SOHO), which has between 2 and 50 users.

To install a Network in a Box solution, Sam first has to make a number of decisions. He has to decide if he wants a system with NICs that plug into ISA or PCI slots in his computers, so he opens the cases on both to see what slots are available. The ISA bus was an early technology that allowed all the components within an IBM personal computer to talk to one another. The PCI bus is a newer technology that allows faster bus transmission speeds. Most IBM-compatible microcomputers offer one or two card slots of both ISA and PCI compatibility. Since Sam also wants to share an Internet connection, he has to make sure he orders a system that has a combination hub and router, since it is the router that allows you to interconnect to the Internet via some form of telephone line. Once he has made all these decisions, he orders the Network in a Box and waits for its arrival.

The Network in a Box includes network interface cards, a router/hub, cabling, and the necessary software to install the system. Installation of the NICs is straightforward, as is installing the cabling and the router/hub. (More details about hubs and routers will be introduced in Chapter Eight.) Installing the software is a little more challenging but not at all difficult. In a relatively short time, Sam has created a system that allows both of his workstations to access the Internet, a high quality printer, and a common disk storage area (Figure 7-24).

Figure 7-24
Sam's Network in a Box solution for his home computer system

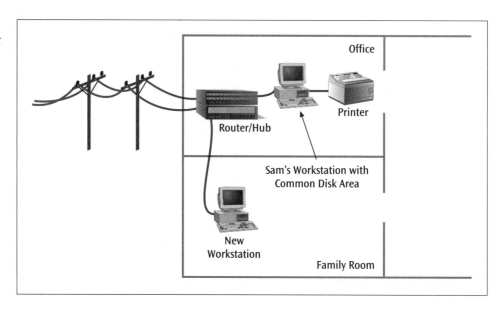

SUMMARY

▷ A local area network is a communication network that interconnects a variety of data communicating devices within a small area and transfers data at high transfer rates with very low error rates.

▷ A local area network enables the sharing of data, software, and peripherals.

▷ Local area networks can be costly and time consuming to maintain.

▷ Several common functions of local area networks include file serving, print serving, support for electronic mail, and process control and monitoring.

▷ A local area network can be configured as a bus/tree topology, a star-wired bus topology, or a ring topology.

▷ A baseband bus topology local area network uses digital signaling and supports one channel.

▷ A broadband bus topology local area network uses analog signaling and can support hundreds of simultaneous channels.

▷ Both baseband and broadband buses can be difficult to expand by adding a new workstation when there is not an available tap.

▷ The star-wired bus topology is a variation on the bus topology, but because it includes a hub has the advantage of easier installation and maintenance.

▷ The ring topology is a circular connection of workstations in which data is passed from workstation to workstation around the ring.

▶ The wireless topology allows a highly flexible placement of workstations and requires no wiring to transmit and receive data.

▶ For a workstation to place data onto a local area network, the network must have a medium access control protocol.

▶ There are three basic forms of medium access control protocols: contention-based (such as CSMA/CD), round robin-based (such as token passing), and reservation-based (such as demand priority).

▶ The most common example of a contention protocol is CSMA/CD, which can be found on bus and star-wired bus local area networks.

▶ CSMA/CD works on a first-come first-served basis and supports half-duplex connections.

▶ The most common example of a round robin protocol is a token passing protocol such as token ring.

▶ The most common example of a reservation protocol is the demand access protocol used to support 100VG-AnyLAN.

▶ CSMA/CD is clearly the most popular access protocol, but it suffers from collisions of data frames during high usage periods.

▶ To standardize the medium access control protocols, IEEE created the 802 series of network standards.

▶ The most popular brands of local area network systems include Ethernet (CSMA/CD), Token Ring, FDDI, and 100VGAnyLAN.

▶ Ethernet LANs have the most variations of product and continue to dominate the local area network market.

▶ The local area network is an indispensable tool for supporting practically any type of computing environment.

KEY TERMS

baseband	hub	tap
basic service set	IEEE 802 suite of protocols	token bus
bidirectional	local area network (LAN)	token ring
broadband	logical design	topologies
bus topology	logical link control (LLC) sublayer	trees
carrier sense multiple access with	medium access control (MAC)	wireless topology
collision detection (CSMA/CD)	sublayer	10Base5
cascade port	multi-station access unit (MAU)	10Base2
collision window	nomadic	10BaseT
contention-based protocol	network interface card (NIC)	10Broad36
demand priority protocol	non-deterministic protocol	100BaseTX
deterministic protocol	passive device	100BaseT4
distributed processing	physical design	100BaseFX
Ethernet	print server	100VGAnyLAN
extended service set	repeater	1000BaseSX
Fast Ethernet	ring topology	1000BaseLX
Fiber Data Distributed Interface	round robin protocol	1000BaseCX
(FDDI) ring protocol	runts	1000BaseT
file server	shared network	
gigabit Ethernet	star-wired bus topology	

REVIEW QUESTIONS

1. What is the definition of a local area network?
2. List the common functions of a local area network.
3. List the advantages and disadvantages of local area networks.
4. What are the basic topologies of local area networks? List two advantages that each topology has over the others.
5. What is meant by a *passive device*?
6. What is meant by a *bidirectional signal*?
7. What are the primary differences between baseband technology and broadband technology?
8. What purpose does a hub serve?
9. What purpose does a MAU serve?
10. How are hubs and MAUs different? How are they alike?
11. What is meant by *nomadic operation*?
12. What is a medium access control protocol?
13. What are the basic operating principles behind CSMA/CD?
14. What is the purpose of the token in a token ring?
15. List three examples of what can go wrong with a token.
16. What is meant by a *non-deterministic* protocol?
17. What does the term *10BaseT* stand for?
18. What is the difference between Fast Ethernet and regular Ethernet?
19. List three advantages of FDDI over token ring.
20. What are the principle characteristics of 100VGAnyLAN?

EXERCISES

1. What properties set a local area network apart from other forms of networks?
2. You have a broadband bus local area network with four stations, numbered 1 through 4, left to right. Because of cable lengths, there are amplifiers at various points along the cable. Station 3 wants to send a copy of a data frame to both station 2 and station 4. Show a cable configuration and a sequence of events that will allow this transmission to take place.
3. Describe another example of a broadband bus system besides cable television.
4. Of all the local area networks introduced in this chapter, is any system capable of supporting a full duplex connection?
5. Why is a transmitting station responsible for removing its own data from a ring?
6. Physically, a hub looks the same as a MAU. Logically they are different. Explain how they are different.
7. In the IEEE 802.5 token ring protocol, the ED field has two copies of the A bit and the C bit. Why are there two copies?
8. Explain the statement that a ring is "composed of segments of one-way point to point cables strung between pairs of repeaters."
9. If, on a token ring LAN, a workstation is turned off, does the entire network stop? Explain.
10. If there is only one active workstation on a demand priority local area network, the demand priority protocol acts like what other kind of medium access control protocol?

11. If a network is described as 10BaseT, list everything you know about that network.

12. A FDDI network has been described as having multiple tokens. What does this mean? Would a multiple token concept work on 16 Mbps token ring?

13. The Internet model doesn't have a data link layer. Does it have a medium access control sublayer? Explain.

14. Suppose workstation A wants to send the message HELLO to workstation B. Both workstations are on an IEEE 802.3 local area network. Workstation A has the binary address "1" and workstation B has the binary address "10." Show the resulting MAC sublayer frame (in binary) that is transmitted. Don't calculate a CRC; just make one up.

15. What is the difference between the physical representation of a ring LAN and the logical representation? What is the difference between the physical representation of a star-wired bus LAN and the logical representation?

16. Are collisions possible in token ring local area networks? Explain.

THINKING OUTSIDE THE BOX

1 A retail department store is approximately square, 35 meters on a side. Each wall has two entrances equally spaced apart. Located at each entrance are four point-of-sale cash registers. Suggest a local area network solution that interconnects all eight cash registers. Draw a diagram showing the room, the location of all cash registers, the wiring, and the server. What type of wiring would you suggest?

2 You work for a small advertising company with approximately 200 employees. Scattered around the company are a number of separate computer workstations that perform operations such as word processing, graphics design, spreadsheeting, and market analysis. Your boss has asked you to consider installing some form of local area network to support computer operations. Create a list of possible computer applications that could operate over a local area network and would support employee and business daily functions. What type of local area network might you suggest? What would be the topology? The medium access control method? What kind of support equipment (hubs, servers) might you need? Where would that support equipment be located?

3 You have three computers at home that you want to network together. Two computers are on the main floor of the house, but the third is upstairs in a bedroom. List as many possible ways to interconnect the three computers so that they could operate on one local area network.

PROJECTS

1. Find the IEEE web site and report on the latest advances on the 802 standards. Are there new standards for supporting 1000 Mbps CSMA/CD or wireless LANs? Any new proposals for systems not mentioned in this chapter?

2. In token ring LANs, there is a device called the bridge that can interconnect multiple hubs. What are the functions of this bridge?

3. Besides the CSMA/CD protocol, there is a CSMA protocol. In what kind of systems is the CSMA protocol used? Why CSMA and not CSMA/CD?

4. What happens next in a token ring LAN when a token disappears?

5. Besides the transmission speed, is gigabit Ethernet the same as 10 Mbps Ethernet?

8

Local Area Networks: Internetworking

◆ ◆

Barnes and Noble learned its lesson the last time it built a computer network for its headquarters. After just 18 months the traffic generated by inventory, purchasing, and other business applications saturated the corporate network to the point of user dissatisfaction. So when Barnes and Noble moved into new headquarters in Manhattan, the company created a brand new Gigabit Ethernet backbone anchored by Ethernet switches. Switches easily won out over traditional devices such as routers because switches tend to be less expensive, are very fast, and don't require the extensive software protocols that routers require. What was the end result of the new system? Room for the system to expand as user requirements grow. With respect to response times, some users swear certain applications are moving twice as fast as before.

Caruso, J., "Barnes & Noble's new net provides plenty of room to grow," *Network World*, July 19, 1999, p.18.

Why did Barnes & Noble decide to use switches?

Can switches really make networks operate faster?

Objectives ▶

After reading this chapter, you should be able to:

▶ List the reasons for interconnecting multiple local area networks and interconnecting local area networks to wide area networks.

▶ Identify the functions and purpose of a bridge.

▶ Distinguish a transparent bridge from a source routing bridge.

▶ Outline how a transparent bridge learns.

▶ Describe what a switch is and how it differs from other devices.

▶ Describe the types of situations in which a switch is advantageous.

▶ Define a hub and describe the situations in which a hub is used.

▶ Describe what a router is and how it differs from other devices.

▶ Describe the types of situations in which a router is used.

▶ Identify the basic features of a network server.

Introduction ▶

Local area networks are a large and growing field of study. Many professionals dedicate their careers to supporting local area networks and their hardware and software systems. Chapter Seven introduced the local area network by discussing its advantages and disadvantages, its functions and applications, its basic topologies and medium access control techniques, and some actual LAN systems. To continue the discussion, this chapter introduces an exciting and often confusing area of study—the internetworking of local area network to local area network and local area network to wide area network.

internetworking

As the popularity of local area networks grows and as more companies come to depend on LANs for everyday operations, employers realize that for employees to operate efficiently and effectively, it is necessary to give the users of a local area network access to as many resources and tools as possible. Having this kind of access to resources and tools often means that users on one local area network need to access the resources on another local area network.

As access to more resources becomes available and as more employees log on to the network—a natural part of organizational growth—the network itself must also grow. As local area networks grow in size, network performance decreases and employee performance follows. It is sometimes necessary to break a local area network into smaller, multiple local area networks. Some type of interconnection among these smaller local area networks is necessary for employees to maintain access to the wide range of resources. Access to resources may also mean that users will need to connect to an external network, such as the Internet. In both cases, technical support must provide a means to interconnect a local area network to other networks. Both the hardware and software needed to interconnect local area networks improve everyday. However, there are so many tools available that technicians assigned to support a company's local area networks must keep abreast of constantly changing technology.

This constantly changing technology, and the terminology that accompanies it, creates confusion among users and network professionals. For example, the term gateway used to be synonymous with the term router. Today, many industry experts use the term **gateway** as a more generic term for a device that interconnects two networks.

To understand the interconnection of computer networks, this chapter examines the four most common interconnection tools: the bridge, the hub, the switch, and the router. We will also examine some of the devices that support the operation of a local area network. Finally, an example of a local area network interconnection design will show how many of the concepts function in a realistic setting.

Why Interconnect Local Area Networks?

Many times a single local area network is not sufficient to support the needs of its users. It may not be capable of supporting a large number of users or the necessary resources may be being used on another internal local area network or an external wide area network. In these situations, multiple local area networks or access to a wide area network might provide better service. If access to multiple local area networks or to a wide area network is desired, steps must be taken to interconnect the multiple

— performance
— security

networks. Interconnecting multiple networks is called **internetworking** and is complex due to the large diversity of networks.

Why would someone want to connect two or more networks? Several valid reasons exist for internetworking. Consider, for example, a company that has two local area networks, one for the research department and one for the design department. If the company could interconnect the two networks, it would be possible for the employees in research to share data and resources with the employees in design, and vice versa. Multiple repositories of data could be created to which both research and design have access, to which only research has access, and to which only design has access.

Another company might have two local area networks that use differing protocols. One LAN uses a token ring protocol, and the second uses a CSMA/CD protocol. By interconnecting the two networks, users on both networks will be able to share data, software, and hardware—in spite of the fact that a token ring network has a completely different frame format than a CSMA/CD network has.

Yet another company might want to interconnect a local area network to a wide area network, such as the Internet. With this interconnection, users of the local area network can gain access to the World Wide Web, electronic mail, remote logins, and file uploads and downloads to any point in the world. Also, workers who travel will be able to access the internal local area network from any point in the world.

Suppose a company has a sufficiently large LAN that is very busy with high demands on a file server and printer. Response time for retrieving files from the server is very slow and employees have to wait long periods of time for a print job. The company would like to divide the network into two LANs, thus dividing the traffic and work load among two different networks. Once this division is made, the response time should improve and wait time for print jobs should drop to an acceptable level. Division of a large network can also provide a level of security. If research is placed on one network and sales is placed on a second network, employees in sales who request data files from research will have their requests monitored and possibly denied. Likewise, employees in research may have their requests for sales records monitored.

Consider a company in which the computer workstations are spread across a number of adjacent buildings. To better manage the number of workstations and the distances between buildings, each building can receive at least one local area network and all the buildings can be interconnected. This concept of interconnecting smaller LANs can be implemented on floors within a building or even within divisions on a floor within a building.

Suppose a company maintains a mainframe computer that processes a number of crucial computer programs or "legacy" applications (older programs that have been used by the company for many years). By internetworking a local area network to the mainframe, all workstation users will have access to the applications on the mainframe. These programs can be transmitted from the mainframe to the workstation (downloaded) and executed on the user workstation. This downloading relieves the amount of work on the mainframe computer since the workstations are now the computers running the programs.

Internetworking isn't just for large corporations. Suppose you have two or more personal computers in your home for family use. Also suppose that you have installed a network-in-a-box so that the computers can share data, software, and peripherals such as a printer. The next step would be to provide the home network with an interface to the Internet so that all the home computers could access Internet resources.

Many devices are available (and more are becoming available every day) for performing the interconnection of two or more networks. These devices can be classified into four basic categories: bridges, hubs, switches, and routers. Although in the past these terms have meant different things to different people, the last several years have seen most people reach a common ground on definition and functionality. You should be aware, however, that these terms are a little flexible. Some manufacturers that produce network interconnection devices are particularly guilty of stretching the naming conventions. Let's discuss each of these devices in turn and examine examples that demonstrate their use.

Bridges

Originally, a bridge was a device that interconnected two local area networks that used the same protocol. For example, if you had two 10BaseT local area networks, you could interconnect the two networks using a bridge. More recently, however, bridges have been gaining computing power. Now, equipment catalogs show bridges that can interconnect two dissimilar local area networks, such as a bridge that interconnects a CSMA/CD LAN to a token ring LAN. These devices that interconnect two dissimilar local area networks are sometimes called "brouters"—a cross between a bridge and a router. It is debatable how much the term brouter is still used in the industry. To reflect the fact that bridges will continue to expand in computing capabilities, let's define a **bridge** as a device that interconnects two local area networks that both have a medium access control sublayer. CSMA/CD and token ring local area networks both have a medium access control sublayer, but as was shown in Chapter Seven, they have slightly different frame formats. The CSMA/CD frame has an eight-byte header and a length field, while the token ring frame has a one-byte starting delimiter, an access control byte, a frame control byte, an ending delimiter, and a frame status byte. To interconnect these two networks, a device must be capable of converting one frame format to another. Bridges exist that can perform this conversion.

A bridge has two basic functions interconnecting two local area networks that have medium access control sublayers. To understand these functions, consider the example of a CSMA/CD local area network connected to a token ring local area network. First, if a data frame originates on the CSMA/CD network and is destined to arrive at a workstation on the same network, the bridge needs to realize this destination information and *not* forward the data frame to the token ring network. Thus, the bridge acts as a filter. A **filter** examines the destination address of a frame and either forwards or does not forward the frame based on some address information stored within the bridge. The bridge will reduce the amount of traffic on the interconnected networks by dropping frames that do not have to be forwarded.

Second, as the data frame passes from one network to a second dissimilar network, the bridge converts the frame from the format of the first network to the format of the second network. For example, if the data is passing from a CSMA/CD network to a token ring network, the bridge converts the frame from an IEEE 802.3 frame to an 802.5 frame. The bridge has to perform a small amount of data manipulation to delete, convert, and create the appropriate fields. A simple example of this data manipulation is shown in Figure 8-1.

Figure 8-1

A bridge interconnecting two dissimilar local area networks

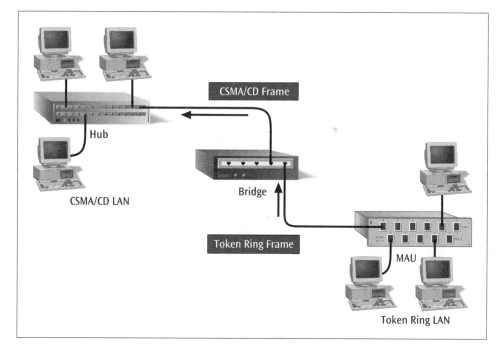

Let's examine the filtering functions of a bridge that interconnects two CSMA/CD networks (Figure 8-2). As a frame of data moves across the first CSMA/CD network and enters the bridge, the bridge examines the medium access control sublayer source and destination addresses. A MAC sublayer address is the address assigned to the network interface card (NIC) when the NIC is manufactured. All companies that produce NICs have agreed to a formula so that every NIC in the world has a unique MAC address. The bridge, using some form of internal logic, determines if a data frame's destination address belongs to a workstation on the current network. If it does, the bridge does nothing more with the frame, since it is already on the appropriate network. If the destination address is not an address on the current network, the bridge passes the frame onto the next CSMA/CD network, assuming that the frame is intended for a station on that network. Additionally, the bridge can check for transmission errors in the data by performing a cyclic checksum computation on the frame and set any necessary error bits before re-transmitting the frame.

Figure 8-2

A bridge interconnecting two identical local area networks

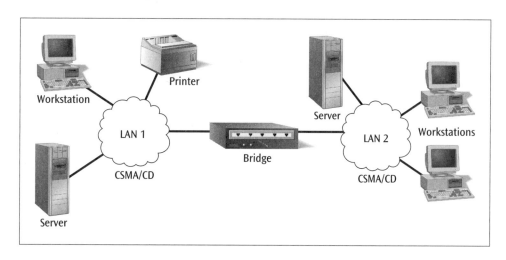

How does the bridge know what addresses are on which networks? Did a technician sit down and type in the address of every NIC on each interconnected network? The answers to these questions depend on whether a bridge is one of these three basic types: a transparent bridge, a source-routing bridge, or a remote bridge.

Transparent bridge

The *transparent bridge*, designed for CSMA/CD LANs, observes network traffic flow and uses this information to make future decisions regarding frame forwarding. Upon installation, the bridge begins observing the addresses of the frames in transmission on the current network and creates an internal port table to be used for making future routing decisions. The bridge creates the internal port table by using a form of **backward learning**, or observing from where a frame has come. If a frame is on the current network, the bridge assumes that the frame originated from somewhere on that network. The bridge takes the source address from the frame and places it into an internal table. After watching traffic for awhile, the bridge has a table of workstation addresses for that network. Then, if a frame arrives at the bridge with a destination address other than an address in the table, the bridge assumes the frame is intended for a workstation on some other network, and passes the frame on to the next network.

Transparent bridge learning network addresses

As an example of how the transparent bridge learns, examine Figure 8-3 and the following scenario. The bridge has two ports, one for CSMA/CD LAN A and the second for CSMA/CD LAN B. When the bridge is first activated, the internal port tables, one for Port A and one for Port B, are empty. Figure 8-4(a) shows the two tables as initially empty. Workstation 1 transmits a frame to Workstation 4. The bridge extracts Workstation 1's address and puts it in Port A's table. It has just learned that Workstation 1 is on network A (Figure 8-4(b)). The bridge still doesn't know the address of Workstation 4, however. Even though Workstation 4 is also on LAN A, the bridge does not know this fact because there is no entry in Port A table for Workstation 4. Consequently, the bridge forwards the frame out Port B onto LAN B.

[handwritten margin notes: ✱Test; The TO address does nothing to update the internal table. The From Address updates the internal table. MAC Address]

Figure 8-3
A bridge interconnecting two CSMA/CD networks has two internal port tables

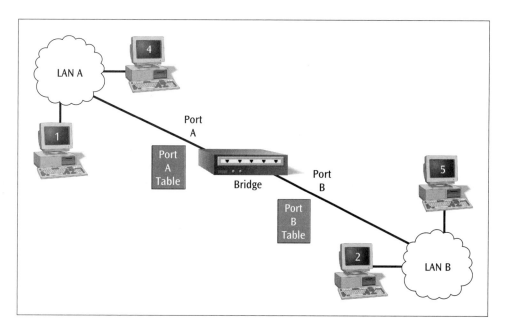

Figure 8-4
Two internal port tables and their new entries

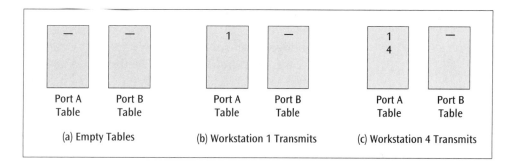

(a) Empty Tables   (b) Workstation 1 Transmits   (c) Workstation 4 Transmits

CBT: For a visual demonstration of how a bridge creates its internal tables and filters frames, see the enclosed CD-ROM.

Eventually, Workstation 4 returns a frame to Workstation 1. The bridge extracts the address of Workstation 4 and places it in the Port A table (Figure 8-4(c)).

The frame is destined for Workstation 1, and the bridge sees that there is entry for Workstation 1 in Port A's table. Now, the bridge knows Workstation 1 is on LAN A and *does not* forward the frame onto LAN B. In addition, if Workstation 1 sends another frame to Workstation 4, the bridge will see that Workstation 4 is on LAN A (because of the entry in the Port A table) and *will not* forward the frame onto LAN B.

If Workstation 1 sends a frame to Workstation 5, the bridge will not recognize the address of Workstation 5 because there is no entry in Port A's table and will forward the frame onto LAN B. The bridge will perform the same learning function for LAN B and update Port B's table accordingly. Thus, the bridge learns where workstations are and then uses that information for future routing decisions.

Transparent bridge converting frame formats

If a bridge interconnects two dissimilar local area networks, how does the bridge convert the frame format from one local area network to the proper frame format for the second local area network? Figure 8-5 shows a data packet as it travels from a CSMA/CD LAN to a token ring LAN. Recall the four layers of the Internet model introduced in Chapter One. The data packet originates in an upper layer application on the CSMA/CD LAN (Figure 8-5(a)). When the data packet is passed from the application layer to the transport layer, the transport layer adds the necessary transport header information (Figure 8-5(b)). The process of adding transport header information on the front of the data packet is called **encapsulation**. By encapsulating an application layer packet with a transport header, the packet has become a transport layer packet. The application layer information is essentially hidden until the time when the transport layer header is removed. Next the data packet is given to the Internet layer where the necessary network header information is applied (Figure 8-5(c)). The transport layer packet is now encapsulated within the network layer packet. Finally, the packet is passed to the interface layer, which contains the medium access control sublayer, and a frame is created using the appropriate CSMA/CD header information (Figure 8-5(d)). The network layer packet is now encapsulated within the CSMA/CD frame. The data frame passes over the CSMA/CD network and arrives at the bridge. Because of the addressing, the bridge determines that the data frame should pass onto the token ring LAN. Before the bridge passes the frame forward, it must convert the CSMA/CD frame to the proper token ring frame format. The bridge removes the CSMA/CD medium access control sublayer information (Figure 8-5(e)), creates the proper token ring medium access control sublayer information, and places this information on the front of the frame (Figure 8-5(f)). The data frame is now in the proper format to traverse the token ring.

Note how the bridge removes only the medium access control sublayer information and does not touch any information from a higher layer (Internet, transport, and application layers). The bridge went only as far as the medium access control sublayer. According to the OSI model, the medium access control sublayer is the second layer in the OSI model. Thus, a bridge is a layer two device. (This is another good example of why you should be familiar with both the OSI and Internet models.)

Figure 8-5

A data frame as it moves from a CSMA/CD LAN to a token ring LAN

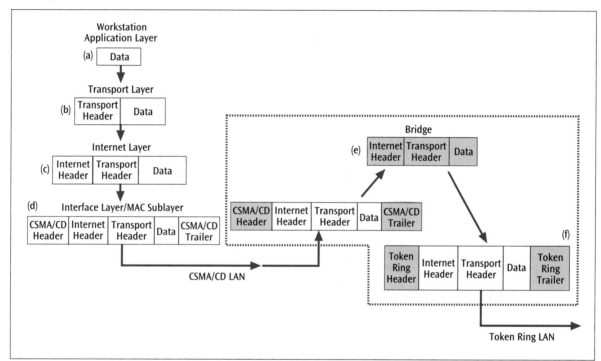

Source-routing bridges

A second form of bridge is the source-routing bridge, which was designed for IBM Token Ring LANs. A **source-routing bridge** does not keep any internal tables, but instead relies on information contained within each data frame. With this type of bridge, the workstation sending the frame has to know the address of the final

Details ▶

Spanning Tree Bridge

Although a transparent bridge is simple and elegant, it suffers from one potential problem. Consider the network design shown in Figure 8-6. The network includes two hubs and two bridges. Workstation 1 transmits a data frame destined for Workstation 2, and it arrives at both bridges. Both bridges update their internal tables, noting that Workstation 1 is on LAN X, and

both bridges forward a copy of the frame to LAN Y. Workstation 2 then receives two copies of the frame. Although you might think that receiving two copies of a frame is a potential problem, most networks can recognize a duplicate frame and will discard the duplicate.

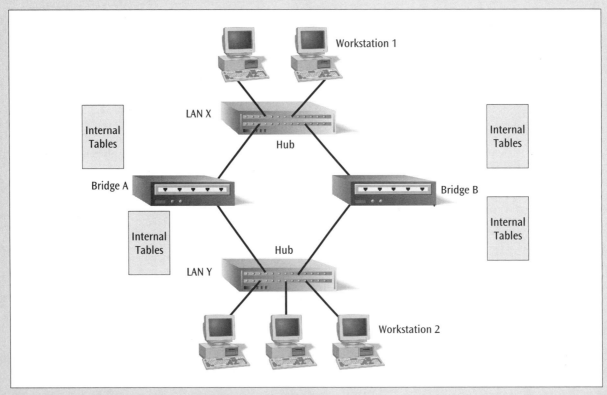

Figure 8-6 *Two bridges on two local area networks causing a loop*

destination and the route necessary to get the frame there. This technique contrasts with a CSMA/CD local area network, in which the workstations do not need to provide the bridges with directions to the destination workstation. Instead, the token ring workstation software has determined the route and inserts this information into the appropriate position within the data frame.

The serious problem lies in what happens next (Figure 8-7). Bridge A's frame will not only arrive at Workstation 2, but will also arrive at Bridge B, and Bridge B's frame will also arrive at Bridge A. Upon receipt of the frame, Bridge A will incorrectly assume that Workstation 1 is on LAN Y (because it received the frame from LAN Y) and update its table in error. Likewise, Bridge B will assume that Workstation 1 is on LAN Y (because it received the frame from LAN Y) and update its table incorrectly. Now if Workstation 2 wants to transmit a data frame back to Workstation 1, neither bridge will forward the frame because both tables incorrectly indicate that Workstation 1 is on LAN Y.

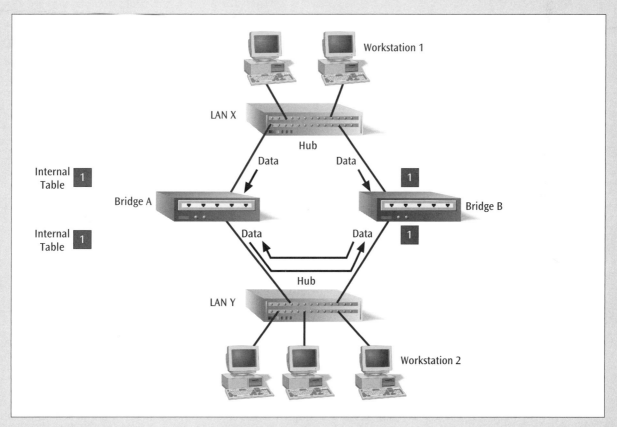

Figure 8-7 *Both bridges receive the frame and update their tables incorrectly*

To avoid this situation, you could simply say "Don't place two bridges in such a way as to create a loop." This is easy to say, but some large networks *want* duplicate paths between networks in case one path fails or becomes congested.

The solution is to install bridges that employ the spanning tree algorithm. The **spanning tree algorithm** looks at all possible paths within a network and creates a tree structure that includes *only* unique paths between any two points. Exactly how this algorithm works is beyond the scope of this text. But in large network configurations, a bridge using the spanning tree algorithm can be essential.

Consider the token ring system shown in Figure 8-8. A data frame originates at Workstation A on LAN1 and is destined for Workstation C on LAN3. For the frame to reach its destination, the frame must traverse LAN1, Bridge1, LAN2, Bridge2, and LAN3. Workstation A inserts into the address portion of the token ring medium access control frame the sequence LAN1—Bridge1—LAN2—Bridge2—LAN3. At each point along the route, this address sequence is examined to determine the next appropriate step. For example, after the frame is forwarded by Bridge1, Bridge2 receives the frame. Bridge2 examines the address sequence and sees the portion "LAN2—Bridge2—LAN3," and knowing that the frame just arrived from LAN2, forwards the frame onto LAN3.

Bridge3 will also receive a copy of the frame as soon as Bridge1 forwards it onto LAN2. Bridge3 will also see the address portion "LAN2—Bridge2—LAN3," and knowing that it is not connected to LAN3 but to LAN4, will not forward the frame. Because of space limitations within the address field of the token ring medium access control frame, the address sequence of LAN—Bridge—LAN can be no more than seven bridges in length.

Figure 8-8
A token ring system composed of multiple token ring local area networks

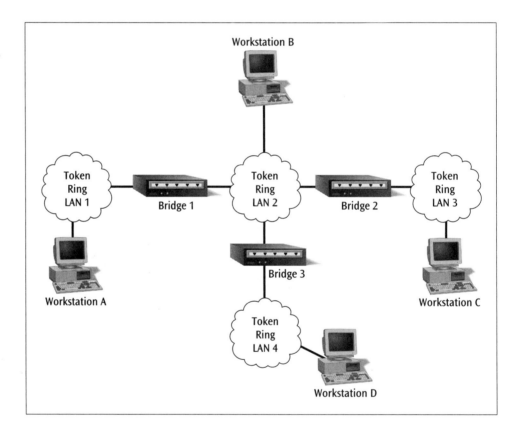

What happens if the sending workstation does not know the exact sequence of addresses between itself and the data frame's destination? To obtain the proper sequence of addresses, the source workstation can dispatch a discovery frame. A **discovery frame** is a special frame that is transmitted from the source workstation for the sole purpose of finding a route to a particular destination workstation. When a discovery frame arrives at a bridge, the bridge automatically forwards the frame onto the next network. If all of the bridges automatically forward this frame, the frame should eventually make its way to the destination workstation. As the discovery frame is forwarded by each bridge, the bridge software records into the discovery frame the name of the bridge and the name of the LAN on which it is being transmitted. When the discovery frame arrives at the destination workstation, the appropriate sequence of LAN—Bridge—LAN addresses has been recorded, and the frame begins its journey back to the source workstation. When the discovery frame arrives at the source, the address sequence is removed from the discovery frame and inserted into the frame to be transmitted.

Remote bridges

A third form of bridge is the remote bridge. A **remote bridge** is capable of passing a data frame from one local area network to another when the two local area networks are separated by a long distance and there is a wide area network connecting the two LANs. In order for a remote bridge to pass a data frame over a wide area network, the frame must be converted into the necessary form for traversing over the wide area network. In Figure 8-9 a frame leaves Workstation 1 destined for Workstation 2, travels across local area network X, and arrives at Bridge A. Remote Bridge A realizes that the frame is not intended for a workstation on LAN X and prepares to forward the frame. But to reach Workstation 2, the frame must traverse an X.25 wide area network. (For a description of X.25, see the Appendix.) The remote bridge adds X.25 header information in front of the existing LAN frame. The X.25 header information is used to get the frame across the X.25 wide area network. Once the frame reaches remote Bridge B, the X.25 header is removed, leaving the original LAN frame. The original LAN frame then travels across LAN Y to Workstation 2.

Figure 8-9

Two LANs with intervening X.25 frame network

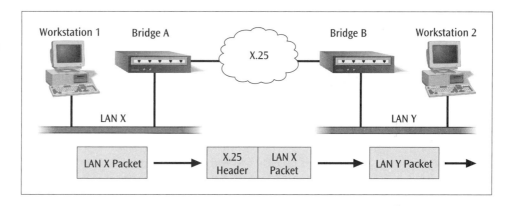

Because of the capability of a remote bridge to transfer local area networks across a wide area network, the remote bridge is a more complicated and thus more expensive device. The situation is quite pronounced since there are many different types of local area networks (10BaseT, 100BaseFX, 1000BaseSX, token ring, and 100VG-AnyLAN to name a few) and many different types of wide area networks (X.25, frame relay, and asynchronous transfer mode to name a few). A remote bridge is unique for each pair of local area network-wide area network combinations. For example, if you want to connect two 100BaseFX local area networks with a frame relay wide area network, you will need to obtain the appropriate 100BaseFX/frame relay remote bridge.

Hubs

Before introducing one of the more exciting innovations for local area networks—the switch—let's discuss the more common local area network interconnecting device that is being replaced by the switch—the hub. Some people consider any device in a local area network that is a collection point for communication lines from workstations to be a hub. Thus, bridges, switches, MAUs, and routers might all be considered hubs. This definition, however, is not held by all professionals in the field. To avoid confusion, this definition will not be used. Instead let's define a **hub** as a device that interconnects two or more workstations in a star-wired bus local area network and broadcasts incoming data onto all outgoing connections. Another way of defining a hub is as a device that works at the physical layer. It is simply a device that accepts a frame and immediately forwards it. The hub does not examine or modify the contents of the frame. Be aware, however, that the distinctions between hub, bridge, router, and switch are becoming more blurry every day.

Although hubs come in many configurations with many different types of options, most hubs can be categorized into one of two categories: managed hubs and unmanaged hubs. A **managed hub** possesses enough processing power that it can be managed from a remote location. The management operations that can be performed on a hub include inventory management (knowing what devices are where on the network), traffic and environmental monitoring, and power management.

An **unmanaged hub** contains little or no intelligence at all and cannot be controlled from a remote location. Although unmanaged hubs are less expensive than managed hubs, they cannot participate in any kind of network management operations. Their sole function is to allow the interconnection of two or more workstations in a local area network.

Hubs continue to become more intelligent, modular, and customizable every day. Some advanced hubs allow for the insertion of bridging, routing, and switching modules within the same unit. Some top of the line hubs can even support a processor board and network operating system, effectively converting the hub into a file server.

Switches

A hub is a simple device and requires virtually no overhead to operate. It is also inefficient. When a network is experiencing a high level of traffic, the hub compounds the problem by taking any incoming frame and retransmitting it out to all connections. Likewise, bridges work well to interconnect local area networks and segments of local area networks when the level of traffic between the segments remains relatively low. But the bridge can become a bottleneck when the traffic between local area network segments increases. A local network administrator can deploy a few tricks to help minimize the bottleneck, but these tricks do not provide a satisfactory solution to the problem. A device is needed that can segment a local area network into separate pieces and still operate efficiently during periods of high traffic.

Recently a new device called a switch has emerged that is useful in high-traffic situations. A **switch** is a combination of a hub and a bridge and can interconnect multiple workstations like a hub, but can also filter out frames providing a segmentation of the network. Switches can provide a significant decrease in interconnection traffic and increase the throughput of the interconnected networks while requiring no additional cabling or rearranging of the network devices. The switch is available for CSMA/CD, token ring, and FDDI networks, but seems to have the most positive effect on CSMA/CD networks. This positive effect is due to the fact that CSMA/CD networks experience collisions, which can significantly affect overall system throughput. Functionally, a switch is very similar to a bridge in that it can act as a filter between two or more segments of a network, but operates in place of a hub.

There are, however, a number of significant differences between a bridge and a switch. First, a switch is a more powerful device than a bridge or a hub. A single switch can be designed to support multiple technologies, including token ring, CSMA/CD Ethernet, Fast Ethernet, and FDDI rings. A switch can also be designed to provide redundant circuits to avoid hardware failures, with the capability to automatically switch to a back-up data circuit or to automatically switch to a back-up controller or management module. These controller and management modules support the operations and control of the switch.

Second, a switch is designed to perform much faster than a bridge. Switches that use cut-through architecture are especially faster than bridges. In a **cut-through architecture**, the data frame begins to exit the switch almost as soon as it begins to enter the switch. A cut-through switch does not store, then forward a data frame. The cut-through capability allows a switch to pass data frames very quickly, thus improving the overall network throughput. A store-and-forward device holds the frame for a small amount of time while various fields of the frame are examined, thus diminishing the overall network throughput. The major disadvantage of cut-through is the potential for the device to forward faulty frames. For example, if a frame has been corrupted, a store-and-forward device will input the frame, perform a cyclic checksum, detect the error, and perform some form of error control. A cut-through device is so fast it is forwarding the frame before the cyclic checksum field is calculated. If there is a cyclic checksum error, it is too late to do anything about it. The frame has already been transmitted. If too many corrupted frames are passed around the network, network bandwidth and time is wasted on transmitting errant frames. The more unnecessary frames transmitted, the greater the likelihood of collisions.

Finally, a switch can have several ports, unlike a bridge, which has only two ports. To support each of these ports efficiently, the main hardware of the switch—called the **backplane**—has to be fast enough to support the aggregate or total bandwidth of all the ports. For example, if a switch has eight 10 Mbps ports, the backplane has to support a total of 80 Mbps. This backplane is similar to a bus inside a microcomputer. It allows you to plug in one or more printed circuit cards. Each circuit card supports one port, or connection, to a workstation or other device. If the circuit cards are **hot-swappable**, it is possible to insert and remove cards while the power to the unit is still on. This capability allows for quick and easy maintenance of the switch.

The switch has one physical similarity to the hub. If you decide to install a switch into a local area network, in many cases it is as simple as unplugging a hub and plugging a switch in the hub's place. Logically, however, the switch is unlike the hub. The switch can examine frame addresses and, based on the contents of an address, direct the frame out the one appropriate path. Whereas the hub simply blasts a copy of the frame out to all connections, the switch uses intelligence (similar to the bridge) to determine the one best connection for outgoing transmission.

Depending on user requirements, a switch can interconnect two different types of CSMA/CD network segments: shared segments and dedicated segments. In **shared network segments,** as shown in Figure 8-10, a switch may be connected to a hub (or several hubs), which then connects multiple workstations. Since the workstations are first connected to a hub, they all share the one channel, or bandwidth of the hub, which limits the transfer speeds of individual stations.

Figure 8-10

Workstations connected to a shared segment local area network

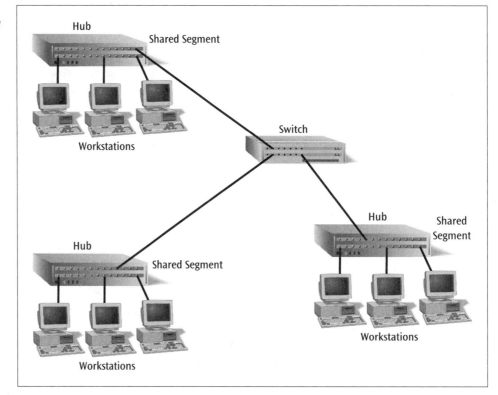

In **dedicated network segments,** as shown in Figure 8-11, a switch may be directly connected to a workstation, and connected to the hub. Each workstation then has a private or dedicated connection. This dedicated connection increases the bandwidth for each workstation over what the bandwidth would be if the workstation were connected to the hub. Dedicated segments are useful for more powerful workstations with high communication demands.

Figure 8-11
Workstations connected to a dedicated segment local area network

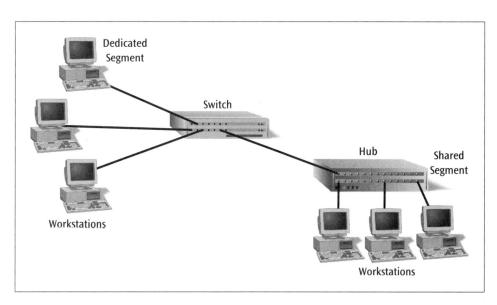

A dedicated network segment can also be used to create a virtual LAN. A **virtual LAN,** or VLAN, is a logical subgroup within a local area network that is created via switches and software rather than by manually moving wiring from one network device to another. For example, if a company wishes to create a workgroup of employees to work on a new project, network support personnel can create a VLAN for that workgroup. Even though the employees and their actual computer workstations may be scattered throughout the building, LAN switches and VLAN software can be used to create a "network within a network." Because the workstations are connected to a dedicated network segment, when workstations within the group transmit to other workstations within the group in the new virtual network, the response time will be faster.

Isolating traffic patterns and providing multiple access

CBT: For a visual demonstration of how a switch transmits packets between workstations and a server, see the enclosed CD-ROM.

Whether shared or dedicated segments are involved, the *primary goal* of a switch is to isolate a particular pattern of traffic from other patterns of traffic or from the remainder of the network. Consider a situation in which two servers, along with a number of workstations, are connected to a switch (Figure 8-12). If Workstation A wishes to transmit to Server 1, the switch forwards the data packetframe directly to Server 1 and to no where else on the network. Workstations B, C, and D do not receive a data frame from Workstation A. Furthermore, Workstation A can transmit to Server 1 and at the same time Workstation B can transmit to Server 2. Finally, the switch can accommodate a peer-to-peer connection between Workstations C and D without sending the data to any other workstations *or servers* on the network.

Broadcast message—often issued by servers to keep the PC's aware of their status and location. A large amount of bdcst messages are sent.

VL1 — 1,2,3,45
V22 — 6,7,8,9,10

Figure 8-12
*A switch with two
servers allowing simul-
taneous access to each
server*

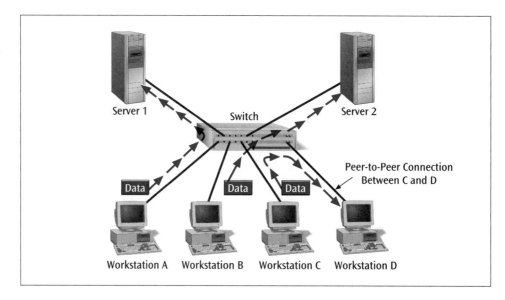

If access to a server is slow because there are many transmissions entering the server through its one cable and network interface card, another network interface card can be added to a server. This new connection can then be attached to an available port on the switch (Figure 8-13). This scenario allows two workstations to transmit to the server simultaneously, thus providing more bandwidth in accessing the server.

Figure 8-13
*A server with two NICs
and two connections to
a switch*

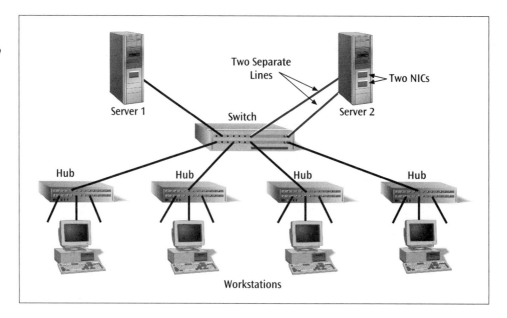

Switches can be used in combination with bridges to further isolate traffic segments in a local area network. In Figure 8-14, notice that Server 1 is located near Workstations A and B and is isolated from the rest of the network via the bridge. If Workstations A and B routinely access Server 1, this traffic will be isolated from everyone else on the larger network. On the other side of the bridge, Workstations C, D and E typically access Servers 2, 3 and 4 through a switch. Thus, their traffic is isolated from each other and from the rest of the network.

Figure 8-14
A bridge and switch combination designed to isolate network traffic

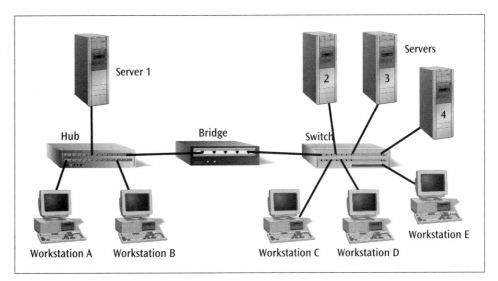

In addition to isolating traffic patterns, switches can provide multiple access to servers. Consider the situation shown in Figure 8-15 in which a number of workstations wish to access an e-mail server. By providing a switch that interconnects multiple segments of workstations with multiple connections to an e-mail server, many users may access the e-mail server in a more efficient manner. Note that each connection to the e-mail server requires its own network interface card. The e-mail server must be powerful enough to handle multiple connections, each performing a query into the e-mail database.

Figure 8-15
Switch providing multiple access to an e-mail server

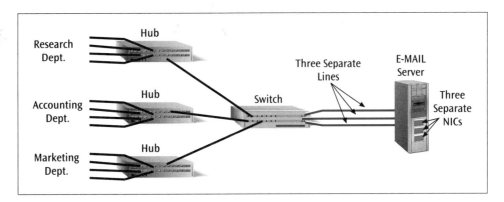

Not on quiz

Full duplex switches

One of the primary disadvantages of a CSMA/CD local area network is the collision. If two workstations talk at the same time, a collision of their signals will occur. Since only one workstation at a time can talk, a CSMA/CD network is a half-duplex system. Recall that a half-duplex system allows both sender and receiver to talk but not at the same time. The bandwidth of a CSMA/CD network would double if it were a full-duplex system in which both sender and receiver could talk simultaneously. In addition, there would be no collisions on an individual segment, which would simplify the CSMA/CD algorithm. The switch allows for a CSMA/CD network to be full-duplex. Thus, a **full-duplex switch** allows for simultaneous transmission and reception of data to and from a workstation.

How can a workstation send and receive signals at the same time? There are two possible solutions: multiplex the send and receive signals over one set of wires or use two sets of wire. It is much simpler to use two sets of wires and transmit the send signal on one set of wires and the receive signal on another set of wires. Figure 8-16 shows the upstream signal traversing one set of wires and the downstream signal traversing the other set.

Figure 8-16
*Full-duplex connection
of workstations to a
LAN switch*

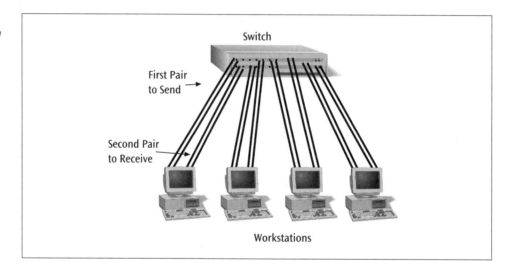

Since the connection between workstation and switch is not a very long connection (usually less than 100 meters) and the wire used is typically twisted pair, it should not be much more expensive to implement a full-duplex connection. In fact, if you are installing a new local area network including the wiring, it is a simple process to run multiple sets of wires to each workstation in the event that you decide to create a full-duplex local area network in the near future.

It is clear the switch is a very powerful tool for segmenting a local area network and thus reducing the overall amount of traffic. Because there is a reduction of traffic, a local area network that employs switches is more efficient and network performance improves. Overall network throughput improves and the response time when a user requests a service is faster. As the switch continues to evolve, do not be surprised to see it replace the hub, the bridge, and perhaps other interconnection devices.

[handwritten: chapter 9]

[handwritten: Not on quiz]

Network Servers

[handwritten left margin: - Processor]
[handwritten: - Memory]
[handwritten: - Disks storage]
[handwritten: - Do not use IDE Drives]
[handwritten: - SCSI]
[handwritten: Small computer Systems Interface]

All of the network devices introduced thus far—bridges, hubs, and switches—are useful and sometimes vital to the proper support of a local area network. One device that has been mentioned numerous times and is essential to a local area network is the network server. The **network server** is the computer that stores software resources such as computer applications, programs, data sets, and databases, and either allows or denies workstations connected to the network access to these resources. A network server is very often the computer that holds and executes the network operating system.

Network servers, however, continue to grow in power and flexibility. As this growth continues, the functions a server can perform also grow. Some of the latest servers can accept additional hardware cards that support bridging, routing, and switching. Conversely, switches are becoming so powerful that they can perform server functions. It is likely that one day there will be a single device that is capable of performing switching, bridging, routing, and server operations. Since servers may also be considered an interconnection device, let's take a few minutes to discuss the main hardware features of a network server. Network servers and the network operating systems that run on them will be covered in more detail in Chapter Nine.

[handwritten left margin: IDE]
[handwritten: Integrated Drive electronics]

Network servers range from small microcomputers to mainframe computers. Typically, however, the network server is a powerful microcomputer workstation. This workstation typically houses hundreds of megabytes of random access memory, one or more large-storage hard disk drives (each drive providing hundreds of gigabytes of storage), and at least one high speed microprocessor. The hard disk drives typically have an interface such as the small computer system interface and are hot-swappable. The **small computer system interface (SCSI)** is a specially designed interface that allows for a very high speed transfer of data between disk drive and the computer. A disk drive that is hot-swappable is capable of being removed from the computer while the power is still on. This makes maintenance and repairs simpler because the power to the workstation does not have to be shut off.

[handwritten margin: TEST]

To protect the workstation from catastrophic disk failure, the disk drives on most network servers support one of the **redundant array of independent drives (RAID)** techniques. A RAID technique, of which there are several, describes how the data is stored on multiple disk drives. For example, one large data record may be broken into pieces with each piece stored on a different disk drive.

[handwritten margin: TEST]

[handwritten left margin: Mirror]
[handwritten: Striping]

Many of the higher power network server workstations support not one but multiple processors. Since the processor is the driving engine within the computer, a workstation with multiple processors can simultaneously execute multiple requests from other workstations. A network server that boasts symmetric multiprocessing implies that requests to the network server are evenly distributed over the multiple processors, thus effectively utilizing the full potential of the multiprocessing system.

[handwritten left margin: Hotswapable - replace equipment without being the system down.]

A number of miscellaneous support devices are associated with a network server. A network server requires at least one connection to the network in the form of a network interface card. To safeguard the contents of the disk drives, a tape backup system is employed to back up hard disk contents on a regular and automatic basis. Also, some form of battery backup system (uninterruptable power supply) is used to maintain power to the computer for various periods of time should electrical power be lost. All of these support devices will be examined in more detail in Chapter Nine.

[handwritten bottom: ADAptec - SCSI Controller]

Routers

So far we have examined bridges for interconnecting two local area networks and hubs and switches for interconnecting workstations. How do you interconnect a local area network to a wide area network, such as the Internet? The most commonly found choice is the router. The most common function of a router is to transmit data frames between two networks in which one network *has* a medium access control sublayer but the second network *does not have* a medium access control sublayer. A very common example is using a router to interconnect a local area network, such as a CSMA/CD LAN, to the Internet. The Internet, as you will see in Chapter Eleven, does not have a medium access control sublayer, but is based on the Internet protocol instead. To perform the routing, the router can not just look at the MAC layer addresses of the two networks. Instead, the router must dig further into the data frames for more information. Since both the CSMA/CD network and the Internet have a network layer (with network layer addresses), the router examines the network layer addresses and uses those to perform the routing. The byproduct of digging deeper into the data frames for network routing information is increased processing time, making the router a slower and more elaborate device than a bridge.

What sequence of events occurs when a CSMA/CD data frame is converted by a router to an Internet data frame? Figure 8-17 shows a data frame passing from a CSMA/CD LAN onto the Internet. A data packet originates in an upper layer application, such as an e-mail program (Figure 8-17(a)). The data packet is passed to the transport and network layer of the CSMA/CD network and the appropriate transport and network headers are inserted (8-17(b)). After the network information has been inserted, the packet is given to the medium access control sublayer, where CSMA/CD header and trailer info is added (Figure 8-17(c)). The packet is then placed onto the medium of the CSMA/CD network, where it travels to the router. Upon receiving the packet, the router does not need the MAC header and trailer and so they are removed (Figure 8-17(d)). Now, the router will extract the destination network address from the network header and determine that the packet has to go out on the Internet. The router then adds the appropriate wide area network header (if any) to the packet (Figure 8-17(e)) and sends the packet out of the router and onto the Internet. From there, the packet will make its way through the Internet to its final destination.

Router is much slower than a bridge or a switch

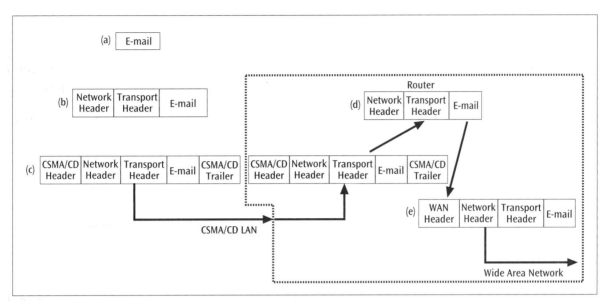

Figure 8-17

A data frame as it passes from a CSMA/CD LAN to the Internet

As you can see from Figure 8-17, the router has to discard the MAC header and trailer information to get at the network header information, which is used for routing across a wide area network. In contrast, a bridge or a switch extracts only the MAC layer information and does not have to dig any deeper than that layer.

Routers, however, are similar to switches in that both are getting more advanced everyday. A very common feature found in many routers is some form of firewall protection. A **firewall** is a system or combination of systems that supports an access control policy between two networks. In the case of a router, a firewall acts as a protection system between the local area network and the wide area network. Some of the newer routers also contain switches, thus combining the services of a router with the advantages of a switch. Modern routers can also accept data in one format and convert the data to another format. For example, when a data packet comes in from the Internet in Internet form (TCP/IP), the router can convert the packet to an appropriate form for the traversal over the local area network.

The more advanced routers can also perform management functions such as monitoring network traffic, providing accounting information, and incorporating quality of service functions. **Quality of service (QoS)** is the concept that data transmission rates, error rates, and other network traffic characteristics can be measured, improved, and, hopefully, guaranteed in advance. Thus, if a user wishes to use an application that incorporates a high bandwidth transmission, such as real-time video, a network connection request containing the appropriate QoS parameters is made. The network examines the request and, based on current network demands and network limitations, either guarantees or does not guarantee the connection.

Routers and routing plays an extensive role in wide area networks such as the Internet. As data packets move across the Internet they pass from network to network. The connection points between these networks are routers. Each router examines the data packet and determines the packet's next step along its journey. Since routers play such an important role in wide area networks and the Internet, we will return to the topic of routers and routing in Chapters Ten and Eleven.

LAN Internetworking in Action: A Small Office Revisited

In Chapter Seven, you met Hannah, the computer specialist for a small business on the west side of Chicago. Her company has approximately 20 workers, each with his or her own computer workstation. They use the computers mainly for word processing and spreadsheets. The owner of the company wanted to update their computer services by offering the following applications to employees:

- ► electronic mail;
- ► a database that contains information on all their past and present customers;
- ► limited access to the Internet; and
- ► one or two high quality color laser printers.

After considering many options, Hannah decided to set up a 100BaseTX local area network with two hubs, Category 5 twisted pair and fiber optic wiring, and a server, as shown in Figure 8-18. Because 100BaseTX systems use CSMA/CD as the medium access control protocol and only one workstation can transmit at a time, the system is half-duplex.

Figure 8-18

Hannah's small business solution from Chapter Seven

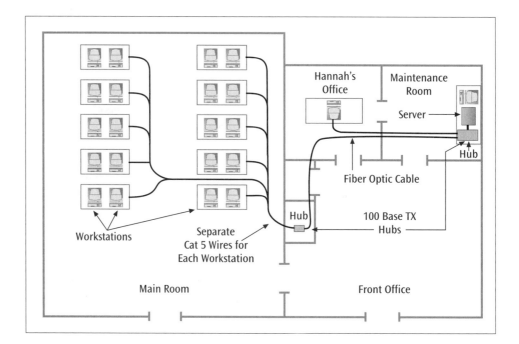

The wires, NICs, hubs, and server were installed, and the appropriate software was loaded on the server and on the individual workstations. After an initial testing period, Hannah created user documentation and trained the employees on how to use a couple of the applications. Despite a few problems and a little bit of employee grumbling for having to learn a new system, daily office functions quickly returned to normal levels.

Now, Hannah's next step is to install a router and get everyone connected to the Internet. Hannah orders a router, then contacts an Internet service provider to acquire approximately 30 to 40 Internet addresses and Internet access. She also calls the telephone company to install a high-speed line such as a digital T1 service to support the connection between the router and the Internet service provider (Figure 8-19).

Figure 8-19
The modified network with a router and high-speed phone line

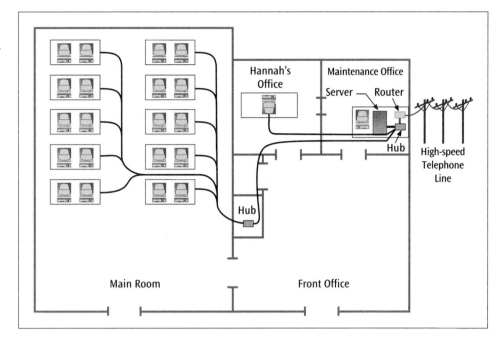

As Hannah fears, after all systems are up and running and employees have access to the Internet, including the World Wide Web, local area network and Internet usage is growing at a phenomenal rate. As the employees began to learn the possibilities of sharing local files, databases, and printers, as well as accessing the vast amount of material available on the Internet, an increasing demand is placed on the local area network and server. Network response time grows slower and slower, and employees begin to complain. It's now time to examine the network and perform an upgrade.

After many hours of observation and talking to employees, Hannah determines that roughly half the employees use the business database quite heavily, and the other half use the Internet heavily. Almost everyone does word processing and some work with spreadsheets, but this work does not significantly contribute to the heavy network load.

To accommodate the heavy network traffic, Hannah decides first that she needs a new server to handle just the Internet traffic. If she can keep the database traffic on the database server and the Internet traffic on the Internet server, traffic flow should improve. Furthermore, if she goes to a full-duplex system instead of the half-duplex system she had originally installed, additional improvements should occur. During the initial installation, Hannah had the technicians install four pairs of Category 5 twisted pair. As a consequence, she can now move up to full-duplex without having to change the original wiring. To create the full-duplex system, the hub adjacent to the servers is replaced with a switch. Since the other hub is connected to this switch, Hannah decides to leave it as a hub and see if her other modifications can improve network performance sufficiently. Figure 8-20 shows the upgraded network.

Figure 8-20
The upgraded network with full-duplex connections, an additional server, and a switch in place of a hub

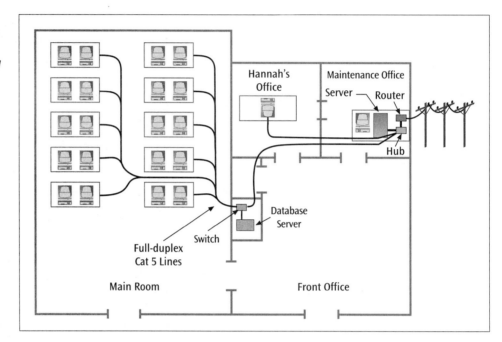

The upgrades Hannah makes to the network work very well. Network congestion is reduced significantly. Both Internet users and database users notice a reasonable reduction in wait time. Since computer usage and corresponding network usage almost always increases over time, Hannah's solutions are more than likely temporary until the next problem is encountered. Nonetheless, Hannah's fellow employees are content for the present.

❖❖❖

SUMMARY

► As local area networks grow in size, and as the need arises to connect a local area network to other local area networks and wide area networks, it is necessary to understand how to interconnect similar and dissimilar networks.

► The reasons for interconnecting networks include interconnecting two local area networks from different departments, interconnecting two local area networks from different geographic locations, providing local area network users with access to the Internet or a legacy system, segmenting a network that is growing too quickly, providing a level of security between two application groups, and providing a means to share software, peripherals, and Internet connections in a home computer network.

► There are four devices that can provide varying levels of interconnection: bridges, hubs, switches, and routers.

► The basic function of a bridge is to interconnect two local area networks with medium access control sublayers. In most cases, the two networks have identical medium access control software.

- ► A bridge acts as a filter as it examines each frame and decides whether the frame should be forwarded onto the next network.

- ► A transparent bridge builds its own routing tables by observing the flow of traffic on the networks. This observation is an example of backward learning.

- ► A source-routing bridge does not create or maintain internal routing tables but instead uses the address sequence in the data frame to decide whether to forward a frame.

- ► If the address sequence to a distant workstation is unknown, the transmitting workstation will issue a discovery frame, which finds a path from the transmitting workstation to the destination workstation.

- ► A remote bridge connects two local area networks that are separated by large distances and a wide area network. As the data frame leaves one local area network, the remote bridge adds the necessary control information for the frame to traverse the wide area network.

- ► Although bridges are relatively quick and simple devices, they can interconnect only two relatively similar networks.

- ► A hub is a device that interconnects multiple workstations within a local area network.

- ► Hubs, as well as most interconnection devices, can be either managed or unmanaged.

- ► A switch can provide a significant decrease in interconnection traffic and increase the throughput of interconnected networks while working with conventional cabling and adapters.

- ► A switch replaces a hub and isolates the traffic flow between segments of the network by examining the address of the transmitted frame and directing the frame to the appropriate port.

- ► A switch that employs a cut-through architecture is the opposite of a store-and-forward device in that the data frame is leaving the switch almost as soon as it begins to enter it.

- ► Switches can create shared segments in which all workstations hear all the traffic or dedicated segments in which other workstations do not hear the local traffic.

- ► Switches can also be made to operate in full-duplex mode if an additional pair of wires are used for the opposite flow of data.

- ► A network server is a vital component in a local area network and often contains a large amount of memory, multiple processors, high-speed disk drives, network interface cards, and a tape backup.

- ► Routers interconnect local area networks to wide area networks.

- ► A router's most common function is to route data packets between two networks in which one network *has* a medium access control sublayer but the second network *does not* have a medium access control sublayer.

- ► Routers provide all users on a local area network with access to outside networks.

- ► Routers perform more slowly than bridges and require more processing because they have to dig deeper into the data frame for control information.

KEY TERMS

backplane
backward learning
bridge
cut-through architecture
dedicated network segment
discovery frame
encapsulation
filter
firewall
full-duplex switch

gateway
hot-swappable
hub
internetworking
managed hub
network server
quality of service (QoS)
redundant array of independent drives
(RAID)
remote bridge

router
shared network segment
small computer system interface (SCSI)
source-routing bridge
switch
transparent bridge
unmanaged hub
virtual LAN (VLAN)

REVIEW QUESTIONS

1. What are the primary reasons for interconnecting two or more networks?
2. What are the basic functions of a bridge?
3. What is meant by a filter?
4. How does a transparent bridge work?
5. What is backward learning?
6. How does a bridge encapsulate a message for transmission?
7. How is a source-routing bridge different from a transparent bridge?
8. What is the purpose of a discovery frame?
9. In what situations does one use a remote bridge?
10. What is the basic function of a hub?
11. What is the difference between a managed hub and an unmanaged hub?
12. What are the basic functions of a switch?
13. How does a switch differ from a hub? From a bridge?
14. What is cut-through architecture?
15. How is a full-duplex switch different from a switch?
16. What are the basic features of a network server?
17. What are the basic functions of a router?

EXERCISES

1. List at least 3 examples of a network to network connection.
2. A transparent bridge is inserted between two local area networks ABC and XYZ. Network ABC has Workstations 1, 2, and 3, and network XYZ has Workstations 4, 5, and 6. Show the contents of the two routing tables in the bridge as the following packets are transmitted. Both routing tables start off empty.

   ▶ Workstation 2 sends a packet to Workstation 3.
   ▶ Workstation 2 sends a packet to Workstation 5.
   ▶ Workstation 1 sends a packet to Workstation 2.
   ▶ Workstation 2 sends a packet to Workstation 3.
   ▶ Workstation 2 sends a packet to Workstation 6.
   ▶ Workstation 6 sends a packet to Workstation 3.
   ▶ Workstation 5 sends a packet to Workstation 4.

- ► Workstation 2 sends a packet to Workstation 1.
- ► Workstation 1 sends a packet to Workstation 3.
- ► Workstation 1 sends a packet to Workstation 5.
- ► Workstation 5 sends a packet to Workstation 4.
- ► Workstation 4 sends a packet to Workstation 5.

3. A CSMA/CD network is bridged to a token ring network. A user on the CSMA/CD network sends an e-mail to a user on the token ring network. Show how the e-mail message is encapsulated as it leaves the CSMA/CD network, enters and leaves the bridge, and arrives at the token ring network.

4. You have three token ring LANs, X, Y and Z, with one source-routing bridge (A) connecting LAN X to LAN Y and a second source-routing bridge (B) connecting LAN Y to LAN Z. A workstation on LAN X wants to send data to a workstation on LAN Z, but does not know the correct path. Show the sequence of messages as a discovery frame makes its way across the networks and returns with the appropriate path sequence.

5. Is the hub the only device on a local area network that can be managed? Explain.

6. a. The local area network shown in Figure 8-21 has two hubs (X and Y) interconnecting the workstations and servers. What workstations and servers will receive a copy of a packet if the following workstations/servers transmit a message:
   - ► Workstation 1 sends a message to Workstation 3:
   - ► Workstation 2 sends a message to Server 1:
   - ► Server 1 sends a message to Workstation 3:

   b. Replace hub Y with a switch. Now what workstations and servers will receive a copy of a packet if the following workstations/servers transmit a message:
   - ► Workstation 1 sends a message to Workstation 3:
   - ► Workstation 2 sends a message to Server 1:
   - ► Server 1 sends a message to Workstation 3:

Figure 8-21

Network example to accompany exercise 6

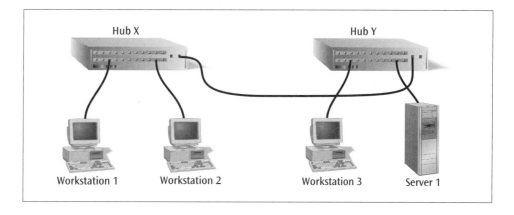

7. What does it mean when a switch or device is cut-through? What is the main disadvantage of a cut-through switch? Is there a way to solve this disadvantage of a cut-through switch without losing the advantages?

8. Give a common business example that mimics the differences between a shared network segment and a dedicated network segment.

9. You currently have a workstation with a half-duplex connection to a switch. What are the possible options that would allow you to create a full-duplex connection to the same switch?

10. A CSMA/CD network is connected to the Internet via a router. A user on the CSMA/CD network sends an e-mail to a user on the Internet. Show how the e-mail message is encapsulated as it leaves the CSMA/CD network, enters the router, and then leaves the router.

THINKING OUTSIDE THE BOX

1 You are working for a very small company with only a dozen employees. Five computer workstations need to be connected to the Internet. List the options that are available. What are the advantages and disadvantages of each? What is your recommendation?

2 You are working for a company that is composed of three departments: marketing, sales, and general support. General support occupies the first floor, marketing the second floor, and sales is on the third floor. General support has 40 workstations, marketing has 12 workstations, and sales has 25 workstations. Some applications require data to transfer between departments, but generally each department has its own applications. Everyone needs access to the Internet and e-mail. Design a local area network solution for this company. Show the location of all interconnecting devices (if used): hubs, switches, bridges, routers, and network servers. What type of wiring would you recommend? What type of local area network topology and protocol would you recommend?

PROJECTS

1. Using the Internet or other external sources, collect literature on a CSMA/CD switch. What are the specifications of the switch (such as port speeds, number of ports, cable types supported, backplane speed, etc.)?

2. Create a map of your company or school's local area network. Where are the hubs located? Are there any switches? If so, where are the switches located? Are there any bridges? Where are they located? Are there any routers? Show where they are located.

3. Some companies produce a device called a brouter. What is a brouter? Find the product description of a brouter and list its characteristics.

4. Locate some literature on a powerful network server. What are its features and its operating characteristics? Determine its approximate cost?

5. What are the different RAID technologies? For each RAID technology, create a figure that shows how it works.

9

Local Area Networks: Software and Support Systems

◆◆◆

"Life after moving Cat's [The Cat's Pajamas, an order fulfillment and inventory system application] to [OS 1, a popular network operating system] was a nightmare. The system was crashing two to three times a day with no reason that I could find. I was on the phone with [the OS 1 company] and Cats constantly, but nobody could figure it out. [The OS 1 company] had me apply [three service packs] and a few [patches], which helped, but the system still was crashing at least twice a week....After many weeks and about $1500.00 in phone support from [the OS 1 company], the technical support rep told me that I should find a better software package than The Cat's Pajamas. This was not the solution I was looking for, since this is the package that a sizeable percentage of presses our size nationwide are running, so I was forced to bring the old [OS 2, another popular network

operating system] server back into production until I could figure something out....Fourteen months later, we are running [OS 3, a third popular network operating system] as our server."

Kirch, J. "Microsoft Windows NT Server 4.0 versus UNIX," [online at *http://www.unix-vs-nt.org/kirch/*] Accessed February 10, 2000.

Is it always this difficult to get an application to run with a particular network operating system?

What choices are available when selecting a network operating system?

Can the choice of a network operating system be so important as to cause this much frustration?

Objectives ▶

After reading this chapter you should be able to:

▶ Identify the main functions of operating systems and network operating systems and distinguish between the two.

▶ Describe the basic functions of a client/server system.

▶ Identify the basic features of Novell NetWare, Windows NT, Unix and Linux network operating systems.

▶ Compare and contrast the Novell NetWare, Windows NT, Unix and Linux network operating systems.

▶ Identify the features of other network operating systems, such as OS/2.

▶ Enumerate the various components of software licenses.

▶ Identify common examples of network utility software, Internet software, programming tools, and network applications.

▶ Identify the different types of support devices commonly found on local area networks.

Introduction ▶

Chapter Seven began the discussion of local area networks by introducing the basic topologies supporting most local area networks and the medium access control techniques that allow a workstation to transmit data onto the network. Chapter Eight continued the discussion, focusing on the interconnection of local area networks to other local area networks and to wide area networks. The bridge, switch, router and gateway are the four basic tools that can perform this interconnection. This chapter will conclude the discussion of local area networks by introducing the software that runs the local area network and that runs on a local area network. We will concentrate on two basic areas: network operating systems and application software.

Although there are many more areas of network software, such as diagnostic and maintenance tools, utilities, and programming development environments, network operating systems and applications are two of the most important. Network operating systems are essential if the network is going to allow multiple users to share resources. The network operating system provides user accounts with password protection, controlled access to network resources, and services to help use and administer the network. Local area network operating systems, much like the hardware that supports them, continue to evolve into more powerful and elaborate tools every day. The more popular network operating systems include Novell NetWare, Microsoft Windows NT, Unix/Linux, and OS/2. This chapter will outline each of these operating system's basic features and capabilities and will compare the advantages and disadvantages of each.

Application software is essential since it is the reason users use networked workstations. The more common applications include word processors, database systems (including client/server systems), web browsers, and office suite software, which often incorporates word processing, spreadsheets, messaging, time management, and project management tools. It is important to understand the difference between how application software works on an individual PC and how it works on a networked workstation. It is equally important to understand the different software licensing agreements that apply to network and application software.

Finally, local area networks require many different types of support devices. As you saw in Chapter Eight, hubs, bridges, routers, and switches are important support devices for the segmentation and interconnection of local area networks. Modems, modem pools, and multiplexors (discussed in Chapters Four and Five) are also useful support devices. Other support devices that will be considered in this chapter include uninterruptable power supplies, tape drives, printers, media converters, and workstations. Each of these devices will be introduced along with examples of how they can be used.

Network Operating Systems

What is a network operating system (NOS)? What functions does it perform? How is a network operating system different from an individual PC's operating system (OS)? The best way to begin answering these questions is by understanding the basic functions of an OS. Once you understand the basic functions of an operating system, you can then begin to understand the additional functions of a network operating system.

An **operating system** manages all the other programs (applications) and resources (such as disk drives, memory, and peripheral devices) in a computer and is the program that is initially loaded into computer memory when the computer is turned on. Even after an application starts and is executing, the application makes use of the operating system by making service requests through a defined **application program interface (API)**. In addition, users can interact directly with the operating system through an interface such as a graphical user interface, command language, or shell.

An operating system can perform a number of services, most of which are crucial to the proper operation of today's computer systems. One of the most important services is determining which applications run in what order and how much time should be allowed for each application before giving another application a turn. A **multitasking operating system**, in which multiple programs can be running at the same time, is an operating system that schedules each task and allocates a small amount of time to the execution of each task. In reality, a multitasking operating system runs only one program at a time, but jumps from one program to the next so quickly that it appears as if multiple programs are being executed at the same time.

An equally important and difficult operating system task is handling the very complex operations of input and output to and from attached hardware devices, such as hard disks, printers, and modems. Hard disk drives are so complex and elaborate that it would be cruel and unusual punishment to require a user to specify the precise location of each record of data as it is written to or read from a hard disk. Equally as complex as controlling the storage space on a hard disk drive is controlling the storage space in the computer's main memory. Since it is likely that multiple applications are running at the same time, the operating system must allocate the limited amount of main memory such that each application has a sufficient amount of memory in which to operate.

A modern operating system must also provide various levels of operating system security, including directory and file security, memory security, and resource security. As applications and their users become increasingly sophisticated, it is becoming more crucial to provide protection from unscrupulous users.

Finally, a service that is not as obvious as multitasking, memory and storage management, or security, but is equally important is communicating about the status of operations. An operating system sends messages to applications, an interactive user, or a system operator about the status of the current operation. It also sends messages about any errors that may have occurred.

A number of popular operating systems exist on different types of computer systems. For example, popular operating systems for microcomputers include Unix, Linux, Windows 95/98/NT/2000, and IBM's OS/2. Popular operating systems for larger computers (minicomputers and mainframes) include IBM's OS/390 and AS/400, DEC's VMS, and again Unix and Linux.

A **network operating system** (NOS) is a large, complex program that can manage the resources common on most local area networks, in addition to performing the standard operating system services listed above. Table 9-1 summarizes the functions of a network operating system. The resources a network operating system must manage include one or more network servers, multiple network printers, one or more physical networks, and a potentially large number of users directly connected to the network and sometimes remotely connected. Among these resources, the network server is critical. The server, as described in Chapter Eight, is usually a high-powered workstation that maintains a large file system of data sets, applications such as word processors and spreadsheets, user profiles, and access information about all the network peripherals.

Table 9-1
Summary of network operating system functions

Network Operating System Functions
Manage one or more network servers
Maintain a file system of data sets, applications, user profiles, network peripherals
Coordinate all resources and services available
Process requests from users
Prompt users for network login, validate accounts, apply restrictions, perform accounting functions
Manage one or more network printers
Manage the interconnection between local area networks
Manage locally connected users
Manage remotely connected users
Support system security
Support client/server functions
Support web page development and web server operations

A network operating system also performs a wide variety of network support functions, including the coordination of all resources and services on the network and the processing of requests from "client" or user workstations. It also supports user interaction by prompting a user for network login, validating user accounts, restricting user access to resources for which access has not been granted, and performing user accounting functions.

Recently, one of the more important network operating system functions has become the ability of the operating system to interact with the Internet. More and more users working on a local area network expect to find a seamless connection between their workstations and the Internet. If they have to create Internet web pages, they like to have the tools available to create and maintain those pages. The network operating system, then, has to provide the means to store the web pages and make them accessible to external users of the Internet. When a network operating system is performing this operation, it is acting as a web server.

One local area network function that is found in many types of business operations is the client/server system. Let's take a little more detailed look at this important network function.

Client/server systems

The network server and network operating system form the heart of client/server systems. A client/server system is a distributed computing system consisting of a server and one or more clients which request information from the server. In a client/server system, a user at a workstation, the client, is running an application program that creates and sends a request to the server. The server performs the request and returns the results to the client (Figure 9-1). For example, to check the balance on your credit card using your personal computer, a client program in your computer forwards your request (the query) to a server program at the credit card company. The computer at the credit card company accesses your account and retrieves the current balance amount. The balance is returned back to the client (the application in your personal computer), which displays the information on your monitor. Many of the additional services of a network operating system revolve around client and server operations.

Figure 9-1
Data flow through a client/server network

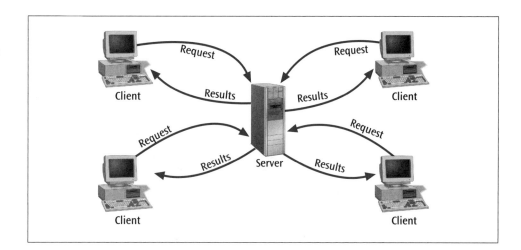

Client/server systems come in two basic architectures: two tier and three tier. A **two tier architecture** is a client/server system in which the user interface is located in the user's workstation environment and the database management system is located in a server. Processing capabilities are divided between the user or client interface environment and the database management server environment. Two tier client/server systems work fine for small numbers of users (less than 100). Above 100 users, the performance of two tier systems degrades. To support a higher number of users, you should consider a three tier architecture. A **three tier architecture** (or multi-tier architecture) adds a middle tier between the client interface environment and the database management server environment. This middle tier, in the most basic form, is a transaction processing monitor. A transaction processing monitor is responsible for performing message queuing and scheduling and prioritization for work in progress. For example, if the middle tier is responsible for queuing requests, the client can deliver its request to the middle tier and then return to prompt a user for another request. The middle tier sends the query to the database and then returns the results to the client's screen.

A strong advantage to a three tier client server system is its ability to handle thousands of client workstations. The main disadvantage is the increased level of complexity involved in creating and supporting a three tier environment.

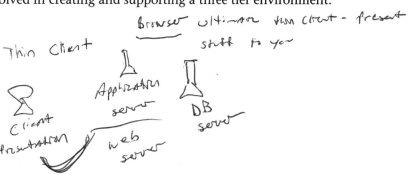

Current Network Operating Systems

A network operating system supports many functions that we, as users, normally take for granted. Having an understanding of these basic functions allows you to examine actual network operating systems more closely, compare these products by how well each supports the basic functions, and evaluate their strengths and weaknesses. Having mastered the basic functions, let's turn our attention to several of the more popular network operating systems on the market and see how each supports these basic functions.

Details ▶

Client/Server Systems and the Remote Procedure Call

In the client/server model, a client performs a request for some operation. The operation could be a database query, a file transfer, or any operation that one computer could call upon a second computer to perform. To support the client/server request, the client must call the remote server and request the data, an operation performed by the remote procedure call. The **remote procedure call** allows a client application to request a service from a server by emulating, or imitating, the call of a local procedure. Thus, similar to one local procedure calling a second local procedure while passing the values necessary for computation (parameter passing), a client procedure calls a remote server procedure, also passing the appropriate parameters.

The actual operation is, of course, a little more complicated. Both the client and server possess a library routine called a **stub**, which handles the details of the actual data transfer. The client calls its stub passing parameters as in a normal procedure call. The stub takes the parameters and packs them into a message (parameter marshalling), which is given to the transport layer. The transport layer passes the message to the network, and the message is transmitted to the server's transport layer. The server's transport layer passes the message to the server stub, which unmarshalls the parameters and gives them to the server. The server performs the requested operation and returns the results in the same manner as the request was transmitted. Figure 9-2 demonstrates a sample client-server interaction.

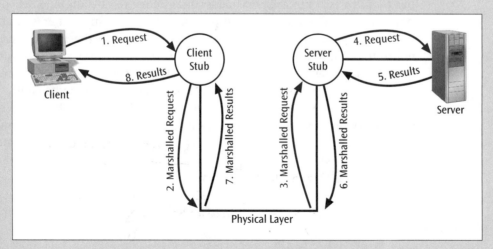

Figure 9-2   *Sample dialogue for a remote procedure call in a client/server system*

Novell NetWare

Novell was founded in 1983 and was one of the first developers of network operating systems. Over the years it has produced a number of successful products including Novell Directory Services (NDS), which is an intelligent system that authenticates users and includes a distributed database of information about every application, user, server and resource on a network. NetWare 5 is the latest version of their network operating system. At one point, over 70% of the LAN market used Novell's network operating system, but this market share has dropped due to serious competition from Windows NT (2000) and other products. Three versions of NetWare are currently in use: NetWare 3, NetWare 4, and NetWare 5.

An interesting feature shared by all versions of NetWare is that the user interface is virtually invisible to the user. When a workstation is part of a Novell NetWare operating system, the workstation must still run a machine-based operating system such as Windows 98 or Macintosh System 9. The user performs Windows-type operations such as opening folders and double-clicking application icons and does not realize that an application is physically located on a network server rather than on his or her own workstation. Novell hides the details from the user and performs all the network functions, such as client/server requests and network printing operations, with little or no knowledge on the part of the user.

For example, if you are using a word processor and wish to print a document, you would click the printer icon in the word processing program. If you have (or a network administrator has) instructed your workstation to redirect all printer output to a network printer, your print job will not print on the printer in your office (assuming you have one) but will be redirected to another printer. Once this **redirection** has been set, all future print requests will automatically be forwarded to the network printer—the operation has become transparent to both you and the word processing application.

Likewise, it is also possible to redirect requests to a hard disk drive. For example, the word processing application may not reside on *your* hard disk but on a *network server's* hard disk. The network administrator has instructed your workstation to redirect certain hard disk requests to a hard disk on the network server. Although the Novell network operating systems were the first to perform this redirection on a local area network, all other network operating systems can now perform the same redirection.

NetWare 3 Most popular

NetWare 3, one of the most popular versions of NetWare, is a 32-bit network operating system that can support large, multi-segmented networks. Even though version 3 is showing its age and is slowly being replaced with newer versions, there are still many networks that run this particular version. Version 3 introduced many concepts that later versions and competing products have incorporated. For example, all directory information is kept in an area called the bindery. The **bindery** is a database that contains the user names and passwords of network users and groups of users authorized to log in to that server; it consists of three non-identical files that are encrypted for security reasons. A bindery contains the data that pertains only to the server in which it resides. It also holds information about other services provided by the server to the client, such as printer, modem and bridge/router access information. Each server on a multi-server network requires and maintains its own bindery, and users are required to login to a particular server. A user requiring access to two servers would need an account on each server and would have to login to each server separately.

NetWare 3 is famous for a number of other features that set it apart from the competition. Three of those features include NetWare loadable modules, disk mirroring, and the IPX/SPX protocol suite. **NetWare Loadable Modules (NLMs)** are software modules that support various network functions and utilities. To save on server memory space, an NLM is loaded only when needed and then is unloaded when it is no longer required. By keeping all NLMs out of main memory until needed, the core of the operating system (called the kernel) is kept as small and fast as possible. For example, if you wanted to format a disk using NetWare, you would first load the NLM that performs disk formatting, then execute the NLM to format the disk. Once you were done formatting disks, you would unload the NLM, thus freeing main memory for other operations.

Disk mirroring and **disk duplexing** are two techniques for saving data onto hard disk drives and are used to protect users from server disk crashes. As data is written to the server disk drive, a duplicate copy is written to a second disk drive. Thus, if one drive crashes, a complete copy, or mirror image, of the disk exists on the second disk.

The **IPX/SPX protocol suite** is NetWare's pair of communication protocols that support network layer and transport layer functions, respectively. All data packets that move throughout a Novell network are created using the IPX/SPX protocols. **IPX**, which stands for Internet packet exchange, is used to transfer data between the server and workstations. Just as network layer packets are then passed to the medium access control layer for encapsulation into frames, IPX packets are encapsulated and carried by the frames at the MAC layer. **SPX**, which stands for sequenced packet exchange, is layered on top of IPX to form a reliable end-to-end transport layer interface. Although IPX and SPX are original Novell creations, they have become a disadvantage, particularly as TCP/IP has become the standard for internetworking protocols.

The future for NetWare 3 is limited. Novell no longer supports versions 3.11 and 3.12. The latest NetWare version, 3.2, is supposed to be the last enhancement to version 3.

NetWare 4

NetWare 4 represented a significant change from NetWare 3. Version 4 introduced Novell Directory Services, which replaced the bindery. The **Novell Directory Services (NDS)** is a database that maintains information on, and access to, every resource on the network, including users, groups of users, printers, data sets, and servers. NDS is based on X.500 which is a well-known standard for directory services. All network resources are objects in the NDS, independent of their actual physical location. NDS is global to the network and is replicated on multiple servers to protect it from failure at a single point. Since NDS is global, users can log in from any point in the network and access any resources for which they have been granted permission. For example, every user that is allowed to log in to the network is entered into the NDS by the network administrator. Likewise, all network support devices such as printers are also entered into the NDS. If user X wants to use printer Y, the network administrator has to assign the appropriate permissions to allow user X access to printer Y. These permissions are also entered into the NDS. Every time user X logs on and tries to send a print job to printer Y, the NDS is referenced for the necessary permissions.

The basic idea underlying NDS is that the network administrator must create a hierarchical tree that represents the layout of the organization. This layout could be physical, such as workstations 1 through 20 on the 3rd floor, workstations

21 through 40 on the 2nd floor, and workstations 41 through 60 on the 1st floor. A more powerful and flexible hierarchical tree can be created based on a logical layout. A logical layout describes the organization in terms of its departmental structure, such as the Engineering department (which could be scattered over floors 1 through 3), the Sales department (which could be situated on floors 2, 6, and 7), and Marketing (which might be physically located in two different buildings).

Thus, a network administrator must give much thought to proper creation of the NDS tree. A good tree design will achieve four goals:

- ▶ A carefully designed NDS tree will structure all the network resources for ease of access.
- ▶ A good design will provide a blueprint for implementation of NetWare 4.
- ▶ A properly crafted tree will provide a flexible layout for future additions.
- ▶ A good design will provide for secure access to shared resources.

Creating a good and appropriate tree design is not a trivial task. However, four basic steps will give the network administrator a good start. First, gather all corporate documents so that you know about available hardware and software and the employee departments and divisions. As part of this first step, you should obtain an organizational chart. Second, design the top section of the tree first. To design the top section of the tree first, you name the tree, name the organization object, and create the first layer of organizational units. An **organizational unit (OU)** is an object that is composed of further objects. For example, a division in a company that is composed of multiple departments is considered an organizational unit. A department is also an organizational unit if it is composed of further objects such as employees. If your network is small, there may not be a need for any organizational units. Third, you design the bottom section of the tree, which includes the remaining hierarchy of organizational units and leaf objects. A **leaf object** is composed of no further objects and includes entities such as the users, peripherals, servers, printers, queues, and other network resources. Figure 9-3 shows an example of an appropriate tree design.

Figure 9-3
Possible design for a Novell NetWare NDS tree

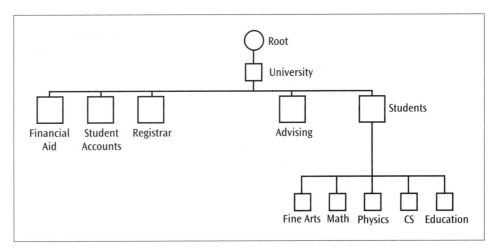

The fourth and final step in creating an NDS tree is to design the security access rights—assign rights to the appropriate NDS objects. Throughout the entire creation process, it is important to review your draft design for errors, flexibility, and completeness. Since an NDS tree is going to be used by every server on the network and for a number of years, it is extremely important to create a well-designed tree. A tree that is designed improperly will lead to difficulties in the future when you try to add new users, user groups, and resources to the network. A poorly designed tree may also create a sluggish system with poor response times.

Test The NDS is one of the most significant features of Netware version 4 and to this day remains one of Novell NetWare's key selling features. One of the few disadvantages of NetWare 4 is that establishing the NDS is much more challenging than creating the bindery found in the earlier versions. Most users of NetWare 4 feel the learning curve necessary to properly create an NDS tree is worth the trouble, as the advantages outweigh the disadvantages. Nonetheless, because of this more challenging setup requirement, a number of NetWare users have been reluctant to move from NetWare 3 to NetWare 4 and higher.

NetWare 5

NetWare 5 is the latest major version of the NetWare network operating system and incorporates some noteworthy features. It retains the NDS distributed database of network resources, which was first introduced in version 4. As NetWare products continue to be used as network operating systems, the NDS continues to grow in popularity and flexibility. The administrative functions of creating and supporting the NDS tree and utilities continue to have a graphical user interface in version 5 as they did in version 4.

Although NetWare 5 maintains compatibility with the IPX and SPX protocols used in earlier versions, it also allows the network administrator to create a network that runs only IP, the internetworking protocol portion of TCP/IP, which supports the Internet. When a network supports IP directly, it is said to "run native IP." This native IP allows a Novell network to interact more quickly with other TCP/IP networks such as the Internet. The NDS has been modified to incorporate some of the directory, file system, and naming protocols found on the Internet, thus enabling NetWare 5 to more easily and quickly interact with the Internet.

NetWare 5 provides a platform for distributed network applications by including Java, database, and a complete set of application programming interfaces. Thus, NetWare 5 can be used to create networked applications that can access databases and the Internet.

Finally, Novell recognized that user and system hard disk requirements are growing at phenomenal rates. In 1985, a microcomputer with a 20 Mb hard disk was a personal computer with a lot of disk storage. Today, microcomputers incorporate disk drives with gigabytes (billions of bytes) of disk storage. And as the number and size of software applications continue to grow, hard disk storage sizes will also continue to grow. NetWare 5 introduces the Novell Storage Services (NSS), which enables the server to handle billions of volumes containing files up to 8 terabytes (8 trillion bytes) in size.

In summary, NetWare has done a very good job of supporting basic network operating system functions. It can easily manage one or more network servers; the NDS directory system is powerful and one of the most popular in the industry at maintaining a file system of data sets, applications, user profiles, and network peripherals; it can coordinate all available resources and services; it can process requests from users and prompt users for network login, validate accounts, apply restrictions, and perform accounting functions; it can successfully manage one or more network printers; it can manage the interconnection between local area networks; and it can manage both locally connected users as well as remotely connected users. As a supporter of client/server functions, NetWare is constantly improving. While NetWare does not have the large base of applications designed for client/server systems that some other network operating systems have, its base of applications continues to grow. The latest release of NetWare also includes the necessary support tools to create and maintain Internet web pages. Finally, network security is strong. NetWare is a stable, robust operating system that has enjoyed success for a number of years.

NT = New Technology = Application, DB, web servers

Microsoft Windows NT

To compete in the local area network operating system market, Microsoft announced a new operating system at Comdex in October 1991. Microsoft had been working on it for roughly three years, and despite a title similar to Windows, it was a significant departure from the personal computer operating system. The new operating system's title was Windows NT version 3 (NT stands for "new technology"). Since its introduction, Windows NT has gone through a number of versions and has become very serious competition for Novell's NetWare operating system.

Similar to NetWare, **Windows NT** is a network operating system designed to run over a network of microcomputer workstations and provide file sharing and peripheral sharing. Also similar to the most recent versions of NetWare, Windows NT was designed to offer the necessary administrative tools to support multiple users, multiple servers, and a wide range of network peripheral devices. NT also supports many of the applications that allow users to create, access, and display web pages, as well as the software that allows a network server to act as a web server. A web server is where users place their web pages such that other users on the World Wide Web can access their pages. One strength that sets NT apart from NetWare is that the Windows NT operating system works seamlessly with the very popular Microsoft applications tools.

Windows NT Version 4

First released in August 1996, Windows NT version 4 contained a number of significant features that very quickly made Windows NT a compelling and competitive network operating system. One of the most noticeable features of NT version 4 is the choice of user interface. The Microsoft Windows 95 operating system user interface for single-user personal computers is incorporated into Windows NT version 4, making the server interface easier to use and consistent with other Microsoft products. Windows NT also offers the Microsoft Management console and administrative wizards. Although NT literature states that this software is supposed to combine the administration tasks into one easy-to-use tool, many network administrators state that administration of an NT system is scattered over many management tools and thus is difficult to perform. The Microsoft Management console allows administrators to create task-based consoles that can be delegated to the appropriate administrator. These tools are assisted by the new network monitor and diagnostic tools, which allow the administrator to follow network traffic, perform traffic analysis, and make network changes and repairs.

NetWare 3 was not a multi server NDS.

NetWare operating system.

Windows NT version 4 has quickly become a successful network operating system. It does a very good job of supporting most, if not all, of the network operating system functions. For instance, it can manage one or more network servers. Although it does not use a directory system as powerful or as popular as NetWare's NDS, it can maintain a file system of data sets, applications, user profiles, and network peripherals. Windows NT can coordinate all available resources and services; it can process requests from users and prompt users for network login, validate accounts, apply restrictions, and perform accounting functions. It can manage one or more network printers. It can manage the interconnection between local area networks, and it can manage both locally connected users as well as remotely connected users. Windows NT is also good at supporting web page creation and administration. Finally, it is highly regarded in its support for client/server functions. Using either Microsoft applications on the user workstations (the front end) or database systems located on the server (the back end), Windows NT provides a smooth marriage between the network operating system and Microsoft applications.

Security is the one area that many people feel is a weakness of Windows NT. Windows NT is a relatively new network operating system, and like most newly released software, it has a number of bugs. NetWare has been available for a longer period of time and has had more opportunities to debug problems. Nonetheless, Microsoft continues to work hard at eliminating most if not all of its security problems, as well as other problems inherent with such a large and complex operating system.

Windows 2000

Windows 2000, released in February 2000, is the next generation of the Windows NT operating system. There are three versions of this operating system:

> ▶ Professional, which supports small to medium-sized businesses;
> ▶ Server, which is a more powerful system designed to support database-intensive business applications; and
> ▶ Advanced Server, which is the most powerful version available and supports systems such as online transaction processing, large-scale data warehousing, web site hosting, large-scale Internet service hosting, and large-scale scientific and engineering simulations.

All three versions incorporate Microsoft's answer to the very popular NetWare NDS directory service—Active Directory. **Active Directory** stores information about all the objects and resources in a network and makes this information available to users, network administrators, and application programs. Like NetWare's NDS, Active Directory creates a hierarchical structure of resources. This hierarchical structure resembles an inverted tree, with the root at the top and the users and network resources—the leaves—at the bottom. Microsoft felt it was in their best interest to create a directory service based on existing standards rather than create a new proprietary directory service. Thus, you may hear network specialists talk about how Active Directory is built around the Internet's Domain Naming Service (DNS) (which is introduced in Chapter Eleven) and a second standard that is growing in popularity—Lightweight Directory Access Protocol (LDAP).

Windows XP - New Release of Windows

To create an Active Directory hierarchy, you create a tree design similar to the tree design of NDS (Figure 9-4). Objects, such as users, groups of users, computers, applications, and network devices, are the leaf objects within the tree. Above the leaf objects are container objects to represent organizations within the business. For example, two of the departments within an organization may be marketing and engineering. Both of these departments are organizational objects. Under both marketing and engineering are the users, their computers, and their trade-specific software applications. There are other users, computers, and applications within the business and they are denoted by other objects within the tree. Grouping objects within the directory tree allows administrators to manage objects and resources on a macro-level rather than on a one-by-one basis. With a few clicks of a mouse, a network administrator can allow all the users of the engineering department to access a new engineering-type software application.

Figure 9-4
Example hierarchical tree design of Windows 2000 Active Directory

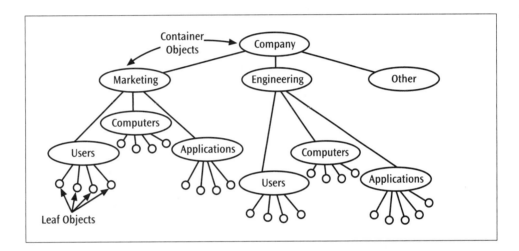

Windows 2000 is a relatively new release and has undergone significant changes from its predecessor Windows NT. It is too soon to tell if it will be a successful network operating system. Like its predecessor Windows NT Version 4, however, it performs the necessary functions a local area network operating system is supposed to perform and has the backing of the largest software company in the world.

Unix - multi user operating system

Unix is a popular operating system that functions well on mainframe computers and network servers, but can also be found to a lesser degree on single-user workstations. It is most often found with a text-based interface, although graphical user interfaces are available. It was initially designed at Bell Labs and was first implemented on a PDP-7 minicomputer in 1970. It is a relatively streamlined system, which explains why it operates quickly. Shortly after its introduction, the software was rewritten in the popular C programming language. Because of the characteristics of the C programming language and because of Unix's design, its internal code is relatively easy to modify. In addition, early versions were given away for free. As a result, Unix became extremely popular in academic institutions.

Because Unix is one of the older operating systems, it is a very stable system that has grown in processing power over the years. It handles network operations well, and a wide variety of applications have been written to run under Unix. Today, many experts consider Unix to be one of the best operating systems for supporting large applications such as multi-user database systems and web-based servers.

One of the biggest drawbacks of Unix is the user interface. Many users today expect a graphical user interface and find Unix's text-based, command line interface old-fashioned and cumbersome. A second disadvantage is that although Unix adequately supports a directory system that can maintain data sets, applications, and user profiles, it is not as powerful as NetWare's NDS.

Unix has many strengths as a network operating system. It can manage one or more network servers. It can coordinate all resources and services available as well as process requests from users, prompt users for network log ins, validate accounts, apply restrictions, and perform accounting functions. Unix is also very capable of managing one or more network printers as well as the interconnections between local area networks. It can support locally connected users as well as remotely connected users and has a very stable security system based on many years of experience. Over the years, many applications have been created for Unix that support client/server functions as well as web page development and web server operations.

Unix is a mature, stable network operating system that will probably be available for many years. It has a devout following that will ensure its evolution in the face of competition from newcomers.

Linux

*Linus
Torvalds*

Since its early days, Unix has evolved into many versions and, as a result, executes on a number of computer platforms. One version that generated a lot of interest in the late 20th century is Linux. **Linux**, while based on the Unix concept, is a complete rewrite of the Unix kernel and borrows additional functions from the well-established Free Software Foundation's GNU toolset and from the even larger free software community. It shares many of the same advantages and disadvantages of Unix and performs similar to Unix as a network operating system. There are, however, a number of features that set Linux apart from Unix and from all other network operating systems. First and foremost is the cost of Linux—it is free if downloaded from the Internet.

Many people, however, purchase Linux from a vendor that specializes in providing the latest version of Linux along with all the supporting applications and

Details ▶

Unix Versions

Currently there are three popular versions of Unix: 4.4BSD, System V Release 4, and Solaris 2.x. 4.4BSD is a Unix system developed by Berkeley University. (BSD stands for Berkeley Software Distribution.) The BSD series of Unix releases are non-commercial and are widely used in academic institutions; they have also served as the basis for a number of commercial Unix products. Many feel it is because of the BSD versions that Unix has become so popular today in the computing arena. System V Release 4 was developed jointly by Sun Microsystems and AT&T. This version of Unix incorporates many useful features and has drawn on the success of earlier releases to strengthen its product, while still featuring an almost completely new core software system (kernel) with updated services. System V Release 4 was designed to operate in a commercial environment and has experienced much success. This version can run on a wide range of computer platforms—from personal computers to supercomputers. Because of this cross-platform flexibility, designers of operating systems feel System V Release 4 is one of the most important operating systems ever developed. Solaris 2.x is Sun's System V Release 4 version of Unix. It is very similar to the System V Release 4 version, but has some additional advanced features.

utilities, such as web page support tools and a graphical user interface. The cost of purchasing Linux from a vendor is insignificant compared to the cost of purchasing either NetWare or NT. Whereas NetWare and NT can cost thousands to tens of thousands of dollars depending upon the number of user workstations, commercial versions of Linux cost less than one hundred dollars.

A second strong selling point for Linux is quick bug fixes. Linux has an extremely devoted group of supporters who have staked their reputations on providing a bug-free operating system. Thus, any errors reported to the various Linux support groups are often fixed in a very short time, and then appear in the next version of the software.

A third advantage of Linux over other network operating systems is that when you purchase or download Linux, you can receive the original source code for free along with the compiled code. Having the original source code gives organizations a great amount of control over the software. In the hands of an experienced programmer, the Linux source code can be modified in almost limitless ways to provide a customized operating system. The ability to customize the software, however, is a double-edged sword. In inexperienced hands, customized source code can be a source of constant trouble.

Linux's final advantage is the size of the code. Since its debut, Linux has been able to operate on a system as small as an Intel 386 processor with only 4 MB of main memory. Although Linux will certainly run better on larger systems, many administrators use a Linux system on an old personal computer that supports simpler network functions such as providing network security (a firewall). These systems apparently run unattended with little or no support.

Linux shares many other advantages with Unix, such as fast execution due to its relatively small size and efficient code, network support functions, and an optional graphical user interface that gives the software the feel of a product with a graphical user interface.

Although many supporters of other network operating systems claim that Linux is not yet ready for "prime-time," articles in industry journals are appearing very frequently that claim the contrary. One article reports that a Linux system running on a network server with four concurrent processors and 2 GB of RAM performed as well as a Windows NT 4.0 system on comparable hardware. Although this data is not the product of a scientific study and deserves closer scrutiny, the sheer volume of such articles should lead you to believe that Linux warrants consideration along with other commercial network operating systems.

Linux, however, has two important negative features. First, since Linux is not sold through a large company like Microsoft or Novell, many network professionals fear that technical support for the system is inadequate. Second, although more and more software companies are announcing that they are producing or will produce a version of their software for Linux, applications that run on a Linux system are not as plentiful as applications that run on other network operating systems.

IBM's OS/2

 Debuting in December of 1987, **OS/2** Version 1.00 was the first personal computer operating system that could perform true multitasking. Remember that multitasking is an operating system's ability to keep two or more programs running at the same time. OS/2 version 1.00 provided only a command line interface and would only allow one program to be visible on the screen at a time, although other programs could be running in the background. The operating system could support only a 32 MB hard disk, and it was designed for the Intel 80286, although it could also run on newer 80386 systems.

By October 1988, IBM released an upgrade to OS/2 Version 1.00 that added a graphical user interface called Presentation Manager. Shortly thereafter IBM upgraded the system again, this time adding Database Manager, a multitasking relational database system, and Communications Manager, which provided interfaces to IBM mainframes and midrange computer systems. More upgrades followed as IBM continued to improve OS/2 by simplifying the administrative chores, adding features such as a new file system, adding Adobe type fonts, and improving the video device drivers to accommodate higher resolution monitors.

Released in early 1992, OS/2 version 2.00 was a true 32-bit operating system for a personal computer. Version 2.00 introduced the Virtual DOS Machine (VDM), which would allow a DOS or Windows application to execute as if it was running on a separate computer (a virtual machine). VDM also supported multiple applications to run at the same time. Since each application was running on its own virtual machine, if the application crashed, it would not cripple the entire operating system. Only the one virtual machine in which the application resided would crash. Subsequent upgrades to version 2.00 included a new and improved 32-bit graphics subsystem, PCMCIA support for laptop computers, and support for sound and video multimedia applications.

Making its debut in the autumn of 1994, OS/2 Warp (Version 3.00) supports TCP/IP and Internet communications and allows users access to the World Wide Web.

Shortly after version 3.00 was introduced, IBM introduced Warp Connect and Warp Server. **Warp Connect** allows a client workstation to access network resources such as files, printers, and other peripheral devices. **Warp Server** allows a network-connected workstation to act as an application server.

The latest version of the OS/2 operating system, OS/2 Warp 4 includes a universal network client that allows an OS/2 machine to connect to virtually any network system, including Windows NT, NetWare, and IBM's mainframe network system SNA. It also includes speech recognition software so a user can issue verbal commands and dictation, support for the Java language, remote access services that allow a user to access an OS/2 system from a remote site, and mobile office services that allow a nomadic user to keep files synchronized with the home office.

Summary of network operating systems

To conclude this section, let's summarize the network operating systems and compare them on several criteria: range of compatible hardware and performance, installed base and corporate acceptance, stability, TCP/IP support, and greatest strength. Table 9-2 compares some of the features of the network operating systems discussed.

For the widest range of compatible hardware, Unix and Linux are hard to beat. These two operating systems run on a large number of processor types, both large and small. Windows NT/2000, NetWare, and OS/2 can also operate on a fairly wide range of platforms. Performance-wise, all NOSs presented here are very good. One NOS may outperform the others in a particular area, however. For example, NetWare is very fast at serving files to requesting workstations. Unix and Linux are also fast operating systems, partly due to their streamlined code. Recent studies, however, have shown that NetWare Version 5 is the fastest network operating system overall.

Except for OS/2, all the NOSs have very large installed user bases and can be found in a wide variety of businesses, both large and small. Microsoft products probably enjoy the widest corporate acceptance rate, although NetWare and Unix are very close behind. Linux lags behind the others partly due to the fact that there is no one large software company standing behind the product, ready to provide support. Linux supporters argue, however, that the software needs little support since it runs so well. Statements such as these are subject to further investigation.

Most NOSs are very stable operating systems. The one exception might be NT, partly due to the fact it is one of the newest players in the market and still experiencing a shake-down.

All contemporary NOSs support TCP/IP protocols. Most operating systems support TCP/IP directly, but NT supports its own proprietary network protocol (Net-BEUI) and layers TCP/IP on top of it.

Finally, each network operating system has a particular strength. NetWare is an excellent file server. In fact, for file system power (directory services), NetWare is hard to beat and clearly holds the advantage. Windows NT/2000 is an excellent application server providing client/server capabilities to a wide range of applications. Unix, Linux, and OS/2 give power and flexibility by providing a wide range of services to many different types of network applications.

Table 9-2
Comparison of network operating system products

Criteria	NetWare	NT	Unix	Linux	OS/2
Range of compatible hardware	Modest	Modest	Very wide	Very wide	Modest
Performance	Highest	Good	High	High	Good
Corporate acceptance	Wide	Wide	Wide	Small, but growing	Modest
Installed base	Millions	Millions	Millions	Millions	Hundred of thousands
File system power	Very high	Modest	Modest	Modest	Modest
Java developer tools	Yes	Yes	Yes	Lagging	Yes
Stability	High	Modest	High	Very high	High
Software cost	Moderate to high ($50 - $200 per workstation)	Moderate to high	Moderate	Free	Moderate to high
TCP/IP support	Yes	Yes	Yes	Native	Yes
Strength	File server, NDS	Application server	Speed, stability	Cost, speed, stability	Flexible

Network Software: Utilities, Tools, and Applications

Even though the network operating system is clearly the most important piece of software on a local area network, the operating system can not work alone. Four additional areas of local area network software that work with and support the network operating system are:

- ▶ utilities — software programs that support one or more functions that help keep the network running at optimal performance and that operate in the background;
- ▶ Internet software — programs that allow access to the Internet;
- ▶ programming tools — software programs and programming environments that allow users to create their own applications; and
- ▶ applications — tools that allow users to perform the daily work functions found in most business, educational, and home environments.

Utilities

To support a local area network and its operating system, a wide variety of utility programs are available. Sometimes these utility programs come bundled with an operating system, but many times they are separate and have to be purchased individually. As with a network operating system, it is important when purchasing utility software (as well as any software product) to pay attention to the licensing agreements. Many utility programs are licensed for a single machine. If there are 200 workstations on your local area network, you may have to purchase 200 licenses before you are within the letter of the law.

Five of the more common groups of network utility software are: anti-virus software, backup software, crash protection software, remote access software, and uninstall software.

Anti-virus software is designed to detect and remove viruses that have infected your memory, disks, or operating system. Since new viruses appear all the time, it is important to continuously update anti-virus software with the latest version. Many times, owners who purchase anti-virus software can download these updates from the software company's web site for no additional charge. The only time it is necessary to purchase a new version of the anti-virus software is when the software has undergone a major revision. Some anti-virus manufacturers offer corporate deals in which a flat annual fee is paid and the company can distribute and update as many copies of the software as desired. See the author's web page for some good websites devoted to anti-virus information (*http://bach.cs.depaul.edu/cwhite*).

Backup software allows network administrators to back up data files currently stored on the network server's hard disk drive. Typically, this backup is written to a tape system, but there are other possibilities, such as performing a backup to a remote site via the Internet. Most backup systems can be completely automated so that the software executes at times when few, if any, users are using the system, such as the early morning hours.

The primary goal of **crash protection software** is to perform crash stalling. To take advantage of this utility it is necessary to have software on both the user's workstation and the network server. When an application is running on a user workstation and is about to crash, that application transmits a signal to the operating system stating that it is experiencing trouble and needs to be shut down. A crash protector intercepts this message and attempts to fix the problem while keeping the application running. Even if the severity of the problem is beyond repair, the crash protector will often create a pause that allows users to save important data and exit the program safely. Since a crash protector runs constantly in the background, you should expect a performance slowdown. Most current crash protectors degrade overall system performance anywhere from 2 percent to 8 percent. Although an extra 8 percent in processing time may not seem like much, if you are a serious computer user (for which this software is designed), you may notice the loss in computing speed.

Remote access software allows a person to access all of the possible functions of a personal computer workstation from a mobile or remote location. The two most common users of remote access software are nomadic users who must access software and data on their computers while traveling on the road or while at home, and support personnel who can enter a user's computer system to troubleshoot problems and suggest or make repairs.

Software applications and the number of support files necessary to execute the programs continue to grow in size. If someone decides to remove an application, it is virtually impossible to locate all the files associated with the application. **Uninstall software** works with the user to locate and remove applications that are no longer desired. As the uninstall software locates each associated file, it prompts the user to decide whether the file should be removed. Since some files may be shared by multiple applications, the user may not want to perform the remove operation. Most uninstall programs keep a backup of removed files to correct problems if a mistake is made and the wrong file is removed. Uninstall programs can also locate and prompt the user for removal of orphan files that no longer belong to any application. Finally, many uninstall programs can locate and remove duplicate files.

Internet server software

One of the quickest growing segments of the software market is the tool set to support Internet-related services. These services and applications include web browsers, server software, and web page publishing software, among other applications.

Browsers allow users to download and view World Wide Web pages. Most users are familiar with the two most popular browsers, Netscape Navigator and Microsoft Internet Explorer. Although a few other browsers exist, such as Opera and Lynx, Navigator and Internet Explorer control almost the entire market.

Server software is the application or set of programs that stores web pages and allows browsers from anywhere in the world to access those web pages. When a user running a browser enters or clicks a web page address (its uniform resource locator, or URL), the browser transmits a page request to the server at the address cited. The server receives the request, retrieves the appropriate web page, and transmits the requested web page back across the Internet to the browser. Server software is capable of supporting secure connections. A secure connection allows the system to transfer sensitive data such as credit card information with the confidence that the integrity of the data will not be violated.

When users wish to create one or more web pages that will be stored on an Internet server, web page publishing software helps them prepare the necessary files. Most web page publishing programs allow users to insert static images, animated images, various forms of script, and Java-based code into html files.

Programming tools

Many business environments, as well as many educational institutions, hire computer programmers to write new programs or modify already existing programs. These programs are tailor-made for specific applications within the organization. When a programmer creates a program, most often the programming language used is a higher level language such as C, C++, Java, COBOL, or Visual Basic. Some programs, specifically designed to function on the Web, use scripting languages, such as JavaScript or VBScript. Programmers spend many hours writing, debugging, and testing the code. To support these common programming operations, an organization's network provides the appropriate compiler, interpreter, or programming environment.

Compilers and interpreters are the complex software programs that translate a program written in a higher level language, such as C or Java, into a form that can execute on a particular processor. A compiler produces machine code which is directly executed by a particular processor. An interpreter produces code that can be interpreted by a number of different processors.

A programming environment is a collection of tools that allow a user to create, edit, debug and test programs, usually all within the same application. Thus, a user can write a program, compile it, read the compiler messages, make changes to the program, re-compile it, link modules together, and execute and debug the program simply by switching between dialog boxes within a single application.

Application software

Most local area networks support a sizeable collection of application software. A user at a workstation requests an application by entering a keyword at a command line prompt or by clicking an icon. The application can be retrieved in two ways. In the first case, the application resides on the user's machine. When the user starts the application, it is loaded from the user's hard disk drive into the user's memory space, where the application executes. In the second case, the network operating system receives the request and sends it to the network server. The requested application is retrieved from the network server's hard disk and transferred across the network to the user's machine, where it is loaded into the user's memory space for execution.

Loading applications locally has the advantage of speed. Since the software is not transferred across the network, but loaded from the user's machine, the application loads much more quickly. The disadvantage to this method is support. If 200 workstations in a business each have their own copy of a word processor and an upgrade to the software is issued, a network administrator has to load a copy of the upgraded word processor onto each of the 200 workstations. If the software is kept on the network server, the administrator has to upgrade only a single copy.

Five of the most common applications found on local area networks include:

- ▶ database software, both stand-alone versions and client/server versions, which allows individuals to store and retrieve small and large volumes of information in an organized, efficient, and timely manner;
- ▶ desktop publishing, which allows individuals to make professional quality documents;

> ► office suites, which are integrated collections of basic office functions, such as word processing, spreadsheets, presentation software, time management software, and electronic mail; and
>
> ► stand-alone spreadsheet, word processing, and presentation software, which allow individuals to perform the office functions necessary to efficiently and effectively support the business.

Software Licensing Agreements

The **licensing agreement** that accompanies a software product is a legal contract and describes a number of conditions that must be upheld for proper use of the software package. Most licensing agreements specify the following conditions:

> ► Software installation and use — Specifies on how many computers a user may legally install and use this software.
>
> ► Network installation — Describes if the package may be installed on a computer network, and if so, if additional licenses are necessary for each machine on the network.
>
> ► Back-up copy — Specifies whether a back-up copy is acceptable.
>
> ► Decompilation — Specifies that a user may not decompile, disassemble, or reverse engineer the software code in an attempt to retrieve the high-level language in hopes of making modifications to the code.
>
> ► Rental statement — States that a user is not allowed to rent or lease the software to a third party.
>
> ► Upgrades — If the software program is an upgrade to a previous version, informs the user that the previous version needs to be purchased before installation of the upgrade.
>
> ► Copyright — Informs the user that all documentation, images, and other materials included in the package are copyrighted and under the protection of copyright laws.
>
> ► Maintenance — Informs user if product support is included in the purchase of the software package.

One of the most important conditions that affects users is software installation and use. When a software package is sold, it is usually intended for a particular type of installation, referred to as the user license. Software companies establish user licenses so that an individual in a company does not purchase one copy of a program and install it on, for example, 200 machines, thus cheating the software company out of 199 purchase fees, or royalties. These royalties pay for the cost of creating the software program and help support the future cost of upgrades and maintenance.

There are several forms of user licenses: single user licenses, interactive user licenses, system-based licenses, site licenses, and corporate licenses. A **single user single station license** is one of the most common user licenses. The software package may be installed on a single machine, and only a single user at one time may use that machine. Some software packages actually count how many times the software has been installed and allow only a single installation. Moving the software to another machine requires running the software's uninstall utility. A **single user multiple station license** is designed for the user who might have a desktop machine at work and a laptop machine for remote sites or another desktop machine at home. It is up to the honor of the user that only one copy of the software is operating at one time. For example, if the user is at work and using a particular word processor program with the single user multiple station license, no one should be running the same program on the user's laptop at the same time.

An **interactive user license**, operating system user license, or controlled number of concurrent users license all refer to essentially the same situation. When a software package is installed on a multi-user system, such as a network server on a local area network, it is possible for multiple users to execute multiple copies of the single program. Many multi-user software packages maintain a counter for every person currently executing the program. When the maximum number of concurrent users is reached, no further users may access the program. If a software counter is not used, the network administrator must estimate how many concurrent users of this software package are possible at any one moment and take the necessary steps to purchase the appropriate number of concurrent user licenses.

System-based licenses, cluster-wide licenses, or network server licenses are similar to the interactive user licenses. However, with system or **network server licenses** there is almost never a software counter controlling the current number of users.

A **site license** allows a software package to be installed on any and all workstations and servers at a given site. No software installation counter is used. The network administrator must guarantee that a copy of the software does not leave the site.

A **corporate license** allows a software package to be installed anywhere within a corporation, even if installation involves multiple sites. Once again the network administrator must guarantee that a copy of the software does not leave a corporate installation.

Every software company may create its own brand of user license and may name it something unique. It is the responsibility of the person installing the software—either the user, system administrator, or network administrator—to be aware of the details of the user license and follow it carefully.

What happens if someone does not behave in accordance with the user license and installs a software package in an environment for which it was not intended? In these situations, some packages simply will not work. Other packages will not work correctly if the installation does not follow the license agreement. For example, some software packages maintain a counter and will not allow more copies than what was agreed upon during the purchase. Some software packages, such as database systems, may install on multiple machines or may give the appearance of allowing multiple users access, but if they were not designed for multiple user access, they may not work correctly. It is important during multi-user database access that the database system lock out any other users while one user is currently accessing data records. If the system is intentionally not designed for concurrent multi-user access, unknown results might occur.

Installing the software on more machines than agreed to in the user license is illegal. If you install more copies than the number for which you have licenses and the software manufacturer discovers this fact, you and your company may face legal consequences. It is not worth taking the risk. Be sure any software has the proper user license before installing it.

LAN Support Devices

A local area network is composed of many pieces, both software and hardware. This chapter, up to this point, has introduced local area network software—network operating systems, utility, application and software tools, and software licenses. The hardware side of local area networks contains many different types of devices. In Chapter Eight, bridges, routers, and switches—the hardware that interconnects devices on the network—were discussed in detail. Servers were introduced because they are becoming an interconnection point within a network and often contain switch and router hardware. Modem pools, which were discussed in Chapter Four, contain multiple modems for employees dialing out or remote users dialing in and are very often a part of local area networks. Let's look at a number of other network support devices that you may encounter on a typical local area network: uninterruptable power supplies, tape drives, printers, media converters, and workstations.

Uninterruptable power supplies (UPS) are valuable devices in the event of a power failure. A UPS is a battery backup device that can maintain power to one or more pieces of equipment for short periods of time (usually less than one hour). In a corporate environment, UPS devices are often employed to backup network servers and mainframe computers. Some people even use UPS devices in the home to protect themselves from the loss of data resulting from a power interruption. And to protect an electronic device from a possible surge in electrical power, surge protectors should be installed on all computer workstations and network components.

Tape drives are excellent backup devices. It is very common to install a tape drive into a network server so that modified files can be backed up one or more times a day. Many companies backup their files nightly when all the employees are at home and network use is at a minimum. As rewriteable CDs become more common place, the tape drive backup system may be replaced by CD technology.

Printers continue to evolve at a rate that rivals other network devices. A number of years ago there was a wide selection of dot matrix printers, daisy wheel printers, chain printers, ink jet printers, and laser printers. Today, the printer market has shrunk to just two basic formats: ink jet and laser. Both are capable of printing black and white and color with very high resolution. When a printer is hooked up to a network, the appropriate print server software is installed on the network, allowing the printer to be shared by multiple users.

Media converters are handy devices if it is ever necessary to connect one type of media with another. For example, if different portions of one network were installed at different times using different media, it may be more reasonable to use a media converter to connect the two type of media rather than remove all the older media and replace it with the newer media. There are many different types of media converters, just as there are many different combinations of media. For example, there are media converters to connect coaxial cable to twisted pair, twisted pair to fiber optic cable, and SCSI cables to non-SCSI cables.

The computer workstation itself has also evolved over the years. Workstations now support hundreds of megabytes of random access memory and have dozens of gigabytes of hard disk storage. A new concept introduced in the late 1990s was the thin client workstation. A **thin client** is a computer with no floppy disk drive or hard disk storage. Any software that operates on the thin client is downloaded from the network server to the thin client. The idea behind thin client is to minimize workstation maintenance and reduce hardware costs. Security is also improved in a thin client system because it is not possible to insert a floppy disk for storing software or data. The idea of saving on hardware costs by using a thin client instead of a fully configured workstation lost some importance when manufacturers of fully configured workstations dropped their prices to under $1000 for entry level machines. Nonetheless, thin clients still address important security considerations, and their maintenance costs are reduced because of the lack of disk drives.

LAN Software in Action: A University Makes a Choice ▶

Recently, Dominican University, a small private college in the Chicago area, was considering whether to replace their Novell NetWare operating system with Windows NT. The computing administration was receiving a large number of inquiries from faculty and students asking when the various Microsoft applications such as Word, Excel, Access, and PowerPoint would be available on the network. Furthermore, a number of departments were beginning to create database applications, most using Microsoft Access. The questions were raised: "If we are going to install the suite of Microsoft applications, would it make sense also to install Windows NT at the same time and create a 'cohesive' enterprise? Or should we stay with NetWare and migrate to the next version when it becomes available?"

At the time of the decision, there were approximately 150 workstations in student labs, faculty offices, and administrative staff offices with three servers providing mostly word processing, computer programming environments, Internet browsing, and electronic mail. Less-used applications included graphics, statistics, spreadsheets, foreign language programs, and biology, chemistry, and physics educational programs. Applications such as billing, registration, and other business applications were handled by a separate mainframe computer system running the VAX VMS operating system. Windows 95 was installed on every machine and Novell's NetWare version 3.12 was installed on all network servers.

To help make the decision, the network administrator asked himself the following questions:

▶ What are the primary uses of the current local area network?

▶ Would the primary uses of the system change if the Microsoft Office suite was installed?

▶ Is the network support staff trained in NetWare or NT?

▶ Can the university afford Windows NT?

Primary uses of current network

Most of the network traffic was estimated to consist of web page transfers from the Internet, email transfers, word processing, and compiling student programs. To handle student programming needs, a C++ programming environment was installed on each machine in the student labs. Network transactions could then be characterized as simple file/message transfers. NetWare handles these types of transactions as well as, if not better than, any network operating system. Thus, the primary uses of the current network did not merit a change from NetWare.

Changes due to Microsoft Office suite

The primary uses of the network probably would not change if the Microsoft Office suite were installed. Users would continue to use both Netscape Navigator and Internet Explorer, the same as before. More people might begin to use Microsoft Word rather than WordPerfect for word processing, and they might possibly switch from the Quattro Pro spreadsheet to Excel, but the application level would remain the same. Even though more users might develop Access database projects, Access would never be the major application on the network. In fact, Access could operate on the NetWare platform with no problem. Although a number of applications (such as Lotus Notes) run only on NT servers, there were no plans to use any of the NT-only applications. Thus, installing the Microsoft Office suite would not require a change from NetWare to Microsoft NT.

Network support staff training

The network support staff at Dominican University is familiar with NetWare and not NT. NT's security, file system, and administration are significantly different from most other NOSs, including NetWare. In moving from NetWare to NT, there would be a significant learning curve.

Cost of NT

Dominican purchased the NetWare system a couple of years ago, and thus the system has been paid for. Switching to a new NT system would involve a significant amount of money. Upgrading from NetWare version 3 to version 4 would also require funds, but upgrades are often less expensive than purchasing new software. Thus the cost of migrating to NT is a significant factor.

When considering the cost of a network operating system, you need to consider the administrative costs of running a system in addition to the initial outlay of funds to purchase the software. If one NOS is not as stable as another, there could be significant downtime costs. When a new system is installed, significant training costs are incurred. Regardless of the particular system, networks require administration, maintenance, and training. Certain network operating systems, however, excel in one area or another. For example, NetWare has very powerful administrative software that, when coupled with its NDS file system, make administration simpler and more powerful.

What was Dominican University's decision? The university decided to stay with NetWare and migrate to version 4. The cost of Windows NT (both the software costs and the expense of training users and network support staff) did not justify the gains. Since most users (student, faculty, and administrative) used the network for basic file and print serving, there was not significant evidence that changing to NT, or any other network operating system, would improve operations or offer benefits.

When considering one network operating system over another, it is most important that a decision be well thought out. All factors must be weighed, and hype and advertising must not be part of the consideration.

◆ ◆

SUMMARY

- ► A network operating system has several additional functions not normally found in an operating system. An NOS: manages one or more network servers; maintains a file system of data sets, applications, user profiles, network peripherals; coordinates all resources and services available; processes requests from users; prompts users for network login, validates accounts, applies restrictions, and performs accounting functions; manages one or more network printers; manages the interconnection between local area networks; manages locally connected users; manages remotely connected users; supports system security; supports client/server functions; and supports web page development and web server operations.

- ► Novell NetWare is a popular network operating system with a powerful directory service (NDS) and is very good at performing file and print serving.

- ► NetWare is currently available in three versions: Versions 3, 4, and 5.

- ► Windows NT is also a popular network operating system and is very good at supporting client/server applications and has a new directory service called Active Directory.

- ► Windows NT is currently available in two versions: Version 4 and Version 2000.

- ► Unix is an older operating system that is stable, fast, and able to run on a variety of platforms.

- ► Linux is a derivative of Unix and shares the Unix features of stability and speed as well as low cost and the capability to run on a variety of platforms.

- ► IBM OS/2 is a modestly popular network operating system that adequately supports the necessary NOS functions.

- ► Many types of software programs support a local area network and include utility programs, Internet software tools, programming tools, and application programs.

- ► There are also many types of hardware devices necessary to support a local area network. These devices include hubs, bridges, switches, and routers, modems and modem pools, uninterruptable power supplies, surge protectors, tape drives, printers, print servers, media converters, workstations, and network servers.

KEY TERMS

Active Directory
anti-virus software
application programming interface (API)
backup software
bindery
client/server system
corporate license
crash protection software
disk duplexing
disk mirroring
interactive user license

IPX/SPX
leaf object
licensing agreement
multitasking operating system
Novell Directory Service (NDS)
NetWare loadable module (NLM)
network operating system (NOS)
network server license
organizational unit (OU)
operating system
OS/2
redirection

remote access
server software
single user single station license
single user multiple station license
site license
thin client
three tier architecture
two tier architecture
uninstall software
Warp Connect
Warp Server
Windows NT

REVIEW QUESTIONS

1. List the six basic functions of an operating system.
2. What separates a multitasking operating system from a non-multitasking operating system?
3. List the primary differences between a network operating system and an operating system.
4. What is the definition of a client/server system?
5. What is the purpose of a NetWare loadable module?
6. What is meant by disk mirroring?
7. What is the purpose of IPX/SPX?
8. What is the function of NetWare directory service NDS?
9. What are the primary features of Windows NT version 4?
10. What are the main advantages of Windows NT 2000 over NT version 4?
11. What is the function of Windows' Active Directory?
12. What are the strengths of Unix?
13. List the reasons for Linux's popularity.
14. What are the disadvantages of Linux?
15. What are the advantages and disadvantages of OS/2?
16. What are the conditions often stated in a software license agreement?
17. List the types of software license agreements.
18. What are the five most common groups of network utility software?
19. What are the primary functions of an Internet web page server?
20. What are the different types of hardware support devices for local area networks?

EXERCISES

1. In a client/server system, a client transmits a request to a server, the server performs a processing operation, and the server returns a result. List all the possible things that can go wrong with transmission in this scenario.
2. What are the primary advantages of NetWare version 4 over NetWare version 3? Are there any disadvantages?
3. What does it mean when NetWare version 5 runs IP in native mode?
4. What is the difference between NetWare's bindery from Version 3 and NetWare's NDS from Versions 4 and 5?
5. You want to create a local area network that protects the contents of the network server's hard disks from disk crashes. List all the different techniques for providing this protection that have been presented thus far.
6. Windows 2000 uses Active Directory for its directory service, and NetWare uses NDS. How are the two directory services alike?
7. When using either Novell's NDS or Windows 2000's Active Directory, an administrator can control resources on a macro level. Explain what controlling resources on a macro level means and show an example.
8. In what ways are Unix and Linux similar? In what ways are they different?
9. Why is the stability of Linux so high when compared to other network operating systems?

10. The problem with anti-virus software is that new viruses are being created every day. If you have anti-virus software installed on your machine or on your network, how do you keep your software up to date?

11. How does a single user single station software license differ from a single user multiple station license? Does one have an advantage over another in the business world? In the home computer world?

12. What kind of software applications might a company consider as likely candidates for a site license?

13. Is a thin client computer more advantageous in a corporate setting or in a user's home? Explain.

THINKING OUTSIDE THE BOX

1 A small business with 100 computer workstations is installing a new local area network. All of the users perform the usual operations of email, word processing, Internet browsing, and some preparation of spreadsheets. Approximately one quarter of the employees perform a large number of client/server requests into a database system. Which local area network operating system would you recommend? State your reasoning.

2 Create an NDS or Active Directory tree structure for one of the following:
   a. A segment of the business in which you are employed.
   b. All of the business in which you are employed (if your company is a smaller company).
   c. A segment of the school at which you are enrolled.
   d. All of the school at which you are enrolled (if your school is a smaller school).
   e. A hypothetical corporation that has multiple departments, servers, peripherals, and users.

3 Consider the following software licensing scenario: Office Suite 1 costs $229 per single-user single-station license, while Office Suite 2 costs $299 per interactive user license. You have 200 users on your network, and you estimate that at any one time only 60 percent of your users will be using a suite application. Determine the best license solution. At what level of interactive use will the cost break even with the single-user single-station licenses?

PROJECTS

1. Who makes thin client machines? What are their specifications, features, and prices?

2. A workstation on a NetWare local area network wishes to send an email to someone on the Internet. Show the encapsulation of packets as the email message works its way down through the layers of a workstation running NetWare version 4. Now show the encapsulation as the email message works it way down through the layers of NetWare 5 running IP natively.

3. A network operating system that holds a very small share of the market is Banyan Vines. What is the current state of this system? Is it still being produced? Is it still being supported? Can you find an estimate of how many networks are using Banyan Vines?

4. Are there any other network operating systems other than those listed in this chapter? Are they holding their own or are they slipping into oblivion?

5. What is the difference between a copyright and a patent when applied to computer software?

10
Introduction to Wide Area Networks

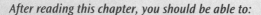

Healthcom is an electronic commerce service that enables wholesale druggists to purchase products from manufacturers and allows wholesale distributors to receive pricing information virtually instantaneously. Healthcom then sends electronic purchase orders from the distributors to the manufacturers. Although a similar service has been available to larger wholesalers for a number of years, it has required expensive hardware and software to implement. Smaller wholesalers, unable to afford this larger system, have relied on U.S. mail, faxes, and costly Western Union mailgrams. Healthcom now gives a more reasonably priced alternative to smaller wholesalers and allows them to compete with larger wholesalers.

Essentially, every drug manufacturer sends every wholesaler an electronic mail message alerting them to new product information. The wholesale distributors can then enter a password-protected site to order products. The service is offered to all wholesalers on a flat subscription rate of $39.95 per month.

"The speed the smaller wholesalers will gain through Healthcom to learn about price increases could work to boost their bottom line. If they get news about manufacturer price increases and act on them even one day in advance, they would save tens of thousands of dollars on each order."

"Health care group offers e-commerce service", *Network World*, August 30, 1999, page 42.

A wide area network allows Healthcom to interconnect all wholesalers. What is a wide area network and how do its characteristics differ from a local area network?

The Internet is the most popular wide area network. Do all wide area networks share a set of features?

Objectives

After reading this chapter, you should be able to:

► Distinguish local area networks, metropolitan area networks, and wide area networks from each other.

► Describe how circuit switched, datagram packet switched, and virtual circuit packet switched networks work.

► Identify the differences between a connectionless network and a connection-oriented network and give an example of each.

► Construct a simple example of Dijkstra's least cost routing algorithm.

► Describe the differences between centralized routing and distributed routing, citing the advantages and disadvantages of each.

► Describe the differences between static routing and adaptive routing, citing the advantages and disadvantages of each.

► Document the main characteristics of flooding and use hop count and hop limit in a simple example.

► Discuss the basic concepts of network congestion including quality of service.

Introduction ▶

Recall that a local area network is typically confined to a single building or a closely located connection of buildings. What happens when a network expands into a metropolitan area, across a state, or across the entire country?

A network that expands into a metropolitan area is called a **metropolitan area network (MAN)**. Many of the same technologies and communication protocols found in local area networks and wide area networks are used to create metropolitan area networks. A good example of a metropolitan area network is the Commonwealth of Pennsylvania's MAN, which is a high speed fiber optic network that was initiated in 1993 to interconnect government agency computers located within the capitol region. All agencies under the Pennsylvania Governor's jurisdiction have now been connected to the MAN, which includes dozens of mainframe and mid-range computers and approximately 25,000 personal computers. In spite of their usefulness, metropolitan area networks are still very small in number compared to other forms of networks.

What happens when a network is larger than a metropolitan area? A network that expands beyond a metropolitan area is a wide area network. Wide area networks share a few characteristics with local area networks: they interconnect computers, they use some form of media for the interconnection, and they support network applications.

More importantly, however, wide area networks differ from local area networks in a number of ways. For example, wide area networks include both data networks, such as the Internet, and voice networks, such as telephone systems, whereas local area networks, in almost all cases, include only data networks. Wide area networks can interconnect thousands, tens of thousands, or more workstations so that any one workstation can transfer data to any other workstation. As the name implies, wide area networks can cover large geographic distances, including the entire earth. In fact, plans are underway to network the planet Mars as more technology is dropped onto the planet and the need for returning signals to earth becomes greater. Thus, wide area networks may someday cover the entire solar system!

Since so many workstations are spread over large distances in a wide area network, a broadcast network, such as a LAN, is not feasible. The opposite situation—connecting every network workstation to every other network workstation—is totally impractical. There would have to be so many connections into each workstation that the technology would be totally unmanageable. Instead, wide area networks use a mesh design and require routing to transfer data across the network. A network that is connected in a mesh is one in which only neighbors are connected to neighbors (Figure 10-1). Thus, to transmit data across a mesh network, the data has to be passed along a route from workstation to workstation.

Figure 10-1
A simple mesh network

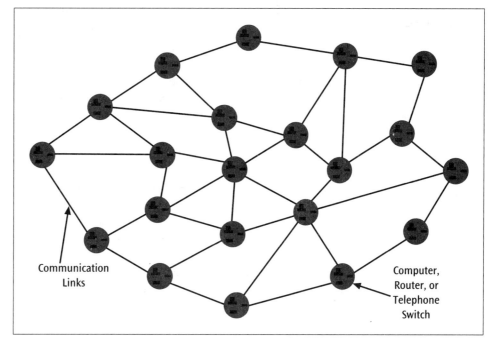

Communication
Links

Computer,
Router, or
Telephone
Switch

Circuit Switched
Packet

Because of the substantial differences between local area networks and wide area networks, wide area networks merit their own discussion. Let's begin with basic terminology and examine the differences between circuit switched networks and packet switched networks. A brief introduction to routing is also in order because wide area networks use routing to transfer data. Subsequent chapters will deal specifically with the Internet as well as many of the latest forms of telecommunication networks.

The topics introduced in this chapter—subnets, routing, and congestion—are topics handled by the network layer of a communications model. Whether it is the OSI model or the Internet model, all matters concerning the connection into and the transfer of data through a network are functions commonly performed by the network layer.

Wide Area Network Basics

A **wide area network** interconnects computers and computer-related equipment in order to perform a given function or functions and typically uses local and long distance telecommunication systems. The types of computers used within a wide area network range from microcomputers to mainframes. The telecommunication lines can be as simple as a standard telephone line or as advanced as a satellite system. Typical functions of wide area networks include the bulk transfer of data between two endpoints, electronic mail services, access to database systems, and access to the Internet. Wide area networks can assist with specialized operations in many fields, such as manufacturing, medicine, navigation, education, entertainment, and telecommunications.

All wide area networks are a collection of at least two basic types of equipment: a station and a node. The **station** is the device with which a user interacts to access a network and contains the software application that allows someone to use the network for a particular purpose. Very often, the station is a microcomputer or workstation, but it could also be a terminal, a telephone, or a mainframe computer.

node
node
Node
Station
Station

Station - most be PC or main frame

Nodes - have multiple
in multiple out

The **node** is a device that allows one or more stations to access the physical network and is a transfer point for passing information through the network. When data or information travels through a network, the information is transferred from node to node through the network. When the information arrives at the proper destination, the destination node delivers it to the destination station. In a network that is designed to transfer computer data, such as the Internet, the node is usually a very fast and powerful router with multiple ports. Incoming data arrives on one port and is retransmitted out a second port depending on the route necessary for the data to reach its destination. In a network that is designed to transfer voice signals, the node is usually a powerful telephone switch that performs functions similar to a router.

Underlying the surface of a wide area network is the subnet. A **subnet** is a collection of nodes, as shown in Figure 10-2. The type and number of interconnections between nodes and how network data is passed from node to node is the responsibility of the subnet. It helps to think of the network and the subnet as being almost two separate entities. A network is the entire entity—the nodes, the stations, the communication lines, the software, and perhaps even the users. The subnet is the underlying physical connection of nodes and communication lines that transfer the data from one location to another. Clearly, a network would not exist without a subnet, but it should not matter to the network what the subnet looks like. The stations, nodes, subnet, communications lines, and software work in concert to create the network. A user sitting at a workstation and running a network application passes his or her data to the network through a station, which passes the data to the subnet. The subnet is responsible for getting the data to the proper destination node, which then delivers it to the appropriate destination station.

Figure 10-2
Network subnet, nodes, and two end stations

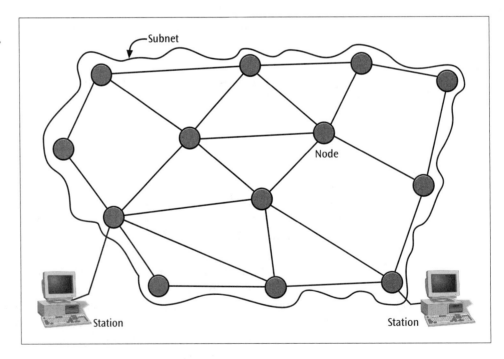

Types of network subnets

A wide area network's subnet may be categorized by the way it transfers information from one end of the subnet to the other. There are three basic types of network subnets: circuit switched, packet switched, and broadcast.

Circuit switched subnet

A **circuit switched subnet** is a subnet in which a dedicated circuit is established between sender and receiver and all data passes over this circuit. One of the best examples of a circuit switched subnet is the dial-up telephone system. When someone places a call on a dial-up telephone network, a circuit, or path, is established between the person placing the call and the recipient of the call. This physical circuit is unique, or dedicated, to this one call and exists for the duration of the call. The information (the telephone conversation) follows this dedicated path from node to node within the subnet as shown in Figure 10-3. A wide area network in which information follows a dedicated path from node to node within the subnet is a circuit switched wide area network.

Figure 10-3

Two people carrying on a telephone conversation using a circuit switched network

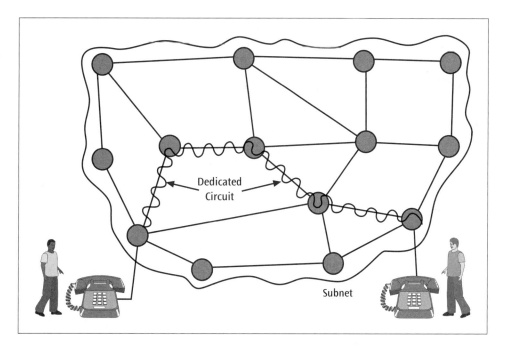

A circuit switched subnet was designed to support voice data, is found in the telephone system, and dedicates a unique circuit for each telephone call. Time is required to establish the circuit and tear down the circuit, but once the circuit is established, all subsequent data travels quickly from node to node. There are two key disadvantages to a circuit switched subnet. First, each circuit is dedicated to only one connection. Second, if the circuit is used for computer data transfer (which is often sporadic), the circuit is not fully utilized.

Packet switched subnet

The packet switched subnet is most often found in networks designed to transfer computer data (such as the Internet). In a **packet switched subnet**, all data messages are transmitted using fixed-sized packages, called packets. No unique, dedicated physical path is established to transmit the data packets across the subnet. To distinguish between a piece of data at the data link layer and a piece of data at the network layer, the term *frame* is used at the data link layer and the term *packet* is used at the network layer. If the message to be transferred is large, it is broken into multiple packets. When the multiple packets arrive at the destination, they are reassembled into the original message.

There are two types of packet switched subnets: the datagram and the virtual circuit. In a **datagram** packet switched subnet, each data packet can follow its own, possibly different, course through the subnet. As each packet arrives at a node, a decision is made as to which path the packet will follow next. This dynamic decision making allows for great flexibility should the network experience congestion or failure. For example, if a number of data packets are currently being routed through Dallas on their way to Phoenix, and Dallas experiences router problems, the subnet can reroute the packets through Denver instead. However, there is a problem with the datagram subnet: despite the fact that a large number of packets may be addressed for the same destination, a node has to examine each packet individually and determine each packet's next path.

To solve this problem, the virtual circuit packet switched subnet was created. In a **virtual circuit** packet switched subnet, all packets that belong to a logical connection can follow the same path through the network. For example, one station may want to transfer a large amount of data, such as the entire electronic contents of a book, across the network to another station. The large amount of data is broken into *n* packets, an optimal temporary path through the subnet is determined, and all the packets follow that path. When the data transfer is complete, the temporary path is dissolved. This type of packet switched subnet is called a virtual circuit because the path followed by the packet is not a physically real circuit like a telephone circuit, although it acts like a circuit.

Although this type of subnet sounds similar to a circuit switched network because all data packets follow a fixed path, it is substantially different. The path exists only in software by creating the necessary routing tables at the appropriate nodes. These routing tables are similar to the port tables used by bridges to determine a path through a local area network. In addition, the path may be shared by other traffic. For example, several workstations may be transferring data across the network at one time. Each workstation has its own virtual circuit, but all the virtual circuits may share one or more communication lines between nodes. Since a virtual circuit is not a dedicated circuit, this type of communication line sharing is common.

Packet switched subnets were designed to support computer data transmissions and break long computer messages into fixed sized packets. In a datagram packet switched subnet, each packet is an entity by itself. If a large message is broken into *n* packets, each packet enters the next node where a unique routing decision is made. Since there is no fixed circuit to follow, there is no time spent creating the circuit. Time is spent, however, when each packet's route is determined. Datagram subnets are the most flexible because they can react quickly to network changes. The virtual circuit packet switched subnet is the marriage of a packet switched

CBT: For a visual example of datagram and virtual circuit subnets, see the CD-ROM that accompanies this text.

network with a circuit switched network. When a large message is broken into multiple packets, all packets follow the same path through the subnet. This path is determined before the first packet is transmitted, which requires circuit setup time. A virtual circuit consisting of internal routing tables move the packets from source to destination.

Broadcast subnet

Similar to the broadcast design of most local area networks, when a node on a wide area network **broadcast subnet** transmits its data, the data is received by all the other nodes. This form of wide area network subnet is, at the moment, relatively rare. Some systems do exist that use radio frequencies to broadcast data to all workstations. For example, there are a few radio broadcast networks that operate in rural areas or in areas where many islands are surrounded by large amounts of water. Some of the new wireless Internet access services such as Local Multipoint Distribution Service (Chapter Three), are based on a broadcast subnet, but they are considered to be more like metropolitan area networks than wide area networks. Since broadcast subnets do not appear as often as circuit switched and packet switched subnets, the remaining discussions do not include broadcast subnets.

In summary, the *physical network design* of a wide area network, or its subnet, has three basic forms: circuit switched, packet switched, and broadcast. Let's now turn our attention to the *logical network design* of a wide area network.

Connection-oriented versus connectionless network applications

The subnet of a network is the physical infrastructure within a wide area network. This infrastructure consists of nodes (routers or telephone switches) and various types of interconnecting media. What about the logical entity that operates over this physical infrastructure? For example, if you are using an e-mail application to send a message to a friend across the country, the e-mail application is the logical entity that uses the network's subnet to deliver the message. Many different types of applications are found on wide area networks including e-mail, web browsing, and many forms of commercial applications. Let's classify all the logical entities—the network applications—into two basic categories: connection-oriented applications and connectionless applications.

A **connection-oriented** application provides some guarantee that information traveling through the network will not be lost and that the information packets will be delivered to the intended receiver in the same order in which they were transmitted. Thus, a connection-oriented network provides what is called a reliable service. To provide a reliable service, the network requires that a logical connection be established between the two end points. If necessary, connection negotiation is performed to help establish this connection. For example, consider the following scenario. A bank wants to electronically transfer a large sum of money to a second bank. The first bank creates a connection with the second bank. During this connection establishment the two banks agree to transfer the funds using data encryption. After the first bank sends the transfer request, the second bank checks the request for accuracy and returns an acknowledgment to the first bank. The first bank will wait until the acknowledgment arrives before doing anything else. All packets transferred during this period are part of this connection and are acknowledged for accuracy. If that is the only electronic transfer, the first bank will say goodbye and the second bank will acknowledge the goodbye. The type of subnet that is used to transfer the funds is not an immediate issue. The subnet could have been circuit switched or packet switched. All the application cared about was that a reliable connection was used to transfer the funds.

A **connectionless** network application does not require a logical connection to be made before the transfer of data. Thus, a connectionless application does not guarantee the delivery of any information or data. Data may be lost, delayed, or even duplicated. If a large message is broken into multiple smaller packets, the order of delivery of the packets is not guaranteed. No connection establishment/termination procedures are followed since there is no need to create a connection. Each packet is sent as a single entity and not as part of a network connection. As a consequence, a connectionless network creates an unreliable service. Because a connectionless network's service is unreliable, the responsibility of providing a reliable service falls to a higher layer, such as a transport or application layer.

A common example of a connectionless network application is the sending of e-mail. When you send an e-mail message, no connection is created between you and the intended recipient. You simply click the Send button and the e-mail is sent. You usually do not know when the e-mail arrives or if the e-mail arrives at all. Connectionless applications, like many e-mail programs, do not negotiate a connection, and the transfer of data is rarely, if ever, acknowledged. If a second e-mail is sent, it has no relationship (network-wise) to the first e-mail. Once again, the underlying subnet is not really an issue. It can be either a circuit switched subnet or a packet switched subnet.

Another illustration of the difference between connection-oriented and connectionless networks is the difference between the telephone system and the postal service. When you call someone on the telephone, if that person is available for conversation, he or she answers the telephone. Once the telephone has been answered, a connection is established. The conversation follows, and when one person has finished with the conversation, some sort of ending statement is issued, both parties hang up, and the connection is terminated (Figure 10-4). Thus when you use the telephone system, you are using a connection-oriented network that provides a reliable service.

Figure 10-4
Connection-oriented telephone call

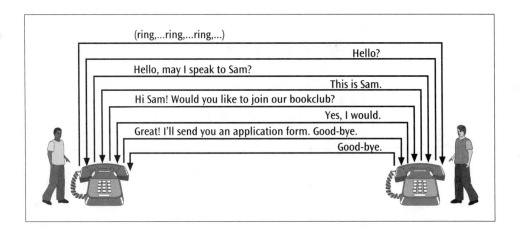

If you send a *standard* letter to someone through the U.S. Postal Service, it will most likely be delivered, although there is no guarantee it will be delivered. There is also no guarantee when it will arrive. If the letter is lost, you will not know until some time passes and you begin to think, "I haven't heard from Kathleen yet; I wonder if she received my letter?" The postal service, then, is similar to a connectionless network that provides an unreliable service. The unreliable service (no offense meant) offered by the connectionless postal "network" requires that you must take further actions if you want to ensure that the letter is delivered as intended. Furthermore, a postal patron never knows for sure which route a letter will take through the postal network. Under some circumstances, mail is transferred by truck; under other circumstances it is transferred by airplane (Figure 10-5).

Figure 10-5
Connectionless postal network

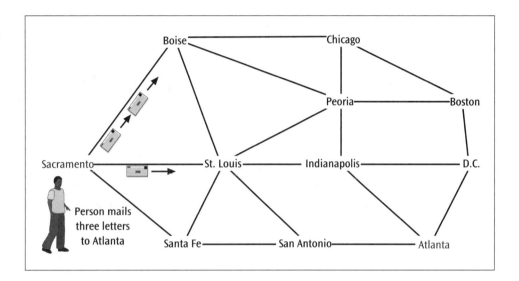

Combinations of network applications with subnet types

Using a wide area network to send e-mail or surf the Internet requires at least two basic ingredients: a connectionless or connection-oriented application and a circuit switched or packet switched subnet to support the connection and transfer the data. It doesn't really matter which type of subnet supports which type of connection. All four combinations—connectionless application over a circuit switched subnet, connectionless application over a packet switched subnet, connection-oriented application over a circuit switched subnet, and connection-oriented application over a packet switched subnet—exist in the real world. Certain combinations of application and subnet, however, make more sense than others and are more efficient.

A circuit switched subnet typically supports a connection-oriented network application. For both the subnet and the application, a circuit is established. In the case of the subnet, the circuit is physical, while in the case of the application, the circuit is logical. In contrast, if the network application is connectionless and is going to transmit only a single packet of data, it is not very efficient to spend the time to establish a physical circuit for the transfer of one or two packets of data. It would be more efficient to use a packet switched subnet which does not require making a connection to transfer one or two packets.

Packet switched subnets can efficiently support either connection-oriented or connectionless network applications. Recall that a datagram packet switched subnet is simply a collection of packet passing nodes. When a packet arrives at a datagram node, the node examines the destination address of the packet and sends it out on the appropriate telephone line based on some routing decision. Virtual circuit packet switched subnets establish a connection before a user transmits any data packets. As the connection is created, a path through the subnet is chosen and each node along the path is notified that it will be part of a virtual circuit. The advantage of creating a virtual circuit is the removal of the routing decision-making process from each node in the subnet. Since the route is determined only once before any data is transmitted, time is not spent determining the route for each and every packet, and the packets move more quickly through the subnet. A disadvantage of a virtual circuit is the loss of each packet dynamically choosing a path, particularly if the subnet is experiencing congestion or node failure along the virtual circuit.

Supporting connection-oriented and connectionless network applications with datagram packet switched subnets and virtual circuit packet switched subnets, provides four more possible combinations. A connection-oriented network with a packet switched virtual circuit subnet makes sense since both require connection and circuit establishment and termination. It is also common to combine a connectionless network with a packet switched datagram subnet since neither require making a connection (either physical or logical).

What about supporting a connection-oriented network application with a packet switched datagram subnet? Although it may sound strange, this combination is quite common. The underlying subnet is a datagram which does not require a physical connection, but the network application creates a logical connection between sender and receiver. In this case, the network does not provide reliability by the creation of a virtual circuit; instead, the responsibility of checking for lost or duplicate packets is performed by the network application in either the transport layer or the application layer.

The final combination, supporting a connectionless network application with a packet switched virtual circuit subnet, makes little sense. It would be wasteful to create a virtual circuit on the subnet if the overall network connection is treated as connectionless, or unreliable. Why take the time to create a virtual circuit on the subnet if the application is only transmitting one or two packets?

Routing

Recall that a wide area network's underlying subnet consists of multiple nodes, each with possible multiple connections to other nodes within the subnet. Each node is a router that accepts an input packet, examines the destination address of the packet, and forwards the packet on to a particular communications line. With multiple-linked nodes there may be one or more paths entering a node as well as one or more paths leaving a node. If most nodes in the subnet have multiple inputs and outputs, numerous routes from a source node to a destination node may exist. Examining Figure 10-6, you can see there are a number of possible routes between Node A and Node G: A-B-G, A-D-G, A-B-E-G, and A-B-E-D-G, to list a few.

Figure 10-6
A seven-node subnet showing multiple routes between nodes

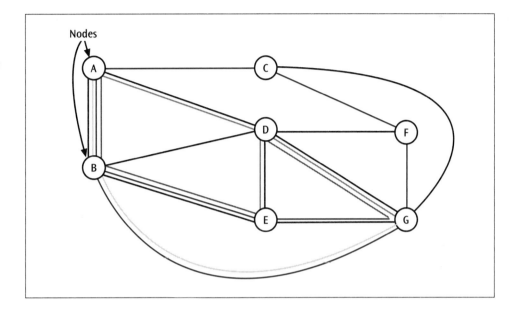

Consider the Internet as an example: it is a massive collection of networks, routers and communication lines (various types of telephone lines). When a data packet enters a router, the router examines the IP address encapsulated in the network layer of the packet and determines where the packet should go next. When there are multiple routes through a network, such as the Internet, how is any one particular route selected? How is routing through a wide area network accomplished? Although routing on the Internet is fairly complex and well beyond the scope of this text, it is possible to examine the basic routing techniques that all types of wide area networks employ. Note that a wide area network does not use only one form of routing. The routing algorithm used by the Internet, for example, is actually a combination of several types of basic routing techniques.

To begin to understand the often complex issue of routing, you need to look at the subnet as a graph, called a network graph, consisting of nodes (computers, routers, or telephone switches) and edges (the communication links between the nodes). Figure 10-7 is a simple example of a network graph. If you assign a weight or cost to the edge between each pair of nodes, the network graph becomes a weighted network graph. A **weighted network graph** is a structure consisting of nodes and edges in which the traversal of an edge has a particular cost associated with it.

Figure 10-7
A simple example of a network graph

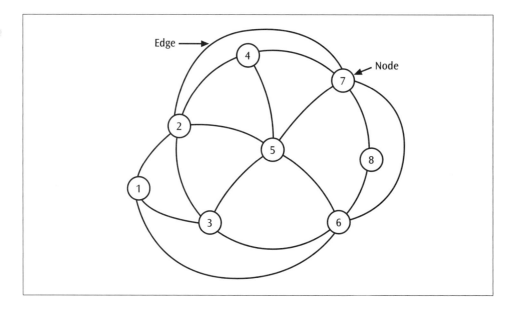

You can assign many meanings to the weights in a weighted network graph. A weight can mean the dollar cost of using the communication link between two nodes. A weight could also mean the time delay cost associated with transmitting data on that link between the source and destination nodes. Another common meaning for weight is the size of the queue that has backed up waiting for a packet to be transmitted onto a link. Each of these weights is useful for determining a route through a network.

Once you look at the subnet as a graph and assign weights to the paths between nodes, you can develop an algorithm—a rule of thumb—for traversing the network. Many algorithms exist for selecting a route through a network. Algorithms often strive for an optimal route through a network, but there are different ways to define optimal. For example, one algorithm might define the optimal route as one that generates the least dollar cost for using a particular path. Another algorithm might determine that the least time delay cost for a particular path is the optimal route. A third algorithm might define the optimal route as the one having the smallest queue lengths at the nodes along the path.

Some algorithms use criteria other than optimality. For example, some algorithms might try to balance the network load over a number of different paths. Another algorithm might favor one type of traffic over another, such as real time traffic over non-real time traffic. A third algorithm may try to remain robust, responding to changing network demands, as nodes and communication links fail or become congested. Yet a fourth algorithm may try to remain static and not switch between possible paths. As you can see, routing is a complicated topic. To get a feel for routing in wide area networks, let's examine several of the most commonly used routing techniques: Dijkstra's least cost algorithm, flooding, centralized routing, distributed routing, isolated routing, adaptive routing, and fixed routing. Most wide area networks use a combination of these routing techniques to achieve a routing algorithm that is fair, efficient, and robust, yet stable.

Dijkstra's least cost algorithm

One possible method for selecting a route through a network is to use a **least cost algorithm**, which chooses a route that minimizes the sum of the costs of all the communication paths along that route. If you could find the one route with the smallest sum, you would find the least cost route. An algorithm such as Dijkstra's is usually calculated by each node. Since this calculation is time consuming, it is only calculated on a periodic basis or when something in the network changes, such as a connection or node failure. So that a node can calculate Dijkstra's algorithm, delay information is often shared among neighboring nodes. The time period of calculation, who performs the calculation, and the sharing of information can vary greatly, as will be shown shortly.

Let's look at an example. Figure 10-8 shows the same subnet as shown in Figure 10-6, but modified to include an arbitrary set of link costs. Path A-B-G has a cost of 9 (2+7), A-D-G has a cost of 10 (5+5), and path A-B-E-G has a cost of 8 (2+4+2). To ensure that you find the minimal cost route, you need to use a procedure that calculates the cost of every possible route starting from a given node. Although the human eye can quickly pick out a path through a network graph, there are a number of drawbacks to "eyeballing" the data to find a solution:

▸ You can easily miss one or more paths.

▸ You may not find the least cost path.

▸ Wide area networks are never as simple as Figure 10-8 depicts; thus eyeballing the data would never work to find the least cost path.

Figure 10-8
Network subnet with costs associated with each link

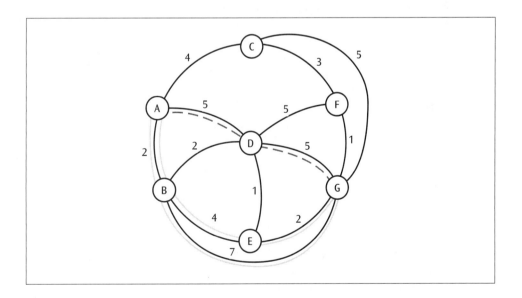

Dijkstra created a **forward search least cost algorithm** to solve this exact problem. Most wide area networks use some form of Dijkstra's algorithm to determine a least cost route through a network, whether that cost is a measure of time or money.

Flooding

Compared to Dijkstra's least cost routing algorithm, the flooding technique seems simple. **Flooding** states that each node takes the incoming packet and retransmits it onto every outgoing link. For example, as Figure 10-9 shows, assume a packet originates at node A. Node A simply transmits a copy of the packet on every one of its outgoing links. Thus, a copy of the packet (the first copy, or copy 1) is sent to nodes B, C and D.

Details ▶

Understanding Dijkstra's Algorithm *Extra credit problem on test*

Understanding how Dijkstra's algorithm works can be beneficial for network designers and analysts. Rather than listing the algorithm in stepwise form, let's walk through an example solution, looking at the algorithm as we go along. Using Figure 10-8, let's find the least cost routes from node A to all other nodes.

To begin, you need to create a table with a column for each node in the network except the starting node A. The table needs to include a row, called Visited, for each path through the subnet, and a row, called Next, to denote the node that should be traversed next after leaving A. For example, if the Next value under column Node G is B, then a packet that is leaving node A and is destined to node G should be transmitted next to node B.

	Node					
Visited		B	C	D	E	F	G
-							
Next							

Then, select the starting node A, Visit it, and add the starting node to the Visited list. Next, locate each *immediate* neighbor of node A (only one link or hop away from node A) that are not yet in the Visited list, calculate the cost to travel from node A to each of these neighbors, and enter these values into the table. For example, node B is one hop away from node A, it has not yet been Visited, and it costs 2 units to travel from A to B. Enter 2 in the table in the column for node B, indicating the cost of the path from node A to node B, and enter B in the Next row noting that to get to B, you go directly

to B on the next hop. You can also go from A to C in one hop with a cost of 4 and a Next value of C, and from A to D with a cost of 5 and a Next value of D. Note that we have not yet Visited B, C or D. We have only Visited A. We are simply examining the costs of the links that run between A and B, A and C, and A and D.

		Node					
Visited		B	C	D	E	F	G
	A	2	4	5	-	-	-
Next		B	C	D			

No more nodes are immediate neighbors of A, and all of node A's immediate neighbor links have been examined, so you need to select the next node to Visit. Let's choose node B, since it has the least cost in our table thus far. By selecting the next node with the least cost, the algorithm will find the least cost in all situations. Locate the immediate neighbors of node B that have not yet been Visited and determine the cost of traveling from node A to each immediate neighbor of B via node B. Note that node A has been Visited, so you exclude it at this stage. The immediate neighbors of node B that have not yet been Visited are D, E and G. The cost of going from node A to node D via node B is 4 (the link from A to B costs 2, and the link from B to D costs 2). Since this cost is less than the cost of going directly from A to D, the current value in the table, replace the value 5 with the new value 4, updating the table. You also replace the D in the Next row under D, since the new least cost path now begins with the packet going to B first after leaving node A.

Figure 10-9
Network subnet with flooding starting from Node A

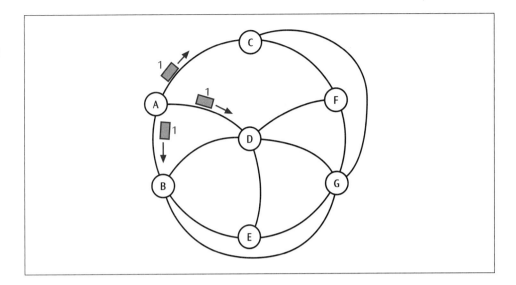

		Node					
Visited		B	C	D	E	F	G
	A	2	4	5	-	-	-
	A B	2	4	4	-	-	-
Next		B	C	B			

The cost of going from A to E via B is 6 (2+4), and the cost of going from A to G via B is 9 (2+7). Enter the values 6 and 9 in the E and G columns, respectively. B is also the Next value for both E and G.

		Node					
Visited		B	C	D	E	F	G
	A	2	4	5	-	-	-
	A B	2	4	4	6	-	9
Next		B	C	B	B		B

Let's Visit node C next. The immediate neighbors of C that have not yet been Visited are F and G. The cost of going from A to F via C is 7 (4+3). Enter the value 7 in the F column and the value C in the Next row. The cost of traveling from node A to G via C costs 9 (4+5). Since the current value is already 9, there is no need to update the table. The new value is not less than the current value in the table.

		Node					
Visited		B	C	D	E	F	G
	A	2	4	5	-	-	-
	A B	2	4	4	6	-	9
	A B C	2	4	4	6	7	9
Next		B	C	B	B	C	B

Let's Visit node D next. The immediate neighbors of D not yet Visited are E, F and G. The cost of going from A to E via D (via B) is 5 (4 + 1). Since this value is less than the current cost from A to E, update the table by entering 5 in the E column. We still get to E by first going to B after leaving A, so the value B in the Next row does not change. The cost of going from A to F via D is 10. The cost of going from A to G via D is also 10. Because the values already entered in the F and G columns are less than 10 (the table already reflects the least cost path for those nodes), you do not update the table.

		Node					
Visited		B	C	D	E	F	G
	A	2	4	5	-	-	-
	A B	2	4	4	6	-	9
	A B C	2	4	4	6	7	9
	A B C D	2	4	4	5	7	9
Next		B	C	B	B	C	B

▶▶

When the packet arrives at node B, B simply transmits a copy of the packet to each of its outgoing nodes (A, D, E, and G). Node C will likewise transmit a copy of the packet to each of its outgoing nodes (A, F and G). Node D will also transmit a copy of the packet to each of its outgoing nodes (A, B, E, F and G) (Figure 10-10). It does not take long to realize that very quickly, the network will be flooded with copies of the original data packet.

The next node to Visit is E. The immediate neighbor of E not yet Visited is G. The cost of traveling from node A to node G via node E (via D via B) is 7 (2+2+1+2). The cost of this path (7) is smaller than the value already entered into column G, so replace the current value in the table with this new, smaller value.

		Node					
Visited		**B**	**C**	**D**	**E**	**F**	**G**
	A	2	4	5	-	-	-
	A B	2	4	4	6	-	9
	A B C	2	4	4	6	7	9
	A B C D	2	4	4	5	7	9
	A B C D E	2	4	4	5	7	7
Next		B	C	B	B	C	B

The next node to Visit is F. The only immediate neighbor of F not yet Visited is G. The cost of traveling from node A to node G via F (via C) is 8. This cost is not less than the current value in Column F, so do not update the table.

The final node to Visit is G. There are, however, no immediate neighbors of G that have not already been Visited, so we are finished. Table 10-1 shows the final results. Now from the table, you can easily look up the least-cost path from node A to any other node. If a data packet originates from node A and is destined for node x, the software in the router would simply consult column x of the table to determine where the data packet should go Next. To find the least-cost route starting from another node, you would need to apply Dijkstra's algorithm again. For example, if you wish to find the least cost path from, say, node C to any other node, you would generate a new table by repeating the least cost algorithm with node C as the starting position.

		Node					
Visited		**B**	**C**	**D**	**E**	**F**	**G**
	A	2	4	5	-	-	-
	A B	2	4	4	6	-	9
	A B C	2	4	4	6	7	9
	A B C D	2	4	4	5	7	9
	A B C D E	2	4	4	5	7	7
	A B C D E F	2	4	4	5	7	7
	A B C D E F G	2	4	4	5	7	7
Next		B	C	B	B	C	B

Table 10-1 *The results Dijkstra's algorithm applied to a seven-node subnet starting from node A*

Figure 10-10
Flooding has continued to nodes B, C, and D

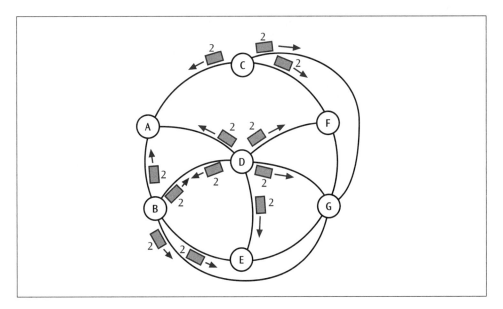

To prevent the number of copied packets from becoming massive, two common sense rules must be established. First, a node need not send a copy of the packet back to the link from which the packet just arrived. Thus, when node A sends a copy of the packet to node C, C does not need to send a copy immediately back to A.

Second, a network limit, called the **hop limit**, can be placed on the number of times any packet can be copied. Each time a packet is copied, a counter, called a **hop count**, associated with the packet is incremented. When the counter reaches the network hop limit, this particular packet will not be copied anymore. For example, suppose the network has a hop limit of 3. When node A first sends copies to B, C and D, each of the three copies has a hop count of 1. When the packet arrives at node C, copies with hop counts of 2 will be sent to F and G (see Figure 10-10). When the copy arrives at node F, two copies with hop counts of 3 will be transmitted to D and G, and the packets that arrive at D and G with hop counts of 3 will go no further.

Although flooding may seem like a strange way to route a packet through a network, it has its merits. If a copy of a packet *must* be sent to a particular node, flooding *will* get it there, assuming, of course, there is at least one active link to the receiving node and the network hop limit is not set at too small a value. Flooding is also advantageous when a copy of a packet needs to get to all nodes, such as when sending emergency information or network initialization information. The major disadvantage of flooding is the large number of copied packets distributed throughout the network.

Centralized routing

The concept of centralized routing is not so much an algorithm for routing data packets through a network as it is a technique of providing routing information. Using a previously calculated method, such as a least cost algorithm, **centralized routing** dictates that the routing information generated from the least cost algorithm is stored at a central location within the network. Any node wishing to transmit a packet to another node consults this centralized site and inquires as to the best route on which to transmit the data. The central site might maintain a least cost routing table such as that shown in Table 10-2. This table was created by performing the least cost algorithm seven times using each node in the network as the starting node.

Table 10-2
Routing table kept at a centralized network site

		Destination Node						
		A	**B**	**C**	**D**	**E**	**F**	**G**
	A	-	B	C	B	B	C	B
	B	A	-	A	D	D	D	D
	C	A	A	-	A	F	F	F
Origination Node	**D**	B	B	F	-	E	E	E
	E	D	D	G	D	-	G	G
	F	C	G	C	G	G	-	G
	G	E	E	F	E	E	F	-

The primary advantage of using centralized routing is that all routing information is kept at one node or site. There is only one routing table, thus minimizing routing table storage and eliminating the possibility of multiple routing tables with conflicting information. A major disadvantage of centralized routing is the "putting all of your eggs in one basket" syndrome. If the node holding the one and only routing table crashes, the entire network will have no routing information. In addition, if all other nodes must consult the one node holding the routing table, network congestion or a bottleneck may result at the routing table's node.

Distributed routing

Distributed routing allows each node to maintain its own routing table. When a data packet enters the network at node x, that node consults its own routing table to determine the next node to receive the packet. Each node needs routing information only for its own locale. As a consequence, the centralized routing table shown in Table 10-3 can be divided by rows, and each row can be assigned to the appropriate node. For example, the routing table for node C would be the table shown in Table 10-4. From the table you can see that if a data packet arrives at node C and is destined for node G, the packet should be sent to node F next.

Table 10-3
Local routing table for node C

	Destination Node						
	A	**B**	**C**	**D**	**E**	**F**	**G**
Origination node C:	A	A	-	A	F	F	F

One of the primary advantages of distributed routing is the fact that no one node is responsible for maintaining a routing table. This situation confers a number of benefits. First, if any node crashes, it will probably not disable the entire network.

Second, a node will not need to send a request to a central routing table since each node has its own table. By eliminating the request packets transmitted to the node holding the routing table and eliminating the resulting packets leaving the routing table node, much data traffic on the network has been eliminated.

One disadvantage of distributed routing is the problems that arise if the routing tables need to be updated. When all the routing information is in one place, it is simple to update a single table. If routing information is scattered throughout a network, getting the appropriate routing information to each node is a complex problem. Additionally, if routing information is stored at multiple locations, there can be one or more routing tables containing old or incorrect information at any given point in time.

Isolated routing

Centralized and distributed routing techniques use routing tables that have been created based on information from some or all of the nodes in the network. In the centralized scheme, all nodes transmit their local routing costs to the central site, which then assembles a centralized routing table. In the distributed scheme, neighbor nodes share routing information, and then each node uses this information to create its own routing table. What happens if there is no sharing of routing information between any nodes, and each node creates its own routing table completely on its own? This situation is exactly what happens in isolated routing.

With **isolated routing**, each node uses only local information to create its own routing table. One technique a node can use to gather routing information is to observe the status information of incoming packets. Suppose all nodes maintain a clock system, and each packet is time-stamped as it leaves a node. Then, when a packet arrives at the receiving node, the receiver can compare the time stamp of the packet with the current time and calculate the time the packet spent in transit. The receiving node can then assume that the time spent coming in on that particular link would be similar to the time a packet would spend going out that same link. Although this information is not always true, it does provide some level of information for making a reasonable routing decision. This information is placed into a local routing table and is consulted each time the node needs to determine the best route for transmitting a data packet.

Isolated routing has one key advantage over centralized and distributed routing. With isolated routing, information does not have to be shared and transmitted to any other nodes, thus reducing the traffic on the network. A disadvantage of isolated routing is that it is possible to gather the wrong or insufficient data. For example, the time spent on a communications link coming into a node may not be the same time spent going out on the same link. Also, this information pertains only to the immediate communications link. What is the state of the network two links away? Or three links away? Each node does not know this information and thus may make a poor routing choice.

Adaptive routing versus static routing

Centralized and distributed routing are methods for distributing routing information. They are typically used in conjunction with some form of least cost routing. Regardless of whether routing information is centralized or distributed, when networks change, routing information needs to change too. Should the routing tables adapt to network changes, or should routing tables stay fixed when network changes occur?

When routing tables adapt to network changes, the routing system is called adaptive. **Adaptive routing** is a dynamic system in which routing tables react to network fluctuations, such as congestion and node/link failure. As a problem occurs in the network, the appropriate information is transmitted to the routing tables, and new routes that avoid the problem areas are created. Adaptive routing raises some questions and issues: how often is information shared, and how often are routing tables updated? How much additional traffic is generated by messages transmitting routing information?

Adaptive routing, unfortunately, can add to network congestion. Each time the network experiences a change in congestion, that information is transmitted to one or more nodes. This transmission of information adds to the congestion, possibly making it worse. In addition, if a network reacts too quickly to a congestion problem and re-routes all traffic onto a different path, the result can be congestion problems in a different area. The network would then recognize the congestion in a different part of the network and possibly re-route all traffic back to the first problem. This back and forth re-routing produces a yo-yo effect.

In contrast to adaptive routing, with **static routing**, routing tables remain the same when network changes occur. Maintaining static routing tables is simpler than maintaining adaptive routing tables. Nodes do not have to share information with other nodes, since the routing tables are not going to change. The down side of static routing is that it does not react quickly (if at all) to network problems of congestion and node/link failure.

Routing examples

Although the Internet is covered in detail in Chapter Eleven, let's take a few minutes to examine two of the routing algorithms that have been used on the Internet over the years. By examining these algorithms, you will see that a real-life routing protocol is actually composed of many of the algorithms and techniques introduced in the preceding sections.

The first routing algorithm used on the Internet (when it was still the Arpanet) was called a distance vector routing algorithm. It was an *adaptive* algorithm in which each node maintained its own routing table called a *vector*. Because each node maintained its own routing table, it was also a *distributed* algorithm. Every 128 milliseconds, each node exchanged its vector with its neighbor. When all the vectors came in from the neighbors, a node would update its own vector with the *least cost* values of all the neighbors. This adaptive, distributed algorithm had the name **Routing Information Protocol (RIP)**. The RIP protocol had the unfortunate side effect of reacting quickly to good news—an identified cheaper route to a destination—but too slowly to bad news—a node or link failure.

The next routing protocol that was implemented on the Internet (in 1979) was called a link state routing algorithm. Link state routing involves essentially four steps. The first step is to measure the delay or cost to each neighboring router. Each router sends out a special echo packet that gets bounced back almost immediately. If a timestamp were placed on the packet as it left and again as it returned, the router should know the transfer time to and from a neighboring router. The second step is to construct a link state packet containing all this timing information. The third step is to distribute the link state packets via flooding. Thus, in addition to using *flooding*, the link state routing algorithm is a *distributed algorithm*. The fourth and final step is to compute new routes based on the updated information. Once a router had collected a full set of link state packets from its neighbors, it creates its routing table, usually using *Dijkstra's least cost algorithm*. **Open Shortest Path First (OSPF)** protocol is a link state algorithm and is still used today by many Internet routers.

Network Congestion

When a network or a part of a network becomes so saturated with data packets that packet transfer is noticeably impeded, **network congestion** occurs. Congestion may be a short-term problem, such as a temporary line or node failure, or it may be a long-term problem, such as inadequate planning for future traffic needs or poorly created routing tables. As with so many things in life, the network is only as strong as its weakest link. If network designers could properly plan for the future, network congestion might exist only in rare instances. But the computer industry, like most forms of communication, is filled with examples of failures to adequately plan for the future. Computer networks are going to experience congestion, and no amount of planning can avoid this situation. Thus it is important to consider effective congestion avoidance and congestion handling techniques.

Preventing network congestion

A node failure in the network subnet can be a difficult problem to handle. The network could require each node site to maintain a backup machine in case the primary machine fails. But this solution is expensive, and many network sites are not likely to comply with it.

As an alternative preventive measure, the network should provide alternate routes such that failed nodes or links could be avoided and routed around. This solution implies that routing tables are adaptive and can change in response to network problems. Almost all currently existing networks provide at least two paths between any pair of nodes, especially if these nodes are main links and carry large amounts of traffic. In the event of a network problem, routing tables are updated and the problem area is avoided.

Handling network congestion

Insufficient buffer space at a node in the subnet can also cause network congestion. It is not uncommon to have from hundreds to thousands of packets arriving at a network node each hour. If the node cannot process the packets quickly enough, incoming packets will begin to accumulate in a buffer space. When packets sit in a buffer for an appreciable amount of time, network throughput begins to suffer. If adaptive routing is employed, this congestion can be recognized and updated routing tables can be sent to the appropriate nodes (or to a central routing facility).

Changing routing tables to reflect congestion, however, might provide only a temporary fix. A more permanent fix would be to increase the speed of the node processor responsible for processing the incoming data packets.

What happens if the buffer space is completely filled and a node can accept no further packets? Packets that arrive after the buffer space is full are usually discarded. Although this solution hardly seems fair, it solves the problem momentarily. A more reasonable solution is to implement error and flow control. (Error and flow control were first discussed in Chapter Six). Flow control allows two adjacent nodes to control the amount of traffic passing between them. When the buffer space of a node becomes filled, the receiving node informs the sending node to stop transmission until later notice.

An alternate solution to controlling the flow of packets between two nodes is buffer preallocation. With **buffer preallocation**, before one node sends a series of *n* packets to another node, the sending node inquires in advance if the receiving node has enough buffer space for the n packets. If the receiving node has enough buffer space, it sets aside the n buffers and informs the sending node to begin transmission. Although this scheme generally works, it introduces extra message passing, additional delays, and possible wasted buffer space if all n packets are not sent. The alternative of discarding packets due to insufficient buffer space, however, is worse.

Flow control and preallocation of buffers helps control congestion between nodes. But what if the problem is congestion between a station and a node? As you saw earlier, one node may have multiple stations connected to it. The node continuously monitors the amount of traffic arriving from each station. If one station begins to transfer data packets at a higher rate than the node can accommodate, the node will inform the station to slow down via a **choke packet**. The concept of choke packets is similar to the concept of flow control between two nodes.

Other techniques have been proposed for limiting the number of packets traveling throughout the network. If the total number of outstanding packets in transit in a network can be limited, it should be possible to reduce or eliminate congestion. For example, some type of **permit system** in which a station may not enter a packet into the network unless it possesses a permit could work to reduce congestion. Although a permit system sounds attractive, many questions need to be addressed for such a system to function well. How do you determine who gets permits and how many? How do you keep the permits equally distributed? What do you do with permits you don't need? What happens if one or more permits are lost? Note that using permits to limit the number of packets in a network does not necessarily eliminate congestion. What happens if all the permits end up at one node? Possible solutions to these problems are difficult, and some are not effective. Although discussing the issues goes beyond the scope of this book, it is important to have an idea of what it would take for a permit system to function well. Congestion may still exist in the area of that one node, while there is little or no traffic anywhere else in the network. All of these potential problems dull the attractiveness of a permit system.

More recent network technologies such as frame relay and asynchronous transfer mode (discussed in more detail in Chapter Twelve) approach network congestion in a more serious fashion. Since these network technologies transfer data at very high speeds, congestion can occur very quickly and be devastating. Thus, it is very important to keep congestion from occurring in the first place.

To achieve congestion prevention, users must negotiate with the network about how much traffic they will be sending or what resources the network must provide to satisfy the user's needs. For example, the following questions could be resolved before any data transfer takes place:

► What is the average (or constant) bit rate at which a user will transmit?

► What is an average *peak* bit rate at which a user might transmit?

► At what rate might a network start discarding packets in the event of congestion?

► What is the average bit rate that the network can provide?

► What is the average peak bit rate that the network can provide?

Many networks relate these issues to **quality of service,** in which a network user and the network subnet agree on a particular level of service. For example, a user who requires a very fast, real-time connection to support live action video will negotiate with the network for a particular quality of service. If the network can provide this level of quality, a contract is agreed to, the user is charged accordingly, and traffic commences. If a second user requires a slower connection for e-mail, a different level of quality is agreed on and the connection is established.

To provide the agreed on level of service, the network may use whatever resources it has available, such as buffer preallocation, choke packets, or permits. These and other techniques are currently used to avoid the potentially serious problem of network congestion.

WANs in Action: Making Internet Connections ►

Accessing the World Wide Web via the Internet is one of the more fascinating technological breakthroughs of recent time. Many people have a personal computer at home and connect to the Internet via an Internet Service Provider (ISP) and a standard telephone line. If you are a student enrolled in a course at a college, you more than likely have Internet access through the college and it's campus computer laboratories. Likewise, many businesses have Internet access over a corporate local area network, allowing employees to download web pages. Although the end result is the same (someone gets to view web pages), the underlying types of subnets and connections can vary depending on whether you are browsing the Web from home, from school, or from work. Let's examine the different types of subnets that support applications commonly found on the Internet.

A home-to-Internet connection

Let's consider the first scenario in which you connect to the Internet from home. Although a number of telecommunication services can connect a home user to an ISP (Chapter Twelve will present several technologies), a majority of people use a standard dial-up telephone service and a modem. The communication software on your personal computer dials a (hopefully) local telephone number, and the local telephone system creates a dedicated circuit switched connection between your modem and your ISP (Figure 10-11).

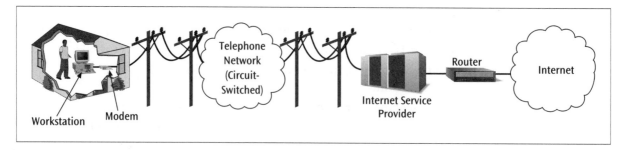

Figure 10-11
User at home using a dial-up telephone line (circuit switched network) to run a connection-oriented application (web browser)

When a user is running a browser and clicks a link, HTTP (Hyper-Text Transport Protocol) creates one or more connections between the user's browser and the server that holds the requested web page. Using these connections, the various pieces (text, graphics, links, JavaScript or Java applications) of the requested web page are transmitted. When the last piece of the requested web page is transmitted, the last connection is dissolved. Thus, a connection-oriented application is running over a circuit switched network.

Suppose instead of web browsing, the user sends an e-mail message to someone. Most e-mail programs are connectionless. The e-mail is transmitted without first informing the recipient that a message is coming, nor does the e-mail software wait for a reply. Now you have a connectionless application running over a switched circuit.

A work-to-Internet Connection

Now let's assume you are at work and using the company computer system. More than likely your workstation is connected to a local area network, which connects to a router, which then connects to an ISP over a leased high-speed telephone line, such as frame relay (Figure 10-12).

Figure 10-12
User at work using a local area network to access the Internet

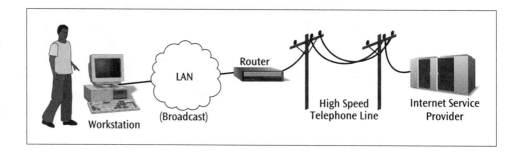

Recall that a local area network is a broadcast subnet. It doesn't use circuit switching or packet switching. Instead, the data is packaged into a frame and broadcast over a medium to a router. The router recognizes the destination address and determines that the frame has to leave the company network and traverse the high-speed telephone line to the Internet service provider. The high-speed telephone line happens to be a frame relay line. Although frame relay isn't discussed in detail until

Chapter Twelve, let's simply say that frame relay is a packet switched network. Thus, whether the user is browsing the Web (a connection-oriented network application) or sending e-mail (a connectionless network application), the application is traveling first over a broadcast subnet and then over a packet switched subnet on its way to the Internet.

These examples show that both connection-oriented network applications and connectionless network applications can operate over packet switched subnets, broadcast subnets, and circuit switched subnets.

◆◆

SUMMARY

► Wide area networks cover larger geographic areas than local area networks.
► Wide area networks are based on potentially different physical subnets: circuit switched, packet switched, and broadcast.
► A circuit switched subnet creates a dedicated circuit between sender and receiver and all data passes over this circuit.
► A packet switched subnet transmits fixed-sized packages of data called packets. Packet switched subnets can be further broken down into datagram and virtual circuit subnets.
► The datagram packet switched subnet transmits each packet independent of each other packet. Each packet is considered a single entity and is not part of a larger grouping of packets.
► The virtual circuit packet switched subnet creates a virtual circuit using routing tables and transmits all packets belonging to a particular connection over this virtual circuit.
► A broadcast subnet transmits its data to all workstations at the same time. Broadcast subnets are more often used in local area networks rather than wide area networks.
► The network application that runs over a network subnet can be of two possible types: connectionless, and connection-oriented.
► A connection-oriented network application provides some guarantee that information traveling through the network will not be lost and that the information packets will be delivered to the intended receiver in the same order in which they were transmitted. To provide this service, a logical connection is established before any data transfer takes place.
► A connectionless network application does not require a logical connection to be made before the transfer of data.
► Connection-oriented network applications can run over either circuit switched or packet switched subnets, whereas it makes more sense to operate connectionless network applications over packet switched subnets.
► Selecting the optimal route for the transfer of a data packet through a network is a common service of many networks. This optimal route is obtained by combining two or more of the many routing algorithms and techniques available today.
► One possible way to select an optimal route through a network is to choose a path whose total path costs have the smallest value. This technique is based on Dijkstra's least cost algorithm and is a common method for determining the optimal route.
► Flooding is a routing technique that requires each node to take the incoming packet and retransmit it onto every outgoing link. If a copy of a packet must be sent to a particular node, flooding will get it there.
► Flooding creates a very large number of copied packets that are distributed throughout the network.
► Centralized routing is a technique for providing routing information and dictates that the routing information generated by a method such as the least cost algorithm be stored at a central location within the network.
► Distributed routing allows each node to maintain its own routing table.

▶ Isolated routing uses only information that can be extracted from watching the connections to neighbor nodes.

▶ Static routing creates a set of routing tables and then does not alter the contents of those tables for long periods of time.

▶ Adaptive routing allows a network to establish routing tables that can change frequently when network conditions change.

▶ When a network or a part of a network becomes so saturated with data packets that packet transfer is noticeably impeded, network congestion has occurred.

▶ Congestion may be the result of network node failure, network link failure, or insufficient nodal buffer space.

▶ Remedies for network congestion include discarding excess packets, preallocation of nodal buffers, flow control, permit systems, choke packets, and pre-network-connection quality of service parameters.

▶ Quality of service parameters can be used by network users and the network to establish minimum guidelines for the proper transfer of data. These guidelines can include transmission speed, level of errors, and overall network throughput.

KEY TERMS

adaptive routing	hop count	packet switched subnet
broadcast subnet	hop limit	permit system
buffer preallocation	flooding	Routing Information Protocol (RIP)
centralized routing	forward search least cost algorithm	static routing
choke packets	isolated routing	station
circuit switched subnet	least cost algorithm	subnet
connection negotiation	metropolitan area network (MAN)	virtual circuit
connection-oriented	network congestion	wide area network
connectionless	node	weighted network graph
datagram	Open Shortest Path First (OSPF)	
distributed routing	quality of service	

REVIEW QUESTIONS

1. What are the main differences between a local area network and a wide area network?

2. How does a metropolitan area network differ from a wide area network? How are they similar?

3. What is a subnet and how does it differ from a network?

4. What is the difference between a station and a node?

5. What are the main characteristics of a circuit switched network? What are its advantages and disadvantages?

6. What are the main characteristics of a datagram packet switched network? What are its advantages and disadvantages?

7. What are the main characteristics of a virtual circuit packet switched network? What are its advantages and disadvantages?

8. How does a connectionless network application differ from a connection-oriented network application?

9. Is a connectionless network application reliable or unreliable?

10. What are the various combinations of circuit switched and packet switched subnets and connection-oriented and connectionless network applications?

11. How does a weighted network graph differ from a network graph?

12. For a weighted network graph, how many different definitions of weight can you list?

13. What are the basic goals of Dijkstra's least cost algorithm?

14. How can flooding be used to transmit a data packet from one end of the network to another?

15. How are the hop count and the hop limit used to control flooding?

16. What are the main advantages and disadvantages of:
    a. centralized routing
    b. distributed routing
    c. isolated routing
    d. adaptive routing
    e. static routing

17. What can cause network congestion?

18. How can network congestion be avoided?

19. What does quality of service have to do with network congestion?

EXERCISES

1. Which type of network application requires more elaborate software: connection-oriented or connectionless? Explain.

2. Create a scenario similar to the telephone call/sending-a-letter scenarios that demonstrates the differences between connection-oriented and connectionless network applications.

3. Explain the difference between a network node and a network station.

4. Does a datagram subnet require any setup time before a packet is transmitted? If so, when and how often?

5. Does a virtual circuit subnet require any setup time before a packet is transmitted? If so, when and how often?

6. List the steps involved in creating, using, and terminating a virtual circuit.

7. You are downloading a file over the Internet. Is the download a connectionless application or a connection-oriented application?

8. (From the Details Section) Given the weighted network graph shown in Figure 10-13, find the least cost route from node A to all other nodes using Dijkstra's least cost algorithm.

Figure 10-13
Sample weighted network graph to accompany Exercise 8

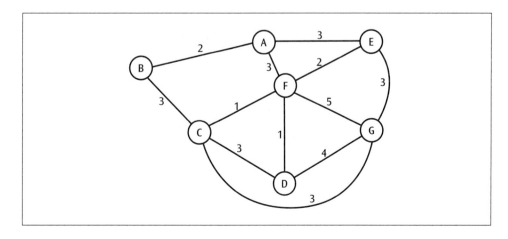

9. Using flooding and the graph from Exercise 8, how many packets will be created if a packet originates at node A and there is a network hop limit of three?

10. How do you determine the hop limit in flooding?

11. Explain how isolated routing may gather the wrong or insufficient data.

12. What can be done to protect a centralized routing network if the computer hosting the central routing table fails?

13. What happens in a virtual circuit packet switched subnet if a node or communications link along the virtual path fails?

14. Does an Ethernet network support quality of service? Explain.

THINKING OUTSIDE THE BOX

1 A large wide area network covers the United States and has multiple nodes in every state. The node in Denver crashes completely. How do the other nodes in the country find out about the crash?

2 Your company is creating a new network application that allows employees to view their payroll information electronically via the Internet. Should this application be connectionless or connection-oriented? Defend your answer, and show an example of a possible payroll information viewing session.

3 A wide area network is using the permit system to control congestion. What happens if, for some unknown reason, all the permits disappear? How can this event be detected? How can this event be repaired?

PROJECTS

1. Given a network with *n* nodes, create a formula that calculates the number of messages transmitted between nodes if one node contains a centralized routing table and each of the other nodes refers to this routing table once a second.

2. Does the Internet use flooding in some situations? Prove your answer.

3. Asynchronous transfer mode is popular for its quality of service capabilities. How does it support quality of service?

4. Write a computer program that inputs the data for a weighted network graph and computes the least cost paths using Dijkstra's algorithm.

5. What is the current routing algorithm used by the Internet? Are any of the basic algorithms and techniques introduced in this chapter used? Which ones?

11

The Internet

◆ ◆

"The Internet will change everything. It's the second Industrial Revolution that will transform every individual, every company and every country." *John Chambers, CEO of Cisco Systems, Inc. 10/1/99.*

"We believe that the online retail market will grow from $2 billion in the value of goods sold during 1997, to more than $17 billion in 2001." *Forrester Research, 1998.*

"As in the PC era, we believe the commercialization of the Internet across industries will represent net wealth creation of hundreds of billions of dollars globally, across several industries, over the next five to 10 years." *Goldman Sach analysts, 1998.*

With so much money at stake, you have to ask, "If something goes wrong with the Internet, who or what is in charge of fixing it?" The answer, surprisingly, is nobody and nothing.

The Internet is managed by a loose confederacy of businesses and non-profit organizations. The players are scattered around the globe and motivated to achieve consensus out of enlightened self-interest. "It has been a highly cooperative effort," states Vinton Cerf, who is credited with creating the TCP/IP protocols that are the backbone of today's Internet. "It's been in everyone's best interest to work together and make sure all this works." Interestingly, this structure makes the Internet the only non-sovereign global association. *Penelope Patsuris, "Who is Running This Joint", Forbes Digital Tool, Forbes.com, Nov. 1998.*

What is this thing called the Internet?

What services are offered on the Internet?

What basic mechanisms keep the Internet working?

Objectives ▶

After reading this chapter, you should be able to:

▶ Describe the major Internet applications and services and outline the advantages and disadvantages of each.

▶ Discuss the business advantages of the World Wide Web.

▶ Cite the basic features of HTML, Dynamic HTML, and XML and describe how they differ from each other.

▶ Describe the importance of e-commerce and the importance of e-retailing, electronic data interchange, micro-marketing, and electronic security.

▶ Recognize that intranets and extranets are business-related Internet systems primarily for in-house use.

▶ Discuss the responsibilities of the Internet Protocol (IP) and how IP can be used to create a connection between networks.

▶ Discuss the responsibilities of the Transmission Control Protocol (TCP) and how it can be used to create a reliable end-to-end network connection.

▶ Identify the relationships between TCP/IP and the protocols ICMP, UDP and ARP.

▶ Describe the parts of a Uniform Resource Locator (URL).

▶ Describe the responsibility of the Domain Name Service and how it converts a URL into a dotted decimal IP address.

▶ Recognize that the Internet is constantly evolving and that IPv6 and Internet2 demonstrate that evolution.

Introduction ▶

During the late 1960s, a branch of the U.S. government titled the Advance Research Projects Agency (ARPA) created one of the country's first wide area packet switched networks, the ARPANET. Select research universities, military bases, and government labs were allowed access to the ARPANET for services such as electronic mail, file transfers, and remote logins.

In 1983 the Department of Defense broke the ARPANET into two similar networks: the original ARPANET and MILNET, for military use only. Although the MILNET remained essentially the same over time, the ARPANET was eventually phased out and replaced with newer technology. During this period, the National Science Foundation funded the creation of a new high speed cross-country network backbone called the NSFnet. The NSFnet backbone was the main telecommunications line through the network connecting the major router sites across the country. It was to this backbone that smaller regional or midlevel (state-wide) networks connected. A set of access or "campus" networks then connected to these midlevel networks (Figure 11-1).

Figure 11-1
Old NSFnet backbone and connecting midlevel and campus networks

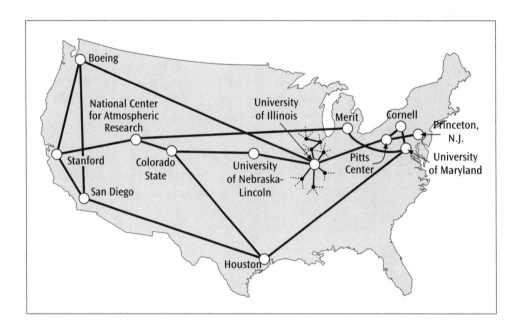

During the late 1980s, the government withdrew all direct support for the NSFnet and turned it over to private industries and universities. Once private industries assumed support of the telecommunications lines and router sites, a network of networks called the Internet was born. Now, there is no longer a single backbone but multiple backbones supported by different businesses and organizations, each in competition with one another. Current estimates suggest that there are more than 72.4 million hosts (computer sites that store and deliver web pages) connected to the Internet and that more than 120 million people around the world access the Internet on a regular basis. Although the number of users cannot be verified, one thing is certain: use of the Internet has grown at a phenomenal rate. Its early creators did not envision the Internet we have today.

Today, the World Wide Web is the Internet application with the broadest appeal. By running a web browser and clicking links, users can examine text and image-based web pages on virtually any topic from virtually any point in the world. To transfer web pages, the Internet uses the HyperText Transport Protocol (HTTP), and to create web pages a combination of HTML, dynamic HTML, and XML are used. The impact of the World Wide Web and other Internet functions on the business community has been monumental. By the late 1990s, an entirely new field of business had been created: e-commerce. You can buy practically anything on the Internet. Entertainment, medicine, groceries, real estate, mortgages, and automobiles are just a few of the products available online through e-commerce.

Many people think the Internet is only the service that allows a person to browse web pages and click links, but it is actually much more. One of the first services, and still one of the more popular offered on the Internet, is the file transfer protocol (FTP), which allows a user to upload or download files. Other services offered by the Internet include remote login, Internet telephony, electronic mail, listservs, Usenet, and streaming audio and video.

Finally, to support the Internet, a host of protocols are necessary. Two of the most common protocols are the Internet Protocol (IP) and the Transmission Control Protocol (TCP). These protocols are supported by a host of secondary protocols, which include Internet Control Message Protocol (ICMP), User Datagram Protocol (UDP), and Address Recognition Protocol (ARP). Let's begin our examination of the fascinating world of the Internet by examining the services available.

Internet Services

When the Internet first came into existence as the Arpanet, most people used it for e-mail, file transfers, and remote logins. To appreciate the many capabilities of the Internet, let's examine several of the more popular services that the Internet provides today. We'll start with File Transfer Protocol for downloading and uploading files and continue with remote login, Internet telephony, e-mail, listservs, Usenets, and streaming audio and video.

File transfer protocol (FTP)

The **file transfer protocol**, or **FTP**, was one of the first services offered on the Internet. Its primary functions are to allow a user to download a file from a remote site to the user's computer and to upload a file from the user's computer to a remote site. These files can contain data, such as numbers, text, or images, or executable computer programs. Although the World Wide Web has become the major vehicle for retrieving text- and image-based documents, many organizations still find it useful to create an FTP repository of data and program files. Using a web browser or specialized FTP software, you can easily access an FTP site. If you desire privacy and wish to restrict access to an FTP site, the site can be designed so that an ID and password are required for entry.

To access an FTP site via a web browser and download a file, you need at least three pieces of information. First, you must know the name of the FTP site. Second, you must know the directory or subdirectory in which to look for the file. Third, you must know the name of the file that you are trying to download. Thus, downloading FTP files is not a "browsing" activity. You must have a good idea of what you are looking for and where it is located.

As an example, let's say you are reading an article in a computer magazine and the article describes a free utility program that organizes your time, keeps you on schedule, and helps you make new friends. You just have to have this program, but it is on some FTP site in the middle of nowhere. No problem. All you need is the utility program's URL and a web browser. As will be shown in more detail in a later section, the URL, or Uniform Resource Locator, is the address of every document on the Internet. Pretending the URL is:

ftp://ftp.goodstuff.com/public/util/

and the name of the file is *perfect.exe*, at the web browser's location or address prompt, you type in the URL. When you arrive at the ftp.goodstuff.com ftp site and are in the public/util subdirectory, you should see the file *perfect.exe* listed. Clicking the filename and specifying the receiving location of the downloaded material starts the download process. The FTP software sends the appropriate FTP request command to the FTP site, which opens a file transfer connection between your workstation and the remote site. The file is broken into packets and the packets are transferred one by one over the Internet to your workstation. When all the packets for the file have arrived the connection is dissolved. Actually, two connections are created. The first connection transfers the control information, which sets up the file transfer, and the second connection transfers the actual data.

Remote login (Telnet)

Telnet is a terminal emulation program for TCP/IP networks, such as the Internet, that allows users to login to a remote computer. The Telnet program runs on your computer and connects your workstation to a remote server on the Internet. Once you are connected to the remote server, or host, you can enter commands through the Telnet program, and those commands will be executed as if you were entering them directly at the terminal of the remote computer.

There are three reasons for using a Telnet program. First, Telnet allows a user to log into a personal account and execute programs no matter where the user is. For example, you might have a computer account at two different companies or two different schools. While physically visiting one site, Telnet allows you to log in to the other computer site. With this login you can check your e-mail or run an application. Second, remote login allows a user to access a public service on a remote computer site. For example, the Colorado Alliance of Research Libraries (CARL) provides a large suite of databases, which include bibliographic indexing and abstracting services as well as full-text files. You can gain access to the CARL libraries through a Telnet connection or a web interface. Third, Telnet enables a network administrator to control a network server and communicate with other servers on the network. This control can be performed from a remote distance, such as from another city or from the network administrator's home.

Internet telephony

One of the most recent services offered by the Internet is sending voice, or telephony. Making telephone calls over the Internet has a number of different names, including **packet voice, voice over packet, voice over the Internet,** and **voice over Internet Protocol (VoIP)**. Internet telephony has been termed voice over Internet Protocol (VoIP) since the Internet Protocol (IP) controls the access to, and transfer of, data over the Internet. Whatever its title, voice over the Internet is beginning to emerge as a new Internet service and has the potential for becoming a popular application.

Because of the newness of this technology, the implementation of voice over the Internet varies greatly depending on your level of involvement. Due to the allure of a potentially large market, many computer-related companies are offering complete packages that can be installed on a local area network system. Some telecommunication companies are also beginning to offer the service. You may also consider establishing your own voice over Internet system, which would require workstation microphones and speakers, workstation software, network server software (called the packet voice agent), and a similar setup on the other end of the connection.

When a reliable and high quality telephone system is already available, why explore new technology that provides the same service? One advantage of Internet telephony is the simple fact that long distance calls, especially overseas calls, cost money, but sending data—or voice—over the Internet is essentially free. Another possible advantage is that a company's data network can be used for both voice and data, possibly eliminating the expense of a separate voice system.

Internet telephony has a number of disadvantages as well. The statement that sending data over the Internet is essentially free is misleading. Nothing, of course, is free. All Internet users must pay for access to an Internet service provider, the interconnecting phone line, and any necessary hardware and software. Furthermore, additional hardware and software are necessary to handle the transmission of voice packets over a corporate data network.

A second disadvantage is the fact that transmitting voice over a corporate network can be resource intensive. If the current corporate network system is straining to deliver data, adding voice to this system can cause severe service problems. These service problems can be compounded because voice systems require networks that can pass the voice data through in a relatively small amount of time. A network that delays voice data by more than 20 milliseconds from end to end will introduce a noticeable echo into the transmission. If the delay becomes longer than 250 milliseconds (that's only a quarter of a second), the system will be basically unusable.

Compounding this delay problem is the fact that the Internet is essentially a large datagram packet switched wide area network. Recall from Chapter Ten that in a datagram packet switched network each packet of data is routed individually. This individual routing can introduce a routing delay for each packet. In contrast, a circuit switched network (for example, the telephone system) first establishes a route (a circuit) and then transmits all the following packets down that established route. Since the route is fixed and dedicated, each packet does not experience a routing delay. To maintain a telephone conversation over the Internet, the voice packets must arrive at their destination in a continuous stream. If they do not, the voice conversation breaks up and is of poor quality. At the present moment, Internet telephony sometimes works and sometimes doesn't work.

One more disadvantage of Internet telephony is the currently slow acceptance of VoIP protocols and systems. Although VoIP may someday be one of the more common ways for establishing telephone connections, it is currently in its infancy and requires serious analysis before a company decides to invest time and resources in Internet telephony.

Note, however, that voice over IP does not have to involve the Internet. Although currently rare, a company can use the Internet Protocol for transmission of data *within* its own network but use traditional telephone lines outside the company network. Because these systems do not use the Internet but instead remain internal, packet delays are minimal, making voice over IP attractive. New systems are beginning to appear that can support both telephone operations and computer data operations over the same set of wires. This area of study—computer telephony integration—is an exciting new marriage of computer systems and telephone systems.

Interestingly, there are other techniques of packet voice that offer better quality and faster throughput than Internet telephony. **Voice over frame relay (VoFR)** and **voice over Asynchronous Transfer Mode (voice and telephony over ATM, or VToA)** are also available. However, both cost more than VoIP. As you will see in Chapter Twelve, frame relay and ATM are two telecommunication technologies that offer very high speed data transfers over connection-oriented networks.

Electronic mail

Electronic mail, or e-mail, is the computerized version of writing a letter and mailing it at the local post office. Many people are so committed to using e-mail that if it were taken away tomorrow there would be some serious repercussions throughout the United States and the world. Many commercial e-mail programs exist, as well as a number of free programs that can be downloaded from the Internet. Although each e-mail program has its own unique feel and options, most e-mail programs offer the following services:

- ▸ creating an e-mail message;
- ▸ sending an e-mail message to one recipient, multiple recipients, or a mailing list;
- ▸ receiving, storing, replying to, and forwarding e-mail messages; and
- ▸ attaching a file, such as a word processing document, a spreadsheet, an image, or a program to an outgoing e-mail message.

Most e-mail systems consist of two parts: the User Agent, which is the portion of the e-mail program that allows a user to create, edit, store, and forward e-mail messages; and the Message Transfer Agent, which is the portion of the e-mail program that prepares and transfers the e-mail message. Furthermore, each transmitted e-mail consists of two basic components: an envelope, which contains information describing the e-mail message; and the message, which is the contents of the envelope.

Most messages consist of plain text and are simple ASCII characters. What if you want to send (or attach) a non-text based item, such as a spreadsheet, a database, or an image? The e-mail program then creates a **Multipurpose Internet Mail Extension (MIME)** type and attaches it to the e-mail message.

Once the e-mail and optional attachment have been created, it is time to transmit the message. The Internet protocol **Simple Mail Transfer Protocol (SMTP),** which is a protocol for sending and receiving e-mail, is used to perform the transfer. To send the e-mail message, the source computer establishes a TCP connection to port number 25 (typically) on the destination computer. The destination computer has an e-mail daemon—a program that is always running in the background and waiting to perform its function—which supports the SMTP protocol. The e-mail daemon watches port 25, accepts incoming connections, and copies messages to the appropriate mailbox. The European and Canadian equivalent to SMTP is the X.400 protocol.

How many times do you receive an e-mail message even though your machine is not turned on? When you turn on your computer and run your e-mail program, a message informs you that you have *x* new e-mail messages waiting. What software performs this operation? Version 3 of Post Office Protocol (POP3) is the software that allows the user to save e-mail messages in a server mailbox and download them from the server when desired. POP3 is useful if you do not have a permanent connection to a network and must dial in using a temporary Internet connection. POP3 will hold your e-mail messages until the next time you dial in and access your mailbox. Thus, POP3 software is commonly found on mobile laptop computers or home computers without a permanent network connection, as well as computer systems that have a permanent connection.

An alternative to POP3 is the more sophisticated Internet Message Access Protocol (IMAP). IMAP (the latest version is IMAP4) is a client/server protocol in which e-mail is received and held for you at your Internet server. You can view just the heading of the e-mail or view the sender of the message and then decide if you want to download the mail. You can also create and manipulate folders or mailboxes on the server, delete old e-mail messages, or search for certain parts of an e-mail message.

Many e-mail packages allow a user to encrypt an e-mail message for secure transfer over an internal local area network or over a wide area network such as the Internet. When an e-mail message is encrypted, it is virtually impossible to intercept and decode the message without the proper encryption algorithm and key. Important e-mail messages can also have a digital signature applied so that, in the future, the owner of an e-mail message can prove that the e-mail message belongs to no one else but the original owner. Encryption techniques will be discussed in more detail in Chapter Thirteen.

Details ▶

How the MIME System Works

There are five different MIME type headers that provide information about the attachment: MIME version, content description, content ID, content type, and content transfer encoding. The MIME version states the version number of the MIME protocol being used. The content description is the actual document being attached. The content ID identifies the MIME attachment and its corresponding e-mail message. The content-type header describes whether the attachment is a text, image, audio, video, application, message, or a multipart file.

The content-transfer-encoding header describes how the non-ASCII material in the attachment is encoded so it doesn't create havoc when transmitted. An e-mail program can use five different schemes when encoding non-ASCII material for transmission:

► Plain 7-bit ASCII

► 8-bit ASCII, which is not supported by many systems

► Binary encoding, which is machine code. Binary encoding is dangerous to use because a particular binary value may signal a special control character, such as carriage return.

► Base64 encoding, which is the best method. To perform Base64 encoding, the computer breaks 24 bits of binary data into four 6-bit groups, then encodes each 6-bit group into a particular ASCII character (based on the standard ASCII character chart).

► Quoted printable encoding, which uses 7-bit ASCII. In quoted printable encoding, any character with a decimal ASCII value greater than 127 is encoded with an equals sign (=) followed by the character's value as 2 hexadecimal digits.

Listservs

A **listserv** is a popular software program used to create and manage Internet mailing lists. Listserv software maintains a table of e-mail addresses that reflects the current members of the listserv. When an individual sends an e-mail to the listserv address, the listserv sends a copy of this e-mail message to every e-mail address stored in the listserv table. Thus, every member of the listserv receives the e-mail message. Other names for listserv or other types of listserv software include mailserv, majordomo, and almanac.

To subscribe to a listserv, you send a specially formatted message to a special listserv address. This address is different from the address for sending an e-mail to all the listserv members. For example, you could subscribe to the University of Southern Michigan at Copper Harbor's listserv about erasers by sending the message SUBSCRIBE ERASERS to *listserv@copper.usm.edu*. To send an e-mail message to all the subscribers, you might send an e-mail to *erasers@copper.usm.edu*.

Listservs can be a useful business tool if you are looking for information on a particular topic or if you wish to be part of an ongoing discussion on a particular topic. By subscribing to a listserv on your topic of interest, you can receive continuous e-mail from other members around the world who are interested in the same topic. Businesses can also use listservs to communicate with their customers about products and services. To find out what listservs are available, visit the web site http://www.liszt.com.

Usenet

Usenet is a voluntary set of *rules* for passing messages and maintaining newsgroups. It is the Internet equivalent of an electronic bulletin board system. When viewing a particular newsgroup, you may see on your screen a continuous stream of comments, organized by topic, from other newsgroup members around the world. This structure differs from a listserv, in which all correspondence is through e-mails.

Usenet supports thousands of newsgroups on virtually every known subject. Newsgroup topics are ordered hierarchically to aid in finding a particular topic. Some of the main newsgroup titles include comp (computer science), news, rec (recreation), sci (science), talk, and misc. For example, under the recreation newsgroup rec you may find the topic music, and under music you may find the topic folk. The hierarchical address, then, would be rec.music.folk.

Although there is stand-alone software that allows you to access newsgroups, most web browsers support access to newsgroups if the network administrator has enabled this feature on your copy of the browser. Since newsgroups are often used for non-business activity, many companies leave the newsgroup option on corporate browsers disabled. If used properly, however, a newsgroup can provide useful business information on one or more business-related topics. For example, if you are responsible for supporting a particular software application, a newsgroup that addresses the bugs and fixes for that software application may be a valuable resource.

Streaming audio and video

Another relatively new service provided by the Internet is streaming audio and video. **Streaming audio and video** involves the continuous download of a compressed audio or video file, which can then be heard or viewed on the user's workstation. Typical examples of streaming audio are popular and classical music, live radio broadcasts, and historical or archived lectures, music, and radio broadcasts. Typical examples of streaming video include prerecorded television shows and other video productions, lectures, and live video productions. Businesses can use streaming audio and video to provide training videos, product samples, and live feeds from corporate offices, to name a few examples.

To transmit and receive streaming audio and video, the network server requires the space necessary to store the data and the software to deliver the stream, and the user's browser requires a streaming product such as RealPlayer from RealNetworks to accept and display the stream. All audio and video files must be compressed since an uncompressed data stream would occupy too much bandwidth and would not travel in real time.

A popular compression technique that delivers near-CD quality audio is MPEG layer 3, more commonly known as MP3. **MP3** allows users to download acceptable quality music in real time mode and high quality audio in a download-, store- and play-later mode. **Real-Time Protocol (RTP)** and **Real Time Streaming Protocol (RTSP)** are two common application layer protocols that servers and the Internet use to deliver streaming audio and video data to a user's browser. Both RTP and RTSP are public-domain protocols and are available in a number of software products that support streaming data.

The World Wide Web

Although the Internet still offers tried-and-true services, such as file transfer, electronic mail, and remote login, a relatively new service has grown dramatically since its introduction in 1992: the **World Wide Web** (**WWW**). Using a web browser such as Netscape Navigator or Microsoft Internet Explorer, you can download and view web pages on a personal computer. Of all the Internet services, the World Wide Web has probably had the most profound impact on business. Internet retail sales and service support has exploded with the use of personal computers and web browsers. Retail sale of virtually every imaginable product and service is now available on the Web. In a single day, you can purchase clothing and groceries, select and purchase an automobile, prepare a funeral, submit an auction bid on a toy you had as a child, find a new job, rent a video tape, and order pizza for dinner. All the personal computer needs is a connection to the Internet, which we will examine later in this chapter, and a web browser. The web pages downloaded can consist of text, graphics, links to other web pages, sometimes music and video, and even executable programs (Figure 11-2).

Figure 11-2
A typical web page

Web pages are created using HyperText Markup Language (HTML) generated manually with a text-based editor such as Notepad or by using a web page authoring tool. A web page authoring tool is similar to a word processor, except rather than creating text documents you create HTML-based web pages. The web page authoring tool has a graphical user interface that allows you to enter text and insert graphics and other web page elements and arrange them on the page using drag-and-drop techniques; the authoring tool automatically generates the underlying HTML code.

Details ▶

HTTP

The entire purpose of **HTTP** is to send and receive web pages. To perform this operation, **HTTP** can perform a number of different commands, called *methods*. A few of the command methods are:

▶ **GET** retrieves a particular web page, which is identified by a URL;

▶ **HEAD** using a given URL, the **HEAD** method retrieves only the HTTP headers of the web page but no document body;

▶ **PUT** used to send data from a user's browser to a remote web site; an example of this is sending your credit card number to a web merchant when purchasing a product; and

▶ **DELETE** requests that the server delete the information corresponding to the given **URL**.

As an example, suppose you click a link during a web browser session. The browser software creates a **GET** command and sends that command to the web server. The web server then returns the web page to the browser. If you are purchasing a product via the Web and you are filling out a form, when you click the send button, **HTTP** creates a **PUT** command and uses it to send the data to a web server.

Once a web page is created, it is stored on a computer that contains web server software and has a connection to the Internet. The web server software accepts hypertext transport protocol (HTTP) requests from web browsers connected to the Internet, retrieves the requested web page from storage, and returns that web page to the requesting computer via the Internet. The **HTTP** protocol is an application layer protocol. As was shown in Chapter One, a user sitting at a workstation running a browser, clicks a link on a web page. The browser software sees this click and creates an application layer HTTP command to retrieve a web page. The HTTP command is passed to the transport layer, network layer, and interface layer before it is placed on the network medium. When the web page request is received at the web site server, the web page is retrieved and returned across the Internet to the user's browser, where it is displayed on the monitor.

Creating web pages

To transmit a web page, a web browser, web server, and the Internet use the HTTP protocol. HTTP, however, is not used to display the web page once it reaches the intended destination. To control how a web page is displayed, another specification is used—HyperText Markup Language (HTML). Although HTML was the original and still most commonly used method for controlling the display of web pages, two new forms of HTML have emerged that offer more power and flexibility in web page creation—Dynamic HTML and eXtensible Markup Language. Let's examine each of these three specifications and discuss their importance to businesses and consumers.

The question of why you should know the basic methods of constructing a web page is important. The web page is the fundamental element of the World Wide Web. If you understand what is involved in creating a web page, you can better communicate with individuals who create web pages, and you can create web pages yourself if called upon to do so. A simple web page can be an effective tool for demonstrating an idea to fellow employees. And in some small businesses, you may be the employee who not only designs the product and creates the marketing campaign, but also designs and creates the supporting web pages.

HyperText Markup Language (HTML)

HyperText Markup Language (HTML) is a set of codes inserted into a document that is intended for display on a web browser. The codes, or markup symbols, instruct the browser how to display text, images, and other elements on a web page. The individual markup codes are often referred to as tags and are surrounded by brackets (< >). Most HTML tags consist of an opening tag, followed by one or more attributes, and a closing tag. Closing tags are preceded by a forward slash (/). Attributes are parameters that specify various qualities that an HTML tag can take on. For example, a common attribute is HREF, which specifies the URL of a file in an anchor (<A>) tag.

Figure 11-3 shows a small example HTML file. Each line in the figure includes an HTML tag, which tells the browser how to display text or images. The line numbers at the end of each line are placed there for comments and are not part of an HTML document. Line 1 (the opening <HTML> tag) begins every HTML document. HTML documents are broken into HEAD and BODY sections. Line 2 starts the HEAD section. Line 3 generates the title that appears on the title bar of the browser; the title statement always appears within the HEAD section. Line 4 denotes the end of the HEAD section. Line 5 denotes the beginning of the BODY section. Line 6 is a text line with a Break tag at the end. A Break tag inserts a line break, and any text following it starts on the next line. Line 7 begins a new paragraph on a new line. Lines 8 and 9 are Headings, which display text in larger size or bolder typeface than the normal typeface. Headings come in different sizes, with H1 being the largest, H2 the next smaller size, and H6 being the smallest. Line 10 generates a Horizontal Rule, which is a graphic dividing line that reaches across the page. Lines 11 and 12 are examples of bolding and italicizing text, respectively.

Line 13 is a statement that places an image at this point in the web page. It consists of two parts: the image tag <IMG> and the SRC attribute. The SRC attribute specifies the location of the image file to be displayed; in this example the image is in a directory called *images* and has the file name 'banner.gif.' All images must be either .gif or .jpg files.

Line 14 is an example of a link or hyperlink. A hyperlink consists of the link or anchor tag (<A>...</A>), the HREF attribute that specifies the URL of the file you want to link to, and text that will be highlighted. The text "DePaul CS Page" will be highlighted (the default highlight color is blue) and underlined. When a user browsing the web moves his or her cursor over this highlighted text, the cursor turns into a 'clickable hand'. If the user then clicks the highlighted text, the browser tries to load the web page identified by the URL specified in the HREF attribute ('http://www.cs.depaul.edu,' in this case).

Lines 15 and 16 denote the end of the BODY section and end of the HTML document, respectively. Figure 11-4 shows the web page as it would appear in a browser.

Figure 11-3
Example HTML file

<HTML>	1. Begins every HTML document
<HEAD>	2. Begins the head section
<TITLE>DePaul University Home Page</TITLE>	3. Title that appears on browser title bar
</HEAD>	4. Closes the head section
<BODY>	5. Begins the body section
This is the first line. 	6. A text line followed by a line break
<P>Start a new paragraph.</P>	7. Begins a paragraph
<H1>First Heading</H1>	8. Level 1 head
<H2>A second level Heading</H2>	9. Level 2 head
<HR>	10. Inserts a rule
<B>Bold this line.</B> 	11. Bold text
<I>Italicize this line.</I> 	12. Italicized text
<IMG SRC="\images\banner.gif">	13. Inserts an image
<A HREF="http://www.cs.depaul.edu"> DePaul CS Page</A>	14. Link to another web page
</BODY>	15. Closes the body section
</HTML>	16. Ends every HTML

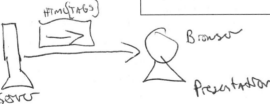

Figure 11-4

*The web page gener-
ated by the example in
Figure 11-3 as it would
appear in a browser*

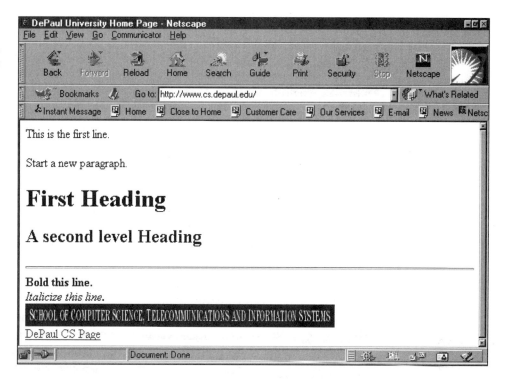

Dynamic HTML

Although HTML is relatively simple to use, it suffers from a number of shortcomings. One of the most serious shortcomings is HTML's inability to place text or an image at a precise position on the web page. Using HTML, it is also difficult to specify and switch between sets of font styles and colors easily and quickly. To better support these types of functionality, Dynamic HTML was created. Rather than being a single specification, **Dynamic HTML** (DHTML) is a collection of new markup tags and techniques that can be used to create more flexible and more powerful web pages. HTML pages are simple, static text documents that browsers read, interpret, and display on the screen. In contrast, Dynamic HTML pages have additional functionality that allows them to be interactive, among other things. Stated simply, Dynamic HTML can grab any element on a web page and change its appearance, content, or location on the page dynamically.

At this time, no one standard exists for DHTML, and its implementation differs between Microsoft and Netscape. Nevertheless, HTML versions 4.0 and 5.0 contain many dynamic elements, and you can incorporate DHTML into your documents using a combination of HTML, style sheets, and scripts.

The range of effects you can achieve with DHTML and the commercial possibilities of DHTML are broad. For example, with DHTML, when a user moves his or her cursor over an area on a page or clicks a particular link (such as a heading), additional text appears on the page (with no help from the server). This pop-up information can be used to provide further details about a particular item. For example, a commercial web site may display several items of merchandise with a simple title under each item. When a cursor moves over or clicks the title, a more detailed description of the merchandise pops up. Since this detailed description is not coming across the Internet from the server, the delivery time is immediate.

A second example of DHTML is the live positioning of elements. You can position text or images anywhere on a page, even on top of other elements, creating

Info DATA

<H1> A Title </H1>

Presentation

[handwritten margin notes:]

TOS

SGML – Meta Language
   HTML – Soph Presentation or DATA
     DHTML
       CSS
XML
    XSL – Not in Book

DATA ⟨ Information / Presentation

XML –

XSL – Extensible Style sheet Language

Unlike CSS XSL allows you to create what ever you want to do.

layers. A **layer** is an HTML page or a section of a page that can be handled separately by the browser from other layers and placed at any position on the web page. You can create a layer, enter any content, specify a text color, font style, and font size. Then by giving the layer X and Y coordinates, you can place it anywhere on the screen. Up to ten layers can be displayed, hidden, and moved, each independent of the others. These layers can overlap, let other layers show through, or hide everything underneath. Businesses and retailers can use this feature to create highly attractive and functional web pages. If a web page is more attractively laid out, the potential customer will linger at the page long enough to, hopefully, find what he or she is looking for and make a purchase. Internet Explorer and Navigator handle layers differently, so what works in one browser may not work in the other.

A third example is the concept of data binding. Internet Explorer enables page elements, such as table cells, to attach themselves to database records, greatly simplifying the chore of sending and retrieving records to and from a server database. Businesses can use this feature to more easily prompt a user for input data and then accept that data and store it in a server database. These transactions can be used for making purchases and answering business queries.

A fourth element of Dynamic HTML is cascading style sheets. **Cascading style sheets (CSS)** allow a web page author to incorporate multiple styles (fonts, styles, colors, and so on) in an individual HTML page. More importantly, cascading style sheets separate the presentation of the web page content from the content itself. By creating a single style sheet, a web page author can use that style sheet for all the web pages at a particular site. This uniformity creates a more pleasing appearance, creates a more unified appearing web site for a business, and makes the life of the author simpler. For example, a style sheet can be created that incorporates a number of font styles, font sizes, and colors. The web page author can then incorporate this style sheet into one or more web pages and use any one of the specified styles, sizes or colors. Both Microsoft and Netscape support cascading style sheets.

Extensible markup language (XML)

HTML and DHTML are members of a family of markup languages called Standard Generalized Markup Language (SGML). SGML is a metalanguage, a description of how to create a markup language; SGML is not a markup language itself.

Another member of this family of languages is eXtensible Markup Language (XML). **Extensible markup language (XML)** is a subset of SGML and is a description for how to create a document—both the definition of the document and the contents of the document. Whereas HTML only determines how the content of a document is to be displayed by a web browser, XML also defines the content of the document. When a document is passed between two entities, the XML document contains the data and a detailed description of the data. This dual construction eliminates the requirement of sending additional earlier documents describing the format of the data.

The syntax of XML is fairly similar to HTML, however there are a number of very important differences. First, XML is extensible, which means a user can define his or her own tags. You can create tags that define entire data structures, such as the entries in an auto parts catalog, which could require tags such as <PARTNAME>, <DESCRIPTION>, and <PARTCOST>. Second, XML is much less forgiving than HTML. Unlike HTML, a document created in XML will not display if there is a mistake in the coding. A mistake in HTML coding is often ignored by the browser and the rest of the document is displayed. Third, XML documents require many more precise rules for the creation of tags and the elements within a document. For example, all tags must be properly nested, all attribute values must have quotation marks around them, and all tags with empty content must end in "/...>".

For a business, XML's biggest advantage is its ability to perform data interchange—the transfer of data records between two companies. Because different companies, and sometimes even different parts of the same organization, rarely standardize on a single set of tools, it takes a significant amount of intercommunication for two groups to reach a point where they can send data records back and forth. XML simplifies sending structured data across the Web so that nothing gets lost in translation. For example, Company A can receive XML-tagged data from Company B, and Company B can receive XML-tagged data from Company A. Neither company has to know how the other company's data is organized, because the XML tags define the data. If another supplier or company wants to be included in this data transfer, no company has to write any computer code to exchange data with their system. The new company simply has to observe the data definition rules spelled out in the XML tags.

If the future growth of the World Wide Web continues at even a fraction of its current growth, the Web is going to remain a vital element in the daily lives of businesses and individuals. To tap that potential, an individual creating a web page must be familiar with HTML, DHTML, and XML. As these markup languages continue to evolve web pages will continue to grow in quality and power.

e-Commerce

Throughout this chapter, numerous references have been made to how the Internet affects a business. The term that has come to represent the commercial dealings of a business using the Internet is **e-commerce**. e-commerce can also be defined as the buying and selling of goods and services via the Internet, and in particular via the World Wide Web. The GartnerGroup predicts that worldwide business-to-business e-commerce will reach $7.29 trillion by the year 2004. Although e-commerce is still a relatively new term and is constantly being redefined, many agree that e-commerce includes each of the following areas:

- ▶ **e-retailing.** e-retailing is the electronic selling and buying of merchandise using the Web. Virtually every product is available for purchase over the Web. Sophisticated web merchandisers can track their customers' purchasing habits, provide online ordering, allow credit card purchases, and offer a wide selection of products and prices that might not normally be offered in a brick and mortar store.

- ▶ **Electronic Data Interchange (EDI).** EDI is the electronic commercial transactions between two or more companies. For example, a company wishing to purchase a large number of mobile telephones may send an electronic request to a number of mobile telephone manufacturers. The manufacturers may bid electronically, and the company may accept a bid and place an order electronically. Bank funds will also transfer electronically between companies and their banks.

- ▶ **Micro-marketing.** Micro-marketing is the gathering and use of potential and current customer browsing habits—critical data for many companies. When a company knows and understands its customers' habits, they can target particular products to particular individuals.

▶ **Electronic security.** The security systems that support all Internet transactions are also considered an important part of e-commerce.

Many analysts predict that e-commerce is in its infancy and will experience tremendous growth in the coming years. There is little doubt that e-commerce has the potential to be an incredible industry and will produce a wealth of jobs.

Cookies and State Information

Web APPLS
"Cookie"
"State"
"Stateless"
"Session"

One web feature that has received a good deal of negative publicity is the cookie. A **cookie** is data created by a web server, that is stored on the hard drive of a user's workstation. This data, called state information, provides a way for the web site that stored the cookie to track a user's web browsing patterns and preferences. A web server can use state information to predict future needs, and state information can help a web site resume an activity that was started at an earlier time. For example, if you are browsing a web site that sells a product, the web site may store a cookie on your machine that describes what products you were viewing. When you visit the site at a later date, the web site may extract the state information from the cookie and ask if you wish to look again at that particular product or at products like it. Other common applications that store cookies on user machines include search engines, the shopping carts used by web retailers, and secure e-commerce applications.

Information on previous viewing habits stored in a cookie can also be used by other web sites to provide customized content. One recent use of cookies involves the storage of information that helps a web site decide which advertising banner to display on the user's screen. With this technique, a web site can tailor advertisements to the user's profile of past web activity.

Although web site owners defend the use of cookies as helpful to web consumers, many users feel that the storing of information by a web site on their computers is an invasion of privacy. Users concerned about this privacy issue can instruct their browsers to inform them whenever a web site is storing a cookie and provide the option of disabling the storage of the cookie. Users can also view their cookies (which may be unreadable) and delete their cookie files. Netscape browsers store cookies in a file called COOKIES.TXT; Internet Explorer keeps each cookie as a separate file in a special cookie subdirectory.

Intranets and Extranets

One of the more powerful advantages of the Internet is that any person anywhere in the world can access your information, be it a web page or an FTP file. For many people, this accessibility is exactly what is desired. A company, however, may not want the entire world to view one or more of its web pages. For example, a company may want to allow its employees easy access to a database, but disallow access to anyone outside the company. Is it possible to offer an Internet-like service inside a company to which only the company's employees have access? The answer is yes, and the concept is called an intranet. An **intranet** is a TCP/IP network inside a company that allows employees to access the company's information resources through an Internet-like interface. Using a web browser on a workstation, an employee can perform browsing operations, but the applications that can be accessed through the browser are available only to employees within the company. For example, a corporate intranet can be used to provide employee e-mail and access to groupware software, a corporate client/server database, corporate human resources applications, or custom applications and legacy applications.

Intranets use essentially the same hardware and software that is used by other network applications such as Internet web browsing, which simplifies installation, maintenance, and hardware and software costs. If desired, an intranet and its resources can be accessed from outside the corporate walls. In these circumstances, an ID and password are often required to gain access.

When an intranet is extended outside the corporate walls to include suppliers, customers, or other external agents, the intranet becomes an extranet. Since an extranet allows external agents to have access to corporate computing resources, a much higher level of security is usually established.

An intranet has often been called a paradigm shift. It is a new way of handling business that has incredible potential yet requires little more hardware and software overhead than what currently is involved. Companies can use an intranet to establish access to internal databases, interface into corporate applications, perform full-text searches on company documents, allow employees to access and download training materials and manuals, perform daily business operations such as filling out a travel expense or requesting a company service, and allow employees to access a database of human resource information such as payroll information, personnel information, and even directions on how to get to work. Whether or not it is a paradigm shift, the intranet provides an exciting electronic tool for businesses and their employees.

Internet Protocols

From simple e-mail to the complexities of the Web, there are many applications available on the Internet. What allows these varied applications to work? What allows the Internet itself to work? The answer to these questions is Internet protocols. Although many protocols are necessary to support the operation of the Internet, several stand out as the most common: Internet Protocol (IP), Transmission Control Protocol (TCP), User Datagram Protocol (UDP), Internet Control Message Protocol (ICMP), and Address Resolution Protocol (ARP).

From the beginning, the ARPANET was designed to be a large interconnection of disparate networks. Data packets would have to travel over major nationwide backbone networks as easily as they would travel over smaller campus networks. The nationwide wide area networks might support many types of telecommunication protocols, while the smaller campus area networks (local area networks) might be Ethernet or token-passing networks. To allow the interconnection of so many types of networks, the Department of Defense needed to create a set of protocol standards or services that could be used to interface one network to another.

Figure 11-5 shows that these services are layered in a hierarchy such that an Internet user requires the use of one or more application services, which require the use of the reliable transport service, which requires the use of the connectionless packet delivery service, which requires the use of the interface services. The application services, as you saw earlier in the chapter, consist of applications such as electronic mail, file transfer, and remote login.

Figure 11-5
Hierarchy of protocol layers as created by the Department of Defense

Application Services
Reliable Transport Service
Connectionless Packet Delivery Service
Interface or Physical Services

The reliable transport service, provided by software called **Transmission Control Protocol (TCP)**, turns an unreliable sub-network into a reliable network, free from lost and duplicate packets. If an application did not need to create a connection to transfer data, the User Datagram Protocol (UDP) could be used instead of TCP. UDP supports a connection-less application in which a connection-oriented stream of packets is not necessary.

The connectionless packet delivery service provides an unreliable, connectionless network service in which packets may be lost, duplicated, delayed, or delivered out of order. Even worse, the sender and receiver of these packets may not be informed that these problems have occurred. This connectionless packet delivery service is called **Internet Protocol (IP)**. Because the connectionless packet delivery service doesn't inform users that problems may have occurred, a reliable transport service is needed "above" the connectionless packet delivery service to cover for its possible shortcomings.

Since TCP/IP is essentially two sets of standards (the TCP set and the IP set), we will examine each independent of the other. We will also examine a number of protocol standards that support TCP, IP and the Internet: User Datagram Protocol (UDP), ICMP, and ARP. Keep in mind that since the use of TCP/IP has proven effective for the Internet, many universities and corporations also use TCP/IP for the internal interconnection of possibly different types of networks.

The Internet Protocol (IP)

The Internet Protocol (IP) provides a connectionless data transfer over heterogeneous networks by passing and routing IP datagrams. IP **datagram** is essentially another name for a data packet. To pass and route IP datagrams, all packets that are passed down from the transport layer are encapsulated with an IP Header (Figure 11-6) that contains the information necessary to transmit the packet from one network to another. The format of this header will be shown in the next section.

Consider once again the example of a workstation sending a message to a distant workstation. Both workstations are on local area networks, and the two local area networks are connected via a wide area network. As workstation A sends a packet down through the layers, the IP header is encapsulated over the transport layer packet, creating the IP datagram. This encapsulation process is similar to the examples of encapsulation presented in Chapters One and Eight. The appropriate medium access control (MAC) sublayer headers are encapsulated over the IP datagram, creating a frame, and the frame is sent to the first router via LAN1. Since the router interfaces LAN1 to a wide area network, the MAC layer information is stripped off, leaving the IP datagram. At this time the router may use any or all of the IP information to perform the necessary internetworking functions. The necessary wide area network level information is applied, and the packet is sent over the WAN to router 2. When the packet arrives at the second router, the wide area network information is stripped off, leaving once again the IP datagram. The appropriate MAC layer information is then applied for transfer of the frame over LAN2, and the frame is transmitted. Upon arrival at workstation B, all header information is removed leaving the original data.

Figure 11-6
Progression of a packet from one network to another

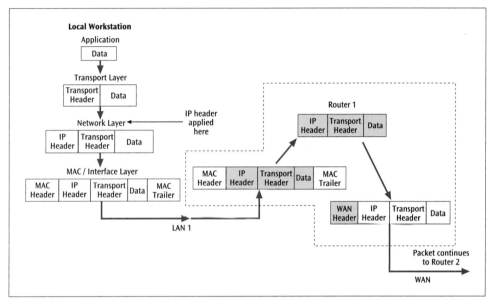

When a router has the IP datagram, it may make several decisions affecting the future of the datagram. The router:

- ► makes routing decisions based on the address portion of the IP datagram;
- ► may have to fragment the datagram into smaller datagrams if the next network to traverse has a smaller maximum packet size than the current size of the packet; and
- ► may determine that the current datagram has been hopping around the network for too long and may delete the datagram.

To perform these functions, the router needs address information, datagram size, and the time the datagram was created. This information is found in the IP header, which is the information applied to the transport packet at the Internet network layer. Each of these functions, along with the IP header fields that support these functions, will be examined in more detail in the next section.

IP datagram format

Figure 11-6 showed that the IP layer of the communications software created an IP datagram consisting of an IP header, before it was passed on to the next layer of software. The information included in this IP header and the way the header is packaged allows the local area networks and wide area network to share data and create inter-network connections.

Exactly what is in this IP header that allows this inter-networking to happen? Figure 11-7 shows in more detail the individual fields of the IP header. Even though there are 14 different fields, let's examine the field that specifies the version of the IP protocol and those fields that affect three of the primary functions of the Internet Protocol: addressing, fragmentation, and datagram discard. By examining these fields, you will begin to discover how IP works and why it is capable of interconnecting so many different types of networks.

Figure 11-7
Format of the IP datagram

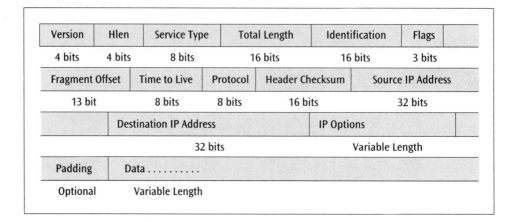

The first field that is of interest is the Version field. The Version field contains the version number of the IP protocol being used, just in case a new version becomes available. Currently, almost all networks involved in the Internet use IP version 4. IP version 6, which was created during the late 1990s, should eventually replace version 4 and is discussed in detail in a later section of this chapter. The Version field is important because two users using two different versions of the IP protocol format at the same time would find that their packets could not talk to each other and would experience a non-functional system.

The next three fields, Identification, Flags, and Fragment Offset are used to fragment a datagram into smaller parts. Why would we want to fragment a datagram? When the IP protocol was created, there existed older networks that had a maximum packet size that was smaller than most other networks. This maximum size was controlled by network hardware, software, or other factors. Since the IP protocol was designed to work over practically any type of network, it had to be able to transfer datagrams of varying sizes. Rather than limit the IP protocol to the smallest maximum packet size in existence (who even knows what that is?), the IP protocol allows a router to break or fragment a large datagram into smaller fragments which will fit onto the next network. Fortunately, most, if not all, modern networks do not have a maximum packet size that is small enough to be a concern. Thus, fragmenting a datagram into smaller packets will some day no longer be an important issue. In fact, as you will see shortly, IP version 6 does not even have a field in the header to perform fragmentation.

The second function, discarding a datagram that has been traveling the Internet for too long, is supported by the Time To Live field. The Time To Live field indicates how long this particular datagram is allowed to live—bounce from network to network—within the system. Each router along the route from source to destination decrements the Time To Live field by 1. When the value of the Time To Live field reaches zero, the router deletes the datagram.

The final two fields of interest, Source IP Address and Destination IP Address, contain the 32-bit IP initial source and final destination addresses of the datagram. A 32-bit address uniquely defines a connection to the Internet—usually a workstation or device—but one workstation or device may support multiple Internet connections. As an IP datagram moves through the Internet, the Destination IP Address is examined by a router. The router, using some routing algorithm, forwards the datagram onto the next appropriate communications link. The details of IP addresses will be covered in the section Locating a Document.

The Internet Protocol is definitely one of the most important communication protocols. Because of its simple design, it is relatively easy to implement in a wide variety of devices. And because of its power, it is capable of interconnecting virtually any type of network. Even though the Internet Protocol is powerful, its primary objective is getting data through one or more networks. It is not responsible for creating an error-free end to end connection. To accomplish this, the Internet Protocol relies on the Transmission Control Protocol, or TCP.

Details ▶

IP Fragmentation

As an example of this fragmentation, consider the IP datagram shown if Figure 11-8(a). The data portion of the datagram is 960 bytes in length. This datagram needs to traverse a particular network that limits the data portion of a datagram to no larger than 400 bytes. Thus, there is a need to divide the original datagram into three fragments of sizes 400, 400 and 160 bytes respectively. Each of the three fragments that are created has an almost identical IP header.

The Fragment Offset field contains the offset, which is the count in units of 8 bytes that this particular fragment is, from the beginning of the original datagram data field. Figure 11-8(b) shows these offsets. The first fragment, which is not offset from the beginning, has an offset value of 0. Since the first fragment is 400 bytes long, the second fragment starts 400 bytes into the original datagram. We are counting in units of 8 bytes, so 400 divided by 8 equals an offset of 50. The first two fragments total 800 bytes, thus the third and final fragment starts 800 bytes into the original datagram. Dividing 800 bytes by an 8 byte unit equals an offset of 100.

The next field that assists in fragmentation is the Flags field. Within the Flags field is a More bit, which indicates if there are more fragments of the original datagram following. Figure 11-8(b) shows the contents of the More bit. The final

field that assists in fragmentation is the Identification field. The Identification field contains an ID unique to these three fragments, which is used to reassemble the three fragments back into the original datagram.

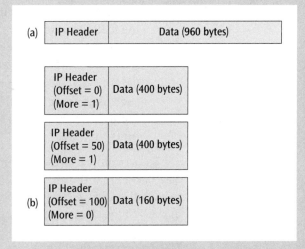

Figure 11-8 *Division of an IP datagram into three fragments*

The TCP Protocol

Perhaps one of the most common examples of a transport layer protocol is the second half of the popular TCP/IP standard: Transmission Control Protocol (TCP). The primary function of **TCP** is to turn an unreliable sub-network into a reliable network, free from lost and duplicate packets. How can a transport layer protocol make an unreliable network reliable? To make a network more reliable, TCP (as well as most transport layers) performs the following six functions:

> ▶ **Creates a connection.** The TCP layer includes a port address that indicates a particular application on a machine. The port address along with the IP address identifies a particular application of a particular machine. When TCP creates a connection between a sender and a receiver, the two ends of the connection use a port number to identify the particular application's connection. This port number is found within the TCP datagram and is passed back and forth between sender and receiver.

> ▶ **Releases a connection.** The TCP layer can also dissolve a connection after all the data has been sent and received.

> ▶ **Implements flow control.** So that the sending station does not overwhelm the receiving station with too much data, the TCP header includes a field, called the Window value, that allows the receiver to tell the sender to slow down. This Window value is similar in operation to the sliding window used at the data link layer. The difference between the two window operations is the data link sliding window operates between two nodes or between a workstation and a node, while the TCP window operates between the two ends of a network connection.

> ▶ **Establishes multiplexing.** Since the TCP header includes a port number and not an IP address, it is possible to multiplex multiple connections over a single IP connection. This multiplexing can be done by creating a connection that has a port number different from a previous connection.

> ▶ **Performs error recovery.** TCP numbers each packet for transmission with a sequence number. As the packets arrive at the destination site, the receiving TCP software checks these sequence numbers for continuity. If there is a loss of continuity, the receiving TCP software uses an acknowledgment number to inform the sending TCP software of a possible error condition.

> ▶ **Establishes priority.** If the sender has to transmit data of a higher priority, such as an error condition, TCP can set a value in a field (the Urgent Pointer) which indicates that all or a portion of the enclosed data is of an urgent nature.

To accomplish these six functions, TCP places a header at the front of every data packet that travels from sender to receiver, or from one end of the connection to the other. As we did with the IP header, let's examine the more important fields in the TCP header.

TCP datagram format

A user is sitting at a workstation and running a network application, such as an e-mail program. When the user wants to send an e-mail message, the e-mail program takes the e-mail message and passes it to the transport layer of the software. Assuming the e-mail is heading out onto the Internet, the transport layer adds a TCP header to the front of the e-mail message. The information in this header is used by the TCP layer at the receiving workstation to perform one or more of the six transport functions. This header contains several fields, which are shown in Figure 11-9. Let's examine only those fields that assist TCP in performing the six functions listed earlier.

Figure 11-9

The fields of the TCP header

Source Port	Destination Port	Sequence Number	
16 bits	16 bits	32 bits	

Acknowledgement Number	Hdlen	Reserved	Flags	Window	
32 bits	4 bits	6 bits	6 bits	16 bits	

Checksum	Urgent Pointer	Options	Padding
16 bits	16 bits	Variable Length	Optional

Data . . . .
Variable Length

The first two TCP header fields, Source Port and Destination Port, contain the addresses of the application programs at both ends of the transport connection. These port addresses are used to identify a connection that can be used in creating and terminating connections. The port number can also be used to multiplex multiple transport connections over a single IP connection.

The Sequence Number field contains a 32 bit value that counts bytes and indicates this packet's data position within the connection. For example, if you are in the middle of a long connection in which thousands of bytes are being transferred, the Sequence Number tells you the exact position of this packet within that sequence. This field can be used to reassemble the pieces at the receiving workstation and determine if a packet of data is missing.

The Window field contains a sliding window value that provides flow control between the two endpoints. The Checksum field, following the window field, provides for a cyclic checksum of the data field that follows the header. The Urgent Pointer is used to inform the receiving workstation that this packet of data contains urgent data.

Like its counterpart IP, TCP is a well-designed protocol. Its primary goal is to create an error-free end-to-end connection across one or more networks. The combination of TCP operating at the transport layer and IP operating at the network layer is found on almost all networks supporting the global Internet. Although TCP and IP will continue to evolve, the basic components should exist for a long period of time.

Internet Control Message Protocol (ICMP)

As an IP datagram moves through a network, any one of a number of things can go wrong. As a datagram nears its intended destination, a router may determine that the destination host is unreachable (IP address is wrong, or host does not exist), the destination port is unknown (there is no application that matches the TCP port number), or the destination network is unknown (wrong IP address). If a datagram has been on the network to long and its Time To Live value expires, the datagram will be discarded. Also, there could be something wrong with the IP header itself. In each of these cases, it would be nice if a router or some device would send an error message back to the source workstation informing the user or the application software of a problem. The Internet Protocol was not designed to return error messages, so some other protocol is going to have to perform these operations.

Internet Control Message Protocol (ICMP), which is used by routers and nodes, performs this error reporting for the Internet Protocol. All ICMP messages contain at least three fields: a type, a code, and the first 8 bytes of the IP datagram that caused the ICMP message to be generated. The type is simply a number from 0 to *n* that uniquely identifies the kind of ICMP message. The code is a value that provides further information about the message type. Together ICMP and IP provide a relatively stable subnet operation that can report some of the basic forms of network errors.

User Datagram Protocol (UDP)

TCP is the protocol used by most networks and network applications to create an error-free end-to-end network connection. TCP, however, is connection-oriented, in that a connection via a port number must be established before any data can transfer between sender and receiver. What if you don't want to establish a connection with the receiver, but simply want to send a packet of data? In this case, User Datagram Protocol is the protocol to use. User Datagram Protocol (UDP) is a no-frills transport protocol that does not establish connections or watch for datagrams that have existed for too long. Although UDP supports error checking and multiple transport connections over a single network connection, it is a connectionless transport protocol. UDP is used by a small number of network applications that do not need to establish a connection before sending data.

Address Resolution Protocol (ARP)

The Address Resolution Protocol is another small but important protocol that is used to support TCP/IP networks. Address Resolution Protocol (ARP) takes an IP address in an IP datagram and translates it into the appropriate CSMA/CD address for delivery on a local area network. As mentioned earlier, every workstation that has a connection to the Internet is assigned an IP address. This IP address is how a packet of data finds its way to its intended destination. There is one problem, however, when that workstation is on an Ethernet or CSMA/CD local area network. Recall that the piece of data that traverses a CSMA/CD LAN is called a frame. This frame consists of a number of fields of information, none of which is an IP address. If the frame is supposed to go to a particular workstation with a unique IP address, but the frame does not contain an IP address, how does the frame know where to go? ARP provides the answer to this question. When an IP datagram enters a CSMA/CD LAN through a router and before the IP header is stripped off leaving only the CSMA/CD frame, ARP broadcasts a message on the LAN asking which workstation belongs to this IP address. The one workstation that recognizes its IP address sends a message back saying, "Yes, that is my IP address, and here is my 48-bit CSMA/CD address. Please forward that IP packet to me via my address." Thus, ARP is the final piece that allows an IP packet to be delivered to a CSMA/CD workstation.

Tunneling protocols

One of the more serious problems with the Internet is its lack of security. Whenever a transmission is performed, it is susceptible to interception. Retailers have solved part of the problem by using encryption techniques to secure transactions dealing with credit card numbers and other secure information. Businesses that want their employees to access the corporate computing system from a remote site have found a similar solution—virtual private networks. A **virtual private network (VPN)** is a data network connection that makes use of the public telecommunication infrastructure, but maintains privacy through the use of a tunneling protocol and security procedures. A **tunneling protocol**, such as the Point-to-Point Tunneling Protocol (PPTP), is the command set that allows an organization to create secure connections using public resources such as the Internet. PPTP is a standard sponsored by Microsoft and other companies, was proposed by Cisco Systems, and is an extension of the Internet's Point-to-Point Protocol (PPP). PPP is a protocol used for communication between two computers using a serial connection. The most common example of a serial connection is a dial-up modem connection between a user's workstation and an Internet service provider. PPP is usually the standard preferred over the earlier Serial Line Internet Protocol, SLIP.

Employees who are located outside a company building can use PPTP to create a tunnel through the Internet into the corporate computing resources. Since this connection runs over the Internet and it may transmit confidential corporate data, it has to be a secure connection. Besides the secure connection, a tunnel is also a relatively inexpensive connection, since it is using the Internet as the primary form of communication. An alternative to a tunnel is a dial-up telephone line or leased telephone line, both of which can be expensive.

Locating a document on the Internet

When a user is running a browser on a workstation and clicks on a link, the browser attempts to locate the object of the link and bring it across the Internet to the user's workstation. This object can be a document, a web page, an image, an FTP file, or any of a number of different types of data objects. How does the Internet locate each object? Stated simply, every object on the Web has a unique English-like address called its Uniform Resource Locator (URL). The Internet, however, does not recognize URLs directly. In order for the Internet to find a document or object, part of the URL has to be translated into the IP address that identifies the web server that contains the document or object. This translation from URL to IP address is performed by the Domain Name Server (DNS).

The IP address itself is not as simple as it appears. The assignment of IP addresses is complex and requires an understanding of the different classes of addresses and of a concept called submasking. Let's examine each of these concepts—URLs, DNS, IP addresses, and submasking—to gain a better understanding of how the Internet finds one document from among an ocean of documents.

Uniform Resource Locator (URL)

To locate something on the Internet, the network's creators decided that every object that is somewhere on the Internet must have a unique "address." This address, or **uniform resource locator** (URL), uniquely identifies files, web pages, images, or any other type of electronic document that resides on the Internet. If you are using a web browser and click a link to a web page, you are actually sending a command out to the Internet to fetch a particular web page at a specific location based on the web page's URL.

All uniform resource locators consist of four parts, as shown in Figure 11-10. The first part, indicated by a 1 in the figure, is the service type. The service type identifies the protocol that is used to transport the requested document. For example, if you are requesting a web page, then HyperText Transport Protocol (HTTP) is used to retrieve the web page (Figure 11-10(a)). If you are requesting an FTP document, then the FTP protocol is used (Figure 11-10(b)). Other service types include telnet:// to perform a remote login, news:// to access a usenet group, and mailto:// to send an electronic mail message.

The second part of the URL, indicated by a 2 in Figure 11-10, is the host name or domain name. Whatever the name, this portion of the URL specifies a computer network server that contains the requested item. Referring to the example in Figure 11-10(a), cs.depaul.edu is one of the servers supporting the computer science program at DePaul University.

The third part of a URL is the directory or subdirectory information. For example, the URL http://cs.depaul.edu/public/utilities/ada specifies that the requested item is located in the subdirectory ada under the subdirectory utilities under the subdirectory public.

The final part of the URL is the filename of the requested object. If no filename is specified in the URL, then a default file, such as default.htm or index.html, will be retrieved.

Figure 11-10

The parts of a uniform resource locator (URL) for HTTP (a) and FTP (b)

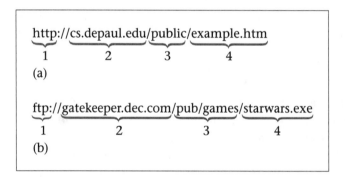

Domain name server and IP addresses

When referencing an Internet site, we often refer to its domain name. A **domain name** is the address that identifies a particular site on the Web. For example, the domain name for The School of Computer Science, Telecommunications, and Information System's web site at DePaul University is *www.cs.depaul.edu*. Starting on the right, *edu* is the top-level domain and indicates that the web site is an educational site. Other top-level domains are com (commercial), gov (government), mil (military), org (non-profit organization), and net (network-based). Sometime in the near future there will be seven additional top-level domains: firm (business/firms), store, web (WWW related activities), arts, rec (recreation), info (information services), and nom (individual/personal nomenclature). Each country has its own top-level domain name. For example, Canada is CA and the United Kingdom is UK. The domain name at the next level—called the mid-level domain name—is usually the name of the organization (often a company or school), or host, such as *depaul*. Any further-level domains are further subdivisions of the host and are usually created by the host. For example, a company, FiberLock, applies to the appropriate agency handling domain name registration, asking for the mid-level domain name *fiberlock*. Since it is a commercial business, the top-level domain name will be .com. If no one else is using *fiberlock*, the company will be granted *fiberlock.com* as its domain name. The company may then add further domain levels, such as *www.fiberlock.com*, or *email.fiberlock.com*.

Computers, however, do not use domain names. They use 32-bit binary addresses called IP addresses. For example, a valid IP address has the following form:

1000 0000 1001 1100 0000 1110 0000 0111

To make IP addresses a little easier for humans to understand, 32-bit binary addresses are represented by dotted-decimal notation, for example 128.156.14.7. This dotted decimal notation is created by converting each 8-bit quantity into its decimal equivalent:

1000 0000 1001 1100 0000 1110 0000 0111

128   156   14   7

Since computers use 32-bit binary addresses, and almost all humans use the domain name form, the Internet converts the English-like domain names into the binary forms using the **Domain Name System (DNS)**, which is a large, distributed database of Internet addresses and domain names. This distributed database consists of a network of DNS servers. The network consists of local DNS servers, mid-level DNS servers, and higher-level DNS servers. To keep the system manageable, management of the DNS database is distributed by domains: edu, gov, com, mil, and so on.

Converting a domain name into a binary IP address can be simple or complicated. The level of complexity depends on whether or not a network server on the originating local area network recognizes the domain name. If a network server cannot resolve an address locally, it will call upon a higher authority. A local DNS server will send a DNS message to the next higher DNS server until the address is found or it is determined that the address does not exist. If the address does not exist, an appropriate message is returned.

Consider an example in which a local server recognizes a domain name. An application program such as a web browser calls a library procedure called the resolver. The resolver sends a DNS message to a local DNS server, which looks up the name and returns the IP address to the resolver. The resolver then returns the IP address to the application program.

What happens if the local DNS server does not have the requested information? It may query other local DNS servers, if there are any. The information concerning the existence of other local servers and remote server locations is kept in a file on the local computer network. If no answer is returned, then the local DNS server tries the top level name server for the domain requested. If no answer is returned again, the top level domain server will query a local DNS server with the appropriate mid-level domain name. If once again no answer is returned, the mid-level domain server will query the appropriate local DNS server. If the answer is found, the results are then returned.

Consider a scenario in which a user at *cs.depaul.edu* wants to retrieve a web page from Dominican University (*www.dom.edu*). The message originates from *cs.depaul.edu* and goes to the *depaul.edu* name server. The *depaul.edu* name server does not recognize *www.dom.edu* because the domain name *www.dom.edu* is not in a list of recently referenced web sites. So the *depaul.edu* name server sends a DNS request to *edu-server.net*. Although the *edu-server.net* does not recognize *www.dom.edu*, it does recognize *dom.edu*, so it sends a query to *dom.edu*. The *dom.edu* name server recognizes its own *www.dom.edu* and returns the result to *edu-server.net*, which returns the result to the *depaul.edu* name server, which returns the result to the user's computer, which converts the domain name address to the appropriate 32-bit binary IP address.

Now, let's talk a little more about the IP address. As you have already seen, IP addresses are currently 32-bit addresses. The addresses, however, are not simply 32-bit integers, but consist of three pieces of information. The size and value of these three pieces of information depend on the basic form of the address. There exist four basic forms for an IP address (Table 11-1).

CBT: For a visual example of how DNS works, see the CD-ROM that accompanies this text.

Table 11-1

Four basic forms of an IP 32-bit address

Address Type	Beginning Bit Pattern	Network Address	Host Address
Class A	0	NetID (128 addresses)	HostID (16,777,216 addresses)
Class B	10	NetID (16,384 addresses)	HostID (65,536 addresses)
Class C	110	NetID (2,097,152 addresses)	HostID (256 addresses)
Class D	1110	Multicast Address	

Each IP address consists of three parts:

▶ A 1-, 2-, 3- or 4-bit identifier field.
▶ A NetID, which indicates a particular network.
▶ A HostID, which indicates a particular host, or computer, on that network.

There are 128 Class A addresses, or particular networks, in existence. Each Class A address can have 16,777,216 hosts, or computers. Clearly, 128 networks is not very many; in fact, all the 128 Class A addresses were assigned a long time ago. And for each network to allow 16,777,216 computers is beyond imagination. As we have seen, many local area networks rarely have more than a few hundred computers attached. Class B addresses allow for 16,384 NetIDs, or networks, each supporting 65,536 HostIDs. Class C addresses allow for 2,097,152 NetIDs, or networks, and 256 HostIDs, meaning that each network can have 256 host computers. The number of host computers allowed by a Class C address is too small except for the smallest networks.

Class D addresses are available for networks that allow multicasting of messages. **IP multicasting** is the ability of a network server to transmit a data stream to more than one host at a time. Consider a scenario in which a company wants to download a streamed video training exercise to 20 users sitting at separate workstations. If a server transmits 20 individual copies to the 20 workstations (unicast), a very high bandwidth signal will be necessary. If the server could instead multicast one copy of the video stream to the 20 workstations, a much smaller bandwidth is necessary. When multicasting, the server can insert a Class D address into the IP

[handwritten margin note:] How many PC's can a Class C license deal with — Less than 256 or 256

datagram and each of the twenty workstations could tell their IP software to accept any datagrams with this Class D address.

Although IP multicasting has some very promising advantages, it suffers from a lack of security. It is relatively easy for any workstation to tell its IP software to accept a particular Class D address, thus allowing workstations that should not be receiving the multicast to receive it anyway. Nonetheless, new applications that address the shortcomings of IP multicasting were announced at the beginning of 2000, opening the door for future possibilities.

When a company applies for a set of Internet addresses, a number of options are possible. First, the company might apply for a Class B address. Assuming any Class B addresses are left, it would be able to allocate 65,536 workstations on its Class B network. This number of workstations might be too large, so the company might consider a Class C address. A Class C address allows for 256 computers on a single network. But if the company has 400 to 500 users, a Class C address is too small. Therefore, the company might apply for two Class C addresses.

More than likely, a company will get its IP addresses from an Internet service provider (ISP). Most ISPs have already applied for a number of IP addresses and are willing to lease some of those addresses to companies. So a company contacts an ISP, which leases 500 IP addresses to the company.

The ISP, however, can present an interesting option depending on whether all 500 users are using their IP addresses at the same time. If all 500 users won't be using their IP addresses at the same time, a company might be able to lease a smaller number of IP addresses and use dynamic IP addressing. With **dynamic IP addressing**, an IP address is temporarily assigned when a workstation invokes an Internet application, such as a web browser. Some routers can perform this dynamic IP addressing, which will save a company from leasing a large number of addresses. The disadvantage is that, despite the fact that the company has 500 employees, only some smaller number of employees can perform an Internet operation at the same time.

If a company receives a large number of IP addresses, rather than have one large IP address space, the company can break their IP addresses into subnets using subnet masking. Breaking a large set of IP addresses into subnets allows for easier network management. The basic idea behind **IP subnet masking** is to take the hostID portion of an IP address and further divide it into a subnet ID and a host ID. Using this technique, an ISP and a company can take a large number of host IDs and break them into subnets (not to be confused with the physical subnets presented in Chapter Ten). Each subnet can then support a smaller number of hosts.

The Future of the Internet

The Internet is not a static entity, but continues to grow by adding new networks and new users every day. People are constantly working on updating and revising the Internet's myriad pieces. The driving force behind all changes, as well as all Internet protocols, is the self-regulating government of the Internet. Based on a committee structure, this government consists of many committees and groups:

- ▶ **The Internet Society (ISOC)**. A volunteer organization that decides the future direction of the Internet.
- ▶ **The Internet Architecture Board (IAB)**. A group of invited volunteers that approves standards.

► **The Internet Engineering Task Force (IETF).** A volunteer group that discusses operational and technical problems.

► **The Internet Research Task Force (IRTF).** The group that coordinates research activities.

► **The World Wide Web Consortium (W3C).** A web industry consortium that develops common protocols and works to ensure interoperability among those protocols.

There are many more steering groups, research groups, and working groups governing the Internet.

The best place to look for recently completed changes and changes in progress is the Internet Engineering Task Force Working Groups. Their web page is *http://www.ietf.org*, and it contains all the information about the Internet and its protocols that is currently available.

For example, the following list presents a few of the many topics recently considered or currently being considered by Working Groups:

► **Internet Printing Protocol (IPP).** This new protocol will allow users on the Internet to send a print job to virtually any printer. IPP will allow a user to learn about a printer's capabilities, submit a print job to a printer, inquire about the status of a printer or print job, and cancel a previously submitted print job.

► **Internet Fax.** A working group is considering a review and specification for enabling standardized messaging-based fax over the Internet. The working group will design informal requirements for fax-Internet gateways as a first step toward devising a set of standards for session-based fax over the Internet.

► **Extensions to FTP.** A working group is examining many changes and additions to FTP, including better security, the addition of standardized checkpoint and restart protocols, and the addition of a mechanism by which FTP clients and servers can exchange information regarding the file extensions supported and not supported.

► **Common Name Resolution Protocol.** A proposal is under consideration that will allow a user to access a web site by using a common name, rather than a possibly long and cryptic URL. For example, a fictitious URL such as http://somewhere.business.com/webs/printers/ribbons/index.html might be replaced with the common name "ribbons," making it much easier for a user to access a particular website.

► **WWW Distributed Authoring and Versioning.** The working group considering this topic is trying to define a standard that allows multiple authors at different, possibly remote, sites to have access to one set of web pages on which they are currently working. These authors can share ideas and work together to create a new product. The standard will allow multiple, simultaneous authors and will automatically track versions as they are created.

One of the biggest changes, however, to affect the Internet will be the adoption of a new version of the Internet Protocol, IPv6. Currently the Internet is using IPv4, which was the version presented earlier in the chapter. In case you are wondering, IPv5 was not a version open to the public, but was used for testing the new concepts. Let's take a closer look at the details of version 6 and see how they compare to the current version 4.

IPv6

When the Internet Protocol was created, the computing climate was not the same as it is today. There was nowhere near the number of users currently using the Internet, and the telecommunication lines used to support the high speed networks were not as fast and as error-free as they are today. Also, the applications transmitted over the Internet involved smaller data packets and there was not such a demand to transmit them in real-time. Since the demands are so much greater today, the designers of the Internet decided it was time to create a more modern Internet Protocol that took advantage of the current technology. Thus, IPv6 was created.

There are several notable differences between IPv6 and the current IPv4. IPv4 uses 32-bit addresses. With the explosive growth of the Internet over the past five years, under the current system there is a concern that we could run out of IP addresses in the near future. IPv6 calls for addresses to be 128 bits long. 128-bit long addresses allow for 7×10^{23} IP addresses per square meter of land and water! Needless to say, we will never run out of addresses using IPv6.

Significant changes were also made to the header. The IP header in version 4 contains 14 fields. In IPv6, the IP header will contain no more than 8 fields (Figure 11-11). As with the IPv4 header, the first field is the version number (6 for IPv6 and 4 for IPv4). The second field is Priority, which allows values 0 through 7 for transmissions that are capable of slowing down, and values 8 through 15 to indicate real time data. The third field is Flow Label, which is experimental and allows a source and destination to set up a pseudo-connection having particular properties and requirements. The fourth field is Payload Length, which identifies how many bytes follow the 40 byte header. The fifth field is called the Next Header and tells which, if any, of 6 extension headers follow. If there are no extension headers following Next Header, then Next Header simply tells which transport layer header (TCP, UDP, and so on) follows. The sixth field is Hop Limit, which says how long this particular datagram is allowed to live—bounce from network to network—within the system; it provides the same information as the Time to Live field in IPv4. The last two fields are the Source Address and the Destination Address.

Figure 11-11
The fields in the IPv6 header

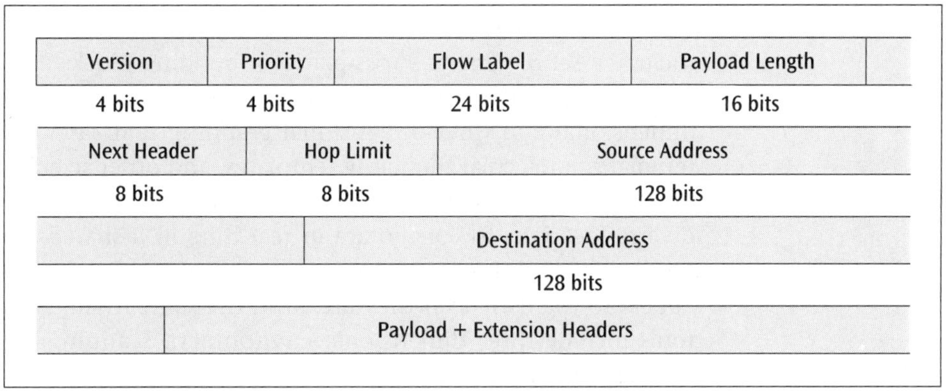

Version	Priority	Flow Label	Payload Length	
4 bits	4 bits	24 bits	16 bits	

Next Header	Hop Limit	Source Address
8 bits	8 bits	128 bits

Destination Address
128 bits

Payload + Extension Headers

What fields that were present in IPv4 are no longer present in the IPv6 header? The first field not present is the Header Length. Since the header is a fixed length in the new version, a length field is unnecessary. Another field not present is the field that allows fragmentation. The creators of IPv6 felt that most networks can handle packets of very large sizes and that it would no longer be necessary to break (or fragment) a packet into smaller pieces. Thus, version 6 does not allow a packet to be fragmented. A final field that is surprisingly missing is the checksum field. Doing away with checksum fields appears to be a trend in many modern networks, as both network quality improves and higher level applications assume the role of error detector.

Although changes to the IP header represent profound changes in the IP protocol, there are even more differences between IPv4 and IPv6. IPv6 has:

- ▶ better support for options; the new Extension Headers allow for much more elaborate support
- ▶ better security; two of the extension headers are devoted entirely to security
- ▶ more choices in type of service

This last improvement to the IP protocol relates to Quality of Service (QoS), which is an important part of modern networks. It is very important that a user can specify a particular level of service and that the network can support that level. IPv6 delivers much improved Quality of Service than IPv4.

Internet2

If changes to the IP protocol are not enough, work is progressing on a new, very high speed network that will cover the U.S., interconnecting universities and research centers at transmission rates up to a gigabit per second (1000 Mbps). The new high-speed network is called Internet2, and its links are currently being built and tested on top of existing Internet links.

Internet2's creators aim to provide high speed access to digital images, video, and music, as well as the more traditional text-based items. In particular, Internet2 has targeted four primary application areas: the digital library, LearningWare, tele-immersion, and virtual laboratories. A digital library is an electronic representation of books, periodicals, papers, art, video, and music. A patron accessing a digital library can quickly retrieve any document based on a powerful query language. Even art, video and music can be retrieved by specifying one or more keywords that describe the contents of the work.

The second application area, LearningWare, is a distributed environment for instruction and research. To provide a proper environment, LearningWare will include a set of generic, cross-platform building blocks and protocols that can be used to create effective learning tools. Some of these building blocks include two-dimensional and three-dimensional graphing templates, mathematical modeling templates, molecular modeling templates, and other science-specific operations.

Tele-immersion, the third application area, enables users at geographically diverse locations to collaborate in real time in a shared, simulated environment. Because of high-power interactive audio and video capabilities however, it will appear to the users as if they are all in the same room. Some of these high-power tools include three dimensional environment scanning, and projective, tracking and display technologies.

Virtual laboratories are the fourth application area. Internet2 will be able to create realistic lab surroundings without the expense of brick and mortar facilities. Through virtual laboratories, students and researchers will be able to conduct experiments that cannot be conducted in the real world, such as atomic explosions. And with the application area of tele-immersion, students and researchers from around the world will be able to collaborate on one or more projects.

In terms of the technology, Internet2 is not a physical network as is the current Internet and it will not replace the Internet. Instead, the Internet2 is a partnership of universities (more than 170 participating schools), industry (dozens of leading companies), and the government in an attempt to develop new technologies and capabilities that will eventually be accessed by a wide number of users. Both the business and academic communities will benefit from the creation and application of these new technologies and capabilities.

The Internet in Action: A Company Creates a VPN ▶

CompuCom, a workstation reseller, was looking for a way to save money. So they sent 3500 workers home. No, they did not fire them. CompuCom told the 3500 workers to do their computing work from home. By having them work from home, the company would not need to house the 3500 workers in corporate offices, and would save money on rent, heating, cooling, and electricity. This move, however, cost more money than CompuCom anticipated. With 3500 workers now remotely dialing in to the corporate computing center daily (see Figure 11-12), CompuCom was spending more money on modems and telephone costs. It had to increase the number of corporate-based modems from 100 to 500, and the monthly 800 number telephone bill increased by as much as 300 percent.

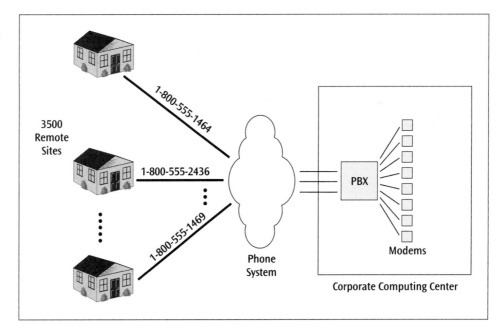

Figure 11-12
CompuCom employees dialing directly into the corporate computing center

After considering several alternatives, CompuCom decided to create a virtual private network. A virtual private network (VPN) is a data network connection that makes use of the public telecommunication infrastructure but maintains privacy through the use of a tunneling protocol and security procedures. A tunneling protocol, such as the Point-to-Point Tunneling Protocol (PPTP), is the command set that allows an organization to create secure connections using public resources such as the Internet. Note that this communication link is not a completely private connection, such as a leased telephone line would be. With a leased line the telephone company installs a special circuit between you and whoever it is that you wish to connect to. No one else can use that circuit. In contrast, a VPN over the Internet uses the public circuits that thousands of users share every second. However, since special routing information and high security measures are used, a VPN is created that only you can access and understand.

By using the Internet as the transfer medium, a VPN achieves a number of advantages. Since the Internet is practically everywhere, a VPN can be created between any two points. Because the Internet is relatively economical, the VPN is relatively economical. Once a connection to the Internet has been achieved (via an Internet service provider), data transferred over the Internet—and over the VPN—is free. The Internet—and thus the VPN—maintains a relatively stable environment with reasonable throughput.

With CompuCom's VPN, an employee at home dials into his or her local Internet service provider at the typical home-to-Internet service provider rate. Once connected to the Internet, tunneling software creates a secure connection across the Internet to the corporate computing center (Figure 11-13).

Figure 11-13
CompuCom's employees creating a tunnel across the Internet into the corporate computing center

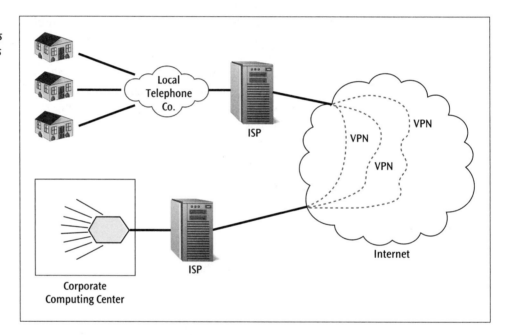

By using VPN connections, CompuCom has netted a savings of $30,000 per month in long distance telephone charges. A VPN is one example of a creative use for the extremely popular and powerful Internet. As businesses continue learning to use the Internet, they will be able to branch out into new territories and improve on existing communication strategies.

◆ ◆

SUMMARY

- The Internet is one of the most important technological advances, if not the most important technological advance, of the late 20th century.
- File transfer protocol (FTP) is useful for uploading or downloading files across the Internet.
- Remote login using Telnet allows an individual to login to a remote computer site and perform operations as if the user were physically located at the remote site.
- Internet telephony (VoIP) offers an inexpensive alternative to long distance calling, but its quality often suffers, which renders the service virtually unusable. The Internet was not designed to transfer real-time data, which is necessary to support interactive voice. Other technologies such as voice over frame relay appear more stable.
- Electronic mail, or e-mail, is a standard requirement for most business operations and can transfer standard text messages and include MIME-encoded attachments.
- Protocols such as Simple Mail Transfer Protocol (SMTP), Post Office Protocol version 3 (POP3), and Internet Message Access Protocol (IMAP) support the operations of e-mail.
- A listserv is a popular software program used to create and manage Internet mailing lists.
- Usenet newsgroups are the equivalent of electronic bulletin boards. Both listservs and newsgroups can be used by the business person to keep abreast of one or more areas of interest.
- A cookie is a data file created by a web server that is stored on the hard disk drive of a user's workstation. Information in the cookie can be used by a commercial company for marketing, observing customer profiles, and tailoring advertisements.
- Streaming audio and video is the continuous download of a compressed audio or video file, which is then heard or displayed on the user's workstation. Streaming audio and video require support protocols such as Real-Time Protocol (RTP) and Real Time Streaming Protocol (RTSP).
- The World Wide Web is a vast collection of electronic documents containing text and images that can be accessed by simply clicking a link within a browser's web page. The browser uses HTTP to transmit and receive web pages, and HTML to display those web pages.
- HTML and Dynamic HTML are markup languages used to define how the contents of a web page will be displayed by a web browser.
- XML is a set of rules for defining your own markup language. By creating your own markup language, you can create a document that contains both the data and a definition of that data.
- E-commerce, a rapidly growing area of the Internet, is the buying and selling of goods and services electronically. Many companies are investing heavily in e-commerce, hoping that it will increase their market share and decrease their costs.
- Cookies store state information on a user's hard drive and provide a way for web sites to track a user's web browsing patterns and preferences.
- An intranet is an in-house Internet with web-like services available only to a business' employees or to customers and suppliers through an extranet.

- ► To support the Internet, several protocols are necessary: IP, TCP, ICMP, UDP, and ARP.
- ► The Internet Protocol (IP) provides a connectionless transfer of data over a wide variety of network types.
- ► Transmission Control Protocol (TCP) resides at the transport layer of a communications model and provides an error-free, end-to-end connection.
- ► Internet Control Message Protocol (ICMP) performs error reporting for the Internet Protocol.
- ► User Datagram Protocol (UDP) provides a connectionless transport layer protocol, in place of TCP.
- ► Address Resolution Protocol (ARP) translates an IP address into a CSMA/CD MAC address for data delivery on a local area network.
- ► Tunneling protocols allow a company to create virtual private network connections into a corporate computing system.
- ► To locate a document on the Internet, you usually refer to its Uniform Resource Locator (URL).
- ► The URL uniquely identifies each document on every server on the Internet. Part of the URL is the 32-bit IP address.
- ► Each 32-bit IP address consists of a network ID and a host or device ID.
- ► The URL is converted to a 32-bit IP address via the Domain Name Server (DNS).
- ► The Internet continues to evolve with a new Internet Protocol version 6 on the horizon as well as a completely new, higher speed Internet2.

KEY TERMS

Address Resolution Protocol (ARP)
ARPANET
cascading style sheets (CSS)
cookies
datagrams
domain name
domain name system (DNS)
Dynamic HTML (DHTML)
dynamic IP addressing
e-commerce
electronic mail
extensible markup language (XML)
extranet
file transfer protocol (FTP)
hypertext markup language (HTML)
hypertext transport protocol (HTTP)
Internet

Internet Control Message Protocol (ICMP)
Internet2
Internet Protocol (IP)
intranet
IP multicasting
IP subnet masking
IPv6
layers
listserv
MIME
MP3
packet voice
Real-Time Protocol (RTP)
Real Time Streaming Protocol (RTSP)
remote login (Telnet)
Simple Mail Transfer Protocol (SMTP)

streaming audio and video
Telnet
Transmission Control Protocol (TCP)
tunneling protocol
uniform resource locator (URL)
Usenet
User Datagram Protocol (UDP)
virtual private network
voice over Asynchronous Transfer Mode (VToA)
voice over frame relay (VoFR)
voice over Internet Protocol (VoIP)
voice over packet
voice over the Internet
World Wide Web (WWW)

REVIEW QUESTIONS

1. What was the precursor to the present day Internet?
2. What is the file transfer protocol (FTP) used for?
3. What is the main function of Telnet?
4. Is voice over the Internet as reliable as either FTP or Telnet?
5. What other options are available for transmitting voice via Internet technology besides voice over Internet?
6. List the basic features of a common electronic mail system.
7. What are the duties of SMTP, POP3, and IMAP when referring to e-mail systems?
8. What is a listserv used for?
9. What tools are necessary to support streaming audio or video?
10. What is the relationship between the Internet and the World Wide Web?
11. What is the HTTP protocol used for?
12. How is a Web markup language different from a programming language?
13. What features have been added to HTML to produce dynamic HTML?
14. What can a cascading style sheet add to a web page?
15. How does XML differ from HTML and dynamic HTML?
16. What is the relationship between Electronic Data Interchange and e-commerce?
17. What is a cookie? Who makes cookies and where are they stored?
18. How do intranets and extranets compare with and differ from the Internet?
19. List the main responsibilities of the Internet Protocol.
20. List the main responsibilities of the Transmission Control Protocol.
21. Explain the relationship of the port to an IP address.
22. Is UDP the same thing as TCP?
23. What is the relationship between IP and ICMP?
24. How can a business use a private virtual network and tunneling to support an off-site connection?
25. What is the purpose of the uniform resource locator (URL)?
26. List the four basic parts of the URL.
27. How does the domain name server translate a URL into a 32-bit binary address?
28. What are the different classes of IP addresses?
29. How will IPv6 differ from the current version (4) of IP?
30. What are the main features of Internet2?

EXERCISES

1. Locate and label the service, host name, directory, and filename in the following URL: *http://www.reliable.com/listings/pages/web.htm*
2. Using a web browser, go to site gatekeeper.dec.com and download the file *brownies-1* from subdirectory *pub/recipes/*.
3. If you use your computer primarily for Telnet into a remote computer, will you have a large long distance telephone bill?
4. What is the difference between voice over the Internet and voice over IP?
5. With respect to e-mail, what is the relationship between SMTP and POP3? Can one operate without the other?

6. Your company has asked you to keep abreast of work place ethics. Which Internet service(s) introduced in this chapter would best help you accomplish this task?

7. Are HTTP and HTML two protocols with the same function? Explain.

8. Given the following section of HTML code, what will be displayed on the screen when the HTML code is read by a browser:

   ```
   <HTML>
   <BODY>
   <H1> Chapter Eleven </H1>
   <HR>
   <IMG SRC="amerflag.gif" ALIGN=RIGHT>
   <A HREF="http://www.everyone.edu"> Everyone College Home Page </A>
   </BODY>
   </HTML>
   ```

9. Which dynamic HTML concept will print the phrase *Computer Network* multiple times on the screen, each one larger than the previous one and each one slightly overlapping the previous one.

10. Does the U.S. government support the Internet? Explain.

11. Is HTML a subset of XML, or is it the other way around? Explain.

12. Given an IP packet of size 540 bytes, and a maximum packet size of 200 bytes, what are the IP Fragment Offsets and More flags for the appropriate packet fragments?

13. The *Time To Live* field in IP version 6 is smaller than in IP version 4. Since this count is decremented each time an IP datagram enters a router, what are the implications of a smaller field size?

14. In your town there is a small commercial retail building with essentially one room. On one side of the room is a real estate agency and on the other side of the room is a guy who sells tea. How does this situation serve as an analogy for the concept of IP addresses with TCP port numbers?

15. Somewhere in the middle of the United States are two Internet routers with routing tables that are all messed up. The two routers keep sending their packets back and forth to each other, nonstop. Will an error message ever be generated from this action? If so, who will generate the error message and what might it look like?

16. Why is ARP necessary if every workstation connected to the Internet has a unique IP address?

17. If your computer workstation has an IP address of 167.54.200.32, is it a Class A, B, C, or D IP address?

18. What are the advantages and disadvantages of using dynamic IP address assignments?

19. Some of the new protocols such as Internet Protocol version 6 are not including any kind of error detection scheme on the data portion of the packet. What significance does this trend indicate?

THINKING OUTSIDE THE BOX

1 You are working for a company that wants to begin electronic data interchange (EDI) with two other companies that supply parts. Your company produces mobile telephones, and the two other companies produce batteries and telephone keypads. When your company places an order with the battery company, it specifies battery size, battery type, battery power, and quantity. When your company places an order with the keypad company, it specifies keypad configuration, power consumption, keypad dimensions, and keypad color. How might XML be used to support EDI between the companies?

2 Two banks want to establish an electronic link between themselves over which they can transmit money transfers. Can they use a virtual private network and a tunneling protocol, or is a better technique available?

3 You are working for a company that sells high-end bicycles, and it has decided to create a series of web pages to promote sales. What kind of services might you offer in your web pages? Will your web pages use HTML, dynamic HTML, or some other language? How might your company use cookies to its advantage?

PROJECTS

1. Using either your computer at home or at work, locate all the cookies stored on your hard disk drive. If you examine the cookies, can you tell who put them there?

2. Locate on the Web a free copy of a streaming audio or video player and download it to your machine. Then locate a web site that delivers a streaming audio or video program. How difficult was the download and installation of the player? How good is the quality of the audio or video stream?

3. Create a web page, using either a web page editor or by writing HTML code manually, which has an H1 heading, some text, one or more links, and one or more images.

4. Find an example of a company that performs Electronic Data Interchange. What companies are involved with the interchange? What data is transferred during the interchange? What other details can you describe?

5. If you have access to the Internet at work, school or home, does your workstation have a Class B IP address or a Class C IP address? Is it statically assigned or dynamically assigned? Did your IP address come from an Internet service provider or from somewhere else?

6. Who are the "players" involved with the new Internet2? What does it take to become one of these players?

12

Telecommunication Systems

◆◆

UNIFIED MESSAGING, many professionals agree, is the management of voice mail, fax, and e-mail through a single user interface. In other words, a system that supports unified messaging is capable of sending, receiving, storing, retrieving, and manipulating voice mail messages, faxes, and e-mail messages from a user's workstation and using a single system. To achieve this combination of services, the single system has to look and act like a local area network (for sending and receiving e-mails) but be able to support voice transactions (the voice mail and faxes). This integration of voice and data into one system is one of the hottest trends in communication systems today and is called computer telephony integration.

Computer Telephony magazine wanted to conduct a test of the various unified messaging systems available, but found there were an "immense" number of systems from which to choose. After examining this quickly growing market, the magazine managed to whittle the list of systems down to a representative four. All four systems adequately provided the features of unified messaging to the user desktop. *Computer Telephony*'s conclusion was: "The days of dealing with three different interfaces to manage voice, fax, and e-mail are dead—or should be. Unified Messaging is a mature technology, ready to implement in even the most technologically conservative infrastructure."

Computer Telephony, September 1999 and October 1999

Have you heard of unified messaging?

Why is it important for someone to know both data communications and voice communications?

Objectives ▶

After reading this chapter, you should be able to:

▶ Identify the basic elements of a telephone system.

▶ Differentiate between the composition of the telephone companies before the 1984 Modified Final Judgment and after the 1984 judgment.

▶ Describe the difference between a local exchange carrier and an interexchange carrier.

▶ Differentiate between the role of the local telephone company before the Telecommunications Act of 1996 and after the 1996 act.

▶ Outline the features of ISDN and distinguish a basic rate interface from a primary rate interface.

▶ List the advantages of using frame relay for transmitting data.

▶ Describe a frame relay permanent virtual circuit and how it is created.

▶ Outline the basic features of frame relay's committed information rate and how it is used to establish a level of service.

▶ List the differences between data over frame relay and data over the Internet.

▶ Compare a frame relay switched virtual circuit to a permanent virtual circuit.

▶ Identify the main characteristics of asynchronous transfer mode including the roles of the virtual path connection and the virtual channel connection.

▶ Discuss the importance of the classes of service available to asynchronous transfer mode.

▶ Identify the main characteristics of digital subscriber line and know the difference between a symmetric system and an asymmetric system.

▶ Recognize the characteristics and possibilities formed by the merger of the data communications field with the voice communications field in the area of computer telephony integration.

Not so long ago, a person preparing to pursue an education in communication systems had to make a choice—was he or she going to study voice communications or data communications? Until recently, the two fields were almost completely separate. Today, computers are starting to take over the voice portion of communications, and the line between data networks and telecommunication systems is blurring at an alarming rate. Just knowing data networks is no longer sufficient. For example, many businesses are installing single systems that can support both voice and data over the same circuits. One such system is computer telephony integration (CTI), a combination of local area networks and the telephone system. CTI is capable of providing telephone services, such as voice mail, and typical computer applications, such as database systems and electronic mail over a single set of wires. In addition, systems that were designed to transfer data, such as frame relay and the Internet, are now transmitting voice with various degrees of success. Asynchronous transfer mode, which was originally designed for both voice and data, is another example of a system that is adept at transferring both forms of signals. The well-informed business network professional must now learn some basic telecommunication systems along with computer networks.

In this chapter we will examine the most common telecommunication systems available today. All of these systems are capable of supporting both voice and data traffic, and almost all are exclusively wide area networks. Although mobile telephone systems (see Chapter Three) are capable of supporting voice and data traffic, they are more regional or metropolitan area networks and do not fit into this chapter's discussion of wide area telecommunication systems. As the quest for higher and higher transmission speeds intensifies, the telephone companies continue to introduce new technologies and services to support these speeds. Before examining these new services, however, let's examine the basic telephone system, which is called **plain old telephone system**, or **POTS**, and learn a number of essential terms for understanding telecommunication systems. The chapter will conclude with a summary of the numerous technologies and a comparison of the advantages and disadvantages of each.

Basic Telephone Systems

The basic telephone system, or POTS, has been in existence since the early 1900s. During most of those years, POTS was an analog system capable of supporting a voice conversation. It wasn't until the 1980s that POTS began carrying computer data signals as well as voice signals. The growth of data became so large that near the end of the 20th century, POTS carried more data than voice. Even though POTS has seen a number of technological changes, such as the increasing use of digital signals in place of analog signals, POTS is still basically a voice carrying medium. Because of the voice requirements, POTS has limitations. Let's examine these limitations before we introduce the main characteristics of basic telephone systems.

Limitations of telephone signals

The average human can hear sounds in the range of 20 to 20,000 Hz. As a person ages, however, the higher frequencies become increasingly difficult to hear. The average human voice is no where near as flexible as the ears. It can generate sounds only from approximately 200 Hz to 3500 Hz (a range of 3300 Hz). Since telephone systems were originally designed to transmit the human voice, the telephone network was engineered to transmit signals of approximately 3000 Hz. Actually, the telephone system allocates 4000 Hz to a channel and uses filters to remove frequencies that fall above and below each 4000 Hz channel. The resulting extra frequencies are used to give a little space between channels when multiple channels are transmitted over the same wire. The channels and their frequencies are demonstrated in Figure 12-1.

Figure 12-1
The various telephone channels and their assignment of frequencies

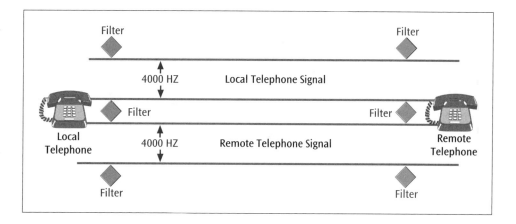

If the telephone system were capable of transmitting CD-quality music, each channel would need to support 22,000 Hz signals. A 4000 Hz channel for voice is a long way from a 22,000 Hz channel for CD-quality music. All this talk about frequencies and voice transmission leads to two important points:

- ▶ The more information you wish to send over a medium, the higher the frequency of the signal you need to represent that information. (See Chapter Two for a detailed discussion of this point.)
- ▶ If you want to send data over a telephone line, it has to travel in a 4000 Hz channel.

When you put these two statements together, a painful fact emerges—regardless of the desired speed of transmission, we have to perform any data transmission over a standard telephone line in a fairly narrow band of 4000 Hz. With various analog modulation techniques, it has been possible to push data transmission speeds to 33,600 bps. But until the telephone system moves away from 4000 Hz channels (which is not likely for quite a while), analog transmission speeds will not exceed 33,600 bps. (Recall from Chapter Two and Shannon's equation—the frequency, noise level, and power level of an analog signal determine the maximum rate of data transmission.) Newer digital transmission techniques have pushed data transmission speeds up to 56,000 bps, but this speed works only under ideal conditions.

Telephone lines, trunks, and numbers

The telephone line that leaves your house or business is called the **local loop** and consists of either four or eight wires. Although it takes only two wires to complete a telephone circuit, more and more customers are demanding that multiple lines and services be connected to their premises. These extra lines and services require additional pairs of wires. So that the telephone company does not have to send out a technician to install additional sets of lines, it often anticipates future user demand and installs extra wires during the initial installation (Figure 12-2). On the other end of the local loop is the local telephone company's central office. The **central office (CO)** contains the equipment that generates a dial tone, interprets the telephone number dialed, checks for special services, and connects the incoming call to the next point.

Figure 12-2
The local loop as it connects your house to the telephone company's central office

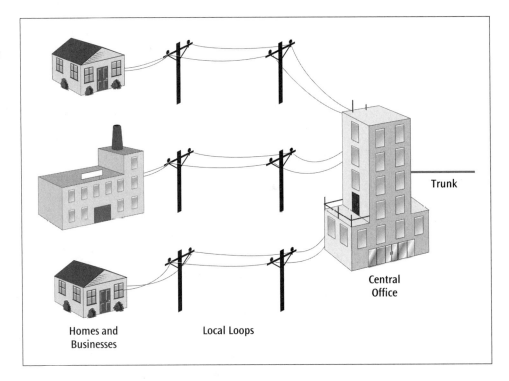

Trunk

Central
Office

Homes and
Businesses

Local Loops

From the central office, the signal can go to one of three places:

▶ over a local loop to another house in your neighborhood (if the number dialed does not start with a 1 and has the same area code and first three digits as your telephone number);

▶ to another central office in a neighboring community (if the number dialed does not start with a 1 and the first three digits are not the same as your telephone number); or

▶ to a long distance provider if the telephone call is a long distance call.

How does the telephone network know when a telephone call is long distance? The United States is divided into local access transport areas. A **local access transport area (LATA)** is a geographic area, such as a large metropolitan area or part of a large state. As long as a telephone call remains within a LATA (intraLATA), the telephone call is local and is handled by a local telephone company. If the telephone call passes out of a LATA and into another (interLATA), the telephone call is long distance and must be handled by a long distance telephone company.

The telephone network consists of two basic types of telephone lines: a standard telephone line, or subscriber loop, and a trunk. A standard telephone line, such as the wire that runs between a house and the central office, has a unique telephone number associated with it. A trunk, unlike a standard telephone line, does not have a telephone number associated with it. It does not have a telephone number associated with it because the trunk can carry hundreds of voice and data channels. These hundreds of voice and data channels are typically carried between switching centers and central offices. Finally, most trunks transmit digital signals, which require repeaters every few miles. In contrast, telephone lines usually transmit analog signals, which require amplifiers every few miles.

There are several different types of telephone trunks. Digital trunks carry digital signals, use time division multiplexing techniques, and require periodic regeneration of the signals. Analog trunks carry analog signals, use frequency division multiplexing, and require amplification at regular intervals. Inter-office trunks, which run from central office to central office, can carry either analog or digital signals. Trunks that run from a central office to a business are used when the business must support a large number of telephone lines, and are usually digital trunks.

A telephone number consists of three parts: the area code, the exchange, and the subscriber extension (Figure 12-3). The area code, which identifies a portion of a state or a portion of a very large city, begins with the digits 2 through 9 (0 and 1 are used for the operator and long distance, respectively). The second and third digits can be any digit between 0 and 9. The exchange has the same composition as the area code: the first digit can be 2 through 9, and the second and third digits can be 0 though 9. Each digit in the subscriber extension can be 0 through 9.

Figure 12-3
Composition of a telephone number

708	555	1212
Area code	Exchange	Subscriber extension

The telephone network before and after 1984

Prior to 1984, AT&T owned all the long distance telephone lines in the United States, a majority of local telephone systems, and Bell Laboratories. In the 1970's, the federal government took AT&T to court citing anti-trust violations. AT&T lost the case, and in 1984 the **Modified Final Judgment**, which was the court's ruling on the breakup of AT&T, required the divestiture of AT&T into separate companies. This breakup of AT&T allowed AT&T to keep the long distance lines and Bell Labs, but it had to divest itself of all local telephone companies.

At the time, AT&T consisted of 23 Bell Operating Companies (BOCs), which provided local telephone service across the country. There were also as many as 14,000 other local telephone companies not owned by AT&T, which typically served very limited areas. With the divestiture, the 23 BOCs were separated from AT&T and, to survive, were reorganized into seven Regional Bell Operating Companies (RBOCs) (Figure 12-4). These seven RBOCs then competed with the remaining 14,000-plus local telephone companies to provide local service to each home and business in the country. Because of the power and size of the seven RBOCs and through competition, the 14,000-plus local telephone companies were reduced to fewer than 1400. Through mergers, by 2000 there were only 4 RBOCs left: Southwestern Bell Company (SBC), US West, BellSouth, and Bell Atlantic.

Figure 12-4
The original 23 Bell companies and the newer 7 regional telephone companies

A number of other results of the 1984 divestiture completely changed the landscape of the U.S. telephone system. The United States was divided into local access transport areas (LATAs), which determined when a telephone call was local distance or long distance. The breakup also allowed long distance telephone companies other than Bell Telephone, such as MCI and Sprint, to offer competing long distance services. To support this competition in long distance providers, the local phone companies had to give all long distance telephone companies equal access to the telephone lines running into each home and business. Finally, the local telephone companies became known as **local exchange carriers** (LEC) and the long distance telephone companies became known as **interexchange carriers** (IEC or IXC).

A local exchange carrier can offer a number of services, including Centrex, private and TIE lines, and many other telecommunications services such as call waiting and conference calls. A **Centrex (central office exchange service)** is a service from local telephone companies in which up-to-date telephone facilities at the telephone company's central (local) office are offered to business users so that they don't need to purchase their own facilities. Businesses are spared the expense of having to keep up with fast-moving technology changes since the telephone company is providing the hardware and the services; the business simply pays a monthly fee. If a business does not want to lease a Centrex service from a local telephone company, it can instead purchase its own private branch exchange, the main competition to Centrex. A **private branch exchange (PBX)** is a large computerized telephone switch that sits in a telephone room on the company property. A PBX handles all in-house calls and places calls onto outside telephone lines. A PBX can also offer many telephone services such as voice mail, call forwarding, and dialing plans that use the least expensive local and long distance telephone circuits.

Private lines and TIE lines are leased telephone lines that require no dialing. They are permanent direct connections between two specified points. Consider a company that has two offices in the same city. It is always transferring data between the two offices. The company could use a dial-up telephone line with two modems, but a dial-up line would be very expensive. A leased line might offer a less expensive alternative and will always be connected—you never have to dial a telephone number.

Finally, an interexchange carrier or long distance telephone company can also offer a large number of services, including credit card and calling card services; 700, 800, 888, and 900 services; international access; and operator and directory assistance.

Telephone networks after 1996

A second major ruling in the telecommunications industry was made in 1996—the Telecommunications Act of 1996. The Telecommunications Act of 1996 paved the way for *anybody* to offer a local telephone service to homes and businesses. New providers of local telephone services are called competitive local exchange carriers (CLECs), and they may include interexchange carriers (long distance telephone companies), cable television operators, small companies with virtually no equipment (called non-facilities based CLECs), and even the power company.

For example, in some areas of the country, local cable television companies are offering local telephone service. Since a majority of homes and businesses in the U.S. are already wired for cable television, it seemed like a reasonable idea to also offer local telephone service over those cable TV lines. Termed cable-telephone or cable telephony, most of the major cable television companies, including Cox Communications, Comcast, and AT&T, as well as many smaller cable companies, offer telephone services over cable television lines in most major cities across the country. Large cable companies in the U.S. are achieving penetration rates as high as 10 to 20 percent in their particular markets. Of those new subscribers, 95 percent are dropping their old local telephone service in favor of the new cable service. In fact, some reports indicate that cable telephony subscribers will grow from 50,000 today to almost 11 million by the year 2005. The network that provides cable telephony is a hybrid of fiber and coaxial cable (Hybrid Fiber Coax, or HFC). This technology is the same system that can deliver cable modem service into homes and businesses.

There is a problem with allowing all these new local telephone companies into a market—the telephone lines. It is extremely expensive for each new telephone company to install new telephone lines into each home and business. Although cable television companies already have the luxury of cable lines running into each house, non-cable companies trying to break into the local telephone market do not have this advantage. To compensate for this high cost of installation, the currently existing local telephone companies, now called Incumbent Local Exchange Carriers (ILECs), by law must give CLECs access to their telephone lines. Furthermore, by law, ILECs must give competitors all services at wholesale prices, access to telephone numbers, operator services, and directory listings, access to poles, ducts, and right-of-ways, and physical co-location of equipment within ILEC buildings.

A very interesting provision of the 1996 telecommunications act is that an ILEC must sell access to the local dial tone at 17 to 28 percent less than the standard price. Thus, a CLEC buys the dial tone for a discount and can offer services to users for less than the ILEC but still more than the CLEC paid the ILEC for access.

The ILEC still has to provide the wire and the local loop, but the CLEC can make a profit off it. Needless to say the ILECs are not too crazy about this arrangement and are resisting. Time will tell if the Telecommunications Act of 1996 will actually work or will collapse into a series of legal battles.

Leased Line Services

Most home computer users use plain old telephone service (POTS) lines to connect their computers to services such as the Internet or online services such as America Online. They have the local telephone number of an Internet service provider or the local America Online port, software instructs the modem to dial the number, and a low-speed connection (less than or equal to 53,000 bits per second) is established. This set up, however, is not good enough for some home users. Businesses also require permanent high-speed connections to connect themselves to the Internet and to access other businesses, remote databases, and high-speed data transfer services such as public data networks. These users desire a faster circuit, perhaps one that does not need to be dialed at all but is constantly connected. One possible solution is to lease a dedicated line from the telephone company.

One leased line service is the **56K leased line**. This service is available for both business and homes, but due to the cost is found almost only in businesses. The telephone company will install a private line between you and whoever you wish. This connection will always be active and will transfer data at speeds up to 56,000 bits per second. The monthly cost of such a service typically depends on the distance between end points in the connection. For example, an intraLATA 56 Kbps leased line averages around $150 per month. An interLATA (long distance) 56 Kbps leased line can cost as much as $500 plus $1.00 per mile.

A second option for companies requiring a data rate faster than 56,000 bits per second is the T1 (or T-1) service. A **T1 service** is an all digital connection that can transfer either voice or data at speeds up to 1.544 Mbps (1,544,000 bits per second). If the user wishes, the T1 line can support up to 24 individual telephone circuits, 24 individual data lines at 56,000 bits per second each, or a variety of combinations of these options. If you don't need 1.544 Mbps to transmit your computer data, you can request a quarter-T1 or a half-T1, which, as the names imply, give 1/4 and 1/2 of the 1.544 Mbps data rate, respectively. These 1/4 and 1/2 T1 lines are called fractional T1 services. Customers who require transmission speeds higher than 1.544 Mbps can consider the T3 service, which transmits data at 45 Mbps. Unfortunately, the speed and cost differences between T1 and T3 are dramatic.

Similar to a 56K leased line, a T1 connection is a point to point service and is always active. The user pays a monthly charge, which is based on the transfer speed requested and the distance between end points. IntraLATA T1 lines typically cost approximately $350 to $400 per month. InterLATA T1 lines can cost as much as $1200 per month plus $2.50 per mile for the connection.

Interestingly, the T1 service, which was created in 1963, was designed to support only the telephone circuits that the telephone companies used to transmit data between switching centers. When businesses heard about the T1 line, they began to ask for it to connect their businesses to a telephone company switching center (central office). Demand for the T1 line has grown steadily ever since. Recently however, the T1 line has started to receive stiff competition from new services that offer equal or faster transmission speeds with smaller price tags. Next, we will examine some of these new services and see how they compare to the 56K leased line and the T1.

Integrated Services Digital Network (ISDN)

Integrated Services Digital Network (ISDN) was designed in the mid-1980s to provide a worldwide public telecommunication network that would support telephone signals, data, and a multitude of other services for both residential and business users. ISDN provides services such as facsimile, teletex, and alarm systems, and modern telephone services including call transfer, caller identification, caller restriction, call forwarding, call waiting, hold, conference call, and credit-card calling. Users connect to ISDN via a "digital pipe." This digital pipe, which consists of only one or two pairs of twisted pair wire, provides users access to circuit switched networks, such as telephone networks, packet switched networks, such as the Internet or public data networks, and a host of local area network services. A public data network is a long distance service designed to transfer data between two geographic locations. (A description of the first public data network, X.25, is provided in the Appendix.) All of these services are provided simultaneously, at varying data transfer rates, and users are charged only according to the capacity used, not the connect time.

The digital pipe that connects users to the ISDN central office carries a combination of various types of channels. Three kinds of channels have been defined: B channel, D channel, and H channel. The B channel, which stands for basic user channel, is a 64 Kbps full-duplex channel capable of carrying digital data, pulse code modulated voice, or a mixture of lower-rate traffic such as home burglar alarm systems and emergency services. Three types of connections can be established using a B channel: circuit switched connections (dial-up telephone), packet switched connections into the Internet or a public data network, and semi-permanent connections, or leased lines.

The second type of channel supported by ISDN is the D channel, which stands for data traffic channel. It is a 16 or 64 Kbps full duplex channel that can carry the signaling information for the B channel. The D channel may also be capable of carrying packet switched data or low speed data such as burglar alarms and emergency services information.

The third channel is the H channel, which stands for high speed data channel. The H channel is a full duplex 384 Kbps channel (H0), 1.536 Mbps channel (H11), or 1.92 Mbps channel (H12) that is capable of carrying high speed data for applications such as facsimile, video, high quality audio, or multiple lower speed data streams.

A user, wishing to subscribe to ISDN, receives a "package," which consists of combinations of B, D, and H channels. Different combinations can be created by the ISDN provider for varying user requirements. For example, a **basic rate interface (BRI)** package can be created, which consists of two full-duplex 64 Kbps B channels plus one full-duplex 16 Kbps D channel. A BRI package is standard for most residential services and many small businesses. It allows the simultaneous transfer of voice as well as a number of data applications. Since most ISDN subscribers already have a standard voice telephone line, it is common to see these users tie together the two B channels for a combined data transfer rate of 128 Kbps. This transfer rate is substantially faster than a standard telephone line using a modem, which might achieve 53 Kbps over a noise-free connection. The pricing for a BRI package is typically $40 per month flat rate plus $0.10 per minute for usage.

For users requiring a larger data capacity, a primary rate interface package is available. The **primary rate interface (PRI)** package consists of 23 64 Kbps B channels plus one 64 Kbps D channel, for a combined data transfer rate of 1.544 Mbps (T1 speed). Additional services have also been defined, which include the H channels in various configurations with the B and D channels.

Even though ISDN has been in existence for many years, it didn't become popular until the mid-1990's. In fact, before the mid-1990s, it was fairly common to call your local telephone company, inquire about getting an ISDN line installed, and be placed on hold while the phone company's customer service agent called around asking who knew about ISDN. Since that time the telephone companies have spent many hours on training personnel and promoting ISDN. Now, a call placed to your local telephone company will result in an immediate response.

Unfortunately, due in large part to earlier lack of preparation by the telephone companies, ISDN's acceptance has been slow, and other technologies are quickly leaving it behind. A 128 Kbps digital telecommunications line into the home is looking fairly old fashioned when compared with newer technologies such as Digital Subscriber Line and cable modems.

Frame Relay

Consider the following scenario. A business wants a fast, permanent connection to another site. It wants to transmit data, so it wants a system that was designed for data and not voice. What should they use? A dial-up telephone line? Definitely not, since a dial-up telephone line is not a permanent connection, was not designed for data, is not fast, and would be very expensive. What about leased or private lines? Leased or private lines, too, would not be acceptable solutions, since they are not fast enough and can be very expensive.

The answer to the question, "what should they use," is frame relay. **Frame relay** is a packet switched network that was designed for transmitting data over fixed lines (not dial-up lines). If someone is interested in leasing a frame relay circuit, he or she contacts his or her local telephone company or frame relay service provider. Most long distance telephone companies, such as AT&T, Sprint, and MCI offer frame relay service over most of the country. Once the service is established, the customer only needs to transmit his or her data over a local link to a nearby frame relay station (Figure 12-5). The frame relay network is responsible for transmitting the user's data across the network and delivering it to the intended destination site.

Figure 12-5
A business connecting to the frame relay network via a local connection

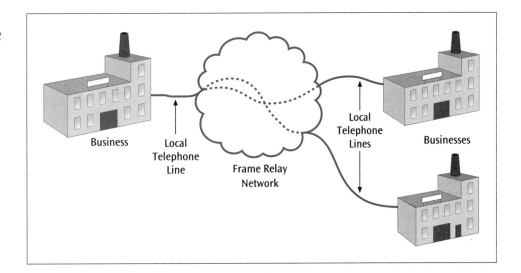

A frame relay service provides many attractive alternatives to leased lines. One of the first noticeable characteristic of a frame relay network is its very high transfer speeds. The data transfer speeds can be very fast, ranging from 56 Kbps up to 45 Mbps. Along with the high data transfer rates is high throughput—the network as a whole is very fast. Frame relay switches quickly transfer data as it travels from one end of the network to the other in a relatively short period of time using fiber optic cables. Frame relay networks also provide very good security. Because of the encryption techniques used to transmit data between frame relay switches, it is extremely difficult to intercept and decode the transmitted data. Frame relay connections are also permanent connections. Since the connection is fixed, it is always available.

Another advantage that is associated with frame relay and other modern high speed wide area networks is low error rates during transmission. In fact, the error rate is so low that the frame relay network does not have any form of error control. If an error occurs, the frame relay network simply discards the frame. It is the responsibility of the application, and not the frame relay network, to perform error control.

Finally, frame relay networks have reasonable pricing. Fixed monthly pricing is based on three charges: a port charge for each access line that connects a business into the frame relay network; a charge for each permanent connection that runs through the frame relay network between two endpoints (companies using the service); and the charge for the access line, which is the high speed telephone line that physically connects the business to the frame relay port. Since a port and access line are capable of supporting multiple permanent connections, a user may pay one port charge, one access line charge, and several permanent connection charges.

The permanent connection that is necessary to transfer data between two endpoints is called a **permanent virtual circuit (PVC)**. When a company wants to establish a frame relay connection between two offices, the company will ask the frame relay provider to establish a PVC between the two offices (Figure 12-6). For example, if a company has offices in Chicago and Orlando, the company can ask a frame relay provider to create a PVC between the Chicago and Orlando offices. Each office is connected to the frame relay network by a high-speed telephone access line through a port. This port and high-speed telephone line combine to create the physical connection. The PVC—the logical connection—runs from the Chicago office, through the frame relay network, to the Orlando office.

Figure 12-6
A frame relay connection between Chicago and Orlando showing access lines, ports and PVC

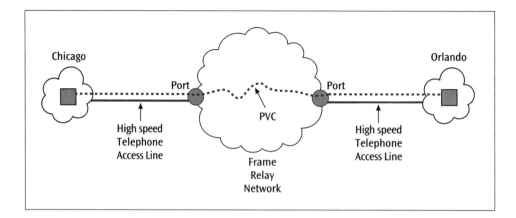

If the two offices in Chicago and Orlando have one application running between them, then only one PVC is necessary. However, if two different kinds of applications are running between the two offices, it will probably be necessary to create two PVCs. Even though there are two PVCs, it may be possible to support both PVCs over the same access line and port. Supporting both PVCs over the same access line and port would be possible if the sum of the two transmission speeds of the PVCs is not greater than the capacity of the port and access line.

To illustrate this concept, let's assume that the Chicago – Orlando application required a 256 Kbps connection. The company asks the frame relay provider for a 256 Kbps PVC. If the company also requests a 512 Kbps port, and orders a telephone line capable of supporting at least 512 Kbps (such as a T1), then the company can add an additional 256 Kbps PVC in the future. The 512 Kbps port and telephone line can support two-256 Kbps PVCs.

Frame relay is called a layer 2 protocol. Recalling the OSI model, **layer 2 protocol** means that frame relay technology is only one part of a network application and resides at the data link layer. When creating a network solution for a particular application, the choice of frame relay is only one part of the solution. You still have to choose network and transport layer software, such as software supporting the TCP and IP protocols, to run on top of the frame relay network, as well as the type of high speed telephone line at the physical layer. The intended application, for example a database retrieval system, then runs on top of TCP/IP.

Frame relay setup

Typically frame relay is used to replace leased lines connecting a business to other businesses or when a business is scattered over multiple locations. For example, consider a company that has four locations and wishes to make a number of connections between those locations. Using leased telephone lines, if one location wants to have a connection to three other locations, there needs to be three leased lines coming from that first location. Figure 12-7, an interconnection diagram, shows that a combination of six separate leased lines are required to make the necessary connections.

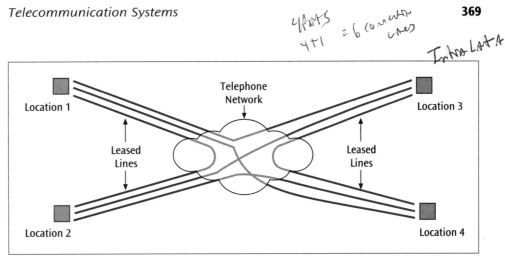

Figure 12-7
Interconnection diagram showing six leased lines used to interconnect a business' four locations

If the business decides to replace the leased lines with frame relay connections, the interconnection diagram now looks like that shown in Figure 12-8. Only four connections—one for each location—into the frame relay network are necessary. The frame relay network has its own interconnections for connecting all the frame relay switches. These internal connections are not the concern of the business. All the business needs to do is get each of its offices connected to the frame relay network, or cloud. The optimal scenario would involve selecting a frame relay provider that has a network cloud connection in each city where the business has an office. If this local network access is feasible, which it should be due to the wide spread availability of frame relay, then the telephone connections between the business' offices and the frame relay network should be local telephone connections, not long distance telephone connections. Local high speed telephone connections are almost always less expensive than long distance high speed telephone connections.

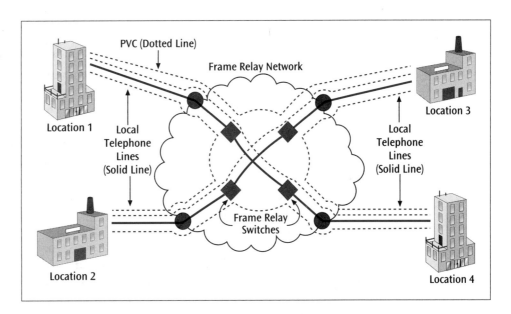

Figure 12-8
Interconnection diagram showing a frame relay network solution to the business interconnection problem

Since each line between a business location and the frame relay network is capable of supporting multiple PVCs (the dotted lines in Figure 12-8), only one physical line and port may be needed to support three virtual connections leaving a business location. Money is saved since frame relay PVCs are reasonably priced and only four telephone lines are needed, whereas in the earlier example using only leased telephone lines 6 telephone lines were needed.

In conclusion, a frame relay solution for a company that has four locations and wishes to make a number of connections between those locations would require the following:

- ► four high speed telephone access lines into the frame relay network;
- ► four port charges for connecting into the frame relay network; and
- ► six PVC charges providing connectivity between sites.

The monthly costs for the frame relay setup would be: 4 x (high speed local telephone line cost) + 4 x (port charge) + 6 x (PVC cost). A non-frame relay solution would require 6 x (high speed long distance telephone line costs), which is very often more expensive. The In Action example at the end of this chapter demonstrates these costs using representative dollar values.

All the various frame relay costs depend on one major factor—the speed of the connection. The customer wants the frame relay carrier to provide a certain connection speed, and the frame relay provider is willing to provide various levels of connection speed, each with a different cost. Let's see how the customer and provider agree on a connection speed and cost—the committed information rate.

Committed Information Rate (CIR)

When a customer establishes a permanent virtual circuit with a frame relay carrier, both customer and carrier agree on a committed information rate. The committed information rate (CIR) is the data transfer rate that is agreed on by both customer and carrier. For example, if a customer seeks a CIR of 256 Kbps, the carrier agrees to transfer the customer's data at this rate. The customer also agrees that it will not exceed this rate. If the customer does exceed its CIR, and the frame relay network becomes saturated, the carrier will drop any of the customer's frames that exceed the committed information rate. An additional part of the agreement is the burst rate, which allows the customer to exceed the committed information rate by a fixed amount for brief moments of time.

For each PVC, the customer can specify a guaranteed throughput and a limited data delivery time. A guaranteed throughput means that the frame relay provider guarantees that a specified percent of all frames will be delivered, assuming that the data is not transferred faster than the committed information rate. A limited data delivery time means that all data frames will be delivered in an agreed upon amount of time, if, once again, the customer does not exceed CIR. The price of the PVC will then be directly proportional to the CIR level. The higher the CIR level, the higher the price of the PVC.

Consider a simple example. A business in New York has a high speed access line (1.544 Mbps) into the frame relay network. It asks its carrier to create a PVC from New York to Dallas with a CIR of 512 Kbps. If the business stays within its CIR, the carrier guarantees 99.99 percent throughput and a network delay of no longer than 20 milliseconds. Now, what happens if the customer exceeds its CIR by transmitting data at 800 Kbps? Most carriers will allow customers to exceed CIR up to a fixed burst rate for up to 2 seconds with no penalty. Let's assume for this example that the customer has a burst rate of 256 Kbps. That means the customer can transmit at 512 Kbps plus 256 Kbps, or 768 Kbps for 2 seconds. If the customer continues to exceed its CIR beyond 2 seconds, the carrier sets a Discard Eligible (DE) bit in the frame header of the customer's frames. If network congestion occurs, DE-marked frames are *discarded* by network switches. It is then the responsibility of the customer's application to watch for discarded frames.

What is the tradeoff between paying for a low CIR and a high CIR? If you request a low CIR and your data rate continuously exceeds that CIR, some of your frames may be discarded. But if the network is not congested, you won't lose frames. If you pay for a high CIR, the data rate is guaranteed and you won't lose frames no matter if the network is congested. The question comes down to how much of a gambler you are. You can save money, go with a lower CIR, and hope the network doesn't become congested. Or you can pay for the higher CIR and be fairly confident that no frames will be discarded. If you are supporting an application that should not lose any frames under any circumstances, it might be worth paying for the higher CIR. Interestingly, many customers choose to pay the lowest cost by selecting a *Zero CIR* option that provides no delivery guarantees.

Frame relay vs. the Internet

Both frame relay and the Internet are packet data networks with wide-spread availability. How does transmitting data over frame relay compare to transmitting data over the Internet? Unlike the Internet, frame relay guarantees throughput and minimum delay. In contrast, when data is transferred over the Internet, there is no guarantee when the data will arrive or if it will arrive at all. In addition, frame relay has better security. Because of the signaling technology used, hackers cannot break into PVCs between corporate sites. On the other hand, frame relay is more expensive than the Internet. In addition, frame relay is relatively inflexible when compared with the Internet. You cannot send data to a new site unless a PVC is already in place, and creating a PVC requires that the customer contact the carrier and make arrangements. An application using the Internet is often more flexible. Depending on the application, such as a virtual private network (introduced in Chapter Eleven), it may be possible for a user to dynamically create a virtual connection through the Internet.

Finally, even though frame relay service is very popular and continues to grow, it is not as pervasive as the Internet. One of the attractive advantages of frame relay is being able to connect to it with a local telephone connection. If no frame relay service providers offer the service in your local area, you will have to pay for a long distance high speed telephone line, which might make frame relay a less attractive alternative.

Voice over Frame Relay (VoFR)

Frame relay was originally designed and used to transfer packets of data between two sites more cost-effectively than leased lines could do. **Voice over frame relay (VoFR)**, as defined in the published standard FRF.11, allows the internal telephone systems of companies to be connected using frame relay PVCs. For example, a large company that supports an internal PBX may use voice over frame relay to connect its PBX (and thus its internal telephone system) to another location, such as another office or to the telephone company's central office.

Transferring telephone calls using frame relay has a number of advantages over using the leased line service of a standard telephone system. First of all, frame relay reduces the cost of a telephone call. Frame relay uses network resources more efficiently by combining a number of channels of voice traffic with data, and reliably transmitting the result over an existing frame relay network. The cost of equipment that connects a company's PBX to the frame relay network is quickly recovered from the savings of avoiding conventional telephone lines. Because VoFR does not significantly complicate network architecture, increase link speeds or CIR, voice, fax and data traffic can be combined effectively over a single network of wires. Finally, up to 255 voice sub-channels can be multiplexed on a single frame relay circuit.

There are disadvantages of course. Most importantly, a data network, which is now being additionally called upon to transmit frame relay voice, may experience congestion problems. Many corporate networks are straining to simply deliver data. To add voice to an already congested network is asking for more trouble. One minor disadvantage is that voice compression is necessary in frame relay applications to help ensure high quality audio while maximizing bandwidth usage. Sometimes this voice compression can affect the quality of the signal, thus affecting the sound quality of the voices on the telephone line.

Frame relay switched virtual circuits

One of the latest developments in frame relay networks removes the disadvantage that a frame relay connection is not dynamic. To create a frame relay connection between two points requires much planning and paperwork between the customer and the carrier. A **Switched Virtual Circuit (SVCs)** enables frame relay users to dynamically expand their current PVC networks and establish logical network connections on an as-needed basis to end points on the same network or through gateways to end points on other networks.

Details ▶

How SVCs Work

The protocol to establish an SVC connection is very similar to other connection-oriented protocols that require a connection be established before any data is transferred. The SVC protocol includes the following steps:

1. An SVC connection is initiated when a customer (the calling user) sends a SETUP message to the network.

2. This call SETUP includes the destination address (the address of the receiving user), the bandwidth parameters, and other information about the call and its use.

3. The network sends a CALL PROCEEDING message back to the calling user, along with the Data Link Connection Identifier (DLCI) designation for the connection.

4. If the receiving party accepts the call, a CONNECT message is sent from the receiving user back through the network to the calling user, indicating acceptance.

5. Once the connection is set up, the two end points can transfer information.

6. When both parties have completed the call, either end point can send a DISCONNECT message to the network.

7. The network releases the resources of the call and sends a RELEASE message to both end points.

8. When the network receives a RELEASE COMPLETE message from both end points, the call is terminated and the connection no longer exists.

From the steps in this protocol, you can see that creating a switched virtual circuit on a frame relay network is quite similar to a person placing a telephone call. However, since humans are not making the connection, a more formal method—the protocol—of transferring messages is required.

Several characteristics make switched virtual circuits attractive. SVCs and PVCs can co-exist on the same network. Thus, if you have an existing system using PVCs, SVCs can be added with relatively few system changes. It is also possible to mix applications that work better with PVCs with applications that work better with SVCs. For example, PVCs are better suited to end points that communicate frequently, while SVCs are ideal for occasional-use applications or end points that need to communicate only periodically. SVCs allow users to establish reliable connections to any destination at any time for any duration desired. And unlike PVCs, service providers are not required to pre-configure SVCs, because the user creates an SVC connection simply by placing a call. Finally, SVCs can also be used as a backup in the event of a transmission system failure because they are available on demand, similar to dial-up modems and ISDN.

Asynchronous Transfer Mode (ATM)

Asynchronous transfer mode (ATM), similar to frame relay, is a relatively new high speed, packet switched service that is offered by the telephone companies. A business that wants to send data between two points (either within a building or across the country) at very high transfer rates might consider using ATM. The transfer rates are as fast as 622 Mbps with even faster speeds possible.

ATM has several unique features that set it apart from frame relay and other packet services. In ATM, all data is sent in small 53-byte packages called cells. The cell size is kept small so that when a cell reaches a node (a switch point) in an ATM network, the cell will quickly pass through the node (switch) and continue on its way to its destination. This fast switching capability gives ATM its very high transfer rates. Furthermore, ATM is a fully switched network. As the cells hop from node to node through the network, the cell is processed and forwarded by high speed switches (recall the concept of the network switch from Chapter Eight).

ATM networks were designed to simultaneously support voice, video and data. As you will see shortly, ATM can handle a wide range of applications, including live video, music, and interactive voice. In addition, ATM can be used as the transfer technology for local area networks, metropolitan area networks, and wide area networks. Since ATM is a layer 2 protocol similar to frame relay, ATM can be supported by many different types of physical layer media, such as twisted pair and fiber optic cable. Finally, ATM can support different traffic classes to provide different levels of service (Quality of Service QoS). ATM traffic classes will be described in detail in the next section.

Similar to frame relay, in which private virtual circuits are established over an existing telecommunication service, to use ATM you must create a logical connection called a **virtual channel connection (VCC)** over a virtual path connection. A **virtual path connection (VPC)** is a bundle of VCCs that have the same endpoints. VCCs are established for data transfer, but they are also established for control signaling, network management, and routing. A VCC can be established between two end users, in which case the VCC would transmit data and control signals. A VCC can also be established between the end user and a network entity, in which case the VCC would transmit user-to-network control signaling. When a VCC transmits user-to-network control signaling, it is called the **User-Network Interface**. Finally, a VCC can be established between two network entities, in which case the VCC would transmit network management and routing signals. When a VCC transmits network management and routing signals, it is called the **Network-Network Interface**.

ATM classes of service

One of the unique features of ATM, which sets it apart from most other data and voice transmission systems, is the ability to specify a class of service. A **class of service** is a definition of a type of traffic and the underlying technology that will support that type of traffic. For example, the real-time interactive class of service defines live video traffic. A real-time interactive application is one of the most demanding with respect to data transmission rate and network throughput. Transmitting a video stream over a medium requires a significant amount of bandwidth, and to provide a real-time service requires the network to be fast enough to send and receive video so that a live application can be supported.

If a network does not provide varying levels of class of service, then all applications share the network bandwidth equally, which may not provide a fair distribution of bandwidth. As an example, let's examine the Internet, which currently does

Details ▶

ATM Cell Composition

ATM transmits all data and control signals using 53 byte cells (Figure 12-9). The cells are kept small so that cell switching within the nodes can be performed by hardware, which is the fastest technique for cell switching. The ATM header is always 5 bytes long, which leaves 48 bytes of the cell for data. Cell fields and their functions include:

▶ Generic flow control, which is applied only at the user-network interface. Generic flow control is used to alleviate short-term overload conditions by telling the user to slow down on sending data into the ATM network. Generic flow control is not currently used by ATM systems.

▶ Virtual path identifier, which is an 8-bit field (allowing for 256 combinations) that identifies a particular VPC.

▶ Virtual channel identifier, which is a 16-bit value that identifies a unique VCC. When a company creates an ATM connection between two locations, a virtual path supports one or more virtual channels. To identify the path and channels, the path and the channels have unique ID numbers.

▶ Payload type, which consists of three bits. These three bits identify the type of payload as either user data or network maintenance and report whether congestion has been encountered in the network.

▶ Cell loss priority, which is a single bit, is set by the user. A value of 0 means this cell is of high priority, so the ATM network should think twice about discarding it in the event of network congestion.

▶ Header error control, which consists of an 8 bit cyclic redundancy checksum ($x^8 + x^2 + x + 1$). It provides error checking for the first 4 bytes of the header, but not for the data. The cyclic checksum can also provide error correction by fixing single-bit errors in the header.

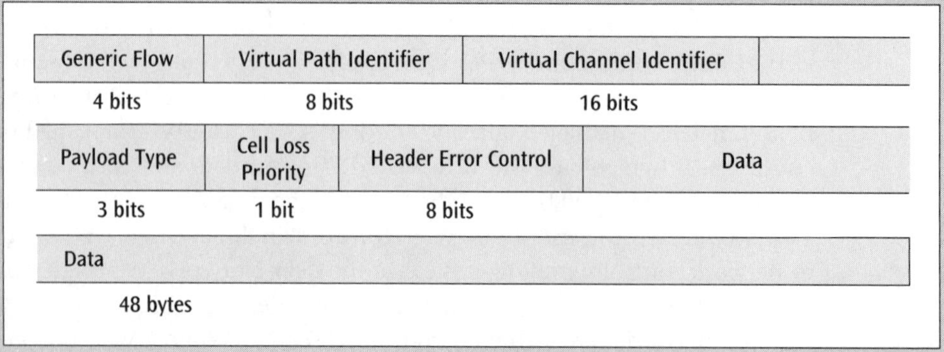

Generic Flow	Virtual Path Identifier	Virtual Channel Identifier	
4 bits	8 bits	16 bits	

Payload Type	Cell Loss Priority	Header Error Control	Data
3 bits	1 bit	8 bits	

Data
48 bytes

Figure 12-9 *The 53-byte ATM cell with its individual fields*

not offer a class of service. Suppose you and a coworker are sitting at computer workstations in adjacent cubicles. You are running an e-mail program. Your coworker is participating in a live videoconference with a third coworker in another state. Both the e-mail and videoconferencing applications are using the Internet for data transmission. Even though your e-mail application often transmits very little data and time is not at all important, your e-mail has exactly the same priority on the Internet as your coworker's application, which is trying to run a very time intensive and data consuming video conference. Although your e-mail program may work fine via the Internet, your coworker's video conferencing is undoubtedly performing poorly. If the Internet could support different classes of service, your e-mail program could request a low class of service, and your coworker's video conference program could request a high class of service. With a high class of service, the video conference program would perform much better.

With ATM, for every VCC that is set up, the customer specifies a desired Class of Service. As you might expect, the desired Class of Service determines the price for using a particular VCC. ATM has defined four classes of service that are functionally equivalent to leased line service, frame relay service, and Internet service. These classes of service determine the type of traffic an ATM network can carry:

> ▶ **Constant Bit Rate (CBR)** Similar to a current telephone system leased line, CBR is the most expensive class of service. CBR acts very much like a time division multiplexed telephone service, but is not a dedicated circuit such as a leased telephone line. CBR delivers a high speed continuous data stream that can be used with transmission-intensive applications.

> ▶ **Variable Bit Rate (VBR)** Similar to frame relay service, VBR is used for real time applications, such as compressed interactive video, which is time dependent, and non-real time applications, such as sending e-mail with large, multimedia attachments, which is not time dependent.

> ▶ **Available Bit Rate (ABR)** ABR is used for traffic that may experience bursts of data, called "bursty" traffic, and whose bandwidth range is roughly known, such as a corporate collection of leased lines. Negotiation establishes a low and high transfer range, but getting bit space for sending data through the ATM network is similar to flying standby. You may get a chance to transmit your data, or you may have to wait. ABR is good for traffic that does not have to arrive in a certain amount of time. ABR also provides feedback that indicates if a part of the ATM network is experiencing congestion.

> ▶ **Unspecified Bit Rate (UBR)** UBR is also capable of sending traffic that may experience bursts of data, but there are no promises as to when the data may be sent, and if there are congestion problems, there is no congestion feedback (as ABR provides). With even worse performance than ABR, data transmitted over a UBR connection may not make it to the final destination. Part way through the network connection, if the bandwidth necessary to transmit your data is required by a higher class of service, your data may be discarded. You are flying standby, but if halfway through the flight, they need your seat, you are tossed out the open door.

By offering different classes of service, the ATM service provider can offer tailor-made transmission services and charge the customer accordingly. Likewise, the customer can request a particular class of service that is appropriate for a particular application.

Advantages and disadvantages of ATM

Because of the various ATM features such as high transfer speeds, various classes of service, and the ability to operate over many types of media and network topologies (LANs as well as WANs), ATM has a number of very significant advantages. ATM can support a wide range of applications with varying bandwidths and at a wide range of transmission speeds. Cell switching, which is performed by ATM's high speed hardware-based switches that route cells down the appropriate path, is so fast that it provides low delays and high bandwidths. ATM's different classes of service allow customers to choose service type and pricing individually for each data connection (VCC). Finally, ATM is extremely versatile. It can carry voice, packet data, and video over the same facilities.

As you might expect, ATM also has a number of disadvantages. It is more expensive than other data transmission options. The cost of the equipment is high because the cell switching equipment is so fast and relatively complex. Due to the complexity of ATM, there is a high learning curve for setting up and managing the network. Finally, compatible hardware and software is still not widely available.

ATM is now being widely used by the large telecommunications carriers (AT&T, MCI, Sprint, and so on) to provide voice and Internet services. Because of its complexity and cost, however, smaller businesses have been reluctant to use ATM. Alternatives that can offer reasonably comparable speeds at much lower costs have attracted much attention. One possible alternative to ATM on a local area network is high speed Ethernet. With transfer speeds hitting 100 Mbps to 1000 Mbps, high speed Ethernet is a very attractive alternative to ATM. For longer distances, newly emerging technologies hold the promise of equaling ATM. One of those promising new players is digital subscriber line, the next topic.

Digital Subscriber Line

Digital subscriber line (DSL) is a relatively new technology that allows *existing* twisted pair telephone lines to transmit multimedia materials and high-speed data. Transfer speeds can range from hundreds of thousands of bits per second up to several million bits per second. Although many of the larger local telephone companies are hesitant to offer DSL service, the newly emerging telephone companies (competitive local exchange carriers or CLECs) are eagerly offering DSL. The hesitation of the larger telephone companies is partially based on the fact that they currently hold onto many customers by offering leased-line services such as T1. These larger telephone companies would have to invest a large amount of money in DSL technology and training just to merely hold onto those customers who might defect from T1 to another provider's DSL. For these companies, investing in DSL just to stay even doesn't make sense. However, they may have to change their plans eventually. Many professionals in the telecommunications arena predict that DSL will play a crucial role in the next decade as a tool for delivering a wide range of information in video and multimedia formats.

Success of DSL will hinge on the telephone companies' ability to offer these services to a wide audience and at reasonable prices. Thus far the reasonable price component is in place—DSL pricing is very competitive with the pricing of both ISDN and cable modems. As you shall see, however, DSL has a number of features that surpass both ISDN and cable modems, making the price even more attractive.

DSL basics

As with any technology that can deliver multimedia data, transmission speed is an important issue. DSL is capable of a wide range of speeds. The transfer speed of a particular line depends on one or more factors: the carrier providing the service, the distance of your house or business from the central office of the local telephone company, and whether the DSL service is a symmetric connection or an asymmetric connection. Each carrier that offers DSL has a choice of a particular form of DSL technology and the supporting transmission formats. The form of DSL and its underlying technology determine the speed of the service.

The transfer speed of a line also depends on distance. The closer your house or business is to the central office, the faster the possible transmission speed. This dependency on distance is due to the fact that copper-based twisted pair is fairly susceptible to noise, which can significantly affect a DSL transmission signal. The longer the wire (without a repeater), the more noise on the line, and the slower the transmission speed. The maximum distance a house or business can be from the central office without a repeater is currently 5.5 kilometers.

The third factor affecting transfer speed is the type of connection: symmetric or asymmetric. A **symmetric connection** is one in which the transfer speeds in both directions are equivalent. An **asymmetric connection** has a faster downstream transmission speed than its upstream speed. For example, an asymmetric DSL service can provide download speeds into the hundreds of thousands of bits per second while upload speeds may be less than 100,000 bps. An asymmetric service is useful for an Internet connection in which the bulk of the traffic (in the form of web pages) comes down from the Internet to the workstation. Usually the only data that goes up to the Internet is a request for another web page, relatively small e-mail messages, or an FTP request to download a file.

Another characteristic of DSL is that it is an always-on connection. Users do not have to dial and make a connection. Thus, a home or business workstation can have a 24 hour a day connection to the Internet, or some other telecommunication service. Furthermore, the connection is charged a flat, usually monthly, rate. The user does not pay a fee based on distance of connection and how long the connection is established.

Because DSL is an always-on connection, it uses a permanent circuit and not a switched circuit. This permanent circuit is not dynamically configurable, but must be created by the DSL service provider. Since most current home users of DSL use the service to connect their computers to an Internet service provider, the user-unchangeable connection is not a serious issue. However, if the user wishes to move the DSL connection to another recipient, a new DSL service has to be contracted.

What does a business or user need to establish a DSL connection? At the present time, four components are required. The local telephone company (LEC) must install a special router called a DSLAM (digital subscriber line access multiplexer) within the telephone company's central office. This device creates and decodes the DSL signals that transfer on the telephone local loop (Figure 12-10). Next, the local telephone company also installs a DSL modem on its premises, which combines and splits the DSL circuits (the upstream and downstream channels) with the standard telephone circuit (POTS). Recall that DSL transmits over the same telephone line that runs from a central office to a home or business. Since it is the same telephone line, DSL must share the line with a POTS signal.

If the DSL is of a particular form, such as DSL.Lite (described shortly), then the DSL line does not carry a standard telephone circuit. Thus, there is no need to split the DSL signal from the telephone signal at either the sending or receiving end. When there is no splitter used to separate the DSL signal from the POTS signal, then the service is called **splitterless** DSL.

The user requires an equivalent DSL modem connected to his or her workstation. Like the DSL modem at the telephone company's central office, this modem splits the downstream DSL from the upstream DSL, and also splits off the standard telephone circuit. If the DSL form is splitterless, there is no telephone circuit to split off.

Finally, the DSLAM router at the telephone company's central office must be connected to an Internet service provider via a high speed line. Since this high speed line will be supporting the Internet service requests from multiple users, the high speed line needs to be a very fast service, such as ATM. All these components are shown in Figure 12-10.

Figure 12-10
The four necessary components of a DSL connection

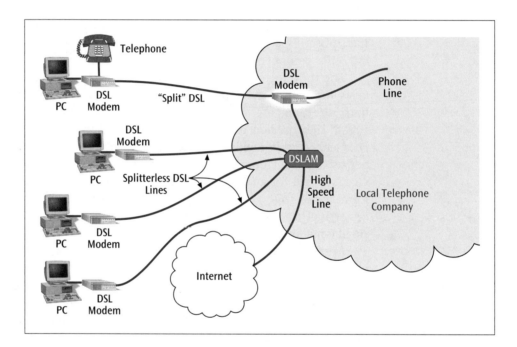

Let's consider a telephone company and the types of DSL service it provides. In late 1999, Bell Atlantic was offering two types of DSL service. The first service, called "Personal Infospeed DSL," provided a downstream speed of 640 Kbps and an upstream speed of 90 Kbps for $39.95 per month, or $20 more with an Internet connection. For users requiring more speed, the second service, called "Power Infospeed DSL," provided a downstream speed of 7.1 Mbps and an upstream speed of 680 Kbps for $109.95 per month, with an additional charge of $80.00 with an Internet connection. The speeds and prices listed here are only one example of hundreds possible across the United States.

DSL formats

Digital subscriber line comes in a variety of formats. Although this variety provides carriers and users with a wide selection of technologies and speeds, the variety can also split users into numerous camps, making it more difficult to settle on one format as a standard. No doubt over time a number of these formats will fall by the wayside, but there is no knowing how long that will take. For now, an informed business network user should be aware of the various formats. (Please note that the data transmission rates provided will change often as the technology evolves.) Often referred to as xDSL, seven DSL formats are in use today:

- ▶ **Asymmetric Digital Subscriber Line (ADSL)** A popular format that transmits the downstream data at a faster rate than the upstream rate. Typical downstream data rates range from 300 Kbps to 800 Kbps, while upstream data rates range from 128 Kbps to 400 Kbps.

- ▶ **Consumer DSL (CDSL)** A trademarked version of DSL that is a little slower than typical ADSL speeds. Rockwell owns this technology and will sell the integrated circuits to support CDSL.

- ▶ **DSL.Lite** A slower format compared to ADSL; also known as Universal DSL, G.Lite, and splitterless DSL. Typical transmission speeds are in the 50 Kbps to 200 Kbps range.

- ▶ **High-bit rate DSL (HDSL)** The earliest form of DSL, which provides a symmetric service with speeds usually equivalent to a T1 service (1.544 Mbps).

- ▶ **Very high data rate DSL (VDSL)** A very fast format (between 51 and 55 Mbps) over very short distances (less than 300 meters).

- ▶ **Rate-Adaptive DSL (RADSL)** A format in which the transfer rate can vary depending on noise levels within the telephone line's local loop.

Comparison of DSL to ISDN, Frame Relay, ATM, and cable modems

To summarize DSL, frame relay, ATM, and ISDN, let's compare them on the following criteria: maximum transmission speed; typical cost per month; if the connection is switchable or fixed; if the technology is scalable (capable of increasing in speed in the future); whether the service provides quality of service choices; if the service can support data or voice or both; and how available the service is within the U.S. Let's also include cable modems in the comparison since, in a number of cases, they may be an alternative technology to services such as ISDN and DSL. Table 12-1 shows the summary results.

Note that when comparing maximum transmission speeds, ATM is the clear winner. However, ATM is clearly the most expensive and the most complex, making it an unlikely candidate for small business and home connections. If you consider DSL systems that transmit around 1.5 Mbps, DSL compares favorably to cable modems, is much faster than ISDN for roughly the same price, and is less expensive than frame relay systems at comparable speeds.

A noticeable shortcoming of both cable modems and ISDN is their inability to scale to higher speeds. While ISDN BRI is definitely locked in at 128 Kbps, cable modems might be able to increase in speed if the cable providers are willing to invest a significant amount of funds in updating their technology.

The last issue worth noting is the availability of each service. Although availability is certain to change for each type of service, currently cable modems and DSL have moderate availability across the country, while ISDN, ATM and frame relay enjoy fairly wide availability.

Table 12-1

Summary of different high speed wide area connection technologies

	Maximum Speeds	Cost per Month ($=100s, 1000s, 10,000s)	Switchable or Fixed	Scalable	QoS	Data or Voice	Availability
Cable modem	1.5 Mbps	~$40	Fixed	No	No	Data	Limited
ISDN BRI	128 Kbps	~$40	Both	No	No	Both	Wide
Frame relay			Both	Yes	No	Both	Wide
ATM	1.5 Mbps	$$$$	Fixed	Yes	Yes	Both	Wide
	622 Mbps	$$$$-$$$$$	Fixed	Yes	Yes	Both	Wide
DSL	640 Kbps	$40	Fixed	Yes	No	Both	Moderate
	7 Mbps	$$$-$$$$	Fixed	Yes	No	Both	Growing

Computer Telephony Integration

Computer Telephony Integration (CTI) is an exciting, newly emerging field that combines more traditional voice networks with modern computer networks. CTI integrates the PBX phone switch with computer services, thus creating modern voice and data applications that run on computer systems. Recall that a PBX is the internal telephone system many businesses use to support all their telephone operations, including in-house calling, in-house to outside calling, voice mail, call transfers, and conference calling. Traditionally these services require a PBX switch, a set of telephones, and standard telephone wiring. In contrast, CTI combines the power of computer systems with the services of a telephone network. Running the appropriate telephony operations on a workstation computer, a user can perform typical telephone operations by clicking in a window of a program.

Consider the following scenario. A business sells a product and provides customer service to support that product. To obtain customer service, a customer calls the company's 800 number. Since the company has a CTI system that combines the power of a local area network with modern telephone services, the caller's telephone number appears on the screen of the customer service representative's computer monitor as the telephone rings (caller ID). Before the customer service representative answers the telephone, the customer's telephone number is used as the key to a database query, and a summary of the customer's account appears on the screen. The customer service representative answers the telephone and has customer information on the screen as they say "Hello." As the representative talks to the customer and learns about the customer's problem, they can click icons on the computer screen to transfer the customer's call to another party, put the customer on hold, or retrieve additional information from the computer database. Thus, the distinction between computers and telephone systems is greatly blurred. The computer and local area network are performing telephone operations as easily as they perform data operations.

CTI can also integrate voice cabling with data cabling. Early CTI systems, despite a strong marriage of data and voice operations, still maintained separate wires for the local area network and its workstations and another set of wires for the telephone lines. The PBX may have talked directly to the local area network, sending data and telephone commands back and forth, but the actual computer data and telephone data remained separate. Recently developed systems support computer data and voice data on the same set of wires. Although this arrangement may save on wiring, it places a larger demand on the single wiring system.

Currently, there are two possible CTI configurations: first party control and third party control. First party control provides individual connections from the PBX to each workstation. Each office has a telephone connected to the company PBX, and the workstation in an office is also connected to the PBX. Thus, the PBX acts as a translator between telephone operations and workstation functions. In third party control, the more popular configuration as of late, a PBX connects to a computer telephony server. All workstations can be controlled by this server through a LAN. In this configuration, the PBX is not connected to each workstation. When a PBX wants a workstation to perform a certain function, the PBX talks to the computer telephony server, which then sends the information to the workstation.

There are three advantages to using CTI. First, it creates new voice/data business applications that can save businesses time. Second, it makes optimal use of current resources. Third, it saves money. These advantages mean that businesses can realize many benefits from CTI applications. For example:

- ▶ **Unified messaging** Users utilize a single desktop application to send and receive e-mail, voice mail, and fax.
- ▶ **Interactive voice response** When a customer calls your company, his or her telephone number is used to extract the customer's records from a corporate database. The customer's records are displayed on a service representative's workstation as the representative answers the telephone.
- ▶ **Integrated voice recognition and response** A user calling into a company telephone system provides some form of data by speaking into the telephone and a database query is performed using this spoken information.
- ▶ **Fax processing** and **fax-back** In fax processing, a fax image that is stored on a LAN server's hard disk can be downloaded over a local area network, converted by a fax card, and sent out to a customer over a trunk line. An incoming fax can be converted to a file format and stored on the local area network server. With fax-back, a user dials into a fax server, retrieves a fax by keying in a number, and sends that fax anywhere.
- ▶ **Text to speech and speech to text conversions** As a person speaks into a telephone, the conversion system can digitize the voice and store it on a hard disk drive as computer data. The system can also perform the reverse operation.
- ▶ **Third party call control** Users have the ability to control a call without being a part of it, such as setting up a conference call.
- ▶ **PBX Graphic User Interface** Different icons on a computer screen represent common PBX functions such as call hold, call transfer, and call conferencing, making the system easier for operators to use.
- ▶ **Call filtering** Users can specify telephone numbers that are allowed to get through. All other calls will be routed to an attendant or voice mailbox.
- ▶ **Customized menuing systems** Organizations can build customized menuing for an automated answering system to help callers find the right information, agent, or department. Using drag-and-drop, the menuing system can be revised daily. The menu system can be interactive, allowing

callers to respond to voice prompts by dialing different numbers. Different voice messages can be associated with each response, and voice messages can be created instantly via a PC's microphone.

As you can see, some very interesting applications are emerging from the convergence between voice and data systems. Only time will tell if CTI is a passing fad or the beginning of a whole new area of computer networks.

Telecommunications Systems in Action – A Company Makes a Service Choice ▶

Bringbring Corporation manufactures telephone sets. Its administrative headquarters is in Chicago, IL, and it has regional sales offices in Seattle, WA, San Francisco, CA and Dallas, TX. Bringbring is expanding its data networking capability and has asked for your help in designing its new corporate network. It wants to add new hardware and software and provide new data applications. Although the company has enough funds to support the new applications, it wants to seek the most cost-effective solution for the long run.

Current data network

Bringbring currently has only a few PCs in use. The Marketing group in Chicago runs applications on 25 standalone PCs. The other sites currently have no PCs.

Details ▶

CTI Configurations

For CTI to work, CTI software is required on the PBX, all workstations, and the computer telephony server, if the configuration involves third party control.

To allow a smooth integration between the desired voice applications and the computer systems they will run on, all CTI functions and messaging are defined by the CTI Applications Programming Interface (API). An Applications Programming Interface is a precisely defined interface that allows a user writing an application program to call on the more technical resources in an operating system or hardware product. Thus, an API frees the application programmer from knowing the precise technical details of the telephone system and lets the programmer concentrate on writing a good, useful, user-friendly interface.

Currently, there are three different groups of CTI APIs:

▶ **Telephony Services API (TSAPI)** Developed by AT&T and Novell, this API provides third party control using NetWare servers.

▶ **Telephony API (TAPI)** Developed by Intel and Microsoft, this API was originally a first party control configuration, but now provides third party control using Windows NT systems.

▶ **Java Telephony API (JTAPI)** Developed by Sun, JTAPI is a Java-based set of APIs. JTAPI is the least used API of the three and is based on third party control.

New data network

Bringbring Corporation plans to add 10 additional PC workstations to the Chicago office and plans to add 35 PCs each to the Seattle, San Francisco and Dallas offices. In its new network, it will also install the following 6 servers:

> ► Web Server (Chicago) — an HTTP server to store web pages for public access to corporate marketing information.

> ► Inventory Server (Chicago) — a Database server that stores information about available product inventory.

> ► Site Servers (Chicago, Seattle, San Francisco, Dallas) — Four servers, one at each site, that provide e-mail services so employees can communicate about important projects, store site-specific management files and sales files, and store information for clerical workers and staff software.

New data applications

Once the new PCs, servers, and network equipment is in place, Bringbring would like to use the network for the following applications:

> ► Local Site Server Access — Each employee needs to be able to access data files on the Site Server at his or her location. Each employee accesses enough local files that at least 1 Mbps of bandwidth is needed for each PC on any shared LAN. To provide this level of service, for example, a 16 Mbps shared Token Ring should have no more than 16 workstations connected to it.

> ► E-mail Services — Each Site Server needs to be able to exchange e-mail messages with any other Site Server.

> ► Internet Access — Connectivity needs to be provided from the Web Server to the Internet so current and potential customers around the world can use web browsers to access company information.

> ► Network Access to Sales Records — Sales people at the regional sales offices must be able to access data on the Inventory Server in Chicago. Typically about 20,000 sales records per month will be uploaded from each of the regional sales offices (Seattle, San Francisco, and Dallas). Each sales record is 400 Kbytes in size. It should take no longer than 20 seconds to upload a single sales record.

Prices

Table 12-2 shows the typical prices of different WAN services.

Table 12-2
Typical prices of different WAN services

Type of WAN	Service and Speed	Per Month Cost	Usage Fee
Dial-up Access Lines	POTS Line	$20.00 plus usage fee	10 cents per minute
	ISDN Line	$40.00 plus usage fee	10 cents per minute
Leased Access Lines	IntraLATA 56 Kbps	$150	
	IntraLATA T1 (1.5 Mbps)	$350-400	
	InterLATA 56 Kbps	$500 plus usage fee	$1.00 per mile
	InterLATA T1 (1.5 Mbps)	$1200 plus usage fee	$2.50 per mile
Frame Relay (port speed and price)	56 Kbps	$220	
	128 Kbps	$400	
	256 Kbps	$495	
	512 Kbps	$920	
	T-1 (1.5 Mbps)	$1620	
Frame Relay PVC (CIR and price)	16 Kbps	$25	
	32 Kbps	$40	
	48 Kbps	$50	
	56 Kbps	$60	
	128 Kbps	$110	
	256 Kbps	$230	
	384 Kbps	$330	
	512 Kbps	$410	
	1024 Kbps	$1010	
	1536 Kbps	$1410	

Given the nature of the data transmission requirements, Digital Subscriber Line would have to be symmetric. Because of its spotty availability across the United States at this time, it is not being considered as an option.

Making the choice

To choose the right WAN service, you need to choose the appropriate data transmission speeds for each service, keep the cost low, and meet Bringbring's requirements. Let's examine the total WAN network service cost per month that Bringbring Corporation might pay to a telecommunications carrier to connect all four sites using a technology. Remember that Frame Relay has three cost components: (a) the access line charges (b) the frame relay port charges (c) the PVC charges.

To begin, you need to reread Bringbring's plans and discover their transmission requirements. Separating the WAN information from the LAN information, you focus on "Network Access to Sales Records" and see the requirement that 400 Kbyte records should take no longer than 20 seconds to download. What speed transmission line will meet this requirement?

400 Kbytes = 400,000 bytes

400,000 bytes * 8 bits/byte = 3,200,000 bits

3,200,000 bits / n bps = 20 seconds

Solving for n:

3,200,000 bits / 20 seconds = 160,000 bps = 160 Kbps

Thus, Bringbring needs a telecommunications line connecting to the outside world with a transmission speed of at least 160 Kbps.

Using a dial-up modem, the best case transmission speed is 56 Kbps (actually 53 Kbps). ISDN Basic Rate provides a transmission speed of only 128 Kbps. A 56 Kbps leased line also would not be fast enough. Bringbring needs to use either a T1 line (1.544 Mbps) or a frame relay service.

Considering a T1 scenario, you recognize that Bringbring would need three T1 lines, one from Seattle to Chicago (2050 miles), one from San Francisco to Chicago (2170 miles), and one from Dallas to Chicago (920 miles). Each of these lines is long distance, or interLATA. Using the mileage between these points, the three interLATA T1 lines would cost approximately $6,325 ($1200 + $2.50 per mile from Seattle to Chicago), $6,625 ($1200 + $2.50 per mile from San Francisco to Chicago), and $3,500 ($1200 + $2.50 per mile from Dallas to Chicago) respectively, for a monthly total of $16,450. This scenario does not even allow Seattle to talk to San Francisco, or Seattle to talk to Dallas; it only provides communication between Chicago and each of the regional sites (Figure 12-11).

Figure 12-11
Three T1 lines connecting the three regional offices to Chicago

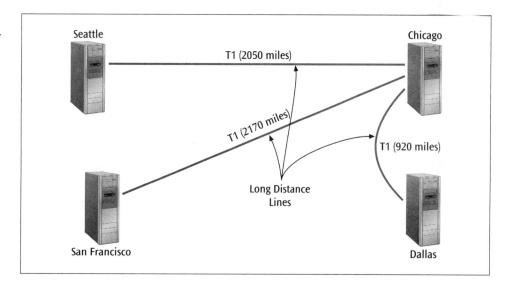

A frame relay scenario means you don't have to worry about running a separate line between every pair of cities. You only have to make a frame relay connection into the frame relay network (Figure 12-12). Once data is in the frame relay network, it can be routed to any of the four cities. Thus, four connections allow us to send data between any two cities, and each connection has to support the 160 Kbps transmission speed. Thus a 256 Kbps frame relay connection is the closest size that will support the desired speed.

Figure 12-12
*Four cities connected
to the frame relay
network*

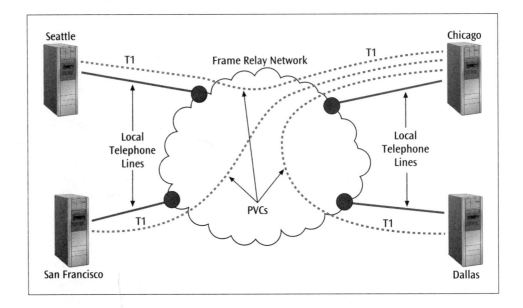

The frame relay charges would include:

▶ Three port charges, one for each 256 K connection into the frame relay network = 3 x $495; plus one 1024K port in Chicago = $1010.

▶ Three PVC charges, for 256 K connection (dotted lines in Figure 12-12) = 3 x $230.

▶ Four IntraLATA T1 telephone lines to connect the four cities to the frame relay network = 4 x $600.

The total charges for frame relay would be $5585 per month, which is much less expensive than using straight T1 lines between the four cities.

Although other charges are generated from the frame relay scenario (for example, Internet access charges and e-mail charges), the WAN connection charges are the most costly portion. Thus, it appears that the frame relay solution will be the least expensive over the long run. The monthly recurring charges of frame relay are significantly lower than the monthly recurring charges of T1 lines.

◆ ◆

SUMMARY

▶ The basic telephone system that covers the U.S. is called Plain Old Telephone System (POTS) and is a mixture of analog and digital circuits.

▶ Telephone channels are designed to carry voice signals with 4000 Hz bandwidth signals.

▶ The telephone line is the set of wires that runs from the central office to a house or business, transfers analog signals, and it is called the local loop.

▶ A telephone trunk carries multiple telephone signals and typically runs between central offices and switching centers.

▶ The divestiture of AT&T in 1984 opened the long distance telephone market to other long distance providers, forced AT&T to sell off its local telephone companies, and divided the country into local access transport areas.

▶ A PBX is an on-premise computerized telephone switch that handles all internal and out-going telephone calls and offers a number of telephone services.

- A Centrex offers the same services as a PBX but the equipment resides on the telephone company's property and the business leases the service.
- Private lines or leased lines are permanent telephone lines that run between two locations and provide constant access without dialing a telephone number.
- The Telecommunications Act of 1996 opened local telephone service to new competitors and required the existing local telephone companies to allow access to the local telephone lines.
- Different types of leased line service include 56 Kbps leased lines and the all digital 1.544 Mbps T1.
- Integrated Services Digital Network (ISDN) is an all digital data and voice service that can service homes and businesses over standard twisted pair telephone lines.
- ISDN comes in two basic services: Basic Rate Interface, which supports two 64 Kbps channels of either voice or data, and Primary Rate Interface, which supports 24 64 Kbps channels of voice or data.
- Frame relay is a relatively new service that provides digital data transfer over long distances and at high data transfer rates.
- To use a frame relay service, a customer contacts a frame relay provider who creates a permanent virtual circuit (PVC) between the customer and to whomever the customer wants to be connected. The customer pays for the PVC, the access to the frame relay network (a port charge), and the telephone line that gives them access to the port.
- Frame relay is a layer 2 service, which means that it can run over a variety of physical media and supports many different types of applications at layers 3 and up.
- The committed information rate is an agreement between the frame relay customer and the service provider. The customer requests a particular transmission speed and if the customer does not exceed that rate, the service provider will guarantee accurate and timely delivery of data frames.
- Frame relay is more reliable than the Internet and provides a better level of security, but the Internet is less expensive and available essentially everywhere.
- Frame relay is now also capable of supporting voice, and can also create switched virtual circuits in addition to permanent virtual circuits.
- Asynchronous transfer mode (ATM) is also a packet switched service, but supports all types of traffic and operates over LANs as well as WANs and MANs.
- ATM, because of the use of very fast switches, can transfer ATM cells at very high rates (over 600 Mbps).
- To use ATM, the ATM service provider creates a virtual path connection between two points and the user then creates one or more virtual channel connections over that virtual path.
- ATM provides a number of different classes of service to support a wide range of network applications requiring varying transmission speeds and network throughputs.
- Digital subscriber line (DSL) is a new digital data service that operates over the local telephone loop and can provide homes and businesses with multi-megabit connections.
- There currently are many different types of DSL service around the country, but most can be classified as either splitterless (only DSL service) or splitter (DSL service and a POTS telephone service on the same set of wires).
- Computer telephony integration is the convergence of data communication networks and voice systems. This convergence provides a number of new voice/computer applications such as workstation-controlled telephone services and data retrieval services.

KEY TERMS

asymmetric connection
asymmetric digital subscriber line
(ADSL)
asynchronous transfer mode (ATM)
available bit rate (ABR)
basic rate interface (BRI)
burst rate
cable telephony
call filtering
central office (CO)
Centrex (central office exchange
service)
class of service
committed information rate (CIR)
competitive local exchange carrier
(CLEC)
computer telephony integration (CTI)
constant bit rate (CBR)
consumer digital subscriber line
(CDSL)
customized menuing systems
digital subscriber line (DSL)
DSL.Lite

fax-back
fax processing
frame relay
high bit-rate DSL
incumbent local exchange carrier
(ILEC)
integrated voice recognition and
response
interactive voice response
interexchange carrier (IEC, IXC)
integrated services digital network
(ISDN)
layer 2 protocol
local access transport area (LATA)
local exchange carrier (LEC)
local loop
modified final judgement
network-network interface
PBX graphic user interface
permanent virtual circuit (PVC)
plain old telephone system (POTS)
primary rate interface (PRI)
private line

private branch exchange (PBX)
rate adaptive DSL (RADSL)
splitterless
switched virtual circuit (SVC)
symmetric connection
T1 service
Telecommunications Act of 1996
Text to speech and speech to text
conversions
third party call control
TIE lines
unified messaging
unspecified bit rate (UBR)
user-network interface
variable bit rate (VBR)
very high data rate DSL
virtual channel connection (VCC)
virtual path connection (VPC)
voice over frame relay (VoFR)
xDSL
56K leased line

REVIEW QUESTIONS

1. What is the plain old telephone system (POTS)?

2. What is the typical frequency range for the human voice?

3. The local loop connects what to what?

4. What is a LATA?

5. How does a trunk differ from a telephone line?

6. List the most important results of the Modified Final Judgment of AT&T.

7. What is the difference between a local exchange carrier and an interexchange carrier?

8. List the most important results of the Telecommunications Act of 1996.

9. How does a leased line service differ from POTS?

10. What are the basic features of a T1 line?

11. What are the basic services of ISDN?

12. What is the difference between a basic rate interface and a primary rate interface in ISDN?

13. What kind of traffic does an ISDN B channel carry? What kind of traffic does a D channel carry?

14. What are the basic characteristics of frame relay?

15. What features make frame relay so attractive?

16. How do you create a frame relay permanent virtual circuit?

17. How does a frame relay switched virtual circuit differ from a permanent virtual circuit?

18. Is it possible to have more than one PVC over a physical line?
19. What is agreed on when a customer and a frame relay service agree on a committed information rate?
20. How is it possible to send voice over frame relay?
21. How does frame relay compare to sending data over the Internet?
22. What are the basic features of ATM?
23. What is the relationship between an ATM virtual channel connection and a virtual path connection?
24. What is meant by the ATM classes of service?
25. What are the basic features of digital subscriber line?
26. What is the difference between a symmetric connection and an asymmetric connection?
27. Describe some applications that incorporate computer telephony integration.

EXERCISES

1. The telephone line that connects your house or business to the central office (the local loop) carries your conversation and the conversation of the person to whom you are talking. What do you estimate is the bandwidth of a local loop?
2. If you play a CD for a friend over the telephone, will the friend hear high quality music? If not, why not?
3. If you place a telephone call and it leaves your LATA and enters another LATA, what kind of telephone call have you placed? What kind of telephone company handles this telephone call?
4. For each of the following scenarios, state whether a telephone line or a trunk should be used:
   a. The connection from your home to the local telephone company
   b. The connection between a large company's PBX and the telephone company
   c. The connection between two central offices
5. How many different area codes are possible? Show your reasoning.
6. In any one area code, how many different telephone numbers are possible? Show your reasoning.
7. You want to start your own local telephone company. Do you have to install your own telephone lines to each house and business? Explain.
8. Why would anyone want to use an ISDN line to connect to the Internet when a modem over a POTS telephone line seems to work just fine? List as many advantages as possible. Then list the potential disadvantages.
9. If you have a frame relay service installed, do you just pick up the telephone and dial the number of the party you want to connect to? Explain.
10. Suppose you want to have a frame relay connection between your Chicago office and your New York office. Itemize the different charges that you will have to pay for this connection.
11. You have established a frame relay connection with a committed information rate of 256 Kbps and a burst rate of 128 Kbps. Several times a day your computer systems transmit in excess of 512 Kbps. What will happen to your data?

12. For each of the following activities, state which is the better transmission medium, frame relay or the Internet:
    a. sending e-mail
    b. sending high speed data
    c. interactive voice communications
    d. receiving a live video stream
    e. participating in a chat room or newsgroup

13. One of the disadvantages of ATM is the "53-byte cell tax." Explain what the 53-byte cell tax means.

14. You have an ATM connection that goes from your location to a network entity, on to a second network entity, and then to your desired destination. Draw a simple sketch that shows each virtual channel connection.

15. State which ATM class of service would best support each of the following applications:
    a. e-mail with image attachments
    b. interactive video
    c. simple text e-mail
    d. voice conversation

16. What do you foresee as the best application use for digital subscriber line? Explain.

17. What is the main advantage of asymmetric DSL over symmetric DSL?

18. Using CTI, how can a local area network support telephone operations?

THINKING OUTSIDE THE BOX

1 A company wants to connect two offices located in Memphis, Tennessee and Laramie, Wyoming. The offices need to transfer data at 512 Kbps. Which is the less expensive solution: a T1 connection or a frame relay connection? What happens if an office in Baton Rouge, Louisiana must also be connected to both Memphis and Laramie and at 512 Kbps. Which solution is cheaper now?

2 You are consulting for a hospital that wants to send three-dimensional high resolution color ultrasound images between an out-patient clinic and the main hospital. A patient will be receiving the ultrasound treatment at the clinic while a doctor at the hospital observes in real-time and is talking to the ultrasound technician and patient via a telephone connection. What type of telecommunication service presented in this chapter might support this application? State the telecommunication service and be sure to list the particular details concerning the service. Explain your reasoning.

3 Your company wants to create an application that allows employees to dial in from a remote location and, using a single connection, access their voice mail, e-mail, and data files. What kind of system would allow this? Describe the necessary hardware and software components.

PROJECTS

1. What kind of hardware is needed to support a T1 connection to your business?

2. Go to the web page www.telcoexchange.com and enter your home or business telephone number. What kind of telephone services are available at that location?

3. What newly emerging telecommunication technology or technologies may replace the technologies introduced in this chapter? Give a brief description of each technology.

4. Using the Bringbring Corporation example, what further steps will be needed and what will be the resulting costs, to include:
   a. E-mail access for all locations
   b. Internet server capabilities for the Chicago location.

5. If you are interested in more technical details of DSL, investigate the different modulation techniques that are used to transmit DSL signals. Although these techniques are quite complex, they are an interesting study in the technological advances necessary to provide high speed data streams over ordinary copper twisted pair. Using either the Internet or hardcopy sources (they have to be fairly recent), look up the following four technologies. All four are being used somewhere on DSL circuits:

   ▶ Discrete Multitone Technology (DMT)
   ▶ Carrierless Amplitude Modulation (CAP)
   ▶ Multiple Virtual Line (MVL)
   ▶ Echo Cancellation

13

Network Security

◆◆

AS THE INTERNET becomes more popular and wide-spread, it opens the door to more attempts at hacking and more creative attempts at hacking. Since virtually anyone can access the Internet from any location, web sites can be particularly vulnerable targets for individuals bent on destruction. In the early part of 2000, one or more hackers managed to use a denial of service attack to cripple a number of very popular e-commerce web sites, including eBay, Amazon, and e*Trade. This feat was accomplished by sending bogus messages to a compromised set of sites, each of which then sent a flurry of messages to the targeted popular web sites. When the popular web sites received the avalanche of messages, they were overwhelmed resulting in their inability to keep up with valid messages. Thus individuals trying to conduct legitimate business at the popular sites were blocked (denial of service). The sites were shut down for hours, and the businesses lost significant amounts of income.

No doubt, one or more hackers are responsible for committing this attack and should be held responsible for their actions, but some experts are wondering if the fault lies partially with the Internet's inability to adequately protect itself and its legitimate users. The creators of the Internet did not envision the world-wide acceptance and stellar growth that have been its hallmarks for a number of years. Few, if any, security techniques were incorporated into the early versions of the Internet.

Even if a corporation is not connected to the Internet, it is still vulnerable to attack from the inside or through dial-up access. Although most corporate networks include some kind of security, as hackers become more inventive and practiced, better security techniques will become a necessity.

What different types of network security exist today?

If the Internet is not capable of protecting itself, is there something you can do to protect yourself while using the Internet?

Objectives ▶

After reading this chapter, you should be able to:

▶ Recognize the concepts of basic security measures.

▶ Cite the strengths and weaknesses of passwords.

▶ Explain the difference between a substitution-based cipher and a transposition-based cipher.

▶ Outline the basic features of public key cryptography.

▶ List the three basic techniques used to discover and eliminate viruses.

▶ Recognize the basic forms of denial of service attacks.

▶ Discuss the advantages and uses of digital certificates.

▶ Cite the uses and basic operating characteristics of the Data Encryption Standard.

▶ List the characteristics and common applications areas of public key infrastructure.

▶ Explain the basic characteristics of a certificate and a certificate authority.

▶ Recognize the importance of a firewall and be able to describe the two basic types of firewall protection.

Computer network security has reached a point at which it can best be characterized by two seemingly conflicting statements: Never has network security been better than it is today; and never have computer networks been more vulnerable than they are today. How both these statements can be true is an interesting paradox. Network security, as well as operating system security, has come a long way from the early days of computers. During the 1950s and 1960s, security amounted to trust—one computer user trusted another computer user not to destroy each other's data files. Modern computer systems have grown in complexity, supporting many simultaneous users with increasingly demanding requests such as database retrievals and graphic-intensive web page downloads. To protect users from one another, computer system security also had to increase.

Today, the Internet allows anyone worldwide to access or attempt to access any computer system that is connected to the Internet. This interconnectivity between computer systems and networks is both a boon and a bane. It allows us to download web pages from Europe and order toys for the kids (or ourselves) from electronic toy stores, but it also opens up all Internet-attached systems to invasion. And unfortunately, some users have a single goal in mind: to access forbidden systems and steal or destroy anything they can get their hands on.

Internet systems alone are not the only systems experiencing security problems. Any system with dial-in capabilities is also open for vandalism. Likewise, any corporate office center or educational facility is a candidate for someone walking in and stealing or destroying computer files. Even building a 30 foot wall and a moat around your company and severing all connections to the outside world will not create a secure environment. Many studies show that employees working at the company are responsible for a majority of business thefts. Carrying a floppy disk home in a briefcase or pocket is a very convenient way to remove data files from a corporate computer network. In this environment, computer network security is an all-encompassing, never ending job.

This chapter's discussion of network security begins by examining the basic security measures that all systems need to use, and is followed by an introduction to the basic concepts of encryption, decryption, and public key technology. Using the concept of public keys, we can then take a look at digital signatures and public key infrastructure. The chapter concludes with an introduction to firewall systems and the basic principles of network security design.

Basic Security Measures

All computer systems need a set of basic security measures. Whether the system is a simple personal computer in your home or a major computer network such as the Internet, it is necessary to protect the hardware and software from theft, destruction, and malicious acts of vandalism. Security measures can be as simple as locking the door or as advanced as applying virtually unbreakable encryption techniques to data. Let's break the basic security measures into eight categories: external security, operational security, surveillance, passwords, auditing, access rights, standard system attacks, and viruses.

External security

External security of a computer system or a computer network consists of protecting the equipment from physical damage. Examples of physical damage include fire, floods, earthquakes, power surges, and vandalism. Common sense damage prevention techniques are usually enough in many cases of external security. Rooms containing computer equipment should always be locked. Unauthorized persons should not be allowed into rooms containing computing equipment. Cabling, and the devices that cables plug into, should not be exposed if at all possible.

If the equipment needs to be in the open for public access, the equipment should be locked down. Many kinds of anti-theft devices exist for locking cabinets, locking cables to cabinets, locking down keyboards, and locking peripheral devices. For example, one manufacturer makes a device that transmits a wireless signal to a pager should a computer cabinet be opened. The person carrying the pager will know immediately which cabinet is being opened so that security can be sent to the appropriate location.

It is also fairly common knowledge not to place expensive computer systems in the basements of buildings. Basements can flood and are often high humidity locations. Rooms with a large number of external windows are also not advisable. Windows can let in sunshine, which can increase the temperature of a room. Computer equipment typically heats up a windowless room. With the addition of sunlight, the increase in temperature may strain the capacity of any existing air conditioning equipment. As temperatures rise, the life expectancy of computer circuits decreases. Also, external windows can increase the probability of vandalism.

To prevent electrical damage to computing equipment, high-quality surge protectors should be used on all devices that require electrical current. The electrical circuits that provide power to devices should be large enough to adequately support the device without placing a strain on the electrical system. Electrical devices that power up and down causing power fluctuations, such as large motors, should be on circuits separate from the computer devices. Finally, devices that are susceptible to damage from static electricity discharges should be properly grounded.

Operational security

Operational security of a computer network involves deciding, and then limiting, who can use the system and when they can use the system. Consider, for example, a large corporation in which there are many levels of employees with varying job descriptions. Employees who do not normally come in contact with sensitive data areas should not have access to sensitive data. For example, if an employee simply performs data entry operations, more than likely he or she should not be allowed access to payroll information. Likewise, employees working in payroll need access to the payroll database, but more than likely do not need access to information regarding corporate research programs. A manager of an area would probably have access to much information in his or her department, but his or her access to information in other departments would likely be limited. Finally, top-level executives often have access to a wide range of information within a company. However, many companies even limit information access to top-level management.

Local area networks and database systems provide much flexibility in assigning access rights to individuals or groups of individuals, as you will see shortly. Computer network specialists, along with database administrators and someone at the top levels of management such as the Chief Information Officer (CIO), often decide how to break the company into information access groups, decide who is in each group, and determine what access rights each group has. As you might recall from Chapter Eleven, network operating systems, such as Novell's NetWare and Windows NT, are very good applications for creating workgroups and assigning rights.

It is also possible to limit access to a system by the time of day or the day of the week. If the primary activity in one part of your business is accessing personnel records, and this activity is only performed during working hours by employees in the personnel or human resources department, then it might be reasonable to disable access to personnel records after working hours, such as from 5:30 p.m. until 7:00 a.m. the next morning. Likewise, the network administrator could also deny access to this system on weekends. Figure 13-1 shows an example of a dialog box that is used to set limits within a network operating system.

Figure 13-1
Sample dialog box from a network operating system for setting time of day restrictions

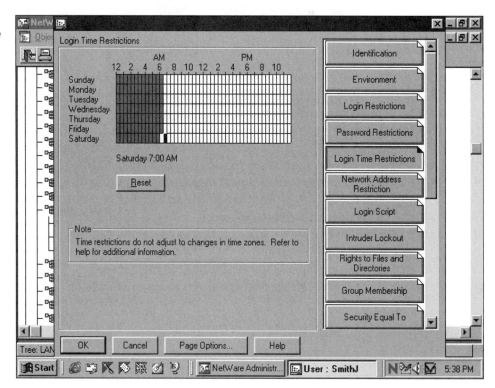

It may also be wise to limit remote access to a system during certain times of the day or week. With appropriate limits set, someone dialing in at 2:30 a.m. to transfer funds from one account to another may signal an illegal activity. If all corporate fund transfers can only occur during typical business hours, this restriction would be reasonable to place on dial-in activity.

Surveillance

Although many individuals feel surveillance is an intrusion into an individual's privacy, many network administrators feel it is a good deterrent to computer vandalism and theft. The proper placement of video cameras in key locations can both deter criminals and be used to identify criminals in the event of vandalism or theft.

There are, however, other forms of surveillance in addition to capturing live action with a video camera. For example, placing a transmitter in each computer that sends a signal to a pager if the computer cabinet is opened, is a wireless form of surveillance. Using a form of surveillance called **intrusion detection**, many companies electronically monitor data flow and system requests into and out of a system. If unusual activity is noticed, protective action can be taken immediately. Intrusion detection is a growing field of study in network security. Companies that accept merchandise orders using the telephone often monitor each telephone call. Companies claim this form of surveillance can improve the quality of customer service and help settle future disputes.

Passwords and ID systems

Almost every system that stores sensitive or confidential data requires an authorized user to enter a password, personal identification number (PIN), or some other form of ID before gaining access to the system. Typically, this password or ID is something either remembered by the user or a physical feature of a user, such as a finger print. Technology in this area is improving rapidly as companies try to incorporate systems that are less vulnerable to fraud.

Perhaps the most common form of protection from unauthorized use of a computer system is the password. Anyone accessing a computer system, banking system, or a long distance telephone system is required to enter a password or PIN. Some common examples of systems that require passwords (or PINs) are:

- national online computer accounts;
- computer network and mainframe computer access at work and school;
- long distance telephone credit card use;
- twenty-four hour automatic banking machines;
- access to retirement accounts and banking services;
- access to email and voicemail systems; and
- access to Internet web sites at which a customer profile is created and stored for future transactions.

Although the password is the most common form of identification, it is also one of the weakest. Too often passwords become known, or "misplaced," and fall into the wrong hands. Occasionally a password is written on paper, and the paper is discovered by the wrong people. More often, however, the password is too simple and someone else guesses it. Standard rules that an individual should follow when creating or changing a password include:

- Change your password often.
- Pick a good password by using at least eight characters, mixing upper and lower case if the computer system is case sensitive, and mixing letters with numbers.
- Don't choose passwords that are similar to first or last names, pet names, car names, or other choices that can be easily guessed.
- Don't share your password with others; doing so invites trouble and misuse.

Figure 13-2 shows how the Novell NetWare operating system allows the network administrator to require a user to use a password, give the user the ability to change his or her own password, and require the user to select a password of a particular size.

Figure 13-2
Controlling a user password with Novell NetWare

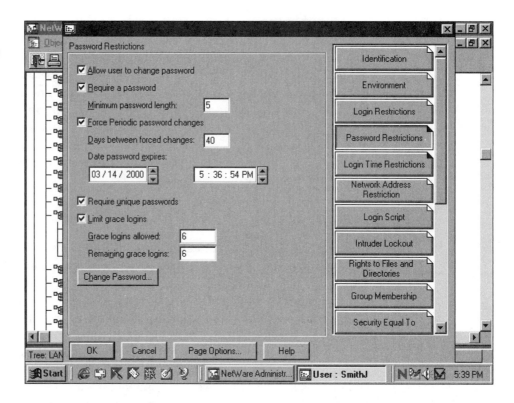

test

Why can't the Admin tell you your password?

test
What Biometric technique does England use?
– EARprint

Some computer systems generate random passwords that are very difficult to guess, but are also hard to remember. Often, the user who is given a randomly generated password either changes it to something simpler, making it easier to guess, or writes it down on a piece of paper, defeating the whole purpose of a secret password. Some systems also disallow obvious passwords or already used passwords, thus requiring the user to be creative and select a password that is difficult to guess.

A common fallacy among computer system users is that the internal operating system file that stores the logon IDs and passwords is susceptible to intrusion. Interestingly, most computer systems store passwords in an encrypted form for which there is no known decryption. How then does the system know when you have entered the correct password? When a user enters his or her login ID and password, the password is encrypted and compared with the entry in the encrypted password file. If the two encrypted passwords match, the logon is allowed. Anyone who gets access to this encrypted password file will discover only unreadable text. This encryption technique explains why, when you forget your password, a computer operator cannot simply read a file and tell you what it is. The computer operator can only reset the password to something new.

Since there are so many weaknesses to the password, other forms of identification have emerged. Biometric techniques that scan something about the user, such as voiceprints, fingerprints, eyeprints, and faceprints, appear to be the wave of the future. For example, England has a large database of earprints. Research shows that no two ears are the same, thus an earprint is useful in helping to identify an individual. Fingerprints have long been used to distinguish one individual from another. Now, desktop devices the size of a computer mouse exist that can scan a thumbprint and allow or disallow access to a computer system. Likewise, systems exist that can record and digitize your voice. The digital voice pattern is compared to a stored sample and the software determines if the match is close enough for validation. Although retina scans (the retina is the inside back lining of the eye) have been portrayed in movies and exist in the real world as security techniques, another eye feature that is unique among all individuals is the iris, or the colored portion of the eye. Some new security devices use the iris to identify people who are allowed to access a system. Other research has been aimed at digitizing the features of the entire face and comparing this digital representation with a stored image. Companies that manufacture twenty-four hour automatic teller machines are interested in replacing the credit card-sized automatic teller machine (ATM) card and corresponding PIN with something that cannot be stolen, such as a fingerprint, faceprint, or eyeprint.

Auditing

Auditing a computer system is often a good deterrent to crime and is useful in apprehending a criminal after a crime has occurred. **Computer auditing** usually involves a software program that monitors every transaction within a system. As each transaction occurs, it is recorded into an electronic log along with the date, time, and owner of the transaction. If an inappropriate transaction is suspected, the electronic log is scanned and the appropriate information is retrieved. In a classic computer crime case that was thwarted because of auditing, a New York man discovered that invoices for an amount under $500 that were sent to local government agencies would be routinely paid without requesting further details. It wasn't until many months later that a lawyer examining the computer audit trail of payments noticed the pattern of checks under $500 sent to the same individual.

Figure 13-3 shows an example of an audit log from the Windows NT operating system's Event Viewer window. Note that each transaction includes the date, time, and source of the event that is recorded. Typical recorded events include the failure of a driver or other system components to load during system startup, any possible breaches in security, and any program that might record an error while trying to perform a file operation.

Figure 13-3
Windows NT Event Viewer example

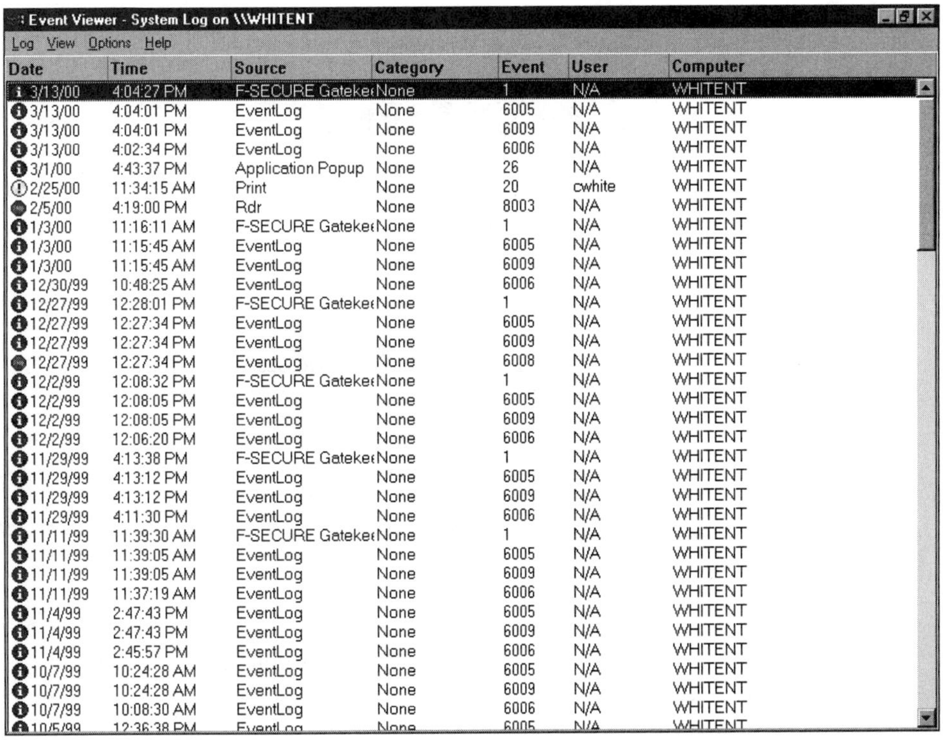

Date	Time	Source	Category	Event	User	Computer
3/13/00	4:04:27 PM	F-SECURE Gatekee	None	1	N/A	WHITENT
3/13/00	4:04:01 PM	EventLog	None	6005	N/A	WHITENT
3/13/00	4:04:01 PM	EventLog	None	6009	N/A	WHITENT
3/13/00	4:02:34 PM	EventLog	None	6006	N/A	WHITENT
3/1/00	4:43:37 PM	Application Popup	None	26	N/A	WHITENT
2/25/00	11:34:15 AM	Print	None	20	cwhite	WHITENT
2/5/00	4:19:00 PM	Rdr	None	8003	N/A	WHITENT
1/3/00	11:16:11 AM	F-SECURE Gatekee	None	1	N/A	WHITENT
1/3/00	11:15:45 AM	EventLog	None	6005	N/A	WHITENT
1/3/00	11:15:45 AM	EventLog	None	6009	N/A	WHITENT
12/30/99	10:48:25 AM	EventLog	None	6006	N/A	WHITENT
12/27/99	12:28:01 PM	F-SECURE Gatekee	None	1	N/A	WHITENT
12/27/99	12:27:34 PM	EventLog	None	6005	N/A	WHITENT
12/27/99	12:27:34 PM	EventLog	None	6009	N/A	WHITENT
12/27/99	12:27:34 PM	EventLog	None	6008	N/A	WHITENT
12/2/99	12:08:32 PM	F-SECURE Gatekee	None	1	N/A	WHITENT
12/2/99	12:08:05 PM	EventLog	None	6005	N/A	WHITENT
12/2/99	12:08:05 PM	EventLog	None	6009	N/A	WHITENT
12/2/99	12:06:20 PM	EventLog	None	6006	N/A	WHITENT
11/29/99	4:13:38 PM	F-SECURE Gatekee	None	1	N/A	WHITENT
11/29/99	4:13:12 PM	EventLog	None	6005	N/A	WHITENT
11/29/99	4:13:12 PM	EventLog	None	6009	N/A	WHITENT
11/29/99	4:11:30 PM	EventLog	None	6006	N/A	WHITENT
11/11/99	11:39:30 AM	F-SECURE Gatekee	None	1	N/A	WHITENT
11/11/99	11:39:05 AM	EventLog	None	6005	N/A	WHITENT
11/11/99	11:39:05 AM	EventLog	None	6009	N/A	WHITENT
11/11/99	11:37:19 AM	EventLog	None	6006	N/A	WHITENT
11/4/99	2:47:43 PM	EventLog	None	6005	N/A	WHITENT
11/4/99	2:47:43 PM	EventLog	None	6009	N/A	WHITENT
11/4/99	2:45:57 PM	EventLog	None	6006	N/A	WHITENT
10/7/99	10:24:28 AM	EventLog	None	6005	N/A	WHITENT
10/7/99	10:24:28 AM	EventLog	None	6009	N/A	WHITENT
10/7/99	10:08:30 AM	EventLog	None	6006	N/A	WHITENT
10/5/99	12:36:38 PM	EventLog	None	6005	N/A	WHITENT

Many good computer programs that can audit all transactions on a computer system are available. The price paid to purchase, install, and support an audit program can be well worth it if the program can help catch a person performing unauthorized transactions.

Access rights

Modern computer systems and computer networks allow multiple users to access resources such as files, tapes, printers, and other peripheral devices. Many times, however, the various resources are not supposed to be shared, or they should be shared only by a select group. If resource sharing is to be restricted, then a user or network administrator should set the appropriate access rights for a particular resource. Most access rights have two parameters: *who* and *how*. The *who* parameter lists who has access rights to the resource. Typical examples of *who* include the owner, a select group of users, and the entire user population. The *how* parameter can specify how a user may access the resource. Typical examples include read, write, edit, execute, append (add data to the end), and print. For example, a user may create a file and allow all users to access the file, but only with read access rights. The usual system defaults grant the owner of the file full read, write, and execute access so that the user can modify or delete the file at any time. Figure 13-4 shows an example of how Novell's NetWare assigns access rights to a resource. The network administrator can assign supervisor, read, write, create, erase, modify, file scan, and access control rights to a particular user.

Figure 13-4
Novell NetWare assigning access rights to a resource

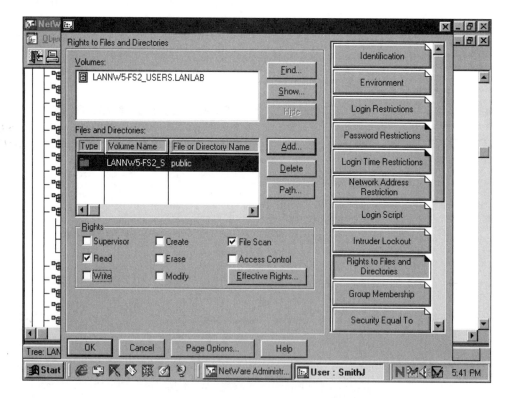

Modern network operating systems allow network administrators to create workgroups. These workgroups are defined by the network administrator and can contain any form of user grouping as desired. For example, one workgroup might consist of all the employees from marketing and engineering that are currently working on a particular project. Once the workgroup is defined, it is then possible to assign a unique set of access rights for this workgroup.

Guarding Against Viruses

A computer **virus** is a small program that attaches itself to another program or computer file and when loaded or executed causes problems with a computer. These problems can consist of deleting and corrupting data and program files, or altering operating system components so that computer operation is impaired or even halted. A computer that has been invaded by a virus program is said to be infected. Since a virus is a computer program, it has a binary pattern that is recognizable. Early model virus scanners from the 1980s looked for the unique bit pattern of a virus. More recent virus scanners watch for the actions of a virus, such as unusual file changes or directory activity. Some common types of viruses include:

- **Parasitic virus** The most common type of virus, a parasitic virus attaches itself to files and then replicates itself once it is in the host computer's memory.

- **Boot sector virus** A boost sector virus is stored on a floppy disk. When the disk is loaded into a new machine, the virus moves from the disk into the host system.

- **Stealth virus** A stealth virus hides itself from antivirus software by assuming a binary form that is not detectable by a virus scanner that is looking for unique binary patterns.

- **Polymorphic virus** These viruses mutate with every infection, thus making them difficult to locate.

- **Macro virus** Macros are found in spreadsheets and word processing documents. A macro virus is one that hides within an application's macro and is activated when the macro is executed.

A worm is similar to a computer virus. A computer **worm** is a special type of virus that manages to store itself in a location that can cause the most damage to the resident computer system. Worms usually replicate themselves by transferring from computer to computer via e-mail. Typically a virus or a worm is transported in a Trojan horse. In computer terminology, a **Trojan horse** is a destructive piece of code that hides inside a harmless looking piece of code, such as an e-mail or an application macro.

To guard against viruses, you can purchase antivirus software that typically checks all of your files periodically and can remove any viruses that are found. As files and e-mail are downloaded and as applications are opened, you may get a message warning of a new virus. To provide effective protection against virus attacks, you need to use antivirus software that includes signature-based scanning, terminate-stay-resident monitoring, and multi-level generic scanning.

The **signature-based scanner** works by recognizing the unique pattern of a virus. All viruses have a unique bit pattern, much like a strand of DNA. Antivirus product developers and virus researchers catalog the known viruses and their signatures. Signature-based scanners then use these catalog listings to search for viruses on a user's computer system. Because new viruses are created daily, it is necessary for a user to update the known virus catalog frequently. This updating is usually accomplished by downloading new virus information from the Internet.

Stealth viruses and polymorphic viruses are designed to elude the detection of signature-based scanners. For these types of viruses, a second antivirus technique is necessary—terminate-and-stay-resident antivirus software. **Terminate-and-stay-resident antivirus software** runs in the background while an application that a user is executing runs in the foreground. Terminate-and-stay resident programs can provide a combination of protective services, including real-time monitoring of disk drives and files, intelligent analysis of virus-like behavior, and stealth and polymorphic virus detection. An advantage of terminate-and-stay-resident antivirus software is its automatic nature—a user does not have to activate the software each time a new file is opened or downloaded. One disadvantage of this software is that an always-running virus checker consumes memory and processing resources. An additional disadvantage is that this type of antivirus software can create false alarms. After a number of false alarms, users often disable the terminate-and-stay resident antivirus software.

The third antivirus technique that is used to assist scanners and terminate-and-stay resident software is **multi-level generic scanning**. The technique is a combination of antivirus techniques including intelligent checksum analysis and expert system virus analysis. Intelligent checksum analysis calculates and applies a checksum to a file and data set at two major times in a file's lifetime—at the beginning when a file is new (or in a known safe state) and at a later time after the file has existed for awhile. Much like cyclic checksum, the checksum from the file when new is compared with the later checksum to determine if the file has changed due to the actions of a virus. Expert system virus analysis involves a series of proprietary algorithms that perform millions of tests on your software and examine the flow of program code and other software functions. The antivirus software then assigns a number of points to the software in question based on the results of these tests, and indicates a virus if a certain point score is achieved.

Standard System Attacks

Malicious computer users who try to break into a computer system often start with a standard set of system attacks. They hope that the system administrator has not properly secured the system and has left it vulnerable to attack. One category of common system attacks is denial of service. **Denial of service attacks** bombard a computer site with so many messages that the site is incapable of performing its normal duties. Some common types of denial of service include e-mail bombing, smurfing, and ping storm. In **e-mail bombing**, a user sends an excessive amount of unwanted e-mail to someone. If the e-mail has a return address of someone other than the person sending the e-mail, then the sender is **spoofing**. Test

Smurfing is the name of a particularly nasty automated program that attacks a network by exploiting Internet Protocol (IP) broadcast addressing and other aspects of Internet operation. Simply stated, an attacker sends a packet to an innocent third party—the "smurf amplifier"—who then multiplies it hundreds or thousands of times and sends the packets on to the intended victim.

Ping storm is a condition in which the Internet ping program is used to send a flood of packets to a server to make the server inoperable. Ping is used to verify if a particular IP address of a host exists and to see if the particular host is currently available. Support for ping is most commonly found in UNIX-based systems, which are very popular systems used to support Internet web servers.

As an example of spoofing, smurfing and Ping storms, let's consider the denial of service attacks that hit a number of commercial web sites in early February 2000. The sites attacked included popular sites such as Yahoo, eBay, Amazon, CNN, and E*Trade. In one type of attack, a hacker somehow takes control of a number of servers on the Internet and instructs them to make contact with a particular web server (the server the hacker wishes to disable). The compromised servers send TCP/IP SYN packets (synchronize/initialization) with bogus source IP addresses (*spoofing*) to the intended target. As each SYN packet arrives, the target server tries to return a legitimate response, but cannot because of the bogus IP address (similar to a *Ping storm*). While the target server waits for a response from the bogus IP address, resources are consumed as more bogus SYN commands arrive.

In a second example, the hacker instructs a number of compromised web servers to send packets to a second set of servers (Figure 13-5). These packets contain the destination address of the target web site. The second set of servers, called bounce sites, receives multiple spoofed requests and responds by sending multiple packets to the target web site at the same time. The target site is then overwhelmed and essentially crippled by the incoming flood of illegitimate packets, leaving no room for legitimate packets.

Figure 13-5
An example of smurfing intended to cripple a web server

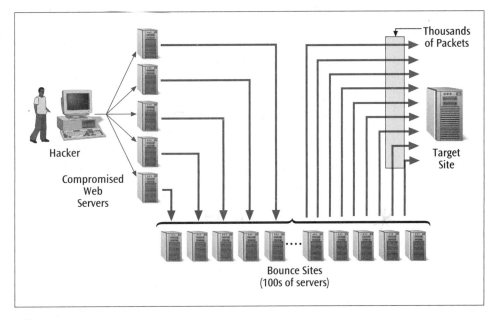

Another common attack, which is not considered a denial of service attack, is the line disconnect attack. With a **line disconnect** attack, the perpetrator attempts to gain access to a user's job after a line disconnect—when a user instructs his or her workstation to disconnect from a dial-up application, but before the system acknowledges the disconnect.

Also popular with hackers is the Trojan Horse attack in which a malicious piece of code is disguised within a seemingly harmless piece of code, such as an e-mail attachment. Once the user reads the e-mail and opens the attachment, the malicious code is released and the damage occurs.

Hackers will also try to steal passwords by guessing simple combinations or eavesdropping on transmissions in which a password is transmitted. Some hackers will even go so far as to create a bogus application that appears to be a legitimate application and prompts users for an ID and password. Once an individual enters his or her ID and password, the software prints some form of message giving the appearance of system failure. The user moves on, not knowing that he or she has just given his or her ID and password to a bogus program.

Professionals who support computer networks, as well as average computer users who have a computer at home or work, should be aware of these common attacks so that they may best protect their systems from vandalism and intrusion.

Basic Encryption and Decryption Techniques

Many times when transferring data from one point to another in a computer network, it is necessary to insure that the transmission is secure from anyone eavesdropping on the line. The term secure means two things. First, it should not be possible for someone to intercept and copy an existing transmission. Second, it should not be possible for someone to insert false information into an existing transmission. Financial transactions and military transmissions are two good examples of data that should be secure during transmission.

Fiber optic cable represented a major improvement in providing secure transmission media for sensitive data. Metal conducted media (twisted pair and coaxial cable) transmitting electrical signals are relatively easy to tap, but fiber optic wire transmitting non-electromagnetic pulses of light are virtually impossible to tap. To illegally tap a fiber optic system you must physically break the fiber optic cable, which is immediately noticed.

No matter how secure the media however, sensitive data also moves through computers, is stored on hard disk drives, and is often transmitted over standard telephone systems. Under these conditions sensitive data is susceptible to interception and additional security measures are necessary. One such additional measure is to encrypt the data before it is transmitted using encryption software, transmit the encrypted data over secure media, then decrypt the received data to obtain the original information. **Cryptography** is the study of creating and using encryption and decryption techniques. Many volumes can be filled on cryptographic techniques, but a knowledge of the basic principles is sufficient to give you an understanding of today's encryption techniques.

Delving into the topic of cryptography requires understanding a few basic terms. **Plaintext** (always shown in lower case characters in the examples) is data before any encryption has been performed (Figure 13-6). An **encryption algorithm** is the computer program that converts plaintext into an enciphered form. The **ciphertext** (shown as upper case characters) is the data after the encryption algorithm has been applied. A **key** is the unique piece of information that is used to create ciphertext and then decrypt the ciphertext back into plaintext. After the ciphertext is created, it is transmitted to the receiver, where the ciphertext data is decrypted.

Figure 13-6
Basic encryption and decryption procedure

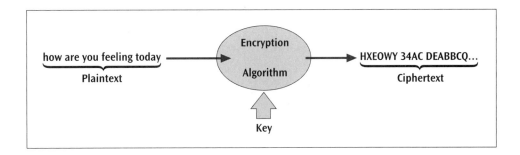

Early cryptography algorithms used the same key for both encryption and decryption. Experts, however, were greatly concerned about the use of a single key. To be able to send and receive encrypted data between local and remote locations, the key had to be given to both local and remote parties. If this key was intercepted and fell into the wrong hands, encrypted data could be decrypted, and false data could be encrypted and sent to either party. Newer techniques, as you will soon see, solve this problem by allowing the use of two different but mathematically related keys. One key is for encrypting the data, and the second key is for decrypting the data.

Monoalphabetic substitution-based ciphers

Despite its daunting name, monoalphabetic substitution-based cipher is actually a fairly simple encryption technique. A **monoalphabetic substitution-based cipher** replaces a character or group of characters with a different character or group of characters. Consider the following simple example. Each letter in the plaintext row maps onto the letter below it in the ciphertext row.

```
Plaintext:  a b c d e f g h i j k l m n o p q r s t u v w x y z
Ciphertext: P O I U Y T R E W Q L K J H G F D S A M N B V C X Z
```

The ciphertext chosen is simply the letters on a keyboard, scanning right to left, top to bottom. To send a message using this encoding scheme, each plaintext letter of the message is replaced with the ciphertext character directly below it. Thus, the message

how about lunch at noon

would encode to

EGVPO GNMKN HIEPM HGGH

A space has been placed after every five ciphertext characters to help disguise obvious patterns. This example is monoalphabetic since one alphabetic string is used to encode the plaintext. It is a substitution cipher because one character of ciphertext is substituted for one character of plaintext.

Polyalphabetic substitution-based ciphers

The **polyalphabetic substitution-based cipher** is similar to the monoalphabetic cipher, except it uses multiple alphabetic strings to encode the plaintext rather than one alphabetic string. Possibly the earliest example of a polyalphabetic cipher is the Vigenére Cipher created by Blaise de Vigenére in 1586. A 26 x 26 matrix of characters is created, as shown in Table 13-1.

Table 13-1
Vigenére Cipher 26 x 26 ciphertext character matrix

Row	Letters																									
A	B	C	D	E	F	G	H	I	J	K	L	M	N	O	P	Q	R	S	T	U	V	W	X	Y	Z
B	C	D	E	F	G	H	I	J	K	L	M	N	O	P	Q	R	S	T	U	V	W	X	Y	Z	A
C	D	E	F	G	H	I	J	K	L	M	N	O	P	Q	R	S	T	U	V	W	X	Y	Z	A	B
...																									

A key is chosen, such as COMPUTER SCIENCE, which is repeatedly placed over the plaintext message. For example:

Key: COMPUTERSCIENCECOMPUTERSCIENCECOMPUTERSCIENCECO
Plaintext: thisclassondatacommunicationsisthebestclassever

To encode the message, look at the first letter of the plaintext, *t*, and the corresponding key character immediately above it, *C*. The *C* tells us to use row C of our 26 x 26 matrix to perform the alphabetic substitution for the plaintext character *t*. We then go to column T in row C and find the ciphertext character V. This process continues for every character of the plaintext. The key, COMPUTER SCIENCE, must be kept secret between the encoder and decoder.

To make matters more difficult for an intruder, the standard 26 x 26 matrix with row A, row B, row C, and so on, as described above, does not have to be used. However, if a unique matrix is chosen for the encoding and decoding, it too must remain a secret along with the key.

Transposition-based ciphers

A transposition-based cipher is different from a substitution-based cipher in that the order of the plaintext is not preserved, as it is in substitution-based ciphers. By rearranging the order of the plaintext characters, common patterns become unclear and the code is much more difficult to break. Let's consider a simple example of a transposition cipher. Choose a keyword that contains no duplicate letters, such as COMPUTER. Over each letter in the keyword, write the position of each letter as the letter is encountered in the alphabet. For the keyword COMPUTER, C appears first in the alphabet, E is second, M is third, O is fourth, and so on.

```
1 4 3 5 8 7 2 6
C O M P U T E R
```

Take a plaintext message, such as "this is the best class i have ever taken" and write it under the keyword in consecutive rows going from left to right.

```
1 4 3 5 8 7 2 6
C O M P U T E R
t h i s i s t h
e b e s t c l a
s s i h a v e e
v e r t a k e n
```

To encode the message, read down each column starting with column 1 and proceeding through column 8. Reading column 1 gives us TESV and column 2 gives us TLEE. Encoding all eight columns gives us the following message:

TESVTLEEIEIRHBSESSHTHAENSCVKITAA

Two interesting observations can be made about this example. First, the choice of the keyword is once again very important, and care must be taken so that the keyword does not fall into the wrong hands. Second, you could make the encryption even more difficult by performing an additional substitution-based cipher on the result of the transposition cipher. In fact, why stop there? You could create a very difficult code if you repeated various patterns of substitution- and transposition-based ciphers, one after another.

Public key cryptography and secure sockets layer

All the encoding and decoding techniques shown thus far depend on protecting the key and keeping it from falling into the hands of an intruder. Yet as important as key secrecy is, it is surprising how often keyword or password security is lax or altogether non-existent. Consider the episode of Stanley Mark Rifkin and the Security National Bank. Posing as a bank employee, Stanley gained access to the wire funds transfer room, found the password to transfer electronic funds taped to the wall above a computer terminal, and transferred $12 million to his personal account. He was later apprehended while bragging in a bar.

One technique to protect a key from an intruder is often seen in late-night black and white spy movies: break the key into multiple pieces and assign each piece to a different individual. Rather than simply assign one or two characters to each person, mathematical techniques, such as simultaneous linear equations, are used to divide the key into parts. Other techniques that require a manipulation of the key or masking the key so as to not make it obvious, can be tedious and sometimes do not add any real security to the key.

One of the problems with protecting a key is that only one key is used to both encode and decode the message. The more people who have possession of the key, the more chances that someone will get sloppy and let the key become known to unauthorized personnel. But what if two keys are involved—one key is public, and the second key is private? Data encrypted with the public key can only be decoded with the private key, and data encrypted with the private key can only be decoded with the public key. This concept of two keys, public and private, is called **public key cryptography**. When one key encrypts the plaintext and the other key decrypts the ciphertext, even if you have access to one of the keys, it is nearly impossible to deduce the second key from the first key. Furthermore, the encrypted data is also extremely difficult to break using the many techniques that experts use to break codes.

How does public key cryptography work? Consider a business that has headquarters in New York. A branch office in Atlanta wishes to send secure data to the New York office. The New York office sends the public key to Atlanta and keeps the private key locked up safe in New York. The Atlanta office uses the public key to encrypt the data, and sends the encrypted data to New York. Only the New York office can decode the data, since they are the only ones possessing the private key. Even if other parties intercept the transmission of the public key to Atlanta, nothing will be gained because it is not possible to deduce the private key from the public key. Likewise, interception of the encrypted data will lead to nothing since the data can decoded only with the private key.

For a more realistic example, consider a situation in which a person browsing the Web wishes to send secure information (such as a credit card number) to a web server. The user at a workstation enters a secure connection and sends the appropriate request to the server. The server returns a "certificate," which includes the server's public key, and a number of preferred cryptographic algorithms. The user's workstation selects one of the algorithms, generates a set of public and private keys, encrypts the user's ~~private~~ key with the server's public key, and sends the result back to the server. The server, using its private key, decrypts the user's ~~private~~ key. Now both sides have their own private keys, and they both have each other's public key. Data can now be sent between the two stations in a secure fashion.

This technique is a part of the secure sockets layer and can be used by most web browsers and servers when it is necessary to transmit secure data. **Secure sockets layer (SSL)** is an additional layer of software added between the application layer and the transport (TCP) layer that creates a secure connection between sender and receiver. Since Unix operating systems create connections between end points in a network via sockets, the software has the title of secure socket layer. A second technique similar to SSL is Secure-HTTP or S-HTTP, but it has not experienced the widespread success that SSL has.

Data Encryption Standard

The **Data Encryption Standard (DES)** is a commonly employed encryption method used by businesses to send and receive secure transactions. The standard became effective in 1977 and was reaffirmed in 1983, 1988, and 1993. The basic algorithm behind the standard is shown in Figure 13-7. Using blocks of 64-bits of data, each block is subjected to 16 levels or rounds of encryption. The encryption techniques are based upon substitution- and transposition-based ciphers. Each of the 16 levels of encryption can perform a different operation based on the contents of a 56-bit key. The 56-bit key is applied to the DES algorithm, the 64-bit block of data is encrypted in a unique way, and the encrypted data is transmitted to the intended receiver. It is the responsibility of the involved parties to keep this 56-bit key secret and not let it fall into the hands of other parties.

[handwritten margin note at top:] your DigitAl Sig is your PrivAte Key applied to the HAsh value generated to that specific Document

Figure 13-7
The basic operations of the Data Encryption Standard

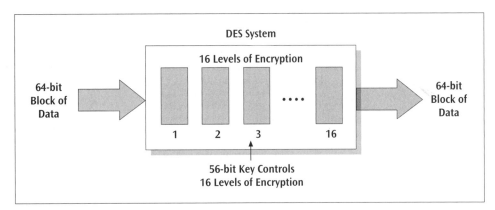

[Figure content:]
DES System

16 Levels of Encryption

64-bit Block of Data → [1] [2] [3] . . . . [16] → 64-bit Block of Data

56-bit Key Controls
16 Levels of Encryption

Even though a 56-bit key provides for over 72 quadrillion possible keys, the Data Encryption Standard was criticized from the beginning for being too weak. Businesses and researchers felt a much larger key should have been used, making it virtually impossible for anyone to break. The government, however, felt that 56 bits was sufficient and that no one would spend the time or resources necessary to try to break a key. Why was the government reluctant to support businesses' concerns about their data security? Perhaps the government's concern was that criminals armed with a key larger than 56 bits could send encrypted materials that no one, including the government, could decrypt thus compromising national security.

In July of 1998, the government was handed a setback when a group called the Electronic Frontier Foundation claimed they cracked a 56-bit key in 56 hours using a single customized personal computer called the DES Cracker. Since that time businesses have increased their pressure on the government to allow stronger 128-bit key encryption. Although many companies use 128-bit keys for domestic transactions, the government has forbidden anything except 56-bit keys for the international transfer of secure documents. At long last, it appears the U.S. government is softening and allowing companies to use techniques better than 56-bit keys.

Digital signatures

Basic encryption and decryption techniques have come a long way since their inception. Today's techniques often employ state-of-the-art encryption and decryption techniques that are based on the earlier basic techniques. Let's examine a number of the more recent advances in cryptography by beginning with the concept of a digital signature.

When participating in financial or legal transactions, people most often identify themselves through a handwritten signature. For example, when you charge something on a credit card, you sign the charge slip, and the salesperson compares the signature on the charge slip with the signature on the back of your charge card. Assuming your card has not been stolen and someone is not posing as you, this system is fairly safe. But what happens when you want to "sign" an electronic document so that later you can prove it is your document and no one else's document? To identify electronic documents as yours, you need to create a digital signature. A **digital signature** uses public key cryptography to assign to a document a code for which you alone have the key.

Digitally signing an electronic document involves sending the document through a complex mathematical computation that generates a large prime number

[handwritten margin notes, left column:]

Rec.

Encrypt message
Private Key

Trans · · · Dig Sig
↳ compute Hash value   User Private Key
Hash value

Transfer 1,000,000
from Acct #123 to Acct #456
↓
Hashing alo.
0123456789 10
↓
encoded with Private Key
↓
9 10 1 2 3 2 1 0
Trans · · · Dig Sig
encrypt with Rec. Public Key =>
encrypt message

[handwritten margin note at bottom:] Digital Sig Does not actually mean that you sign something

Dgital sg.
Is your private key
encryption of ~~~~
the Hash number

called a hash. The original document and the hash are inextricably tied together. If either the document or the hash is altered, the hash will not match the document, and it will not be possible to decode the document.

For a user to digitally sign a document, a hash (the original hash) is generated from a document and then encoded with the user's private key. Any data encoded with the user's private key can only be decoded with the user's public key. If someone wants to verify that this document belongs to this user, the hash is decoded with the user's public key and compared to the original hash. If the two hashes agree, the data was not tampered with and the user's digital signature is valid. Someone could argue that he or she did not *see* the user "sign" the original document, so how do they know that the document is legitimate? This argument is resolved by the fact that only this one user should have this private key. Thus this user is the only person who could have created this precise encoded document.

As you might imagine, if someone discovers the user's private key, a digital signature could be forged. Creating a system that assigns hashes and keys and keeps them secret requires a public key infrastructure, which will be discussed shortly.

Pretty Good Privacy (PGP)

In an effort to create an encryption scheme that could be used by the average person, Philip Zimmermann created encryption software called **Pretty Good Privacy (PGP)**. PGP is very good encryption software that has become the de-facto standard for creating secure e-mail messages and encryption of other types of data files. PGP employs some of the latest techniques of encryption including public key cryptography and digital signatures and is available to anyone in the U.S. for free. It can be downloaded from a number of web sites, FTP sites, and bulletin boards, or can be purchased for commercial installations. Even though the use of PGP and other encryption software is legal in the U.S., it may not be legal in some countries, and it may not be legal to send it from the U.S. to another country. Zimmermann himself ran into much legal trouble because he allowed his PGP software to leave the U.S. in violation of federal export laws that ban the external transportation of encryption software.

PGP can be applied to documents such as e-mail if you wish to send a secure transmission. Since PGP is based on public key encryption techniques, public and private keys are necessary and the receiving workstation must possess the same PGP software to decrypt the message. Because there are many different versions of PGP and various restrictions on their use, you should consult one or more of the popular web sites devoted to supporting PGP and learn about the particular details.

Kerberos

Kerberos is an authentication protocol designed to work on client/server networks and uses secret key cryptography. Unlike public key cryptography in which there are two keys—a public key and a private key—Kerberos uses an encryption key and a decryption key, which are identical. If a client requests a service from a server, and the server wants assurance that the client is who he or she says he or she is, Kerberos can provide the authentication.

To provide this level of authentication, a client presents a *ticket* that is issued by the Kerberos authentication server. In this ticket is a password that is selected by the client. The server accepts the ticket along with the transaction request, examines the ticket, and verifies that the user is who he or she says he or she is. Many application packages found on client/server systems support the use of Kerberos. Although other encryption and decryption techniques are available, Kerberos is one more weapon in the arsenal to protect users and computer systems.

Public Key Infrastructure

Suppose you are working for a company that wants to open its internal local area network to the Internet. The reasons for connecting the company network to the Internet might be to allow company employees access to corporate computing resources—such as e-mail and corporate databases—while at a remote location, allow corporate customers or suppliers access to company records, or allow retail customers to place orders or inquire about previous orders. In each of these transactions, you want to be sure that the person conducting the transaction is a legitimate employee or customer and not a hacker trying to compromise your system. A new technology that assists a company to achieve this goal is public key infrastructure.

Public-key infrastructure (PKI) is the combination of encryption techniques, software, and services that involves all the necessary pieces to support digital certificates, certificate authorities, and public-key generation, storage, and management. A company that adheres to the principles of PKI issues digital certificates to legitimate users and network servers, supplies enrollment software to end-users, and provides the tools necessary to manage, renew, and revoke certificates.

A **certificate**, or digital certificate, is an electronic document, similar to a passport, that establishes your credentials when you are performing transactions on the World Wide Web. It can contain your name, a serial number, expiration dates, a copy of your public key, and the digital signature of the certificate-issuing authority so that a user can verify that the certificate is legitimate. Certificates are usually kept in a registry so that other users may look up a particular user's public key information.

Many certificates conform to the X.509 standard. The X.509 standard, created and supported by the International Telecommunications Union, defines what information can go into a certificate. All X.509 certificates contain the following pieces of information:

- ► Version, which identifies which version of the X.509 standard applies to this certificate.
- ► Serial Number, which is a unique value that identifies a particular certificate. When a certificate is revoked, its serial number is placed in a Certificate Revocation List (CRL).
- ► Signature Algorithm Identifier, which identifies the algorithm used by the certificate authority to sign the certificate.
- ► Issuer Name, which is the name of the entity, normally a certificate authority (defined below), that signed the certificate. In some instances, the issuer signs its own name.
- ► Validity Period, which is the limited amount of time for which each certificate is valid. This period can be as short as a few seconds or as long as a century and is denoted by a start date and time and an end date and time.
- ► Subject Name, which is the name of the entity whose public key this certificate identifies.

▶ Subject Public Key Information, which is the public key of the entity being named, together with an algorithm identifier that specifies to which public key encryption system this key belongs.

▶ Digital Signature, which is the signature of the certificate authority that will be used to verify a legitimate certificate.

All certificates are issued by a certificate authority (CA). A **certificate authority** is either the specialized software on a network or a trusted third-party organization or business that issues and manages certificates. One such example of a third-party business is VeriSign, which is a fully integrated PKI service provider designed to provide certificates to a company that wants to incorporate PKI but does not want to be involved with the elaborate hardware and software necessary to create and manage its own certificates.

Consider a scenario in which a user wants to order some products from a web site. When the user is ready to place the order, the web site wants to make sure the user is really who he or she claims to be. Therefore, the web site requests that the user sign the order with the user's private key, which should be issued by a certificate authority, such as VeriSign. The package consisting of the user's order and digital signature is sent to the server. The server then requests both the user's certificate and VeriSign's certificate. The server validates the user's certificate by verifying VeriSign's signature, and then uses the user's certificate to validate the signature on the order. If all signatures match, the user is who he or she claims to be and the order is processed. Even though most web retailers do not use PKI, but instead rely on secure sockets layer, more retailers in the near future may use PKI to verify their customers.

A **certificate revocation list (CRL)** is a list of certificates that have been revoked before their originally scheduled expiration date. Why might a certificate be revoked? If the key specified in the certificate has been compromised and is no longer valid, or the user specified in the certificate no longer has authority to use the key, a certificate will be revoked. For example, if an employee working for a company has a certificate assigned and the employee quits or is fired, the certificate authority will revoke the certificate and place the certificate ID on the CRL. The employee will no longer be able to send secure documents using this certificate, assuming the software takes the time to check each certificate with the CRL. Whether it is worth the time to perform this check might depend on the importance of the signed document.

Companies that need to conduct secure transactions will often invest in their own public key infrastructure (PKI). Before purchasing a PKI system, a company needs to consider whether the transactions that need to be secure remain internal within the system, or whether they need to traverse external networks. External transactions, such as electronic commerce transactions, require a system that can inter-operate with other companies' systems. Since PKI is relatively new, interoperability is still in its infancy. It may be difficult to find a system that smoothly solves the external problem, or the system found may be very expensive. A company also needs to consider the size of the application requiring secure transactions. If you have a small number of employees (roughly less than 100), it might be more economical to purchase a PKI system that is embedded within an application. If the number of employees is larger (more than 100), then a PKI system that supports a complete system of key and certificate storage and management might be more useful.

What kind of applications or transactions would benefit from public key infrastructure?

- ▶ **World Wide Web access** Ordering products over the Internet is a common activity that benefits from PKI.
- ▶ **Virtual private networks** A PKI system can assist in creating a secure VPN tunnel (see Chapter Eleven) through the insecure Internet.
- ▶ **Electronic mail** PKI can be used to create secure messages when users wish to send secure e-mail messages across an insecure network such as the Internet.
- ▶ **Client-server applications** Client-server systems, such as interactive database systems, often involve the transfer of secure data. PKI can be used to ensure that the data remains secure as it travels from client to server and back.
- ▶ **Banking transactions** Most banking transactions, which traverse external networks, need to be secure. PKI is commonly employed in banking systems to create public/private key pairs and digital signatures.

As good as it sounds, PKI has it problems. Currently, it is expensive. Typical PKI systems cost tens of thousands to hundreds of thousands of dollars for medium sized businesses. Many PKI systems are proprietary and do not interact well with other PKI systems. Due to the newness of PKI, there are still many interoperability issues to be resolved. On the positive side, technology prices for PKI systems, similar to most technologies, continue to drop. In addition, the Internet Engineering Task Force continues to work on standards to which future PKI systems can adhere. Hopefully these new standards will increase the possibility that PKI systems will interoperate.

Firewalls

A **firewall** is a system or combination of systems that supports an access control policy between two networks. The two networks are usually an internal corporate network and an external network, such as the Internet. A firewall can limit users on the Internet from accessing certain portions of a corporate network, and can limit internal users from accessing various portions of the Internet. Figure 13-8 demonstrates a firewall as it blocks both internal and external requests.

Figure 13-8
A firewall as it stops certain internal and external transactions

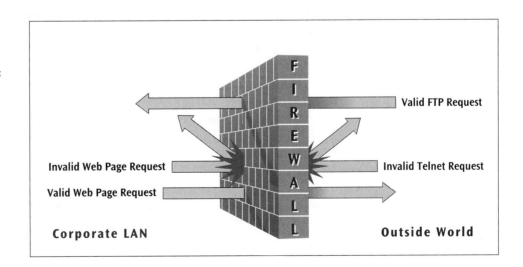

In a corporate environment in which a local area network is connected to the Internet, the company may want to allow external e-mail messages to enter the network, but may not want to allow remote logins into the corporate network. If remote logins are allowed, intruders may learn how to login to a corporate application and vandalize the system. Likewise, the company may want to outlaw file transfers out of the corporate network for fear of losing corporate data or trade secrets. As you will see, a firewall system involves more than just programming a device to accept or reject certain types of transactions.

Firewall efficacy

What types of transactions will a firewall stop, and what types of transactions will it not stop? It is possible for a firewall system to stop inbound or outbound e-mail, remote logins, and inbound or outbound file transfers. It is also possible for a firewall to limit inbound or outbound web page requests. Once a company has created a security policy that decides what should be restricted and what should be allowed, it is possible to create a firewall system that allows or disallows a fairly wide range of transactions.

For example, TCP port numbers indicate which application service is desired. When a transaction enters a system destined for a particular application, the appropriate port number is included as part of the transaction. Some typical port number assignments are:

- ► Telnet (remote login) is usually port 23
- ► Finger (service to find a user ID) is often port 79
- ► USENET (the newsgroup service) is often port 119
- ► E-mail (SMTP) is usually port 25
- ► Web browsing (using HTTP) is port 80

If a company wanted to block anyone from the outside trying to Telnet into the company, it could configure its firewall to stop all incoming transactions requesting port 23. Or a company could stop employees from "wasting" time on USENET by configuring the firewall to block outgoing transactions with port number 119.

Firewalls, unfortunately, do not protect a network from all possible forms of attack. A virus, which can hide within a document, will more than likely not be detected by a firewall if the document it is attached to is allowed into the system. Some firewall systems, however, do boast that they can detect viruses within documents. But these systems are rare and, as you saw earlier, may not detect all current viruses.

A firewall will also not protect the system properly if it is possible to gain entrance to the system through a means other than passing through the firewall. For example, if a network system has one or more dial-up modems that are not part of the firewall system, a perpetrator can bypass the firewall and gain illegal access. A fax machine provides another way for sensitive information to leave a company. The best firewall system in the world will not stop an employee from faxing a confidential document over a standard telephone line to an outside recipient. For a firewall system to work, a comprehensive security policy must be created, installed, and enforced. All possible entrances and exits to the corporate environment, including computer systems, telephone systems, fax systems, and persons physically carrying media into and out of the building, should be part of a security policy.

Basic firewall types

Firewalls come in two basic types:

▶ packet filter, or network level and

▶ proxy servers, or application level.

Let's examine each of these types of firewalls and review their advantages and disadvantages.

The **packet filter** is essentially a router that has been programmed to filter out or allow to pass certain IP addresses or TCP port numbers. Earlier types of routers performed a static examination of the IP addresses and TCP port numbers and, based on information stored in a table, either denied a transaction or allowed it to pass. These types of routers are essentially packet filters, are relatively simple in design, and act very quickly. However, they are a little too simple if a high level of security is desired. For example, it has been relatively easy for someone to fool these early types of static packet filters. An external user can spoof a static packet filter into thinking the external user is a friendly agent and not an external user bent on causing internal network problems.

More modern router models are becoming increasingly intelligent and can follow, to some degree, the flow of conversation between an internal entity and the corresponding external entity. This capacity to follow the flow of conversation allows modern routers to detect spoofing. Modern routers are also beginning to scan for viruses on incoming transactions.

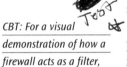

CBT: For a visual demonstration of how a firewall acts as a filter, see the enclosed CD-ROM.

The proxy server is a more complex firewall device. A **proxy server** is a computer running proxy server software that acts as the "doorman" into a corporate network. Any external transactions that request something from the corporate network must enter though the proxy server. The proxy server then creates an "application," called a proxy, that acts as a servant and goes into the corporate network to retrieve the requested information. Since all external transactions must go through the proxy server, it provides a very good way to create an audit log.

A proxy server works in a way similar to how a person gains access to books in a library's rare book room. Library personnel do not want library patrons going into the rare book room for fear of vandalism. Thus, a patron fills out an information request slip, and hands it to a librarian. The librarian enters the rare book room and retrieves the requested volume. The librarian then photocopies the requested information and gives the photocopy to the patron. The patron never comes in contact with the actual rare book.

A proxy server that supports FTP requests provides a realistic example. Recall from Chapter Eleven that the FTP protocol uses two network connections—one for control information and one for data transfer. An FTP proxy examines the control connection at the application layer, decides which commands are allowed or denied, and then logs individual commands. When the proxy encounters a command requesting the second connection—the data connection—it uses company programmed security policy information to decide whether to filter the data packets.

Since the proxy server sits outside the corporate network and its security, the proxy server is wide open to vandalism (Figure 13-9). To protect the proxy server, you need to make sure it is a stripped down version of a network computer. Thus, there is little a vandal can do to harm the proxy server.

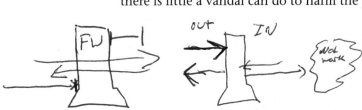

Figure 13-9
The proxy server sitting outside the protection of the corporate network

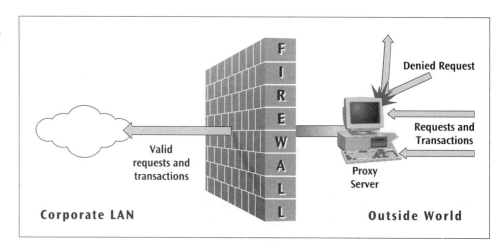

Although the proxy server provides a higher level of security, it is slower than a packet filter since a proxy has to be created for each type of application that may request data from inside the corporate network. As you might expect, proxy technology is improving too. Someday we may see a router-like device that can efficiently and effectively create a proxy, which enters an internal network to retrieve the requested information.

Security Policy Design Issues

When designing a firewall system and its corresponding security policy, a number of questions should be answered. The first question involves the company's expected level of security. Is the company trying to restrict all access to services not deemed essential to the business? Or does the company wish to allow all or most types of transactions, thus asking the firewall system only to audit transactions and create an orderly request for transactions? Restricting all access to services not deemed essential requires a more elaborate firewall system and thus more work and expense. Allowing most types of transactions requires a simpler system that only performs queue management operations and creates an audit trail.

A second question stems from the first decision: How much money is the company willing to invest in a firewall system? Commercially-purchased firewall systems can be powerful, complex, and expensive. It is possible, however, to construct a home-grown firewall system that takes advantage of the capabilities of existing equipment, such as routers and network operating systems. As we saw earlier, it is possible to restrict access into a system based on time of day, day of week, and location. It is also possible to use existing software to create an audit trail of all incoming and outgoing transactions. Depending on the detail of auditing required, additional software can be purchased and installed that will work in concert with network operating system software to provide any desired level of audits.

Similarly, many routers can be programmed to restrict access to certain kinds of traffic. A router can be programmed to accept and reject requests with specific IP addresses or a range of IP addresses. Routers can also be programmed to deny access to certain port addresses at the TCP level.

A third question relates to the company's commitment to security. If the company is serious about restricting access to the corporate network through a link such as the Internet, will the company be equally serious about supporting security on any and all other links into the corporate network environment? Dial-up modem access, wireless network access, and other telecommunication links should also be considered when making security decisions. Fax machines, both stand alone and computer based, as well as removable disk media are two more examples of how data may enter or leave a corporation. Any security policy must take these entrance and exit points, as well as the Internet, into consideration.

Having a well-designed security policy in place will make the jobs of network support staff clearer. The staff employees will know what the network users can and cannot access and where they can and cannot go. A well-designed security policy will make enforcement more straightforward, and it will allow the staff to react properly to specific security requests. The policy will also make clear the goals and duties of network employees in enforcing security with respect to requests from the outside. If there is a good security policy, the users themselves will have a better understanding of what they can and cannot do. This understanding will hopefully assist the network staff in conducting their jobs and will allow the company to maintain security in an increasingly insecure world.

Perhaps because companies have well-designed security policies in place, many people who use the Internet to purchase items online are growing comfortable with the fact that, if they transfer credit card information during a secure session, their data is safe from hackers and other eavesdroppers. This sense of security may change, however, because the Internet Engineering Task Force is considering whether to allow a backdoor entry into all Internet traffic. This backdoor entry would allow authorized persons to intercept any data traffic on the Internet. Since this proposal appears to be a violation of privacy, why would anyone want to create such a backdoor?

At the core of the argument is the fact that standard telephone systems currently allow agencies of the U.S. government to wiretap communications. This wiretap occurs at the telephone central office and is built into central office telephone switches. The act that allows wiretapping (the Communications Assistance for Law Enforcement Act) has been in existence since 1994. Now that the Internet is beginning to carry voice traffic, should it also be possible for the U.S. government to wiretap voice transactions on the Internet? As one critic of the proposal states, if they can tap voice, then they can tap data. Furthermore, if the designers of the Internet create such a backdoor, it is also possible that this knowledge could fall into the wrong hands and be used for criminal intent.

This issue is further complicated by the fact that many businesses presently encrypt all data leaving the corporate network. Most encryption techniques used by businesses are so effective that virtually no one, including the government, can crack them. If the network does the encryption just before the data leaves corporate boundaries, then it would be the responsibility of the corporate network support personnel to provide the U.S. government, if asked, with unencrypted data. If, on the other hand, the encryption is applied at the user workstation before it is inserted onto the corporate network, who will supply the U.S. government with the unencrypted data? Clearly, this issue will be hotly debated for some time to come.

Despite the fact that a company may have a well-designed security policy in place, external events are making this area more complex all the time.

Network Security in Action: Banking and PKI

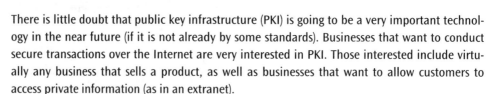

There is little doubt that public key infrastructure (PKI) is going to be a very important technology in the near future (if it is not already by some standards). Businesses that want to conduct secure transactions over the Internet are very interested in PKI. Those interested include virtually any business that sells a product, as well as businesses that want to allow customers to access private information (as in an extranet).

Unfortunately, PKI is off to a rough start. A number of factors are contributing to this rough start:

- ► The cost of a PKI system to support a business can be very expensive.
- ► PKI technology is difficult to understand.
- ► There are many parts to PKI, such as the certificates, the certificate authority, public and private keys, and more.
- ► A user wishing to invoke a secure session using PKI is often required to deal with details that he or she should not have to deal with. This lack of user-friendliness is scaring people away, which concerns businesses trying to implement and support PKI.

There are a number of ways that a technology can help itself remain in a market in spite of initial difficulties. One method is to have good publicity and lots of it. A second method involves marketing. A third method is having a reputable company support your product. Such is the case with PKI and Entrust Technologies, Inc. Entrust Technologies creates PKI systems for other businesses. One of Entrust's best success stories is the creation of a PKI system for ScotiaBank, one of North America's largest financial institutions.

Launched in September 1997, ScotiaBank's Scotia OnLine allows customers to use a web browser 24 hours a day, 7 days a week to access their bank account balances, transfer funds, pay bills, and view their latest VISA transactions and statement. They can also move easily between banking and brokerage transactions without having to enter separate passwords. Naturally, all these transactions had to be secure, thus the need for PKI.

To support PKI, Entrust created a system that required Entrust software clients on the customers' personal computers and on the ScotiaBank web server. Having Entrust software installed on the web server and customers' personal computers helps to ensure that both ends of the connection are conducting secure transactions and that all transactions must pass through these clients to be encrypted and decrypted.

Scotia OnLine issues a digital certificate to each customer, which is used to encrypt and decrypt transactions and generate any necessary digital signatures. These signatures prove that the customer is who the customer says he or she is and the ScotiaBank is really ScotiaBank and not someone trying to spoof a customer into thinking they are ScotiaBank.

Customers choose their own password with help from the customer software client, thus eliminating the need to transfer a customer password over a network. Unlike a typical secure sockets layer session, which transmits customers' unique access identifiers and passwords over the Internet, Scotia OnLine uses only digital certificates to mutually authenticate the customer and the bank.

Customers apparently feel confident that their transactions are secure. As of April 1998, ScotiaBank has issued over 40,000 public-key certificates, making it the largest user of digital certificates in Canada. Other banking institutions are apparently taking notice, too. Beginning in early 2000, CitiBank in the United States will begin offering similar types of services.

◆◆◆

SUMMARY

▶ Network security continues to be an increasingly important topic, particularly with the increase in network interconnectivity. The Internet has helped to open the door to worldwide vandals, making any system connected to the Internet vulnerable to attack.

▶ Network personnel must be aware of external security measures, such as equipment placement, to avoid flooding, vandalism, and other detrimental environmental factors.

▶ Operational security measures exist that allow a network administrator to restrict the time of day, day of week, and location from which someone can log on to a computer system.

▶ Surveillance systems can be a good deterrent of vandalism and equipment theft.

▶ Passwords and other ID systems are very common security techniques, but passwords can be stolen and used by unscrupulous parties.

▶ Newer biometric techniques that use some form of the body for ID are more secure than password systems.

▶ Using software that conducts a continuous audit of network transactions can create an electronic trail that might be useful in catching malicious users.

▶ Most computer systems apply access rights to the resources of the system and the users. By properly setting access rights, computer resources can be made more secure.

▶ Network administrators and users must be aware of standard computer attacks and viruses that can damage computer systems.

▶ Software and hardware exist that can help to protect a system and its users from computer attacks and viruses. Virus scanners have three basic forms: signature-based scanners, terminate and stay resident software, and multi-level generic scanning.

▶ Spoofing, smurfing, and Ping storms are three common examples of a class of system attacks called denial of service.

▶ Encryption technology provides an important set of techniques that can aid in the fight against computer fraud.

▶ Substitution-based ciphers replace one letter or series of letters in a message with a second letter or series of letters.

▶ Transposition-based ciphers rearrange the order of letters in a message.

▶ Public key technology uses two keys—one key to encode messages and a second key to decode messages.

▶ The secure sockets layer is used to encrypt data that travels back and forth between web servers and web browsers.

▶ The Data Encryption Standard was created in 1977 and uses a 56-bit key to encrypt data transmitted between two business locations.

▶ Pretty Good Privacy is free encryption software that allows average users as well as commercial users to encrypt and decrypt everyday transmissions.

▶ Kerberos is a secret key encryption technique that can be used by commercial application programs to verify that a user is who he or she says he or she is.

▶ Digital signatures use public key technology and can be used to verify that a document belongs to a person.

▶ Public key infrastructure uses public key technology, digital signatures, and digital certificates to allow the secure passage of data over unsecure networks.

▶ A firewall is a system or combination of systems that supports an access control policy between two networks.

▶ Firewalls come in two basic types: packet filters, which examine all incoming and outgoing transmissions and filter out those transmissions that have been deemed illegal, and proxy servers, which are computers running at the entrance to a computer network and act as gatekeepers into the corporate network.

► A proper network security design aids network staff personnel by clearly delineating the acceptable network transactions submitted by internal employees and by external users.

KEY TERMS

ciphertext	key	public key cryptography
computer auditing	kerberos	public key infrastructure
cryptography	line disconnect attack	secure sockets layer
certificate	monoalphabetic substitution-based	signature-based scanner
certificate authority	cipher	smurfing
certificate revocation list	multi-level generic scanning	spoofing
Data Encryption Standard (DES)	operational security	terminate and stay resident antivirus
denial of service attack	packet filter	software
digital signature	plaintext	transposition-based cipher
e-mail bombing	ping storm	Trojan horse
encryption algorithm	polyalphabetic substitution-based	Vigenére Cipher
external security	cipher	virus
firewall	Pretty Good Privacy (PGP)	worm
intrusion detection	proxy server	

REVIEW QUESTIONS

1. List three examples of external network security.
2. What is meant by the phrase *operational security* and how does it apply to computer networks?
3. How can surveillance be used to improve network security?
4. What is the major weakness of a password? What is its major strength?
5. How can auditing be used to protect a computer system from fraudulent use?
6. What are the most common types of access rights?
7. What are the most common examples of denial of service attacks?
8. What is a computer virus and what are the major types of computer viruses?
9. What are the three different techniques used to locate and stop viruses?
10. How can a vandal use spoofing to create a denial of service attack?
11. Show a simple example of a substitution-based cipher.
12. Show a simple example of a transposition-based cipher.
13. How can public key cryptography make systems safer?
14. Give a common example of an application that uses secure sockets layer.
15. What is the Data Encryption Standard?
16. What kind of applications can benefit from Pretty Good Privacy?
17. Is Kerberos a public key encryption technique or a private key?
18. What is meant by digital signature?
19. List the basic parts of public key infrastructure.
20. What kind of applications can benefit from Public Key Infrastructure?
21. What kind of entity issues a certificate?

22. Under what circumstances might a certificate be revoked?
23. What is the primary responsibility of a firewall?
24. What are the two basic types of firewalls?
25. What are the advantages of having a security policy in place?

EXERCISES

1. A major university in Illinois used to place the computer output from student jobs on a table in the computer room. This room is the same computer room that housed all the campus' mainframe computers and supporting devices. Students would enter the room, pick up their jobs, and leave. What kinds of security problems might computer services encounter with a system such as this?

2. You have forgotten your password, so you call the help desk and ask them to retrieve your password. After a few moments, they tell you your forgotten password. What has just happened and what is its significance?

3. Using the Vigenére Cipher and the key NETWORK, encode the phrase "this is an interesting class."

4. Using the transposition-based cipher from this chapter and the same key, COMPUTER, encode the phrase "birthdays should only come once a year."

5. You are using a web browser and want to purchase a music CD from an electronic retailer. The retailer asks for your credit card number. Before you transfer your credit card number, the browser enters a secure connection. What sequence of events created the secure connection?

6. You want to write a song and apply a digital signature to it so that you can later prove it is your song. How do you apply the signature, and later on, how do you prove the song is yours?

7. List three examples (other than those listed in the chapter) of everyday actions that might benefit from applying PKI.

8. Can a firewall filter out requests to a particular IP address, a port address, or both? What is the difference?

9. One feature of a firewall is its ability to stop an outgoing IP packet, remove the real IP address, insert a "fake" IP address, and send the packet on its way. How does this feature work? Do you think it would be effective?

THINKING OUTSIDE THE BOX

1 Create a cipher that is composed of either a substitution-based cipher, a transposition-based cipher, or both. Encode the message "Meet me in front of the zoo at midnight." Explain your encoding technique(s).

2 Create a hypothetical business with approximately 50 to 100 employees. Place the employees in two or three different departments. Assign to each department a title and basic job duties. All employees in all departments use personal computers for numerous activities. Identify the computer activities for the employees of each department. Then create a security policy for the employees of each department. Be sure to address when and where the employees have access to computer resources and if any types of transactions should be restricted.

3 You are working for a company that allows its employees to remotely access computing resources and allows suppliers to send and receive order transactions. Your company is considering incorporating PKI. How would you recommend PKI be implemented to support these two application areas?

PROJECTS

1. Find and report on three different security techniques that use some part of the body for verification.
2. Intrusion detection is a growing area of network security. A number of companies offer software systems that perform intrusion detection. Write a one- to two-page paper that summarizes how intrusion detection systems operate.
3. What are the allowable access rights for protecting files and other objects on the computer system at your school or place of employment? List the options for "who" and "how."
4. Create a short list of some of the recent viruses that come inside application macros, such as the macros that work with Microsoft Office applications.
5. Report on the current status of encryption standards for businesses that send data outside the country. What is the businesses' side of the issue? What is the government's side of the issue? What will the government currently allow businesses to export?
6. Pretty Good Privacy (PGP) can be used as a form of public key infrastructure (PKI). Explain how PGP and PKI relate.

14

Network Design and Management

◆◆◆

IT IS COMMON business knowledge that more can be learned from a failure than from a success. If true, then many systems analysts have learned an awful lot. According to Sequent Computer Systems, 76 percent of 500 Information Technology directors surveyed had experienced a major project failure. The top reasons cited for projects failing included requirements that changed as the project progressed, poor planning, unrealistic expectations, ambiguous objectives, lack of project resources, lack of user input, insufficient executive support, technical incompetence of the supplier, and poor quality suppliers.

How do you know when to bail out of a project, such as a computer network upgrade, before time and money are wasted? What are the warning signs of a failing project? You can tell a project is failing or about to fail when:

> it doesn't have a clear owner or sponsor in upper management;

> there is a lack of accountability;

> there is a significant lack of communication;

> missed deadlines become the rule, not the exception;

the project is significantly over budget; and

project spending increases late in the game.

Finally, what can you do to avoid a project that has a high failure potential? First of all, pin down the details up front. Find out what determines project success and under what conditions the project should be terminated. Second, make sure you have clear support from middle management (project manager and business manager), from the unit that will benefit from the project, and from one senior-level executive. Third, make sure you have a clear understanding of objectives before starting. Fourth, don't be afraid to cancel a project immediately after its pilot stage if the pilot stage is a failure. Fifth, adhere to solid project management principles. And sixth, "Check it out, make yourself heard, and if ignored, forget it."

Network World, March 2nd, 1998, pp. 46–47.

How difficult can it be to create a new network or add to an existing network?

Do professionals follow a procedure when they analyze and design computer systems?

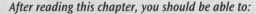

Objectives ▶

After reading this chapter, you should be able to:

▶ Recognize the systems development life cycle and define each of its phases.

▶ Explain the importance of creating one or more location connectivity diagrams.

▶ Outline the differences between technical, financial, operational, and time feasibility.

▶ Create a cost-benefit analysis incorporating the time value of money.

▶ Explain why performing capacity planning and traffic analysis is difficult.

▶ Describe the steps involved in performing a baseline study.

▶ Discuss the importance of a network manager and the skills required of that position.

▶ Calculate component and system reliability and availability.

▶ Describe the importance of a help desk with respect to managing network operations.

▶ List the main features of the Simple Network Management Protocol (SNMP) and distinguish between a manager and an agent.

▶ Describe the use of the Remote Network Monitoring (RMON) protocol and its relationship to SNMP.

▶ Recognize the basic hardware and software network diagnostic tools.

Introduction ▶

For a computer network to be successful, it has to be able to support the current and future amount of traffic, pay for itself within an acceptable amount of time, and provide the services necessary to support users of the system. All these goals are very difficult to achieve. Why are these goals so challenging? First, computer networks continue to increase in complexity. In many business environments, it is extremely difficult for one person to completely understand every component, protocol, and network application. Thus, network management is becoming increasingly formidable.

A second reason is that it is very difficult for an individual or a business to properly define the future of computing within a company. Each company has its own expectation of what computing services a company should provide. Indeed, each user within a company has his or her own idea of the computing services that should be available. It is extremely hard to determine a single service or set of services for an entire company.

Finally, computer network technology changes at a break-neck speed. In some areas, new technologies are emerging almost daily, while in other areas you can expect major developments about every six months. Keeping abreast of new hardware, software, and network applications is a full-time job in itself. Incorporating the new technology into existing technology while trying to guess the needs of users is exhausting. Being aware of this challenging environment makes it clear why designing new systems and updating current systems is filled with dangerous pitfalls. With one wrong decision, a lot of time and money can be wasted.

Most of the topics introduced in this chapter come from very large areas of study. For example, the area of planning, analyzing, designing, and implementing computer-based solutions is often the subject of a 1- or 2-term college course sequence. Nonetheless, it is important to understand the basic concepts used to create computer-based solutions.

Why is understanding basic computer system development concepts so important? These concepts are important because during your computer network career you will either be designing or updating a network system, or you will be assisting one or more persons who are designing or updating a network system. If you will personally design or update a network, you should know how to attack the problem logically and in a proper progression of steps. If you will be working with one or more network professionals, you should be cognizant of the steps involved and how you might be involved. Likewise, conducting feasibility studies, capacity planning, traffic analysis, and creating a baseline are also complex but necessary skills for a person desiring to manage a network. This chapter will introduce each of these topics, with relatively simple examples of their use. The chapter will conclude with an introduction to the skills required of a network manager and a look at some of the tools available for properly supporting a network.

Systems Development Life Cycle

Every business, whether non-profit or for-profit, has a number of major goals. Examples of these goals include:

- ▶ The desire to increase the business' customer base.
- ▶ The need to properly serve customers and keep them happy by providing the company's service as best as possible.

> ► The desire to increase the company's profit level, or in a non-profit organization, to provide the funds necessary to meet the organization's goals and objectives.
> ► The desire to conduct business more efficiently and effectively.

From these major goals, system planners and management personnel within a company try to generate a set of questions that, when satisfactorily answered, will assist an organization in achieving its goals and move the organization forward. For example, someone might ask: Is there a way to streamline the order system? Streamlining the order system will allow the company to conduct business more efficiently and effectively. Can we automate the customer renewal system? Automating the customer renewal system may better serve customers and keep them happy. Can we efficiently offer new products? Offering new products might help the company increase the customer base. Is there a better way to manage our warehouse system? Better management of the warehouse system may help to increase company profits.

Very often in today's business world, the answers to these types of questions require the use of computer systems, and the computer systems used are often large and complex. When large amounts of time, money, and resources are involved in a particular computer solution, you want to make sure that the solution is the best possible solution for the problem.

To properly understand a problem, analyze all possible solutions, select the best solution, and implement and maintain the solution, you need to follow a well-defined plan. One of the most popular and successful plans currently used by businesses today is the systems development life cycle (SDLC). The **systems development life cycle (SDLC)** is a methodology for a structured approach for the development of a business system, including planning, analysis, design, implementation and support. A methodology is a series of steps and tasks that professionals, such as systems developers, can follow to build quality systems faster, with fewer risks, and at lower costs. Although virtually every company that uses SDLC and every textbook that teaches SDLC has a slightly different variation, most agree that the SDLC includes the following phases:

> ► Planning
> ► Analysis
> ► Design
> ► Implementation
> ► Maintenance

Before investigating each of these phases in detail, let's take a closer look at the big picture of the SDLC. The idea of *phases* is critical to the SDLC concept. The intent of SDLC is that the phases are not disjoint steps in a big plan, but overlapping layers of activity. It is quite common to perform two and three phases of a single project at the same time. For example, the design of one component of a system can be in progress while the implementation of another component is being performed.

A second critical concept is *cycle*. After a system has been maintained for a period of time, it is relatively common to restart the planning phase in an attempt to seek a better solution to the problem. Thus, the systems development life cycle is a never ending process (Figure 14-1).

Figure 14-1
The cyclic nature of the phases of the systems development life cycle

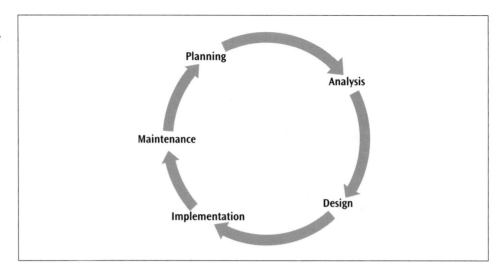

Who is responsible for initiating the phases and shepherding a project through the cycle? A professional called a systems analyst is typically responsible for managing a project and following the SDLC phases, particularly the analysis and design phases. Although systems analyst is a trained professional position, many individuals can learn systems analysis and design methods and incorporate them into designing computer networks. It is also possible that a person who supports computer networks may be called upon by a systems analyst to provide input into a new network solution. In this instance the network professional should be aware of the SDLC and be prepared to provide any requested information or design materials.

For the SDLC to work, a plan must be written and adhered to. Exactly what is written varies from company to company. Regardless, a plan should exist and it should describe in detail each step by including the following information:

> ► Title of the phase to which a step belongs
> ► Title of the step
> ► The objectives or description of the step
> ► Possible data that serve as input to the step
> ► Sub-steps of the step. If the sub-steps are significant, you may want to break out each sub-step as its own separate step.
> ► The deliverables, or results, of the step that will be passed on to the next step.
> ► The participants in the step.

Consider the following example of a step within the analysis phase:

Phase: Analysis

Step: Model the Current System

Objectives: To learn about the current system, including its data, processes, interfaces, and network geography.

Input: Gather all possible data concerning the current system, including interview results, questionnaire results, organization charts, system flow charts, sample business forms, and so on.

Sub-steps: Create an interface model, a data model, a process model, and a network geography model.

Deliverables: The models in electronic and hard copy form.

Participants: Systems analyst, system users, management, owners.

Although the preceding example is simplified, it is important to note that whatever the design plan looks like, it is important that each step is clearly established so that a strong, consistent method is followed for every project a company considers.

Having discussed SDLC in general terms, let's now turn to more precisely defining what happens during each phase. Let's pay particular attention to the analysis and design of the computer networks that might be involved in proposed solutions.

Planning phase

During the planning phase, you identify problems, opportunities and objectives. The systems analyst must honestly look at what is occurring in a business. What is the business trying to do? To effectively carry out this phase, the systems analyst meets with other members of the organization to outline the problems, opportunities, and objectives. The systems analyst interviews managers to learn what their goals are. This information gathering (the input to the phase) is often not the systems analyst's job alone, but is delegated to a cross-functional committee consisting of upper management, end-users, middle management, and systems analysts. Often, upper management assigns problems to a systems analyst who can carry out the planning phase without the need for a cross-functional committee. The output, or deliverables, from the planning phase includes a list of one or more possible problems that have been assigned a priority value.

Analysis phase

During the analysis phase, you determine information requirements for the particular users involved. Information requirements can be gathered by sampling and collecting hard data, interviewing, questionnaires, observing decision makers' behavior and office environments, and prototyping. The systems analyst needs to know who the users are, what is the business activity, what is the work environment, what is the timing, and what are the current procedures. The analyst must also know what operations are performed at what locations. The systems analyst may rely on a specialist called an information analyst during this phase.

During the analysis phase, you also analyze system needs. What are the needs of the system? What is the flow of data within the system? The systems analyst creates data flow diagrams to show the flow of data within the system. What are all the data items in the system? The systems analyst will create data dictionaries to show all the system's data items. What decisions are made with respect to the proper flow of data and execution of the business goal? The systems analyst can create decision tables or decision trees to map the proper flow of data and execution of business goals. The systems analyst should also create a series of maps of the locations that are part of the proposed system. These maps—called network models—can be used to determine computer network topologies and the necessary data flow over those networks.

At the end of the analysis phase, the systems analyst prepares a written systems proposal that summarizes what has been found, provides feasibility studies and cost-benefit analyses of alternatives, and makes recommendations on what (if anything) should be done. This systems proposal is a business proposal and usually does not yet recommend a particular type of computer system or network installation. The recommendation of particular computer hardware and software usually occurs in the next phase, the design phase.

Design phase

During the **design phase**, you design the system that was recommended and approved at the end of the analysis phase. The systems analyst uses the information collected earlier to create a logical design of the information system. The systems analyst will also design accurate data-entry procedures, input and output screens (the user interface), output formats, files and databases, and controls and backup procedures to protect the system. At this point the systems analyst, often assisted by a network specialist, will also design the computer network structure to support the newly created information system. Capacity planning studies will be conducted to determine the necessary size and speed of the proposed network if it is to support the anticipated traffic from the new information system. If the analysis and design call for a modification to an existing system, a network specialist will perform a baseline study to determine the current network capabilities, and then decide what modifications to the current network will be necessary to support the additional services being designed.

The design phase also involves developing and documenting software. The systems analyst works with programmers to develop original software, if necessary. If pre-programmed software is needed, the systems analyst solicits bids for it and all necessary hardware. The creation of all documentation is begun at this time.

At the end of the design phase, the new system is tested. Testing is performed by the systems analyst and end-users, and it can be done on individual pieces and on the system as a whole. Network specialists are, once again, called upon to test the networks and to make sure they perform the necessary transactions, can support the anticipated volume of transactions, and can perform the transactions fast enough to support the anticipated load.

Implementation phase

During the **implementation phase**, the system is installed and preparations are made to move from the old system to the new system. Training of users is performed. Testing continues, evaluations are performed, and feedback is accepted. Any and all networks are monitored continuously by network managers, and all feedback is used to correct network problems or weaknesses. Any system documentation that is needed to support the new system is created during the implementation phase.

Maintenance phase

Maintenance is the largest phase in the SDLC and one that often lasts for years. The **maintenance phase** has been compared to the portion of an iceberg that is hidden under water (Figure 14-2). With respect to time and costs, the planning, analysis, design, and implementation phases are relatively small—like the portion of the iceberg that is above the surface of the water—compared to maintenance. In contrast, the maintenance phase takes more time and money; it is comparable in size to the portion of the iceberg that is hidden below the surface. With careful and accurate analysis and design, the time and cost of the maintenance phase can be kept to a minimum.

Figure 14-2
The iceberg principle of the systems development life cycle

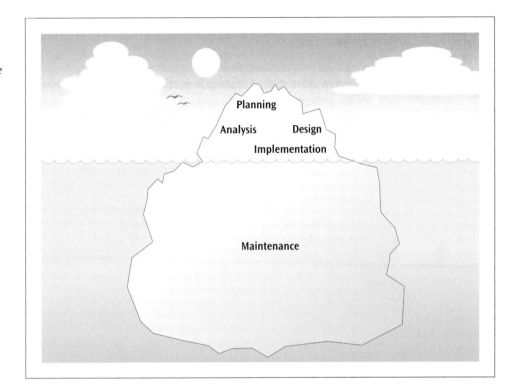

Anyone who is called on to design a computer network or even a portion of a network must follow some analysis and design plan. If the company does not already have an existing plan, it is the responsibility of the network designer to create and then adhere to a plan. If you follow the systems development life cycle to create that plan, you will increase your odds of designing a successful project.

Start Here

Network Modeling

When a systems analyst or a person acting in the role of systems analyst is asked to design a new computer system, the analyst will create a set of models for both the existing system (if there is one), and for the proposed system. Typically these models are designed to show the flow of data through the system and the flow of procedures within the system and help the analyst and other professionals visualize the current and proposed systems. One very important part of most computer systems today is a network. Most businesses have at least one internal local area network, and one or more connections to external wide area networks such as the Internet. Many businesses have multiple office locations scattered over a geographic location. If a systems analyst or even a person such as yourself has been called upon to contribute to the design of a computer network, you will want to create one or more network models. These network models can either demonstrate the current state of the network or can model the desired computer network.

To model a network environment for a corporation, you create a figure called a location connectivity diagram. A location connectivity diagram (LCD) is a network modeling tool that depicts the various locations involved in a network and the interconnections between those locations. Two basic forms of LCDs can be created: an overview LCD and a detailed LCD. An overview LCD depicts the major locations involved in a corporate networking environment. For example, an overview LCD could include the cities in which the company has offices and the communication lines connecting those offices. A detailed LCD depicts the network environment that exists at one of those locations and includes the local area networks within that environment. Let's begin studying how to model a network environment by creating an overview LCD, selecting one of the locations within the overview LCD, and creating a detailed LCD.

Let's assume the company you are working for has offices in Atlanta, New York, Omaha, San Diego, and San Antonio, and deals directly with a separate company in St. Louis. Your company also has employees who travel with laptop computers and dial in to the New York office using telephone lines and modems. Figure 14-3 illustrates an overview LCD showing these locations. Since St. Louis is a separate or external company, we denote the St. Louis location with an X in a circle. The traveling or mobile employees are denoted with an M in a circle. The distance between a pair of locations is shown in miles on the telecommunications line connecting the locations.

Figure 14-3
Overview location connectivity diagram

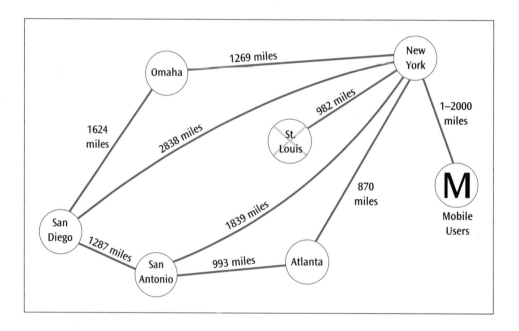

Now let's create a detailed LCD for the San Antonio location. Let's assume that the San Antonio site consists of one building. If San Antonio consisted of multiple locations within the city, a second overview LCD could be created showing the individual sub-locations within the San Antonio location. Then a detailed LCD could be redrawn for each of these sub-locations. Since our example has only one building in San Antonio, only one detailed LCD is needed to model this location.

The detailed LCD in Figure 14-4 shows the logical departments or divisions found in the San Antonio location. There are four departments within the San Antonio office building—marketing, sales, management, and technical support.

The distances between the departments will be used later to determine the type of networking or cabling that can be used to connect them. The connections between the departments indicate the network interconnections that exist or need to exist between the departments. Note that technical support handles the external connections to the outside world.

Figure 14-4
Detailed location connectivity diagram for the San Antonio location

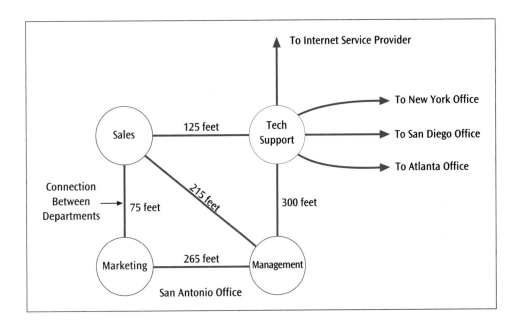

Network models, such as location connectivity diagrams, depict major sites both external and internal to the company, distances between sites, and whether sites are standard, external, or mobile. This information can then be used later in the analysis and design phases to help determine locations for network equipment, types of interconnecting media, and possible locations for data storage.

Feasibility Studies

Analyzing and designing a new computer system can be time consuming and expensive. While the project is in the analysis phase and before a system is designed and installed, a feasible solution must be found. The term feasible has several meanings when applied to computer-based projects. The proposed system must be technically feasible. A **technically feasible system** is one that can be created and implemented using currently existing technology. Does technology exist that can be incorporated into a working solution? If technology from two or more vendors is suggested, will the differing technologies work together? If a vendor claims that a particular hardware or software system will do what is desired, are the vendor's claims accurate or misleading? Does your company have the technical expertise to build, install, or maintain the proposed system?

The proposed system must also be financially feasible. A **financially feasible system** is one that can be created given the company's current financial ability. Can the proposed system solve the current problem and stay within budget? How long, if ever, will it take before the proposed system returns a profit?

In addition, the proposed system must be operationally feasible. An **operationally feasible system** is one that operates as designed and implemented. Will the proposed system produce the expected results? Will users be able to use the proposed system, or will it be so difficult or inconvenient to use that users will not adopt the proposed system?

Finally, the proposed system must be time feasible. A **time feasible system** is one that is installed in a timely fashion that meets organizational needs. Can the proposed system be designed, built, tested, and installed in an agreeable and reasonable amount of time?

All of these feasibility questions are difficult to answer, but they must be answered. Technical, operational, financial, and time feasibility can best be determined with knowledge of computer systems, the state of the current market and its products, and experience. Anyone embarking on designing and installing a new computer network will work much more effectively if they understand analysis and design techniques, project and time management techniques, financial analysis techniques, and other network design techniques. These techniques are an integral part of feasibility studies.

As an example of one technique, let's consider a common financial analysis technique that involves determining the proposed system's costs and benefits: payback analysis. **Payback analysis** charts the initial costs and yearly recurring costs of a proposed system against the projected yearly income (benefits) derived from a proposed system. Systems analysts along with middle and upper management use payback analysis, along with other financial techniques, to determine the financial feasibility of a project.

Before you see how to calculate the payback analysis, you need to review a few common financial concepts as they apply to computer systems. To determine the cost of a system, it is necessary to include all possible costs. First, you need to consider all one-time costs, including:

► Personnel costs, including analysts, designers, programmers, consultants, specialists, operators, administrative staff, and so on.

► Computer usage costs, which reflect the computing needed to perform the analysis and feasibility studies.

► Costs of hardware and software for the proposed system.

► Costs to train the users, support personnel, and management for using the proposed system.

► Supplies, duplication, and furniture costs for the personnel creating the proposed system.

You must also calculate the recurring costs of the proposed system, including:

► Lease payments on computer hardware or other equipment.

► Recurring license costs for software purchased.

► Salaries and wages of personnel who will support the system.

► Ongoing supplies that will keep the proposed system working.

► Heating, cooling, and electrical costs to support the proposed system.

► Planned replacement costs to replace pieces of the system as they fail or become obsolete.

Once the one-time and recurring costs have been established, it is time to determine the benefits that will result from the proposed system. You need to include both tangible benefits and intangible benefits when calculating benefits. With respect to tangible benefits, the most common measurement is monthly or annual savings that will result from the use of the proposed system. Intangible benefits, for which assigning a dollar amount is difficult, include customer good will and employee morale.

Now that the costs and benefits have been determined, you can apply them to a payback analysis. When calculating payback analysis, you should show all dollar amounts using the time value of money. The **time value of money** states that one dollar today is worth more than one dollar promised a year from now because the dollar can be invested now and make interest. Likewise, if something is going to cost one dollar one year from now, you need to put away only 90.9 cents today, assuming a 10 percent discount rate. The discount rate is the opportunity cost of being able to invest money in other projects, such as stocks and bonds. The value of the discount rate is often set by a company's chief financial officer. The value 90.9 cents is the present value of one dollar one year from now. Table 14-1 shows the present value calculations that yield the value 90.9.

Table 14-1

The present value of a dollar over time given 8, 10, 12, and 14 percent discount rates

Present Value of a Dollar for different Discount Rates				
Periods	8%	10%	12%	14%
1	0.926	0.909	0.893	0.877
2	0.857	0.826	0.797	0.769
3	0.794	0.751	0.712	0.675
4	0.735	0.683	0.636	0.592
5	0.681	0.621	0.567	0.519
6	0.630	0.564	0.507	0.456
7	0.583	0.513	0.452	0.400

Figure 14-5 shows an example of a payback analysis. When examining the figure, note that *Development Cost* is a one-time cost that occurs in Year 0. Recurring costs are listed as *Operation and Maintenance Costs* in Years 1–6. Together, Development Cost and Operation and Maintenance Costs total the costs of the project over its intended lifetime. Assuming a discount rate of 12 percent, *Time Adjusted Costs* reflect total costs for each year times the discount rate. The *Cumulative Time Adjusted Costs* are simply the running sums of the Time Adjusted Costs over all the years. *Benefits Derived* values are the benefits, or income amounts, that are expected each year. Still assuming a discount rate of 12 percent, *Time Adjusted Benefits* are the Benefits times the discount rate for each year. The *Cumulative Time Adjusted Benefits* values are the running sums of the Time Adjusted Benefits over all the years. *Cumulative Lifetime Time Adjusted Costs* are the totals of the Cumulative Time Adjusted Costs plus the Cumulative Time Adjusted Benefits for each year.

Figure 14-5
*Payback analysis calcu-
lation for a proposed
project*

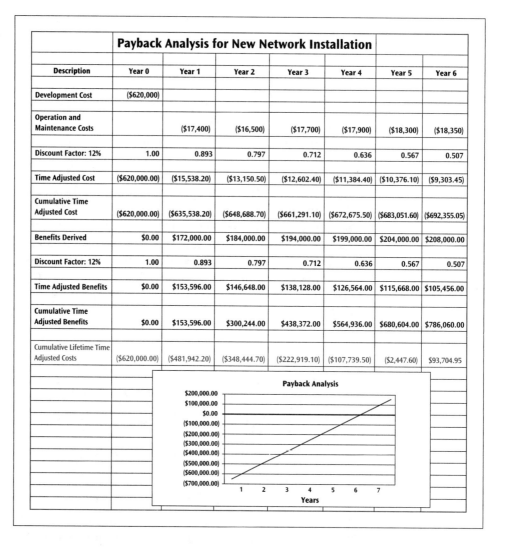

Description	Year 0	Year 1	Year 2	Year 3	Year 4	Year 5	Year 6
Payback Analysis for New Network Installation							
Development Cost	($620,000)						
Operation and Maintenance Costs		($17,400)	($16,500)	($17,700)	($17,900)	($18,300)	($18,350)
Discount Factor: 12%	1.00	0.893	0.797	0.712	0.636	0.567	0.507
Time Adjusted Cost	($620,000.00)	($15,538.20)	($13,150.50)	($12,602.40)	($11,384.40)	($10,376.10)	($9,303.45)
Cumulative Time Adjusted Cost	($620,000.00)	($635,538.20)	($648,688.70)	($661,291.10)	($672,675.50)	($683,051.60)	($692,355.05)
Benefits Derived	$0.00	$172,000.00	$184,000.00	$194,000.00	$199,000.00	$204,000.00	$208,000.00
Discount Factor: 12%	1.00	0.893	0.797	0.712	0.636	0.567	0.507
Time Adjusted Benefits	$0.00	$153,596.00	$146,648.00	$138,128.00	$126,564.00	$115,668.00	$105,456.00
Cumulative Time Adjusted Benefits	$0.00	$153,596.00	$300,244.00	$438,372.00	$564,936.00	$680,604.00	$786,060.00
Cumulative Lifetime Time Adjusted Costs	($620,000.00)	($481,942.20)	($348,444.70)	($222,919.10)	($107,739.50)	($2,447.60)	$93,704.95

In the project's sixth year, it finally turns a profit. Thus, it takes approximately five and one half years for payback. Most companies have established an acceptable payback period. If a company's payback period is, for example, six to seven years, then this payback analysis is one indicator that demonstrates that this proposed project might be financially feasible.

Capacity Planning

Computer networks are mission critical systems, and designing a new computer network or increasing the capacity of a current system requires careful planning. If you design a system that is not capable of supporting the generated traffic, response times will be sluggish and users will not be able to complete the necessary work in the required amount of time. This inability to perform work duties will lead to missed deadlines, projects backing up or even failing, and low employee morale. If the company is selling a product and that product can be purchased electronically, sluggish response times will equate to dissatisfied customers. In the Internet age when a competitor's web site is a simple click away, dissatisfied customers will quickly turn elsewhere.

At the other end of the spectrum, if you over-design a system, you will spend money unnecessarily to create a system that may never reach its capacity. It is arguable that if you are going to err in design, it is probably best to err in the direction of over designing, especially since it is difficult to predict the growth rate of new users and applications. Of all systems designed during the last ten years, a large percentage of systems are more than likely too small and cannot support the demands placed on them. A smaller percentage of network systems are over designed and remain relatively idle.

Capacity planning involves trying to determine the amount of network bandwidth necessary to support an application or a set of applications. Capacity planning is a fairly difficult and time consuming operation, and it is easy to plan poorly and thus design a system that will not support the intended applications. A number of techniques exist for performing capacity planning, including linear projection, computer simulation, benchmarking, and analytical modeling.

Linear projection involves predicting one or more network capacities based on the current network parameters and multiplying by some constant. For example, if you currently have a network of 10 nodes, and the network has a response time of X, using a linear projection, you might conclude that a system of 20 nodes would have a response time of $2X$. Some systems, however, do not follow a linear projection. If a linear projection is applied to these systems, inaccurate predictions may be the result. In these cases, an alternative strategy is required.

A computer simulation involves modeling an existing system or a proposed system using a computer-based simulation tool and subjecting it to varying degrees of user demand (called load). The advantage of a computer simulation is its ability to mimic conditions that would be extremely difficult, if not impossible, to create on a real network. On the negative side, computer simulations are difficult to create, and mistakes are easy to make and hard to discover. One mistake in a simulation can produce false results. Thus extreme care must be taken when creating a simulation.

Benchmarking involves generating system statistics under a controlled environment and then comparing those statistics against known measurements. A number of network benchmark tests exist that can be used to evaluate the performance of a network or its components. Setting up a benchmark test can be very time consuming. Benchmarking is a relatively straight-forward technique when compared to simulation and can provide useful information when analyzing a network. Unfortunately, like simulation, it can also suffer from a weakness due to possible errors. If all variables in the test environment are not the same as all variables in the benchmark environment, inaccurate comparisons will result. For example, if the test environment of the network uses one brand of router or switch and the benchmark network uses a different brand of router or switch, it may be invalid comparing the results of the two networks.

Analytical modeling involves the creation of mathematical equations to calculate various network values. For example, to calculate the utilization of a single communications line within a network (the percentage of time the line is being used), you can use the following equation:

$$U = t_{frame} / (2t_{prop} + t_{frame})$$

in which t_{frame} is the time to transmit a frame of data, and t_{prop} is the propagation time for a signal to transfer down a wire or over the airwaves.

Many experts feel the analytical model is a good way to determine network capacity planning. Like simulation, you can create analytical models that represent network systems that are difficult to create in the real world. Also similar to simulation is the disadvantage of inaccurately creating the analytical models, thus generating results that are invalid.

Let's consider two simple examples of capacity planning for single applications on a computer network. Since most networks support multiple applications, a person who performs capacity planning for a network has to calculate the capacity of each application on the network. Once the capacity of each application has been determined, it should be possible to determine the capacity for the entire network. You can calculate the individual capacities using analytical methods and then estimate the total network capacity using a linear projection.

The first example is a simple mathematical calculation and involves selecting an appropriate telecommunications technology that will support the transfer of very large sales records. (You may remember this example from Chapter Twelve.) Each sales record is 400 Kbyte and has to download in 20 seconds or less. The following equations calculate the capacity needed for one user downloading one file:

400 Kbytes = 400,000 bytes = 3,200,000 bits (8 bits to the byte).

3,200,000 bits / n bps = 20 seconds

Solving for n:

3,200,000 bits / 20 seconds = 160,000 bps (160 Kbps)

Thus, this application requires a telecommunications line capable of supporting a data rate that is at least 160 Kbps.

As another example of capacity planning, this time using a linear projection, consider a company that allows its users to access the Internet. Informal studies performed in the late 1990s show that the average Internet user requires a transmission link of 50 Kbps. As a precaution, let's double that figure to 100 Kbps to allow room for growth. Further measurements have shown that the peak hour for Internet access is around 11:00 am, with about 40 percent of the potential users on the system. If your system has 1000 potential users, 40 percent would yield 400 concurrent users, each at 100 Kbps. To satisfy 400 concurrent users each requiring 100 Kbps connections, network capacity would need to be 40 Mbps (400 X 100 Kbps). How many local area networks are capable of supporting this much traffic? An even more difficult question is how many telecommunication services can provide 40 Mbps between the Internet service provider and the company? Capacity planning in this example indicates that if the company is not willing to install communication links totaling 40 Mbps, restrictions might have to be applied to the employees using the Internet. For example, the company may have to limit the number of simultaneous users, or each user may not be given the full 100 Kbps capacity.

As you can see, capacity planning is difficult and not trivial. Once capacity planning is properly done, however, a network administrator can determine if the company's local area networks and wide area network connections can support the intended applications. Knowing the necessary capacity, however, does not tell the whole story. Other questions need to be answered. For example, how do you know whether the current network can or cannot handle the intended applications? Once the needed capacity is determined, you then need to closely examine the current network to determine its actual capacity. One of the best techniques for determining current network capacities is creating a baseline.

Creating a Baseline

Creating a **baseline** for an existing computer network involves the measurement and recording of a network's state of operation over a given period of time. As you will see shortly, creating a baseline involves capturing many network measurements over all segments of a network, including numerous measurements on workstations, user applications, bridges, routers, and switches. Since collecting this information appears to be such a large undertaking, why would you want to create a baseline? Network personnel create a baseline to determine the normal and current operating conditions of the network. Once a baseline is created, the results can be used to identify network weaknesses and strengths, which can then be used to intelligently upgrade the network.

Very often network managers feel pressure from users and network owners to increase the bandwidth of a network. Without thoroughly understanding whether network problems exist or where network problems are, a network manager may fix the wrong problem in order to improve network operation. Improving the network may involve increasing network bandwidth, but it could just as easily involve something less expensive such as upgrading some older equipment or segmenting a network with the use of a switch. By conducting a baseline study, and preferably a continuing study, a network manager can gain a better understanding of the network and can more effectively improve its overall quality.

Baseline studies can be started at any time, but are most effective when they are started during a time that the network is not experiencing severe problems, such as a node failure or a network interface card that is transmitting non-stop (called a jabber). Therefore, before beginning a baseline study, extinguish all immediate fires and try to get the network into fairly normal operation. Since you will be generating a large number of statistics, you will want to have access to a good database or spreadsheet application to keep the data organized. Once the database or spreadsheet has been set up, you are ready to begin your baseline study.

The next question is on what are you going to collect baseline information? Some options include system users, system nodes, operational protocols, network applications, and network utilization levels. Collecting information on system users involves determining the maximum number of users, the average number of users, and the peak number of users.

To collect baseline information on system nodes, you create a list of the number and type of system nodes, including computer workstations, routers, bridges, switches, hubs, and servers. It is a good practice to have an up-to-date road map of all nodes along with model numbers, serial numbers, and any address information such as Ethernet and IP addresses. This information should also include the products' vendor names and telephone numbers in case technical assistance is ever needed.

Collecting baseline information on operational protocols involves listing the types of operational protocols used throughout the system. Most networks support multiple protocols, such as TCP, IP, IPX, SPX, NetBIOS, and more. The more protocols supported, the more processing time necessary to convert from one protocol to another. If a baseline study discovers an older protocol which can be replaced, network efficiency should improve.

During the baseline study, you should also list all network applications, including the number, type, and utilization level of each application found on the network. A comprehensive list of applications on the network will help to identify old applications that should no longer be supported and should be deleted from the system. A list of applications will also help to identify the quantity of a particular application, which might indicate a violation of a particular software license. When creating a comprehensive list of network applications, don't forget to include the applications stored on both individual user workstations and network servers.

Assembling information on network utilization levels requires that you create a fairly extensive list of statistics. These statistics include many of the following values:

- ▶ Average network utilization (%)
- ▶ Peak network utilization (%)
- ▶ Average frame size
- ▶ Peak frame size
- ▶ Average frames per second
- ▶ Peak frames per second
- ▶ Total network collisions
- ▶ Network collisions per second
- ▶ Total runts
- ▶ Total jabbers
- ▶ Total cyclic redundancy checksum (CRC) errors
- ▶ Node(s) with the highest percentage of utilization and corresponding amount of traffic

Once you have collected and analyzed network utilization data, you can make several important observations. First, if network utilization on a CSMA/CD-based LAN is consistently over 40 to 50 percent, the network may be reaching saturation. If network utilization is at 100 percent, then 100 percent of the useable transmission space on the network is consumed by valid data. Since CSMA/CD networks are contention-based, they suffer from collisions. As the number of transmitting stations increases, the number of collisions increases, thus lowering the network utilization. Therefore, networks experiencing over 50 percent utilization are more than likely experiencing a high number of collisions and might need to be segmented with either bridges or switches.

A second observation you can make is when peak periods of network use occur. Making observations about peak periods of network use is easiest when you graph network activity data. Consider the hypothetical example shown in Figure 14-6. Peak periods occur at approximately 8:30 a.m., 11:30 a.m., 1:00 p.m., and 4:00 p.m. The most likely reason for these peaks would be users logging in and checking e-mail at 8:30 a.m. and 1:00 p.m., and users finishing work before going to lunch (11:30 a.m.) or going home (4:00 p.m.). If peak periods occur at times that appear unusual, it would be advantageous to understand reasons underlying each peak. If you know when peak periods occur and why they occur, network resources can be rearranged to help lessen the load during peak periods.

Figure 14-6
Peak periods of network activity in a typical day

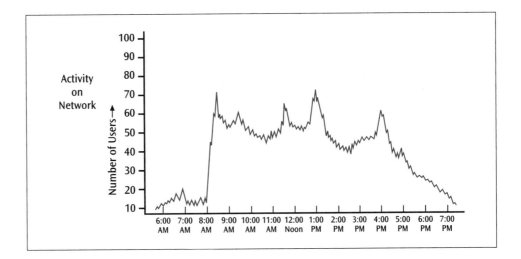

Examining network error statistics allows you to make another set of valuable observations. For example, a high number of runts with cyclic checksum errors indicates a high number of collisions and consequent workstation back offs. If a large number of workstations have to back off and wait random amounts of time, network utilization is going to suffer. A high number of runts with no cyclic checksum errors indicates the frames completely left the workstation before the collision was detected. One possible cause for frames completely leaving the workstation before a collision is detected might be a network segment that is longer than it should be.

Examining the amount of traffic on each node also yields valuable information about network performance. Typically, a small percentage of network nodes accounts for a large percentage of network traffic. It is not unusual to encounter a node, such as a router or server, which is at the center of a great deal of traffic. A user workstation, however, that generates a high amount of traffic is suspect, and should be examined more closely. A classic example is the case of a user who had two workstations at two different locations on one network. Using the e-mail system, the user inadvertently set the e-mail program of each workstation to forward incoming mail to the other workstation. The user had created an infinite loop of forwarded e-mail messages and brought the network to a crawl.

Once you have performed the baseline study, don't stop observing the network. For a baseline study to be really effective, you need to maintain it. An on-going baseline study gives a system manager an effective tool for identifying network problems, repairing the network, responding to complaints, improving the weak spots, and requesting additional funding.

Network Manager Skills

Once the analysis and design phases of network development are completed and the computer network is in place and operating, it is the network manager's responsibility to keep it running. Keeping a network running includes making repairs on failed components, installing new applications and updating the current ones, keeping current system users up to date, and looking for new ways to improve the overall system and service level. It is not an easy job. With the complexity level of today's networks and businesses' dependence on their applications, network managers are highly valuable, visible, and always on the move.

Since many network managers are dealing with both computers and people, they need the skills necessary to work with both. A checklist of skills for the network manager would include a wide platform of technology skills, including but not limited to knowledge of local area networks, wide area networks, voice telecommunication systems, data transmission systems, video transmission, basic hardware concepts, and basic software skills. A network manager should also have people skills, including the ability to talk to users in order to service problems and explore new applications. Along with people skills, a network manager also needs training skills. Training skills include the ability to train users or other network support personnel.

To make effective use of limited funds, a network manager should also possess a number of common management skills, such as budget management skills, which include knowledge of how to prepare a budget to justify continuing funds or to request additional funds. Along with budget management skills basic statistical skills are needed, including the knowledge of how to collect and use system statistics to justify existing systems or to validate the addition of new systems. Time management skills are also a necessity and include the ability to manage not only his or her own time, but also project schedules and any information technology workers' time. Just as valuable as time management skills are project management skills, including the ability to keep a project on schedule and knowledge of how to use project estimating tools, project scheduling tools, and methods for continuous project assessment. Finally, a network manager should possess policy creation and enforcement skills, which include the ability to create policies concerning the use of the computer systems, access to facilities, password protection, access to applications, access to databases, distribution of hardware and software, replacement of hardware and software, and the handling of service requests.

To learn new skills and demonstrate proficiency within a particular area, the network manager can obtain certification. Many network managers become certified on a particular type of network operating system, such as Novell NetWare or Windows NT, or on a particular brand of network equipment, such as Cisco routers or Nortel Networks. The following is a list of the more popular certification programs:

- ► Novell NetWare Certified Novell Engineer (CNE) - This certification addresses the design, installation, and support of Novell network operating systems. You can get certified for a particular version of NetWare.
- ► Novell NetWare Certified Novell Administrator (CNA) - This certification addresses the design of Novell network operating systems.
- ► Microsoft's Certified Network Systems Engineer (CNSE) - This certification addresses the design, installation, and support of the Windows NT operating system.
- ► Cisco Certified Internet Engineer (CCIE) - This certification addresses the Cisco family of routers, switches, and other related Cisco equipment.
- ► Nortel Networks Certification - A certification in Nortel Networks provides an in-depth study of Nortel network systems.
- ► IBM Certified Solutions Expert (CSE) and Certified Specialist (CS) - These certifications demonstrate the ability to successfully plan, install, and support IBM's Networking LAN products.

The position of network manager is demanding, challenging, and always changing. To be a successful network manager requires a wide range of technical, management, and people skills. A good network manager is constantly learning new skills and trying to keep abreast of the rapidly evolving technology. A computer network system could not survive without the network manager.

Generating Useable Statistics

Computer networks are in a constant state of change. New users and applications are added, while former users and applications no longer desired are deleted. A network and its underlying technology is often based on Internet-years, which many experts equate to approximately 90 calendar days. Since the technology changes so quickly and networks are constantly being called upon to support new and computationally intensive applications, a network administrator is constantly working on improving the data transfer speed and throughput of network applications.

To support changes to a network, a network administrator needs funding. Management, unfortunately, is not always receptive to investing more funds in technology. Often management needs to be convinced that services are suffering and response time is not what it needs to be. Statistics on computer network systems can be a very useful tool for demonstrating the need to invest in technology. If properly generated, statistics can be used to support the request for a new system or modifications to an existing system.

Four statistics, or measures, that are useful in evaluating networks are mean time between failures, mean time to repair, availability, and reliability. **Mean time between failures (MTBF)** is the average time a device or system will operate before it fails. This value is sometimes generated by the manufacturer of the equipment and passed along to the purchaser. Often this value is not available, and the owner of the equipment has to generate a value given the equipment's past performance.

Mean time to repair (MTTR) is the average time necessary to repair a failure within the computer network. This time includes the time necessary to isolate the failure. It also includes the time required to either swap the defective component with a working component or the time required to repair a component either on site or by removing and sending the component to a repair center. Finally, mean time to repair includes the time needed to bring the system back up to normal operation. The value of mean time to repair depends on each installation and, within an installation, on each type of component.

The third statistic, **availability**, is the probability that a particular component or system will be available during a fixed time period. A component or network with a high availability (near 1.0) is almost always operational. The value for availability is based on mean time to repair and mean time between failure values. Components with a small MTTR and a large MTBF will produce values very near to 1.0. Availability is defined by the following equation:

$$A(t) = a/(a+b) + b/(a+b) \times e^{-(a+b)t}$$

in which: $a = 1/\text{MTTR}$
$b = 1/\text{MTBF}$
$e = \text{natural log function}$
$t = \text{the time interval}$

To understand how this formula works, let's work through an example. Suppose we want to calculate the availability of a modem that has a MTBF of 3000 hours and a MTTR of 1 hour. The availability of this modem for an 8-hour period is:

$a =$ 1/1
$b =$ $1/3000 = 0.00033$
A(8 hours) = $1/(1 + 0.00033) + 0.00033/(1 + 0.00033) \times e^{-(1 + 0.00033)8}$
      = $0.9997 + 0.00033 \times 0.000335$
      = 0.9997

Since the availability is near 1.0, there is a very high probability that the modem will be available during an 8-hour period.

To calculate the availability of a system of components, calculate the availability of each component and find the product of all availabilities. For example, if a network has three devices with availabilities of 0.992, 0.894, and 0.999, the availability of the network is the product of 0.992 * 0.894 * 0.999.

The fourth statistic, reliability, calculates the probability that a component or system will be operational for the duration of a transaction of time t. Reliability is defined by the equation:

$$R(t) = e^{-b}t$$

in which: $b = 1/\text{MTBF}$
             t = the time interval of the operation

What is the reliability of a modem if the MTBF is 3000 hours and a transaction takes 20 minutes, or 1/3 of an hour (0.333 hours):

$$R(0.333 \text{ hours}) = e^{-(1/3000)(0.333)} = e^{-0.000111} = 0.99989$$

The reliability of the modem is very near to 1.0. A reliability of exactly 1.0 means the network or device is reliable 100 percent of the time.

Let's consider a second example. Suppose a computer terminal is getting old and starting to fail frequently. If we estimate that the MTBF for this terminal is 100 hours, and a terminal transaction takes 30 minutes, what is the reliability of this device?

$$R(0.5 \text{ hours}) = e^{-(1/100)(0.5)} = e^{-0.005} = 0.995$$

Although this value also appears to be near 1.0, there is a difference between the two examples. In the first example, a value of 0.99989 was calculated, and in the second example the value 0.995 was calculated. The difference between these two values is 0.00489. Stated another way, in 1000 repetitions of a trial, a particular event may occur 5 times. On a network many events occur or repeat thousands of times. For example, you wouldn't want to experience 5 terminal failures in a year, let alone one during the transmission of data. Therefore, many network managers strive to maintain system availability and reliability values of 0.9999 to 0.99999.

Managing Operations

To assist network managers and information technologists in doing their jobs, the computing services within a business needs a control center. The control center is the heart of all network operations. It contains, in one easily accessible place, all the network documentation, including network resource manuals, training manuals, baseline studies, all equipment documentation, user manuals, vendor names and telephone numbers, procedure manuals, and forms necessary to request services or equipment. The control center can also contain a training center to assist users and other information technologists. In addition, the control center contains all hardware and software necessary to control and monitor the network and its operations.

One of the more important elements of a control center is the help desk. A help desk answers all telephone calls and walk-in questions regarding computer services within the company. Whether hardware problems, questions about running a particular software package, or computing services providing a new service, the help desk is the gateway between the user and computing and network services. To assist the operations staff, very good help desk software packages exist to support computing services in tracking and identifying problem areas within the system.

A well designed control center and help desk can make an enormous impact on the users within a business. When users know there is a friendly, available person they can turn to for any computing problems, there is much less computer system/computer user friction.

Simple Network Management Protocol (SNMP)

Imagine a network that is composed of many different types of devices, including workstations, routers, bridges, switches, and hubs. The operation of the network is running smoothly, when all of a sudden the network begins to experience problems. It becomes sluggish, and users start calling you complaining of poor network response times. Is there something you can do to monitor or analyze the network without leaving your office and running from room to room or building to building? There is, if all or most of the devices on the network support a network management protocol. A network management protocol facilitates the exchange of management information between network devices. This information can be used to monitor network performance, find network problems, and then solve those problems, all without physically touching the affected device.

Although a number of different protocols exist to support network management, one protocol stands out as the simplest to operate, easiest to implement, and most widely used—Simple Network Management Protocol. Simple Network Management Protocol (SNMP) is an industry standard created by the Internet Engineering Task Force designed originally to manage Internet components but now also used to manage wide area network and telecommunication systems. Three versions of SNMP exist:

- ▶ SNMP Version 1, the original protocol and based on a standard similar to OSI, Common Management Information Protocol/Common Management Information Service (CMIP/CMIS);
- ▶ SNMP Version 2, developed to be independent of the OSI standard;
- ▶ SNMP Version 3 which addresses the problem of management security.

SNMP operates on a network between the application layer and the UDP/IP layers. Note that the SNMP protocol does not operate over TCP/IP, but operates over UDP/IP instead. TCP is a connection-oriented protocol, while UDP (User Datagram Protocol from Chapter Eleven) is a connectionless protocol. Since SNMP is operating at a higher layer and is also a connectionless protocol, it makes sense that the next lower layer—UDP—is also a connectionless protocol.

All three versions of SNMP are based on the following set of principles. Network objects consist of network elements such as servers, mainframe computers, printers, hubs, bridges, routers, and switches. Each of these elements can be classified as either managed or unmanaged. A managed element has management software, called an **agent**, running in it and is more elaborate and expensive than an unmanaged element, which does not have the software agent. A second object— the **SNMP manager**—controls the operations of a managed element and maintains a database of information about all managed elements. A manager can query each agent and receive management data, which it then stores in the database. An agent can send unsolicited information to the manager in the form of an alarm. Finally, a manager can also act as an agent if a higher level manager calls upon the manager to provide information for a higher level database. All this managing and passing of information can be done either locally or remotely—for example, from across the country when the information is transmitted over the Internet.

The database that holds the information about each managed device is called the **Management Information Base (MIB)**. The MIB is a collection of information that is organized hierarchically, much as a network printer is connected to a hub, which is connected to a local area network, which is connected to a router, which is connected to the Internet. A manager can query a managed element (agent) and ask for the particular details that currently exist within that element at that moment in time. For example, a manager might ask a router how many packets have entered the router, how many packets have exited the router, and how many packets were discarded due to insufficient buffer space. This information is then stored for later use by a management program, which might conclude after looking at the statistics that a particular element is not performing properly.

SNMP can also perform an auto-discovery type operation. This operation is used to discover new elements that have been added to the network. When SNMP discovers a newly added element, the information about the element is added to the MIB. Thus, SNMP is a dynamic protocol that can automatically adapt to a changing network. This adaptation does not require human intervention (except for the intervention of connecting the new element to the network).

Managed elements are monitored and controlled using three basic SNMP commands: read, write, and trap. The read command is issued by a manager to retrieve information from the agent in a managed element. The write command is also issued by a manager, but is used to control the agent in a managed element. By using the write command, a manager can change the settings in an agent, thus making the managed element perform differently.

A weakness of the first two versions of SNMP was the lack of security in this write command. Anyone posing as an SNMP manager can send bogus write commands to managed elements, thus causing potential damage to the network. SNMP Version 3 addresses the issue of security so that bogus managers cannot send malicious write commands.

The third command—the trap—is used by a managed element to send asynchronous reports to the manager. When certain types of events occur, such as a buffer overflow, a managed element can send a trap to a manager reporting the event.

Most often the SNMP manager requests information directly from a managed element on the same network. What if a manager wants to collect information from a remote network? **Remote Network Monitoring (RMON)** is a protocol that allows a network manager to monitor, analyze, and troubleshoot a group of remotely managed elements. RMON is defined as an extension of SNMP, and the most recent version is RMON Version 2 (often referred to as RMON2). RMON can be supported by hardware monitoring devices (known as probes), through software, or through a combination of hardware and software. A number of vendors provide networking products with RMON support. For example, Cisco's series of local area network switches includes software in each switch that can record information as traffic passes through and can record it in its MIB. RMON can collect several basic kinds of information, such as number of packets sent, number of bytes sent, number of packets dropped, host statistics, and certain kinds of events that have occurred. A network administrator can find out how much bandwidth or traffic each user is imposing on the network and can set alarms in order to be aware of impending problems.

Network Diagnostic Tools

To support a computer network and all of its workstations, nodes, wiring, applications, and protocols, network personnel need an arsenal of diagnostic tools. The arsenal of possible diagnostic tools continues to grow with more powerful and helpful tools becoming available every day. Diagnostic tools can be grouped into two categories: tools that test and debug the network hardware and tools that analyze the data transmitted over the network. Let's examine the tools that test the network hardware first.

Tools that test and debug network hardware

Tools that test and debug network hardware range from very simple devices to more elaborate, complex devices. Three common testing devices are electrical testers—the simplest, cable testers, and local area network testers—the most elaborate.

Electrical testers measure ac and dc volts, resistance, and continuity. An electrical tester will show if there is a voltage on a line, and if so, how much voltage. If two bare wires are touching each other, the two wires will create a short, and the electrical tester will show zero resistance. The continuity tester is a handy device that shows whether two wires are grounded to each other. Electrical and continuity testers are used to determine if the wires themselves are experiencing simple electrical problems.

Cable testers are slightly more elaborate devices. They can verify connectivity and test for line faults such as open circuits, short circuits, reversed circuits, and crossed circuits. Other handheld cable testers can also test fiber optic lines, asynchronous transfer mode networks, and T1 circuits. For example, if a connector hidden in some wiring closet contains two wires that are switched, a cable tester will detect the problem and point to the source of the problem.

One of the most elaborate devices is the local area network tester. These testers can operate on Ethernet and token ring networks with or without switches. One particular tester from the Fluke Company has a display that graphically shows a network segment and all of the devices attached to it. This tester can troubleshoot the network and suggest possible corrections simply by plugging into an available network jack. A common problem solved by these devices is the identification and location of a network interface card (NIC) that transmits continuously without sending valid data (a jabber). The Fluke device will pinpoint the precise NIC by indicating the 48-bit NIC address. A network administrator simply looks up the particular NIC address in the system documentation and maps it to a unique machine in a unique office.

Tools that analyze data transmitted over the network

The second category of tools analyze data transmitted over the network. These tools include protocol analyzers and devices or software that emulate protocols and applications.

One of the most common tools is the traffic analyzer or protocol analyzer. A *protocol analyzer* monitors a network 24 hours a day, seven days a week, and captures and records all transmitted packets. Each packet's protocol is analyzed and statistics are generated that show which devices are talking to which devices and which applications are being used. This information can then be used to update the network so that it operates more effectively. As an example, if a protocol analyzer sees that a particular application is being used a great deal and is placing a strain on network resources, a network administrator may consider alternatives such as replacing the application with a more efficient one, or redistributing the application to the locations where it is used the most.

Other useful tools include devices or software programs that can emulate protocols and applications. By generating fictitious data, a network manager can simulate conditions that might not normally occur or that occur only under unusual circumstances. This information can then be used to decide if a particular modification to the network would be worth the time and expense. For example, suppose management is considering a new application that allows customers to place orders directly and check the status of previous orders. A network designer can generate fictitious data that might represent these new transactions and observe the additional load that will be placed on the network. This information can then be used to help determine what additional resources, if any, will be necessary to support the new application.

Capacity Planning and Network Design in Action: BringBring Corporation

To see how capacity planning and network design work in a realistic setting, let's return to the BringBring Corporation from Chapter Twelve. Recall that BringBring's administrative headquarters is in Chicago, IL, and its regional sales offices are in Seattle, WA, San Francisco, CA, and Dallas, TX. BringBring is expanding its data networking capability and has asked you to help design its new corporate network. It wants to add new hardware and software and provide new data applications. Although the company has enough funds to support the new applications, it wants to seek the most cost-effective solution for the long run.

BringBring currently has only a few PCs in use. The Marketing group in Chicago runs applications on 25 standalone PCs. The other sites currently have no PCs. BringBring Corporation plans to add 10 additional workstations to the Chicago office and plans to add 35 PCs each to the Seattle, San Francisco, and Dallas offices. The following 6 servers will be installed in its new network:

- ▶ Web Server (Chicago) - an HTTP server to store web pages for public access to corporate marketing information.
- ▶ Inventory Server (Chicago) - a Database server that stores information about available product inventory.
- ▶ Site Servers (Chicago, Seattle, San Francisco, Dallas) – Four servers, one at each site, that provide e-mail services so employees can communicate about important projects, store site-specific management files and sales files, and store staff software and information for clerical workers.

Once the new workstations, PCs, servers, and network equipment are in place, BringBring plans to use the network for local site server access, e-mail services, Internet access, and network access to sales records. Each employee needs to be able to access data files on the site server at his or her location. Each employee accesses enough local files that at least 1 Mbps of bandwidth is needed for each PC on any shared LAN. To provide this level of service, for example, a 16 Mbps shared Token Ring should have no more than 16 workstations connected to it. In addition, each site server needs to be able to exchange e-mail messages with any other site server. Connectivity needs to be provided from the web server to the Internet, so current and potential customers around the world can use web browsers to access company information. Finally, sales people at the regional sales offices must be able to access data on the inventory server in Chicago. Typically about 20,000 sales records per month will be uploaded from each of the regional sales offices (Seattle, San Francisco, and Dallas). Each sales record is 400 Kbytes in size. It should take no longer than 20 seconds to upload a single sales record.

In Chapter Twelve, we found a solution only for the network access to sales records portion of BringBring's networking problem. That solution included 256 Kbps frame relay connections between Seattle and Chicago, San Francisco and Chicago, and Dallas and Chicago. Now let's find a solution for the complete problem—network access to sales records, local site server access, e-mail access, Internet access, and internal local area networks.

To support e-mail access and Internet web page access for each site, each site needs a high speed connection to a local Internet service provider. Each Internet user needs a 100 Kbps capacity. Assuming one third of the 35 users (12 users) at a site will be accessing the Internet at one time, a linear projection of 12 times 100 Kbps per link produces a need for a 1.2 Mbps connection. Since the connection from a site to an Internet service provider should be a local connection, a local T1 line could be added to each site to support Internet access. Recall that T1 lines are capable of supporting a continuous 1.544 Mbps data stream between two locations. Since e-mail access also goes through an Internet service provider and is very low capacity compared to web page browsing, these additional T1 lines should also be fine for supporting e-mail. To allow BringBring's Chicago office to host a web server will require another connection to a local Internet service provider using at least one more T1 line to provide 1.544 Mbps access to the web server.

To help us see the overall physical layout of the four geographic locations, let's create a simple overview location connectivity diagram (Figure 14-7). The overview LCD includes the four locations: Chicago, Seattle, San Francisco, and Dallas. The networking need that involves downloading sales records from the Chicago office requires us to create a connection between Seattle and Chicago, San Francisco and Chicago, and Dallas and Chicago. Note that the mileage figures are included on the connections between the sites. To provide Internet e-mail access and web access, each site requires a local connection to an Internet service provider. Since each Internet service provider is external to the BringBring corporation, each ISP is shown as a circle with an X drawn through it.

Figure 14-7
Overview location connectivity diagram showing Seattle, San Francisco, Dallas, and Chicago sites

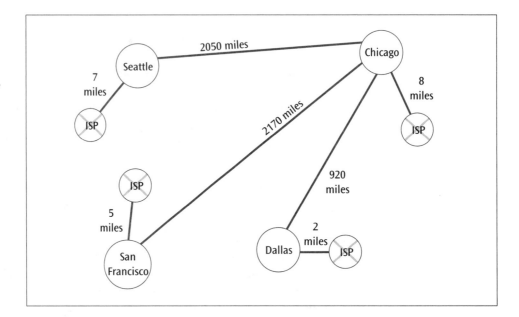

To support local site server access (which includes e-mail services so employees can communicate about important projects, storage services for site-specific management files and sales files, and storage services for staff software and information for clerical workers), each site will need to create a local area network solution. Recall that each user will require at least 1 Mbps bandwidth to connect his or her workstation to the local area network. If each site has 35 workstations, the local area network will need to support at least a 35 Mbps total capacity. Thus, a 100 Mbps CSMA/CD LAN would be prudent. CSMA/CD is a good choice because it is the most popular local area network, which should improve its chances for quality support and reduce its overall costs. A 100 Mbps version has a very reasonable cost, and it should be sufficient to support a total of 35 users per site. Furthermore, it might be worthwhile to use high-speed switches instead of hubs to support the 35 workstations. The use of switches should provide better network segmentation and decrease the likelihood of collisions, thus increasing overall network throughput. In addition, the cost of switches continues to decline, making them extremely cost effective given the high degree of network segmentation they provide.

In conclusion, BringBring will need a variety of network devices to support its increased computing requirements. Frame relay connections provided by a national frame relay provider will create the permanent virtual connections between each of the regional sales offices and the administrative office in Chicago. T1 connections will connect each office to the frame relay provider using local distance services. Each office will also require additional local T1 connections to Internet service providers. These connections will provide Internet e-mail access and World Wide Web access. Finally, CSMA/CD local area networks will provide the on-site glue that allows each workstation to access local servers as well as remote servers via the Internet.

SUMMARY

Read over this carefully

▶ When creating a new network or adding to an existing network, there are many potential pitfalls and opportunities for inaccurate and incomplete assessments.

▶ The Systems Development Life Cycle (SDLC) is one of the most popular techniques used to guide analysts through the difficult decision making phase. Although there are many versions of the SDLC model, most models consist of the following phases: planning, analysis, design, implementation, and maintenance.

▶ Persons designing a new network or upgrading an existing network may want to create one or more network models to help them visualize the system. The location connectivity diagram is one of the common network models used to depict internal and external network structures.

▶ A very important part of the SDLC model is conducting one or more feasibility studies. Feasibility studies can be conducted during the planning phase, the analysis phase, the design phase. They can also be conducted by themselves not in conjunction with SDLC. There are four basic types of feasibility studies: financial, operational, technical, and time.

▶ Payback analysis of a proposed computer network system is one possible financial analysis technique that involves determining the proposed system's costs and benefits.

▶ Capacity planning is a necessary technique that allows a network manager to determine the needed network bandwidth that will support one or more applications within a business. A number of techniques exist for performing capacity planning, including linear projection, computer simulation, benchmarking, and analytical modeling.

▶ A baseline study involves measuring and recording a network's state of operation over a given period of time. Many network administrators feel a baseline study is a must for all network operations, regardless of whether or not more network resources are currently being requested.

▶ The baseline study can serve a number of purposes, such as providing an understanding of the current system, helping to isolate and identify network problems, and providing evidence that more computing resources will be needed in the near future.

▶ Once a network is in operation, good network management is necessary to keep the network operating at peak efficiency. The network manager is responsible for making sure that the network operates at peak efficiency.

▶ A network manager should possess a number of skills, including hardware and software knowledge, people management skills, problem solving skills, and a knowledge of statistics. To develop the technical skills of the manager, a number of certification programs exist from the more popular vendors of network hardware and software systems.

▶ A network manager should be able to create and use basic statistics, such as mean time between failures, mean time to repair, reliability, and availability. These statistics can be used to justify current network resources or to validate the need for additional network resources.

▶ All networks need a command center. It is in this command center that you find help desks, documentation, training centers, and the central nervous system of network operations.

▶ The Simple Network Management Protocol (SNMP) helps network support personnel monitor network performance, find network problems, and then solve those problems without physically touching the affected device.

▶ SNMP defines managers that request information from agent software running on a managed device or element. SNMP managers can also send information to managed elements in order to control operations on the network.

▶ The database that holds the information about each managed device is called the Management Information Base (MIB).

▶ Remote Network Monitoring (RMON) is a protocol that allows a network manager to monitor, analyze and troubleshoot a group of remotely managed elements. RMON is an extension of SNMP and is currently in its second version.

▶ A good variety of diagnostic tools exists that can aid network personnel in troubleshooting and maintaining the complex computer networks that so predominantly exist today.

▶ The more common diagnostic tools include electrical testers, continuity testers, cable testers, local area network testers, protocol analyzers, and devices and programs that can emulate protocols and applications.

KEY TERMS

agent	maintenance phase	Remote Network Monitoring (RMON)
analysis phase	Management Information Base (MIB)	Simple Network Management Protocol (SNMP)
availability	mean time between failures	
baseline	mean time to repair	SNMP manager
benchmarking	network management protocol	systems analyst
capacity planning	operationally feasible system	systems development life cycle (SDLC)
design phase	payback analysis	technically feasible system
financially feasible system	planning phase	time feasible system
implementation phase	protocol analyzer	time value of money
location connectivity diagram (LCD)	reliability	

REVIEW QUESTIONS

1. Describe each phase of the systems development life cycle.
2. What is the primary goal of the planning phase of SDLC?
3. What is the primary goal of the analysis phase of SDLC?
4. What is the primary goal of the design phase of SDLC?
5. What is a location connectivity diagram and how can it assist you in designing a network?
6. Describe the four different types of feasibility studies.
7. What is meant by the time value of money?
8. Describe the four different ways to perform capacity planning.
9. For what reasons might someone perform a baseline study?
10. List the three most important skills a network manager should possess.
11. What is the difference between mean time between failures and mean time to repair?
12. What is meant by the statistical term availability?
13. What is meant by the term utilization?
14. What is meant by the statistical term reliability?
15. What is the function of the Simple Network Management Protocol?

16. What is the difference between a manager and an agent in SNMP?

17. How can Remote Network Monitoring be used to assist SNMP?

18. What should be found in the control center for a network operation?

19. What basic diagnostic tools are used to support a computer network?

EXERCISES

1. State during which phase or phases of the systems development life cycle the following actions are performed:
   a. Create a data dictionary
   b. Install the system
   c. Train users
   d. Write documentation
   e. Perform feasibility studies
   f. Test the system
   g. Create files and databases

2. Using the following data, calculate the payback period.

   Development cost: $418,040

   Operation and maintenance costs (year 0 – year 6): 0; $15,045; $16,000; $17,000; $18,000; $19,000; $20,000.

   Discount rate: 12%

   Benefits (year 0 – year 6): $0; $150,000; $170,000; $190,000; $210,000; $230,000; $250,000.

3. Create a simple analytical model that includes two formulas for calculating the approximate total time T for n terminals using roll call polling and hub polling. Use TD = time to transmit data, TRP = time to transmit a roll call poll, and THP = time to transmit a hub poll.

4. You are performing a baseline study for your company, which is located on the east coast. Your company does a lot of work with businesses on the west coast. You note peak network utilization at approximately noon, when most of your employees are on lunch break. What could be causing this peak activity?

5. During a baseline study, a high number of runts with no cyclic checksum errors were discovered. Explain precisely what this information has to do with network segment length.

6. If a component has a MTBF = 2300 hours and the MTTR = 30 minutes, calculate the availability of the component during an 8-hour period.

7. If a component has a MTBF = 10 hours and a transaction takes 20 minutes, calculate the reliability of the component.

8. In a system with three components, each component has a MTBF = 500 hours, and the MTTR = 3 hours, calculate the availability of the system during a 24-hour period.

9. If a network has four devices with the availabilities of 0.994, 0.778, 0.883, and 0.5, what is the availability of the entire network?

10. Is it possible for an SNMP agent in a managed device to also serve as a manager? Explain how this situation might work.

11. What are the differences between network line continuity testers and network cable testers?

12. You are working for a small company that has a local area network with two hubs. The communications line between the hubs has just been cut, but you don't know that. How can you determine what has happened?

THINKING OUTSIDE THE BOX

1 You have been asked to create a help desk for the computer support division of your company. What services will your help desk provide? How will you provide those services? What type of employees will you hire to work at the help desk?

2 Your company wants to create a web server to promote its business. One of the features of the web server allows remote users to download service bulletins and repair manuals. These bulletins and repair manuals are approximately 240 Kbytes in size. You anticipate approximately 30 users per hour will want to download these documents. What speed communications line do you need to support this demand?

3 The company you are working for sells dolls and their clothing outfits through a retail outlet and through mail order catalogs. It is now considering selling merchandise via the Web. Create an SDLC plan for adding a web merchandising system to the business that shows each step involved in the analysis, design, and implementation phases. The plan should have the appearance of an outline. For example, the analysis phase could begin with the following:

I. Analysis Phase

    A. Interview upper management

    B. Create a questionnaire and present to current employees

    C. Create a model showing the current data flow of the mail order catalog business

    D. Etc.

PROJECTS

1. Create a series of overview and detailed location connectivity diagrams for either your place of work or your school. Try to include as many different external locations as possible. Selecting one external location, create one or more detailed location connectivity diagrams.

2. Perform a baseline study for the network at work or at school. Create a list of all network devices (servers, routers, bridges, switches, hubs, and so on). What protocols are supported by the network? What applications are supported by the network? Try to collect some statistics for one type of device on the network. If it is not possible to get actual values, simply create a list of the statistics that would be valuable for a particular device.

3. What certification programs, other than those listed in the chapter, exist for network administrators? What type of systems are these certificates for?

4. What different benchmark tests are available to test a computer system or computer network? Write a paragraph about each benchmark test stating what the benchmark test actually tests.

5. Scanning the local newspaper want-ads, what percentage of advertisements seeking network support personnel mentioned the requirement of a certification degree?

6. SNMP is only one of several management protocols. What other management protocols exist? Who created them? What types of situations do they manage?

7. How does SNMP Version 3 handle security? Write a one- or two-page report that summarizes the main features.

8. How does the auto-discovery feature of SNMP operate?

Appendix
Pioneering Protocols

◆ ◆

Students studying ancient civilizations and languages often wonder about the point of studying cultures that no longer exist. How will the way people lived hundreds and thousands of years ago benefit you in the here and now? The answer that is often given is that current civilizations are based in many ways upon the ancient civilizations that existed long before the present moment. The same argument, to a lesser degree, can be made for some of the early protocols that shaped and guided the world of computer networks and data communications to where it is today. Unlike ancient civilizations that no longer exist, however, some of the early network protocols still exist today and are actively supporting computer networks. One of the earliest protocols for providing a data link connection between terminals and a mainframe computer is the BISYNC protocol. Although you may be hard pressed to find a current system still using BISYNC, the concepts and principles introduced with BISYNC can be found in more modern protocols.

Synchronous data link control protocol was designed to replace BISYNC and is a more modern data link control protocol. Although many descendants from SDLC have appeared over the years, SDLC is still used in IBM-type systems today. SDLC's close cousin, high-level data link control (HDLC) protocol, is also still found in some non-IBM systems. Many of the concepts introduced in SDLC (and then HDLC) have carried over into many different types of protocols.

Finally, the pioneering protocol for packet switched networks—X.25—will be introduced. When packet switched networks were introduced in the 1960s, the modern day Internet (also a packet switched network) was barely a twinkle in someone's eye. Although X.25 is still offered by some telecommunication carriers across the country, its descendants—frame relay and asynchronous transfer mode—are quickly replacing the original protocol. Once again, however, many concepts were introduced with X.25 that are used in other similar protocols today.

BISYNC Transmission

Introduced in the mid-1960s, IBM's BISYNC protocol was designed to provide a general-purpose data link protocol for point-to-point and multipoint connections. Although the protocol itself is very old and rarely used today, the concepts introduced with BISYNC have carried over into many protocols that are currently in use. Because of this carry over, it is worthwhile to examine this classic protocol.

BISYNC (or BSC) is a half-duplex protocol—data may be transmitted in both directions, but not at the same time. Stated another way, it is a stop and wait protocol, in that one side sends a message, then stops and waits for a reply. Typically one end of the connection is termed the *primary*, and the opposite end is termed the *secondary* (one secondary if point-to-point, multiple secondaries if multipoint). In most situations, the primary controls the dialogue and may perform polls and selects (in which the primary sends to the secondary) of the secondary counterparts.

BISYNC is also called a character-oriented protocol. The messages or packets that are transmitted between primary and secondary are collections of individual characters, many of which are special control characters that drive the dialogue. For example, if the primary wishes to poll the first of multiple secondaries, the following packet is transmitted:

SYN SYN address1 ENQ

The SYN character establishes synchronization of the incoming message with the receiver and precedes all message packets. It can also be inserted into the middle of longer messages to maintain synchronization. The address1 field is the address of the intended secondary. The ENQ character, or inquiry, is used to initiate a poll. Note that, as stated earlier, each control character is an individual character. If you examine the ASCII character set, you will see that the SYN character is a valid ASCII character with the decimal value of 22; the ENQ character is a decimal 5.

If the addressed secondary has nothing to transmit, it responds with:

SYN SYN EOT

The EOT character signifies the End Of Transmission, or that the secondary has nothing to send. If the secondary did have something to send to the primary, it would respond with a message such as:

SYN SYN STX text EOT BCC

The STX character signals the Start of TeXt, and data (text) follows. The BCC character is the Block Check Count, which is an error checksum appended to the end of the data.

If an optional header with control information is included in the packet, the message would look something like the following:

SYN SYN SOH header STX text EOT BCC

SOH indicates the Start Of Header.

Upon receipt of the data at the primary, if there is no checksum error, the primary responds with:

SYN SYN ACK0

The ACK0 character is a positive acknowledgement for even-sequenced packets. If the packet was corrupted during transmission and the primary receives a garbled message, it replies with:

SYN SYN NAK

The NAK character is a negative acknowledgement. It would then be the responsibility of the secondary to retransmit the message.

The more commonly found BISYNC control codes and their meanings are listed in Table A-1.

Table A-1
Commonly found BISYNC characters

BISYNC Control Codes	Meaning
ACK0	ACKnowledgement - Positive acknowledgement to even-sequenced packets of data or as a positive response to a select or bid.
ACK1	ACKnowledgement - Positive acknowledgement to odd-sequenced packets of data.
DLE	Data Link Escape - used to implement transparency (described below).
ENQ	inquiry - used to initiate a poll message or select message; used to bid for the line in contention mode; used to ask that a response to a previous transmission be resent.
ETB	End of Transmission Block - used to signal the end of transmission of a block of data. A single message, due to its length, may be broken into multiple blocks.
ETX	End of TeXt - used to signal the end of a complete text message. If a message is broken into multiple blocks, the ETX is placed only at the end of the message.
EOT	End Of Transmission - used to signal the end of a transmission. If the transmission of data consists of multiple messages, the EOT is placed only at the end of the last message. May also be used as a response to a poll.
ITB	end of Intermediate Transmission Block - used to signal end of data block.
NAK	Negative AcKnowledgement - used to indicate the previously received block was in error; used as a negative response to a poll message.
RVI	ReVerse Interrupt - if a station which is receiving data from another station must send a high-priority message to the sender, the receiver issues an RVI to interrupt the sender and seize control of the line.
SOH	Start Of Header - used to indicate an optional header field follows immediately.
STX	Start of TeXt - used to indicate the start of text and that the data follows immediately.
SYN	SYNchronization - used to provide synchronization as previously described.
TTD	Temporary Text Delay - used by sending station to indicate that it cannot send data immediately but does not wish to relinquish control of the line.
WACK	Wait before transmit positive ACKnowledgement - receiving station is acknowledging previous message but is also saying it cannot accept anymore messages until further notice.

Transparency

When a sender transmits a message (or packet) to a receiver, how does the receiver know when the incoming message has ended? As Table A-1 shows, if it is the end of a block of data, the sender inserts an ETB character at the end of the data. The receiver inputs this ETB character and realizes that the end of the block has been reached and that a checksum should follow. If it is the end of the text, the sender inserts an ETX character, and, once again, the receiver responds accordingly.

Since the receiver is carefully watching for an ETB or ETX, what would happen if an ETB or ETX suddenly appeared in the middle of the data? The receiver would erroneously assume that the end of the text had been reached and that the characters following comprise the checksum, (which they don't) resulting in an incorrect checksum calculation.

Why would a sender cause all this confusion by inserting an ETB or ETX into the middle of the text? Suppose a sender is transmitting a memory listing of a program. A memory listing consists of multiple bytes, each byte having values of hexadecimal 00 to hexadecimal FF. If one of those bytes just happens to have the exact same value as the ordinal value of the ETX character, the receiver will believe it has received the ETX character and act accordingly. To avoid this problem, a technique is needed that allows the binary equivalent of the ETX character to occur in the text, but not be recognized as the control character ETX. Such a technique is termed *transparency*. More generally, transparency is a scheme in which any bit sequence can be included within the text, even those that may appear as control characters.

To enable transparency, the Data Link Escape character (DLE) precedes the STX character. Then, to end the stream of text characters, a DLE ETX character is transmitted, rather than just an ETX character. The DLE is also inserted in front of STX, ETB, ITB, EXT, ENQ, DLE, and SYN control characters.

This system seems fairly straight-forward until you ask: Couldn't the binary equivalent of DLE ETX also appear as data within the text? Yes, it could. To solve this problem, an extra DLE is inserted before each DLE *in the text and only within the text*. Thus, if the sender encounters a DLE ETX sequence in the text during transmission, it inserts an extra DLE creating DLE DLE ETX. The receiver will see the DLE DLE ETX and discard the first DLE. Since it encountered two DLEs before an ETX, the receiver knows it is not receiving the control sequence DLE ETX, but simply text. The only single DLE followed by a TXT should occur after the end of the text. Figure A-1 demonstrates a before and after example.

Figure A-1
Example of transparency before and after DLE insertion

SYN SYN STX DLE ETX DLE ETX ETX BCC

SYN SYN DLE STX DLE DLE ETX DLE DLE ETX DLE ETX BCC

Figure A-2 is a more complete example of two stations engaged in data transfer using the BISYNC protocol. Station A (the primary) is polling Stations B, C, and D (the secondaries), but only Station D has data to return to Station A. After Station D sends it data to Station A, Station A informs Station C that it is going to receive data and to get ready. The data is transmitted but arrives garbled, so Station C asks Station A to retransmit.

Figure A-2
Two stations engaged in data transfer using the BISYNC protocol

Station A	Direction of Transmission	Station B, C, or D	Description
SYN SYN addrB ENQ	→		Poll to station B
	←	SYN SYN EOT	No data to send
SYN SYN addrC ENQ	→		Poll to station C
	←	SYN SYN EOT	No data to send
SYN SYN addrD ENQ	→		Poll to station D
	←	SYN SYN STX text ETX BCC	Send first part of data
SYN SYN ACK0	→		Acknowledge data
	←	SYN SYN STX text EOT BCC	Send remaining data
SYN SYN ACK1	→		Acknowledge data
SYN SYN addrB ENQ	→		Poll to station B
	←	SYN SYN EOT	No data to send
SYN SYN EOT SYN SYN addrC ENQ	→		Primary selects station C
	←	SYN SYN ACK0	OK, I'm ready, send the data
SYN SYN STX text EOT BCC	→		Data is sent
	←	SYN SYN NAK	Data arrives garbled
SYN SYN STX text EOT BCC	→		Data sent again
	←	SYN SYN ACK1	Acknowledge data

Synchronous Data Link Control (SDLC)

Synchronous Data Link Control (SDLC) was created by IBM in the mid-1970s to replace BISYNC. It is a bit synchronous protocol in which the receiver examines individual bits looking for control information. Unlike BISYNC, SDLC is capable of supporting both half-duplex and full-duplex connections. SDLC is a full duplex data link protocol because both sender and receiver may transmit at the same time. You will see shortly the mechanism that allows SDLC to support a full duplex connection.

Another difference between BISYNC and SDLC is that SDLC is code independent whereas BISYNC is code dependent. BISYNC requires a data code, such as ASCII or EBCDIC, that includes control codes such as SOH, STX, or ETX in the code set. SDLC does not rely on character codes to control execution, but relies instead on particular bit patterns to indicate a command.

The basic packet format for SDLC is shown in Figure A-3. The flag field is an 8-bit field with the unique value 01111110. The receiver scans the incoming bit stream looking for the pattern 01111110. When it encounters the pattern, the receiver knows the beginning of the packet has arrived and continues to scan for the remainder of the packet.

Figure A-3
Basic packet format for SDLC

FLAG	ADDRESS	CONTROL	DATA	CHECKSUM	FLAG
8 bits	8 bits	8 bits	$8 \times n$ bits	16 bits	8 bits

Note that the same flag is used to signal the end of the packet. Just as it looked for the beginning of the packet, the receiver scans the incoming bit stream looking for this unique pattern for the ending. When the ending flag has been encountered, the receiver knows to mark the end of the packet.

As with the control codes in BISYNC, if the bit pattern 01111110 occurs in the data, the receiver will interpret it as signaling the end of the packet erroneously. The SDLC needs a way of preventing the bit pattern 01111110 from occurring in the data. As the data and CRC portions of the message are transmitted, the transmitter watches for five 1s in a row. If it encounters five 1s, the transmitter automatically inserts a 0. With this procedure, the data and CRC portions of the packet can never contain 01111110. The receiver also scans the incoming bit stream, and if it finds five 1s immediately followed by a 0 within the data field of the packet, it discards the extra 0. This technique of inserting an extra bit is known as bit stuffing.

The address field contains an 8-bit address that is used to identify sender or receiver. In an unbalanced configuration (the host computer is the primary or master, the terminal is the secondary or slave), the address field always contains the address of the secondary station. In a balanced configuration the host computer and terminal have equal power, more like peers.

The control field describes the type of packet for this particular message. In SDLC there are three different types of packets:

- ▶ I (Information) packets—used to send data, flow control information, and error control information.
- ▶ S (Supervisory) packets—used to send flow and error control, but no data.
- ▶ U (Unnumbered) packets—used to send supplemental link control information.

Figure A-4 shows a further breakdown of the control field. The N(S) and N(R) fields are the send and receive counts for the packets that have been transmitted from and received at a particular station. A station is capable of transmitting packets to another station and receiving packets from another station at the same time. Unlike BISYNC, which sends one packet then awaits a reply, SDLC may send multiple packets to another station before waiting for a reply.

Figure A-4

Bit contents of the control field of an SDLC packet

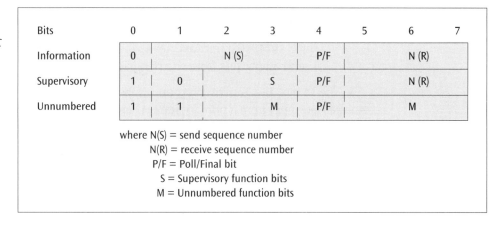

where N(S) = send sequence number
N(R) = receive sequence number
P/F = Poll/Final bit
S = Supervisory function bits
M = Unnumbered function bits

Likewise, a receiving station does not have to acknowledge every single packet. Instead, a receiving station may wait until several packets have arrived before acknowledging any or all of the packets. For example, station A may send seven packets (numbered 0, 1, 2, 3, 4, 5, and 6) to station B. As the packets arrive at station B, station B may wait until the fourth packet arrives (numbered 3) before sending an acknowledgement. Station B would return an acknowledgment to station A with the N(R) count set at 4, implying that packets 0 through 3 were accepted correctly and packet 4 is the next packet expected. The N(R) and N(S) counts always reflect the next packet number expected or sent. At a later time, station B will acknowledge the remaining three packets by transmitting a packet with the N(R) count set to 7.

So that a transmitting station does not overwhelm a receiving station with a flood of packets, SDLC has a technique that limits the number of packets a station may transmit at one time. The window size states the number of packets that may be unacknowledged at any given time. Assume that the window size is 7 and station A sends 6 packets to another station. Station A can still send one more packet, since the window size is 7 and only 6 packets have been sent. If the receiving station acknowledges 4 of those packets, station A can then send up to 5 more packets, since 2 of the original 6 packets transmitted have not yet been acknowledged. Because the number of packets that can be transmitted grows and shrinks with transmissions and acknowledgments, the technique has more accurately been called a sliding window.

The P/F bit is the Poll/Final bit. If a primary is polling a secondary, the P/F bit is a poll bit and is set to 1. If a secondary is sending multiple messages to a primary, the last message will have the P/F bit set to 1, and it will act as a final bit.

Following the control field is the data, which is of variable length but always a multiple of eight bits. After the data is the cyclic redundancy checksum, followed by the ending flag (01111110).

The S and M subfields of the control field are used by SDLC to further define the type of Supervisory or Unnumbered packets. The S bits, which exist only within the Supervisory format, are used to specify flow and error control information. Three types of Supervisory messages are available:

- Receive Ready (RR) 00 – positive acknowledgment; ready to receive an Information packet.

▶ Receive not Ready (RNR) 01 – positive acknowledgment, but not ready to receive Information packets.

▶ Reject (REJ) 10 – negative acknowledgment, go back to the Nth packet and resend all packets from the Nth packet on.

The M bits, used only within unnumbered packets, represent a command when the packet comes from a primary station, and a response when the packet comes from a secondary station. The available unnumbered packet commands are as follows:

▶ Nonsequenced information (NSI) : C/R_1:00 - C/R_2:000
▶ Set Normal Response Mode (SNRM) : 00-001
▶ Disconnect (DISC) : 00-010
▶ Optional Response Poll (ORP) : 00-100
▶ Set Initialization Mode (SIM) : 10-000
▶ Request Station ID (XID) : 11-101
▶ Request Task Response (TEST) : 00-111
▶ Configure for Test (CFGR) : 10-011

The available Unnumbered packet Responses are as follows:

▶ Nonsequenced Information (UI) : 00-000
▶ Nonsequenced Acknowledgement (UA) : 00-110
▶ Request for Initialization (RIM) : 10-000
▶ Command Reject (FRMR) 10-100 : reject packet, cannot make any sense out of it
▶ Request Online (DM) : 11-000
▶ Test Response/Beacon (BCN) : 11-111
▶ Disconnect Request (RD) 00-010 : request for a disconnect

To better understand SDLC and its commands, you need to examine several examples. The first example, shown in Figure A-5, demonstrates the Primary polling station A followed by station A requesting initialization information.

Figure A-5
Example showing polling and request for initialization

→ FLAG, Address-A, 10-00-1-000, CRC, FLAG   Primary polls A

Decoding the Control Field *(10-00-1-000)*: *10*=Supervisory Format, *00*=Receive Ready, *1*=Poll Bit is On, *000*=N(R)=0

← FLAG, Address-A, 11-11-1-000, CRC, FLAG   A Requests online

Nonsequenced Format, Request Online, Final Bit is On

→ FLAG, Address-A, 11-00-1-001, CRC, FLAG   Primary sets A to Normal Mode

Nonsequenced Format, Set Normal Response Mode, Poll Bit is On

← FLAG, Address-A, 11-00-1-110, CRC FLAG   A acknowledges

Nonsequenced Format, Unnumbered Acknowledge, Final Bit is On

In the second example, shown in Figure A-6, the primary polls station A to see if it has data to send, and A replies with three packets of data.

Figure A-6
Primary polls A, and A responds with three packets of data

→ FLAG, Address-A, 10-00-1-000, CRC, FLAG	Primary polls A
← FLAG, Address-A, 0-000-0-000, data, CRC, FLAG	A sends first packet (data packet #0)
←FLAG, Address-A, 0-100-0-000, data, CRC, FLAG	A sends second packet (data packet #1)
← FLAG, Address-A, 0-010-1-000, data, CRC, FLAG	A sends final packet (data packet #2, final bit)
→ FLAG, Address-A, 10-00-1-011, CRC, FLAG	Primary acks packets 0-2
	Primary says packet #3 next expected packet

In a third example, shown in Figure A-7, the primary polls station B, and station B replies with several data packets (). Due to a transmission error, one packet arrives garbled, and the primary requests retransmission.

Figure A-7
Primary polls station B, and B replies with several data packets

→ FLAG, Address-B, 10-00-1-000, CRC, FLAG	Primary polls B
← FLAG, Address-B, 0-000-0-000, data, CRC, FLAG	B sends first packet
← FLAG, Address-B, 0-001-0-000, data, CRC, FLAG	B sends second packet
← FLAG, Address-B, 0-010-0-000, data, CRC, FLAG	B sends third packet
← FLAG, Address-B, 0-011-0-000, data, CRC, FLAG	B sends fourth packet
← FLAG, Address-B, 0-100-0-000, data, CRC, FLAG	B sends fifth packet
← FLAG, Address-B, 0-101-0-000, data, CRC, FLAG	B sends sixth packet
← FLAG, Address-B, 0-110-0-000, data, CRC, FLAG	B sends seventh packet, stops
packet 110 arrives garbled which results in a CRC error	
→ FLAG, Address-B, 10-00-1-110, CRC, FLAG	Primary acks packets 000-101 but says Reject, go back and resend packet 110 and all subsequent packets again.
← FLAG, Address-B, 0-110-0-000, data, CRC, FLAG	B resends seventh packet
← FLAG, Address-B, 0-111-0-000, data, CRC, FLAG	B sends eighth packet
← FLAG, Address-B, 0-000-1-000, data, CRC, FLAG	B sends ninth and final packet
→ FLAG, Address-B, 10-00-1-001, CRC, FLAG	Primary acks packets 110-000

The above examples are greatly simplified and they show a primary communicating with only one secondary. However, you should note that it is quite possible for the primary to carry on concurrent conversations with multiple secondaries.

High-Level Data Link Control (HDLC)

High-level Data Link Control (HDLC) is a data link standard created by ISO that closely resembles SDLC. Typically, anyone dealing with IBM products and software would use SDLC. HDLC is used by anyone dealing with non-IBM products. Note that the two protocols are very similar, but not exactly the same. It is possible to make HDLC behave similar to SDLC, but it is not necessarily possible to make SDLC behave similar to HDLC. Do not assume that the two protocols are interchangeable.

There are a number of major differences between HDLC and SDLC. For example, SDLC allows for only an eight bit address, whereas HDLC can be extended to have an address size that is a multiple of eight bits. To extend the address in HDLC, a 0 is inserted in the high order bit position of each octet, except for the last octet of the address, which has a 1 in the high order bit position.

A further difference is that a balanced configuration is available in HDLC, but not in SDLC. In HDLC's balanced configuration, a command packet contains the destination address and a response packet contains the sending address.

Like the extended address, the control field in HDLC may be 8 bits (as in SDLC) or 16 bits in length. The 16-bit control field allows for 7-bit N(R) and N(S) counts, thus allowing a larger window size for packet transmission. HDLC also allows for an extended checksum. The cyclic checksum field may be either 16 bits (as in SDLC) or an extended 32 bit checksum.

Unique to HDLC is the selective reject command. Supervisory packets in SDLC have Receive Ready, Receive Not Ready, and Reject(go back N). HDLC adds Selective Reject(SREJ), which informs the sender that a message was in error and to go back to that one message and retransmit it, but NOT all the messages that followed it. HDLC also has additional Unnumbered Commands.

Of primary interest here is the fact that multiple modes of dialogue between sender and receiver may be established. The following modes exist in HDLC:

- ▶ Set Normal Response Mode (SNRM) - primary (master) - secondary (slave) type arrangement.

- ▶ Set Normal Response Mode Extended (SNRME) - control field is 16 bits in length as opposed to standard 8-bit length.

- ▶ Set Asynchronous Response Mode (SARM) - the secondary may initiate transmission without explicit permission of the primary, but the primary still retains responsibility of the line (initialization, error recovery, and logical disconnection).

- ▶ Set Asynchronous Response Mode Extended (SARME) - same as SARM, but in extended mode (16 bit control field).

- ▶ Set Asynchronous Balanced Mode (SABM) - either station may initiate transmission without explicit permission from the other station. No station implicitly retains responsibility. This arrangement is a "peer-to-peer" connection, as opposed to the primary-secondary configuration of SDLC.

- ▶ Set Asynchronous Balanced Mode Extended (SABME) - same as SABM, but in extended mode.

Public data networks and X.25

Data transfer over short distances is often performed by local area networks, metropolitan area networks, and dial-up modem transmission. But when the distance covered encompasses a state or a country, LANs and MANs can no longer do the job. Dial-up long-distance transmission using voice-grade telephone lines and modems has improved in quality and reliability over the years, but is still plagued by transmission noise and high telephone line costs. One possible solution presented by a number of companies is to provide a user with a local connection to a long-haul network for a fee. The long-haul company deals with the details of transmitting the data across the network. As shown in Figure A-8, the user at location A (DTE) transmits its data to the network host 1 (DCE). The network has the responsibility of transmitting the data across the subnet to the destination DCE, where the user at site B may receive the information. Networks such as these are termed public data networks, or PDNs.

Figure A-8

Two users accessing a public data network

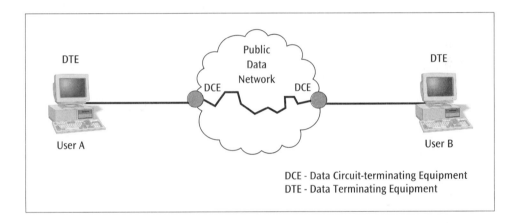

As an analogy, a PDN is similar to a package delivery system such as the United Parcel Service (UPS). You take the package to the UPS depot, and they ship it across the country in the best manner possible. The person sending the package does not necessarily care how the package gets there, just as long as it arrives in a reasonable length of time and undamaged.

One advantage of using a PDN is the fact that User A pays only for the number of characters of data transferred. This situation is like talking on the telephone to a friend and being charged only for the number of words spoken, not the overall connect time.

So that a user may connect a DTE terminal to a PDN DCE, the International Telecommunications Union (ITU) created the X.25 standard in 1974. The X.25 standard allows a uniform approach to making a connection to the network station. Since the use of X.25 involves a high-speed synchronous interface and requires a fair of amount of software and computing power, the device that is the DTE should be more powerful than a simple dumb terminal. If a user does not have a powerful enough workstation, or is using a low-speed asynchronous DTE, X.25 cannot be used. Luckily, CCITT has provided ways to connect low-speed, asynchronous terminals to a PDN. These low-speed, asynchronous methods will be discussed later.

The three levels of X.25

X.25, much like the OSI model, is divided into levels. Since X.25 originated before the OSI model, however, the three levels do not precisely fit into the three layers of the OSI model. The lowest level of X.25, the **physical level**, follows the X.21 standard introduced in Chapter Four. Almost any physical layer protocol can be used since, like any network model, the levels are disjoint from one another. Since X.21 is not widely used, X.25 also allows the use of EIA-232 or V.24/V.28 protocols. When you use EIA-232 or V.24/V.28, the physical level interface is termed X.21 bis (secondary standard).

The second level, the **data link level**, is responsible for creating a cohesive, error-free connection between the user's DTE and the network DCE. Since these responsibilities are virtually identical to the data link layer of the OSI model, X.25 allows two variations on the HDLC data link protocol, Link Access Protocol (LAP) and LAP-B. As a data packet is passed from the network level of the DTE to the data link level, the network level data packet is encapsulated with beginning and ending 8-bit flags, control field, address field, and a packet check sequence (Figure A-9).

Figure A-9
X.25 levels and the flow of data between levels

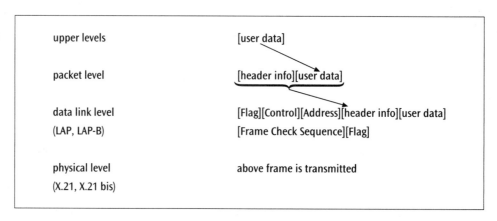

It is important to note that the LAP fields (flag, control, address, packet check sequence and flag) remain with the data packet only for the transmission from DTE to DCE. Once the packet arrives at the DCE, and before it is placed onto the network, all LAP fields are removed.

The third level of the X.25 protocol is termed the **packet level**. This level deviates the most from the OSI's third layer, the network layer. The packet level's main responsibility is to establish a connection between the user (DTE) and the network, and finally to the receiver (DTE) at the other end of the circuit (Figure A-10).

Figure A-10
X.25 connection of user to packet network

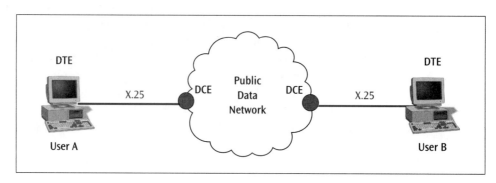

Since it is possible for a user to establish multiple connections simultaneously, each network connection is identified by a 4-bit logical group number plus an 8-bit logical channel number. Although all possible connections are not used, there is sufficient room for many concurrent logical connections over each physical channel.

A user of an X.25 PDN can choose from among four interface options. A user may establish a **permanent virtual circuit**, which is quite similar to a leased line in a public telephone network. Thus, before a session begins, the two users (or sender and receiver) and the network administration reach an agreement for a permanent virtual connection, and a logical connection number is assigned. This assigned logical connection number is then used for all future transmissions between the two users.

A second type of interface is the **virtual call** and is similar to the standard telephone call. The sending or originating DTE issues a call request packet, which is sent to the network. The network routes the call request packet to the destination DTE, which can accept or reject the request for a connection. If the destination DTE accepts, a call accepted packet is sent to the network, which transposes the packet into a call connected packet. The call connected packet arrives at the originating DTE and a virtual circuit is established. The two DTEs are now in a data transfer state until either DTE issues a clear request.

The permanent virtual circuit and virtual call interface designs are connection-oriented, in that a virtual circuit must be created with the consent of both DTEs and the network. After going through all the work to establish a connection, you would imagine that a relatively large number of data packets will be transferred between the two DTEs. But what if one DTE wishes to send only one packet of information to another DTE? Spending so much time making a connection for only one packet of data seems wasteful.

In situations in which only one packet of information is sent, a third connection strategy, the fast select, is available. **Fast select** allows one DTE to transfer data to another DTE without using call establishment and call termination procedures. The originating DTE sends a fast select packet to the network. This packet may also contain up to 128 bytes of user data. When the network delivers this packet to the destination DTE, the destination DTE has two response choices: clear request or call accepted. The first choice is to return a clear request packet, which may also contain 128 bytes of user data. When the originating DTE receives the clear request packet, that DTE realizes "I accepted your packet, thank you very much." The destination DTE could also respond with a call accepted packet, which then invokes the standard X.25 data transfer and clearing procedures of a virtual call.

The final interface option is the **fast select with immediate clear**. This technique is similar to the fast select, but the destination DTE has only one possible response and that is to return a clear request. Upon receipt of the clear request, the originating DTE returns a clear confirmation to the destination DTE. Thus, if a DTE has only one packet of data to send, this fourth and final interface choice is the simplest.

Packet types

So that X.25 may establish connections, transfer error-free data packets, and clear connections, several types of packets exist (Table A-2).

Table A-2
X.25 packet types

Packet Types	
From DCE to DTE	**From DTE to DCE**
Call Setup and Clearing	
Incoming Call	Call Request
Call Connected	Call Accepted
Clear Indication	Clear Request
DCE Clear Confirmation	DTE Clear Confirmation
Data and Interrupt	
DCE Data	DTE Data
DCE Interrupt	DTE Interrupt
DCE Interrupt Confirmation	DTE Interrupt Confirmation
Flow Control and Reset	
DCE Receive Ready	DTE Receive Ready
DCE Receive Not Ready	DTE Receive Not Ready
	DTE Reject
Reset Indication	Reset Request
DCE Reset Confirmation	DTE Reset Confirmation
Restart	
Restart Indication	Restart Request
DCE Restart Confirmation	DTE Restart Confirmation
Diagnostic	
Diagnostic	
Registration	
Registration Confirmation	Registration Request

The call setup and clearing packets are used to create a virtual call. The originating DTE issues a Call Request packet, the network delivers an Incoming Call packet to the destination DTE, the destination DTE responds with a Call Accepted packet, and the network delivers a Call Connected packet to the originating DTE. A similar dialogue follows for clearing a call.

The data and interrupt packets are used for transferring data and sending interrupts. Similar to HDLC commands, the flow control packets are used to acknowledge or suspend the actions of the DTEs. The reset and restart packets are used by the network and DTEs to recover from errors, such as loss of a packet, congestion, loss of the network's internal virtual circuit, or sequence number error. The reset packet can be used to reinitialize a virtual circuit, and the restart packet is reserved for more serious calamities.

Glossary

4B/5B   A digital encoding scheme that takes four bits of data, converts the four bits into a unique five bit sequence, and encodes the five bits using NRZ-I.

10Base2   An encoding scheme for CSMA/CD local area networks that incorporates 10 Mbps baseband (digital) signaling over 185 meter coaxial cable segments.

10Base5   An encoding scheme for CSMA/CD local area networks that incorporates 10 Mbps baseband (digital) signaling over 500 meter coaxial cable segments.

10BaseT   An encoding scheme for CSMA/CD local area networks that incorporates 10 Mbps baseband (digital) signaling over 100 meter twisted pair segments.

10Broad36   An encoding scheme for CSMA/CD local area networks that incorporates 10 Mbps broadband (analog) signaling over 3600 meter coaxial cable segments.

56K leased line   A leased-line service provided by a telephone company that is fixed between two locations and provides data and voice transfer at rates up to 56 Kbps.

100BaseFX   An encoding scheme for CSMA/CD local area networks that incorporates 100 Mbps baseband (digital) signaling over two-pair Cat 5 UTP for a maximum segment length of 100 meters.

100BaseT4   An encoding scheme for CSMA/CD local area networks that incorporates 100 Mbps baseband (digital) signaling over four-pair Cat 3 or higher UTP for a maximum segment length of 100 meters.

100BaseTX   An encoding scheme for CSMA/CD local area networks that incorporates 100 Mbps baseband (digital) signaling over two-pair Cat 5 or higher UTP for a maximum segment length of 100 meters.

100VGAnyLAN   An encoding scheme for local area networks that employs the demand priority access method and transmits data at 100 Mbps over Cat 3 or higher UTP.

1000BaseCX   An encoding scheme for CSMA/CD local area networks that incorporates 1000 Mbps baseband (digital) signaling over fiber optic cable; designed for backbone cabling within a building or across a campus.

1000BaseLX   An encoding scheme for CSMA/CD local area networks that incorporates 1000 Mbps baseband (digital) signaling over fiber optic cable; supports longer distance cabling within a single building.

1000BaseSX   An encoding scheme for CSMA/CD local area networks that incorporates 1000 Mbps baseband (digital) signaling over fiber optic cable; supports relatively close clusters of workstations and devices.

1000BaseT   An encoding scheme for CSMA/CD local area networks that incorporates 1000 Mbps baseband (digital) signaling over Cat 5e UTP.

Active Directory   Network directory structure for Windows 2000 operating system; a hierarchical structure that stores information about all the objects and resources in a network and makes this information available to users, network administrators, and application programs.

adaptive routing   A dynamic system in which routing tables react to network fluctuations, such as congestion and node/link failure.

Address Resolution Protocol (ARP)   An Internet protocol that takes an IP address in an IP datagram and translates it into the appropriate CSMA/CD address for delivery on a local area network.

advanced mobile phone service (AMPS)   The oldest analog mobile telephone system that covers almost all of North America and is found in more than 35 other countries.

agent   The software (or management software) that runs in an element; an element that has an agent is considered a managed element and can react to SNMP commands and requests.

amplitude   The height of the wave above (or below) a given reference point.

amplitude modulation   A modulation technique for encoding digital data using various amplitude levels of an analog signal.

analog data   Data that are represented by continuous waveforms that can be at an infinite number of points between some given minimum and maximum.

analog signals   Signals that are represented by continuous waveforms that can be at an infinite number of points between some given minimum and maximum.

analysis phase   One phase of the systems development life cycle in which an analyst determines the information requirements for the particular users involved.

anti-virus software   Software designed to detect and remove viruses that have infected your memory, disks, or operating system.

application programming interface (API)   Software modules that act as an interface between application programs and technical entities such as telephone switching systems.

ARPANET   One of the country's first wide area packet switched networks; the precursor to the modern Internet; interconnected research universities, research labs, and select government installations.

ASCII   A seven-bit code that is used to represent all the printable characters on a keyboard plus many non-printable control characters.

asymmetric connection   A connection in which one direction has a faster transfer rate than the opposite direction; for example, numerous systems exist that have a faster downstream connection (such as from the Internet) and a slower upstream connection.

asymmetric digital subscriber line (ADSL)   A popular form of digital subscriber line that transmits the downstream data at a faster rate than the upstream rate.

asynchronous connection   One of the simplest examples of a data link protocol and is found primarily in microcomputer to modem connections. Adds one start bit, one stop bit, and an optional parity bit to every character transmitted.

asynchronous transfer mode (ATM)   Similar to frame relay, is a relatively new high speed, packet switched service that is offered by the telephone companies.

AT command set   The most important feature of the Hayes modems. Their use of AT commands simplified microcomputer-to-modem connections and became a worldwide standard.

attenuation   The continuous loss of a signal's strength as it travels though a medium.

availability   The probability that a particular component or system will be available during a fixed time period.

available bit rate (ABR)   A class of service used in asynchronous transfer mode: used for traffic that may experience bursts of data, called "bursty" traffic, and whose bandwidth range is roughly known, such as a corporate collection of leased lines.

backbone   The main connecting cable that runs from one end of the installation to another, or from one end of a network to another.

backplane   The main hardware of a device (such as a LAN switch) into which all supporting printed circuit cards connect.

backup software   Software that allows network administrators to back up data files currently stored on the network server's hard disk drive.

backward learning   A technique in which a bridge or switch creates its routing tables by watching the current flow of traffic.

bandwidth   The absolute value of the difference between the lowest and highest frequencies.

baseband   A single digital signal (such as a Manchester encoding) that is used to transmit data over a medium.

baseband coaxial   A transmission technique that uses digital signaling in which the cable carries only one channel of digital data.

baseline   One of the best techniques for determining current network capacities.

basic rate interface (BRI)   An ISDN service that consists of two full-duplex 64 Kbps B channels plus one full-duplex 16 Kbps D channel.

basic service set   Resembles a cell in a cellular telephone network and is the transmission area surrounding a single access point in a wireless local area network.

baud   The number of times a signal changes value per second.

baud rate   See baud.

Baudot Code   A character encoding technique that uses 5-bit patterns to represent the characters A to Z, the numbers 0 to 9, and several special characters.

benchmarking   Involves generating system statistics under a controlled environment and then comparing those statistics against known measurements.

bidirectional   When a signal is transmitted, it is transmitted from a given workstation and propagates in both directions on the cable away from the source.

bindery   A database that contains the user names and passwords of network users and groups of users authorized to log in to that server. Used in Netware Version 3.

bits per second (bps)   The number of bits that are transmitted across a medium in a given second.

bridge   A device that interconnects two local area networks that both have a medium access control sublayer.

broadband   Uses analog signaling in the form of frequency division multiplexing to divide the available medium into multiple channels.

broadband coaxial   A coaxial cable technology that transmits analog signals and is capable of supporting multiple channels of data simultaneously.

broadband wireless   One of the latest techniques for delivering Internet services into homes and businesses that employs broadband signaling using radio frequencies over relatively short distances.

broadcast subnet   The physical component of a network that transmits data in a broadcast fashion (when one workstation transmits, all other workstations receive that data).

buffer preallocation   An alternate solution to controlling the flow of packets between two nodes.

burst rate   In frame relay systems, allows the customer to exceed the committed information rate by a fixed amount for brief moments of time.

bus topology   The first topology used when local area networks became commercially available in the late 1970s. Essentially a single coaxial cable to which all workstations attach.

cable modem   A communications device that allows high-speed access to wide area networks such as the Internet via a cable television connection.

cable telephony   Telephone services over cable television lines in most major cities across the country.

call filtering   A technology in which users can specify telephone numbers that are allowed to get through. All other calls will be routed to an attendant or voice mailbox.

capacity planning   Involves trying to determine the amount of network bandwidth necessary to support an application or a set of applications.

CardBus A newer 32-bit version of PC card technology; adds memory, mass storage, and I/O capabilities.

carrier sense multiple access with collision detection (CSMA/CD) A contention-based medium access control protocol for bus and star-wired bus local area networks; a workstation wanting to transmit can only do so if the medium is idle, otherwise it has to wait. Collisions are detected by transmitting workstations requiring both workstations to backoff and retransmit.

cascade port In a 100VGAnyLAN network, used to connect a hub to its upstream connection.

cascading style sheets (CSS) Allow a web page author to incorporate multiple styles (fonts, styles, colors, and so on) in an individual HTML page.

Category 1-5 (CAT 1-5) A technique used to label a form of twisted pair wire by how far the wire may transmit and at what data rate.

cellular digital packet data (CDPD) Supports a wireless connection for the transfer of computer data from a mobile location to the public telephone network and the Internet.

central office (CO) Contains the equipment that generates a dial tone, interprets the telephone number dialed, checks for special services, and connects the incoming call to the next point.

centralized routing Dictates that the routing information generated from the least cost algorithm is stored at a central location within the network.

Centrex (central office exchange service) A service from local telephone companies in which up-to-date telephone facilities at the telephone company's central (local) office are offered to business users so that they don't need to purchase their own facilities.

certificate An electronic document, similar to a passport, that establishes your credentials when you are performing transactions on the World Wide Web.

certificate authority The specialized software on a network or a trusted third-party organization or business that issues and manages certificates.

certificate revocation list A list of certificates that have been revoked either before or at their originally scheduled expiration date.

choke packets Similar to the concept of flow control between two nodes.

ciphertext The data after the encryption algorithm has been applied.

circuit switched subnet A subnet in which a dedicated circuit is established between sender and receiver and all data passes over this circuit.

class of service A definition of a type of traffic and the underlying technology that will support that type of traffic. Commonly found in asynchronous transfer mode systems.

client/server system A distributed computing system consisting of a server and one or more clients which request information from the server.

coaxial cable A single wire wrapped in a foam insulation, surrounded by a braided metal shield, then covered in a plastic jacket.

codec A device that accepts analog data and converts it into digital signals. This process is also known as digitization.

collision window The interval during which the signals in a CSMA/CD local area network propagate down the bus and back (the interval during which a collision can happen).

committed information rate (CIR) The data transfer rate that is agreed on by both customer and carrier in a frame relay network.

competitive local exchange carrier (CLEC) New providers (that have been created since the Telecommunications Act of 1996) of local telephone services.

computer auditing A software program that monitors every transaction within a system.

computer network An interconnection of computers and computing equipment using either wires or radio waves over small or large geographic areas.

computer telephony integration (CTI) New telephone services and systems that combine more traditional voice networks with modern computer networks.

connection negotiation The negotiation between the devices at the ends of a communication line that establishes the parameters (such as transfer speed) for transmission.

connection-oriented A type of network application that provides some guarantee that information traveling through the network will not be lost and that the information packets will be delivered to the intended receiver in the same order in which they were transmitted.

connectionless A type of network application that does not require a logical connection to be made before the transfer of data.

constant bit rate (CBR) Used in asynchronous transfer mode, a type of connection that is similiar to a current telephone system leased line. CBR is the most expensive class of service.

consumer digital subscriber line (CDSL) A trademarked version of DSL that is a little slower than typical ADSL speeds.

contention-based protocol A first-come first-served protocol—the first station to recognize that no one is transmitting data is the first station to transmit.

cookies Data created by a web server, that is stored on the hard drive of a user's workstation.

corporate license The agreement that allows a software package to be installed anywhere within a corporation, even if installation involves multiple sites.

crash protection software Software whose primary goal is to perform crash stalling on a workstation or network of workstations.

crosstalk An unwanted coupling between two different signal paths.

cryptography The study of creating and using encryption and decryption techniques.

CSU/DSU A hardware device about the size of an external modem that converts the digital data of a computer into the appropriate form for transfer over a 1.544 Mbps T1 digital telephone line or over a 56 Kbps to 64 Kbps leased telephone line.

customized menuing systems Menus that are custom-created for an automated answering system to help callers find the right information, agent, or department.

cut-through architecture A technology in a LAN bridge or switch that allows the data frame to exit the switch almost as soon as it begins to enter the switch.

cyclic redundancy checksum (CRC) An error detection technique that typically adds either 16 or 32 check bits to potentially large data packets and approaches 100 percent error detection.

data Entities that convey meaning within a computer or computer system.

data code The set of all textual characters or symbols and their corresponding binary patterns.

data communications The transfer of digital or analog data using digital or analog signals.

Data Encryption Standard (DES) A commonly employed encryption method used by businesses to send and receive secure transactions.

data link layer The layer of the OSI model that is responsible for taking the raw data and transforming it into a cohesive unit called a frame.

data network An interconnection of computers that is designed to transmit computer data.

data transmission speed Calculated as the number of bits per second that can be transmitted.

datagram The entity or packet of data transmitted in a datagram network such that each packet can follow its own, possibly different, course through the subnet.

DCE Data circuit terminating equipment—the device, such as a modem, that terminates a data transmission line.

decibel (db) A relative measure of signal loss or gain that is used to measure the logarithmic loss of a signal.

dedicated network segment A portion of a local area network in which a switch is used to interconnect two or more workstations. Since a hub is not involved, if one workstation wants to transmit to another workstation

connected to the same switch, the communication is performed on a dedicated segment.

delay distortion The error that occurs because the velocity of propagation of a signal through a wire varies with the frequency of the signal.

delta modulation A method of converting analog data to digital signal in which the incoming analog signal is tracked and a binary 1 or 0, respectively, is output if the analog signal rises or falls.

demand priority protocol The most common example of a reservation protocol for local area networks, which is based on the IEEE 802.12 standard.

denial of service attack A malicious hacker technique that bombards a computer site with so many messages that the site is incapable of performing its normal duties.

dense wavelength division multiplexing Used with fiber optic systems and involves the transfer of multiple streams of data over a single optical fiber using multiple-colored laser transmitters.

design phase A phase in the systems development life cycle in which the systems analyst uses the information collected earlier to create a logical design of the information system.

deterministic protocol A protocol (such as a LAN medium access control protocol) in which a workstation can calculate (determine) when it will next get a turn to transmit.

Differential Manchester code A digital encoding technique that transmits a binary 0 with a voltage change at the beginning of the bit frame and transmits a binary 1 with no voltage change at the beginning of the bit frame. There is always a voltage transition in the middle of the bit frame.

digital advanced mobile phone service (D-AMPS) The newer digital equivalent of the original analog advanced mobile telephone service.

digital data Entities that are represented by discrete waveforms, rather than continuous waveforms. Between a minimum value X and a maximum value Y, the digital waveform takes on only a finite number of values.

digital signals The electric or electromagnetic encoding of data that is represented by discrete waveforms, rather than continuous waveforms. Between a minimum value X and a maximum value Y, the digital waveform takes on only a finite number of values.

digital signature A technology the uses public key cryptography to assign to a document a code for which you alone have the key.

digital subscriber line (DSL) A relatively new technology that allows existing twisted pair telephone lines to transmit multimedia materials and high-speed data.

DIN connector A family of plugs and sockets used to connect a variety of devices, such as audio and video equipment.

discovery frame   In token ring LANs, a special frame that is transmitted from the source workstation for the sole purpose of finding a route to a particular destination workstation.

disk duplexing   A technique that saves different parts of a record to different disk drives; used to protect users from server disk crashes.

disk mirroring   A technique that saves one record to two or more disk drives creating duplicates; used to protect users from server disk crashes.

distributed processing   A technique in which a task is subdivided and sent to remote workstations on the network for execution.

distributed routing   A wide area network routing technique in which each node maintains its own routing table.

domain name   The address that identifies a particular site on the Web.

domain name system (DNS)   A large, distributed database of Internet addresses and domain names.

downlink   The portion of a communications link that traverses from a satellite to a ground station.

DSL modem   The modem that supports an all-digital phone service that can provide data transmission at speeds of 100s of thousands up to millions of bits per second.

DSL.Lite   A slower format compared to ADSL; also known as Universal DSL, G.Lite, and splitterless DSL. Typical transmission speeds are in the 50 Kbps to 200 Kbps range.

DTE   Data terminal equipment - the device, such as a workstation or a terminal, that connects to a DCE.

Dynamic HTML (DHTML)   A collection of new markup tags and techniques that can be used to create more flexible and more powerful web pages.

dynamic IP addressing   The strategy used when an IP address is temporarily assigned when a workstation invokes an Internet application, such as a web browser.

e-commerce   The term that has come to represent the commercial dealings of a business using the Internet.

e-mail bombing   A malicious hacker technique in which a user sends an excessive amount of unwanted e-mail to someone.

EBCDIC   An 8-bit code allowing 256 possible combinations of textual symbols ($2^8 = 256$).

echo   The reflective feedback of a transmitted signal as the signal moves through a medium.

effective bandwidth   The more realistic bandwidth of a signal, which depends on factors such as environmental conditions.

EIA-232E   An interface standard for connecting a DTE to a voice-grade modem (DCE) for use on analog public telecommunications systems.

electrical component   One portion of an interface that deals with voltages, line capacitance, and other electrical components.

electronic mail   The computerized version of writing a letter and mailing it at the local post office.

encapsulation   The process performed by a layer of a protocol stack which adds control information to the front (or rear) of a data packet.

encryption algorithm   The computer program that converts plaintext into an enciphered form.

Ethernet   The first commercially available local area network system (and currently the most popular). Almost identical in operation to CSMA/CD.

even parity   A simple error detection method in which a bit is added to a character to produce an even number of binary 1s.

extended service set   In a wireless LAN topology, the collection of all the basic service sets attached to a local area network through access points.

extensible markup language (XML)   A subset of SGML that is a description for both the contents of a web-based document and how to display the document.

external modem   A modem with its own power supply and is located outside the computer's cabinet.

external security   Protecting the equipment of a computer system or a computer network from physical damage.

extranet   When an intranet is extended outside the corporate walls to include suppliers, customers, or other external agents.

fall forward   The capability of two modems to negotiate line speeds to a higher level either at the beginning of a connection or during transmission.

fallback   The capability of two modems to negotiate line speeds to a lower level either at the beginning of a connection or during transmission.

Fast Ethernet   100 Mbps Ethernet standards.

fax processing   A fax image that is stored on a LAN server's hard disk and can be downloaded over a local area network, converted by a fax card, and sent out to a customer over a trunk line.

fax-back   The technique in which a user dials into a fax server, retrieves a fax by keying in a number, and sends that fax anywhere.

Fiber Data Distributed Interface (FDDI) ring protocol   A medium access control protocol for a ring local area network topology based on token passing and using fiber optic cables.

fiber optic cable   A thin glass cable approximately a little thicker than a human hair surrounded by a plastic coating. Fiber optic cables transmit light pulses, can transfer data at billions of bits per second, have very low noise, and are not subject to electromagnetic radiation.

file server A workstation on a local area network, with the primary function of providing storage for files, data sets, and network operating system.

file transfer protocol (FTP) The primary functions are to allow a user to download a file from a remote site to the user's computer and to upload a file from the user's computer to a remote site.

filter Examines the destination address of a frame and either forwards or does not forward the frame based on some address information stored within the bridge.

financially feasible system One that can be created given the company's current financial ability.

firewall A system or combination of systems that supports an access control policy between two networks.

FireWire The name of a bus that connects peripheral devices such as wireless modems and high speed digital video cameras to a microcomputer and is currently IEEE standard 1394.

flooding A wide area network routing protocol in which each node takes the incoming packet and retransmits it onto every outgoing link.

forward error correction The process in which a receiver, upon detecting an error in the arriving data, corrects the error without further information from the transmitter.

forward search least cost algorithm A procedure that determines the least cost from one network node to all other network nodes.

frame A cohesive unit of raw data. The frame is the package of data created at the data link layer of the OSI model.

frame relay A packet switched network that was designed for transmitting data over fixed lines (not dial-up lines).

frequency The number of times a signal makes a complete cycle within a given time frame.

frequency division multiplexing Involves assigning non-overlapping frequency ranges to different signals.

frequency modulation Uses two different frequency ranges to represent data values of 0 and 1.

full duplex Data transmitted from sender to receiver and from receiver to sender in both directions at the same time.

full-duplex switch Allows for simultaneous transmission and reception of data to and from a workstation.

functional component The function of each pin or circuit that is used in a particular interface.

gateway A generic term for a device that interconnects two networks.

generating polynomial An industry-approved bit string that is used to create the cyclic checksum remainder.

geosynchronous earth orbit (GEO) Satellites that are found 22,300 miles from the earth and are always over the same point on earth.

gigabit Ethernet An Ethernet specification for transmitting data at 1 billion bits per second.

global system for mobile (GSM) Currently the least popular mobile telephone technology in the United States and the most popular mobile telephone technology in Europe.

go-back-N ARQ Part of the sliding window protocol. When a frame arrives garbled, the receiver returns a go-back-N command, which informs the transmitter to go back to the Nth frame and retransmit it along with all subsequent frames.

half duplex Data is transmitted from the sender to the receiver and from the receiver to the sender in only one direction at a time.

Hayes standards A popular set of communication standards for microcomputer modems.

Hertz (Hz) Cycles per second, or frequency.

high bit-rate DSL The earliest form of DSL, which provides a symmetric service with speeds usually equivalent to a T1 service (1.544 Mbps).

hop count The counter associated with a packet as it moves from node to node within a wide area network. Every time the packet moves to the next node, the hop count is incremented.

hop limit Used with the flooding routing algorithm, the value that is compared to the hop count of each packet as it arrives at a node. When a packet's hop count equals the hop limit, the packet is discarded.

hot-swappable The ability to remove a disk drive unit from a network server without turning off the power to the server.

hub A device that interconnects two or more workstations in a star-wired bus local area network and broadcasts incoming data onto all outgoing connections.

hub polling A polling technique in which the primary polls the first terminal, which then passes the poll to the second terminal, and each successive terminal passes along the poll.

hybrid telephone A mobile telephone that can send and receive D-AMPS signals as well as PCS signals.

hypertext markup language (HTML) A set of codes inserted into a document (web page) that is used by a web browser to determine how the document is displayed.

Hypertext Transport Protocol (HTTP) An Internet protocol that allows Web browsers and servers to send and receive World Wide Web pages.

IEEE 802 suite of protocols   A collection of protocols that define various types of local area networks, metropolitan area networks, and wireless networks. For example, the IEEE 802.3 protocol defines the protocol for CSMA/CD local area networks.

implementation phase   One phase of the systems development life cycle in which the system is installed and preparations are made to move from the old system to the new system.

impulse noise   A non-constant noise and one of the most difficult errors to detect because it can occur randomly.

incumbent local exchange carrier (ILEC)   Local telephone company that existed before the Telecommunications Act of 1996.

infrared transmission   A special form of radio transmission that uses a focused ray of light in the infrared frequency range ($10^{12} - 10^{14}$ Hz).

integrated services digital network (ISDN)   Designed in the mid-1970s to provide an all digital worldwide public telecommunication network that would support telephone signals, data, and a multitude of other services for both residential and business users.

integrated voice recognition and response   A system in which a user calling into a company telephone system provides some form of data by speaking into the telephone and a database query is performed using this spoken information.

interactive user license   An agreement in which the number of concurrent active users of a particular software package is strictly controlled.

interactive voice response   When a customer calls your company, his or her telephone number is used to extract the customer's records from a corporate database.

interexchange carrier (IEC, IXC)   The long distance telephone companies.

interface layer   The lowest layer of the Internet model that defines both the physical medium that transmits the signal and the frame that incorporates flow and error control.

intermodulation distortion   The noise caused when the frequencies of two or more signals mix together and create new frequencies.

internal modem   A printed circuit board that is designed to plug into an available printed circuit card slot within a personal computer.

Internet   A network of networks.

Internet application layer   The final layer of the Internet model, supports the network applications for which one uses a network and presentation services such as encryption and compression that are necessary to properly present or support the application.

Internet Control Message Protocol (ICMP)   Used by routers and nodes, performs error reporting for the Internet Protocol.

Internet model   The communications architectural model that incorporates the TCP/IP protocols, and has surpassed the OSI model in popularity and implementation.

Internet Protocol (IP)   The software that prepares a packet of data so that it can move from one network to another on the Internet or within a set of networks in a corporation.

Internet transport layer   The layer of software in the Internet model that provides a reliable end-to-end network connection; similiar to the transport layer of the OSI model. The Internet transport layer is usually defined as either the TCP protocol or the UDP protocol.

Internet2   A new, very high speed packet-switched wide area network that supplements the currently existing Internet and may eventually replace it.

internetworking   The interconnecting of multiple networks.

intranet   A TCP/IP network inside a company that allows employees to access the company's information resources through an Internet-like interface.

intrusion detection   The ability to electronically monitor data flow and system requests into and out of a system.

IP multicasting   The ability of a network server to transmit a data stream to more than one host at a time.

IP subnet masking   The process of taking an IP address consisting of a network ID and host ID and further breaking the host ID into subnet ID and host ID.

IPv6   A more modern Internet Protocol that takes advantage of the current technology. Currently, most Internet systems are using IPv4.

IPX/SPX   NetWare's pair of communication protocols that support network layer and transport layer functions, respectively.

ISDN modem   The device necessary to convert the digital data from the computer into the proper form necessary for transmission over an ISDN network.

ISDN multiplexing   The ability of ISDN to support multiple digital channels of information. Currently there are two basic forms of ISDN multiplexing: primary rate interface and basic rate interface (BRI).

isochronous connection   Provides guaranteed data transport at a pre-determined rate, which is essential for multimedia applications.

isolated routing   Each node uses only local information to create its own routing table.

jitter   The result of small timing irregularities during the transmission of digital signals that become magnified as the signals are passed from one device to another.

kerberos An authentication protocol designed to work on client/server networks and uses secret key cryptography.

key The unique piece of information that is used to create ciphertext and then decrypt the ciphertext back into plaintext.

lambda The wavelength of each different colored laser.

layer 2 protocol A protocol that operates at the second layer, or data link layer, of the OSI seven layer model.

layers An HTML page or a section of a page that can be handled separately by the browser from other layers and placed at any position on the web page.

leaf object An object in a hierarchical directory structure that is composed of no further objects and includes entities such as the users, peripherals, servers, printers, queues, and other network resources.

least cost algorithm Chooses a route that minimizes the sum of the costs of all the communication paths along that route.

licensing agreement A legal contract that describes a number of conditions that must be upheld for proper use of the software package.

line disconnect attack A hacker technique in which the perpetrator attempts to gain access to a user's job after a line disconnect.

line of sight transmission The characteristic of certain types of wireless transmission in which the transmitter and receiver are in visual sight of each other.

listserv A popular software program used to create and manage Internet mailing lists.

local access transport area (LATA) A geographic area, such as a large metropolitan area or part of a large state. Telephone calls that remain within a LATA are usually considered local telephone calls, while telephone calls that travel from one LATA to another are considered long distance telephone calls.

local area network (LAN) A communication network that interconnects a variety of data communicating devices within a small geographic area and broadcasts data at high data transfer rates with very low error rates.

local exchange carrier (LEC) The local telephone companies.

local loop The telephone line that leaves your house or business, and consists of either four or eight wires.

local loop-back testing A diagnostic technique performed by modems in which the computer or terminal A sends data to its local modem, which immediately returns the data to computer A.

local multipoint distribution system (LMDS) A broadband wireless communication system that can be used to provide digital two-way voice, data, video services, and Internet access.

location connectivity diagram (LCD) A network modeling tool that depicts the various locations involved in a network and the interconnections between those locations.

logical connection A flow of ideas without a direct physical connection—between sender and receiver at a particular layer.

logical design A procedure performed during SDLC that maps how the data moves around the network from workstation to workstation.

logical link control (LLC) sublayer A sublayer of the data link layer of the OSI model. Primarily responsible for logical addressing and providing error control and flow control information.

longitudinal parity Sometimes called longitudinal redundancy check or horizontal parity, tries to solve the main weakness of simple parity in which all even numbers of errors are not detected.

low earth orbit (LEO) The orbit around earth in which satellites are closest to the earth. They can be found as close as 100 miles from earth and as far as 1000 miles from earth.

maintenance phase The largest phase in the SDLC and one that often lasts for years.

managed hub A hub in a local area network that possesses enough processing power that it can be managed from a remote location.

Management Information Base (MIB) The database that holds the information about each managed device in a network that supports SNMP.

Manchester codes A digital encoding scheme that ensures that each bit has a signal change in the middle of the bit and thus solves the synchronization problem.

mean time between failures The average time a device or system will operate before it fails.

mean time to repair The average time necessary to repair a failure within the computer network.

mechanical component One of the four parts of an interface; deals with items such as the connector or plug description.

medium access control (MAC) sublayer Works closely with the physical layer and contains a header, computer (physical) addresses, error detection codes, and control information.

metropolitan area network Networks that serve an area of 3 to 30 miles—approximately the area of a typical city.

middle earth orbit (MEO) The orbit around earth that is above LEO but below GEO. Satellites in MEO can be found 1000 miles to 22,300 miles from the earth.

MIME Multipurpose Internet Mail Extension. The protocol used to attach a document, such as a word processor file or spreadsheet, to an e-mail message.

MNP 1-5 The Microcom Network Protocols that were created for modems to automatically support error correction and data compression.

mobile service area (MSA) When mobile telephone service was first introduced the country was broken into mobile service areas, or markets. It was determined that two cellular carriers could operate within each mobile service area.

modem A device that modulates digital data onto an analog signal for transmission over a telephone line, then demodulates the analog signal back to digital data.

modem pool A relatively inexpensive technique that allows multiple workstations to access a modem without placing a separate modem on each workstation.

modified final judgement The court's ruling on the breakup of AT&T.

modulation The process of converting digital data into an analog signal.

monoalphabetic substitution-based cipher A fairly simple encryption technique which replaces a character or group of characters with a different character or group of characters.

MP3 A compression/encoding technique that allows a high quality audio sample to be reduced to a much smaller sized file.

multichannel multiple distribution service (MMDS) A broadband wireless transmission technique that transfers high quality video into homes and businesses.

multi-level generic scanning An antivirus technique that is used to assist scanners and terminate-and-stay resident software.

multiplexing Transmitting multiple signals on one medium.

multiplexor The device that combines (multiplexes) multiple input signals for transmission over a single medium and then demultiplexes the composite signal back into multiple signals.

multipoint connection A single wire with the mainframe connected on one end and multiple terminals connected on the other end.

multi-station access unit (MAU) A device in a token ring local area network that accepts data from a workstation and transmits this data to the next workstation downstream in a ring fashion.

multitasking operating system An operating system that schedules each task and allocates a small amount of time to the execution of each task.

NetWare loadable module (NLM) A software program that performs a network operating system function, but is only loaded into memory when its execution is necessary. Once the execution has been completed the NLM can be unloaded from memory, thus freeing main memory for some other operation.

network architecture model A template that outlines the layers of hardware and software operations for a computer network and its applications.

network congestion The network characteristic when too many data packets are moving through a network and a degradation of services results.

network interface card (NIC) An electronic device, typically in the form of a computer circuit board, that performs the necessary signal conversions and protocol operations so that the workstation can send and receive data on the network.

network layer A layer in the OSI and Internet models that is responsible for creating, maintaining, and ending network connections.

network management The design, installation, and support of a network and its hardware and software.

network management protocol Facilitates the exchange of management information between network devices.

network operating system (NOS) A large, complex program that can manage the resources common on most local area networks, in addition to performing the standard operating system services.

network server The computer that stores software resources such as computer applications, programs, data sets, and databases, and either allows or denies workstations connected to the network access to these resources.

network server license A license similiar to the interactive user license in which a software product is allowed to operate on a local area network server and be accessed by one or more workstations.

network-network interface One of the types of connections in asynchronous transfer mode (ATM); a network-network interface is created by a network and used to transfer management and routing signals.

node The computing devices that allow workstations to connect to the network and that make the decisions as to which route a piece of data will follow next.

noise Unwanted electrical or electromagnetic energy that degrades the quality of signals and data.

nomadic An application in which a workstation moves around.

non-deterministic protocol A local area network medium access control protocol in which you cannot calculate the time at which a workstation will transmit.

Novell Directory Service (NDS) A database that maintains information on, and access to, every resource on the network, including users, groups of users, printers, data sets, and servers.

odd parity A simple error detection scheme in which a single bit is added to produce an odd number of binary 1s.

Open Shortest Path First (OSPF)   A routing algorithm used to transfer data across the Internet and is based on the link state algorithm.

Open Systems Interconnection (OSI) model   A template that consists of seven layers that defines a model for the operations performed on a computer network.

operating system   Manages all the other programs (applications) and resources (such as disk drives, memory, and peripheral devices) in a computer and is the program that is initially loaded into computer memory when the computer is turned on.

operational security   Deciding, and then limiting, who can use the system and when they can use the system.

operationally feasible system   One that operates as designed and implemented.

organizational unit (OU)   An object in a hierarchical tree structure for a local area network operating system that is composed of further objects.

OS/2   An operating system created by IBM that was the first personal computer operating system that could perform true multitasking.

packet filter   A router that has been programmed to filter out or allow to pass certain IP addresses or TCP port numbers.

packet switched subnet   The physical component of a network that is designed to transfer all data between sender and receiver in fixed-sized packets.

packet voice   One name for telephone calls over the internet.

pager   A wireless communication technology that has grown immensely in popularity within the last decade, it is also often called a beeper.

parallel port   A connection in which there are eight data lines transmitting an entire byte of data at one moment in time.

parity bit   The bit added to a character of data to perform simple parity checking.

passive device   A simple connection point between two runs of cable that does not regenerate the signal on the cable.

payback analysis   Charts the initial costs and yearly recurring costs of a proposed system against the projected yearly income (benefits) derived from a proposed system.

PBX graphic user interface   An interface in which different icons on a computer screen represent common PBX functions such as call hold, call transfer, and call conferencing, making the system easier for operators to use.

PC Card   Credit card-sized, rugged peripherals that add memory, mass storage, and I/O capabilities to laptop computers.

PCMCIA   Personal Computer Memory Card International Association. The organization that creates and maintains the standards for personal computing devices.

period   The length, or time interval, of one cycle.

permanent virtual circuit (PVC)   A fixed connection between two endpoints in a frame relay network. Unlike a telephone circuit, which is not shared by multiple users, multiple PVCs may share the same physical connections, thus making it a virtual circuit.

permit system   A networking technique that requires a workstation to possess one or more permits before transmitting any data; helps to eliminate network congestion.

personal communication system (PCS)   The newest category of mobile telephone technology that is all digital and does rely on the more noisy analog techniques.

phase   The position of the waveform relative to a given moment of time or relative to time zero.

phase modulation   The modulation technique that uses different phase angles of a signal to represent 0s and 1s.

photo diode   A inexpensive light source that is placed at the end of a fiber optic cable to produce the pulses of light that travel through the cable.

photo receptor   The device at the end of a fiber optic cable that accepts pulses of light and converts them back to electrical signals.

physical design   A procedure performed during the systems development life cycle that maps the actual physical components of a network.

physical layer   The lowest layer of the OSI model that handles the transmission of bits over a communications channel.

ping storm   A condition in which the Internet ping program is used to send a flood of packets to a server to make the server inoperable.

plain old telephone system (POTS)   The basic telephone system.

plaintext   Data before any encryption has been performed.

planning phase   A phase of the systems development life cycle in which an analyst identifies problems, opportunities, and objectives.

point-to-point connection   A direct connection between a terminal and a mainframe computer.

polling   The operation in which a mainframe prompts the terminals if they have data to submit to the mainframe.

polyalphabetic substitution-based cipher   Similar to the monoalphabetic cipher, except it uses multiple alphabetic strings to encode the plaintext rather than one alphabetic string.

presentation layer   A layer of the OSI model that performs a series of miscellaneous functions necessary for presenting the data package properly to the sender or receiver.

Pretty Good Privacy (PGP)   A very good encryption software that has become the de-facto standard for creating secure e-mail messages and encryption of other types of data files.

primary   The mainframe computer in a primary—secondary connection between mainframe and terminals.

primary rate interface (PRI)   A service under ISDN that consists of 23 64 Kbps B channels plus one 64 Kbps D channel, for a combined data transfer rate of 1.544 Mbps (T1 speed).

print server   The local area network software that allows multiple workstations to send their print jobs to a shared printer.

private branch exchange (PBX)   A large computerized telephone switch that sits in a telephone room on the company property.

private line   Leased telephone lines that require no dialing.

procedural component   One of the four components of an interface that describes how the particular circuits are used to perform an operation.

propagation delay   The delay between when a signal leaves a transmitter and arrives at a receiver.

propagation speed   The speed at which a signal moves through a medium.

protocol analyzer   A computer program that monitors a network 24 hours a day, seven days a week, and captures and records all transmitted packets.

proxy server   A computer running proxy server software that acts as the "doorman" into a corporate network.

public key cryptography   One key encrypts the plaintext and another key decrypts the ciphertext.

public key infrastructure   The combination of encryption techniques, software, and services that involves all the necessary pieces to support digital certificates, certificate authorities, and public-key generation, storage, and management.

pulse amplitude modulation (PAM)   Tracking an analog waveform and converting it to pulses that represent the wave's height above (or below) a threshold.

pulse code modulation   An encoding technique that converts analog data to a digital signal. Also known as digitization.

quadrature amplitude modulation   A modulation technique that incorporates multiple phase angles with multiple amplitude levels to produce 16 combinations, creating a bps that is four times the baud rate.

quadrature phase modulation   A modulation technique that incorporates four different phase angles, each of which represents two bits: a 45 degree phase shift represents a data value of 11; a 135 degree phase shift represents 10; a 225 degree phase shift represents 01; and a 315 degree phase shift represents 00.

quality of service (QoS)   The concept that data transmission rates, error rates, and other network traffic characteristics can be measured, improved, and, hopefully, guaranteed in advance.

quantizing noise   The noise that occurs during digitization when the reproduced analog waveform is not an accurate representation of the original waveform.

rate adaptive DSL (RADSL)   A form of digital subscriber line in which the transfer rate can vary depending on noise levels within the telephone line's local loop.

Real Time Streaming Protocol (RTSP)   An application layer protocol that servers and the Internet use to deliver streaming audio and video data to a user's browser.

Real-Time Protocol (RTP)   An application layer protocol that servers and the Internet use to deliver streaming audio and video data to a user's browser.

redirection   The technique of moving a data signal to an alternate path.

redundant array of independent drives (RAID) Describes how the data is stored on multiple disk drives.

reliability   Calculates the probability that a component or system will be operational for the duration of a transaction of time.

remote access   Allows a person to access all of the possible functions of a personal computer workstation from a mobile or remote location.

remote bridge   The device that is capable of passing a data frame from one local area network to another when the two local area networks are separated by a long distance and there is a wide area network connecting the two LANs.

remote login (Telnet)   The Internet application that allows you to login to a remote computer.

remote loop-back testing   The modem testing procedure that allows computer A to transmit to its modem, which then passes the signal to the remote modem and back to computer A.

Remote Network Monitoring (RMON)   A protocol that allows a network manager to monitor, analyze, and troubleshoot a group of remotely managed elements.

repeater   A device that regenerates a new signal by creating an exact replica of the original signal.

reservation system   A medium access control procedure that allows users to place a reservation for future time slots.

ring topology   A circular connection of workstations.

roll-call polling   The polling method in which the mainframe computer (primary) polls each terminal (secondary), one at a time, in round-robin fashion.

round robin protocol   A protocol in which each workstation takes a turn at transmission and the turns are uniformly distributed over all workstations.

router   The connecting device between local area networks and wide area networks and between transmission links within a wide area network.

Routing Information Protocol (RIP)   A protocol used to route data across the Internet.

RS-232   An older protocol designed for the interface between a terminal or computer (the DTE) and its modem (the DCE).

RS-449   An interface protocol based on the original RS-232 standard and including several major improvements.

runts   Frames on a CSMA/CD local area network that are shorter than 64 bytes (more than likely due to a collision).

satellite microwave   A wireless transmission system that transmits data using microwave signals from a groundstation to a satellite in space and back to another groundstation.

secondary   A channel in an RS-232 interface in which the data and control lines are equivalent in function to the primary data and control lines, except they are for use on a reverse or backward channel.

secure sockets layer   An additional layer of software added between the application layer and the transport (TCP) layer that creates a secure connection between sender and receiver.

selection   In a mainframe-terminal configuration, the process of a mainframe (primary) transmitting data to a terminal. The primary creates a packet of data with the address of the intended terminal and transmits the packet.

selective-reject ARQ   An error control technique in which a receiver, after detecting an error, requests the transmitter to resend the one packet in error.

self-clocking   A signal that changes at a regular pattern, thus allowing the receiver to synchronize to the changing signal.

serial port   A connection on a computer that is used to connect devices such as modems and mice to personal computers.

server   Computers that store the network software and shared or private user files.

server software   The application or set of programs that stores web pages and allows browsers from anywhere in the world to access those web pages.

session layer   A layer of the OSI model that is responsible for establishing sessions between users and for handling the service of token management.

shared network   A local area network in which all workstations immediately hear a transmission.

shared network segment   A portion of a local area network in which hubs interconnect multiple workstations. When one workstation transmits, all hear the signal and thus are sharing the segment.

shielded twisted pair (STP)   Copper wire used for transmission of signals in which shielding is either wrapped around each wire individually, around pairs of wires or around all the wires together and provides an extra layer of isolation from unwanted electromagnetic interference.

shift keying   A technique in which digital data is converted to an analog signal for transmission over a telephone line.

signals   The electric or electromagnetic encoding of data and are used to transmit data.

signature-based scanner   An anti-virus technique that works by recognizing the unique pattern of a virus.

Simple Mail Transfer Protocol (SMTP)   An Internet protocol for sending and receiving e-mail.

Simple Network Management Protocol (SNMP)   An industry standard created by the Internet Engineering Task Force designed originally to manage Internet components but now also used to manage wide area network and telecommunication systems.

simple parity   A simple error detection technique in which a single bit is added to a character in order to preserve an even number of 1s (even parity) or an odd number of 1s (odd parity).

simplex   A system that is capable of transmiting in one direction only, such as a television broadcast system.

single user multiple station license   An agreement that allows a person to install a copy of a software program on multiple computers, such as a person's home computer and the same person's work computer.

single user single station license   An agreement that allows a person to install a single copy of a software program on one computer only.

site license   An agreement that allows a company to install copies of a software program on all machines at a single site.

sliding window protocol   A protocol that allows a station to transmit a number of data packets at one time before receiving an acknowledgment (a Receive Ready, or RR command).

slope overload noise   The noise that results during analog to digital conversion and the analog waveform rises or drops too quickly; the hardware may not be able to keep up with the change.

small computer system interface (SCSI)   A specially designed interface that allows for a very high speed transfer of data between the disk drive and the computer.

smurfing   The name of a particularly nasty automated program that attacks a network by exploiting Internet Protocol (IP) broadcast addressing and other aspects of Internet operation.

SNMP manager   Controls the operations of a managed element and maintains a database of information about all managed elements.

source-routing bridge   A bridge used in token ring networks that does not keep any internal tables, but instead relies on information contained within each data frame.

spectrum   The range of frequencies that a signal spans from minimum to maximum.

splitterless   A technique used to deliver digital subscriber line signals in which there is no splitter used to separate the DSL signal from the POTS signal.

spoofing   A process by which someone (or something) pretends they are something they are not, such as an e-mail message that has a return address of someone other than the person sending the e-mail. A modem can also perform spoofing by mimicking older protocols that are rarely used today.

spread spectrum   A high security transmission technique that bounces the signal around on seemingly arandom frequencies rather than transmit the signal on one fixed frequency.

star-wired bus topology   The most popular configuration for a local area network; a hub (or similiar device) is the connection point for multiple workstations and may be connected to other hubs.

start bit   Used in asynchronous transmission, a binary 0 that is added to the beginning of the character and informs the receiver that an incoming data character (frame) is arriving.

static routing   A wide area network routing technique in which routing tables remain the same when network changes occur. Maintaining static routing tables is simpler than maintaining adaptive routing tables.

station   The device with which a user interacts to access a network and contains the software application that allows someone to use the network for a particular purpose.

statistical time division multiplexing   A form of time division multiplexing in which the multiplexor creates a data packet of only those devices that have something to transmit.

stop bit   Used in asynchronous transmission, a binary 1 that is added to the end of a character to signal the end of the frame.

stop-and-wait ARQ   An error control technique usually associated with a class of protocols also called stop-and-wait.

streaming audio and video   The continuous download of a compressed audio or video file, which can then be heard or viewed on the user's workstation.

subnet   The collection of nodes and telephone lines into a cohesive unit.

switch   A device that is a combination of a hub and a bridge and that can interconnect multiple workstations like a hub, but can also filter out frames providing a segmentation of the network.

switched virtual circuit (SVC)   A connection that enables frame relay users to dynamically expand their current PVC networks and establish logical network connections on an as-needed basis to end points on the same network or through gateways to end points on other networks.

symmetric connection   A type of connection in which the transfer speeds in both directions are equivalent.

synchronization point   Some form of backup points that are inserted into a long transmission in case of errors or failures.

synchronous connection   A technique for maintaining synchronization between a receiver and the incoming data stream.

synchronous time division multiplexing   A multiplexing technique that gives each incoming source a turn to transmit, proceeding through the sources in round-robin fashion.

systems analyst   A professional that is typically responsible for managing a project and following the SDLC phases, particularly the analysis and design phases.

systems development life cycle (SDLC)   A methodology for a structured approach for the development of a business system, including planning, analysis, design, implementation and support.

T1 multiplexor   The device that creates a T1 output stream that is divided into 24 separate digitized voice/data channels of 64 Kbps each.

T1 service   An all digital telephone service that can transfer either voice or data at speeds up to 1.544 Mbps (1,544,000 bits per second).

tap   A passive device that allows you to connect a coaxial cable to another continuous piece of coaxial cable.

technically feasible system   One that can be created and implemented using currently existing technology.

telecommunications   The study of telephones and the systems that transmit telephone signals.

Telecommunications Act of 1996   A major ruling in the telecommunications industry which, among other things, opened the door for businesses other than local telephone companies to offer a local telephone service.

Telnet   A terminal emulation program for TCP/IP networks, such as the Internet, that allows users to login to a remote computer.

terminate and stay resident antivirus software   Antivirus software that is activated and then runs in the background while users perform other computing tasks.

terrestrial microwave A transmission system that transmits tightly focused beams of radio signals from one ground-based microwave transmission antenna to another.

Text to speech and speech to text conversions Telephone systems that can digitize human speech and store it as a text file, and take a text file and convert it to human speech.

thick coaxial cable A coaxial cable that ranges in size from approximately 6 to 10 mm in diameter.

thin client A workstation computer connected to a network that has no floppy disk drive or hard disk storage.

thin coaxial cable A coaxial cable that is approximately 4 mm in diameter.

third party call control A telephone feature in which users have the ability to control a call without being a part of it, such as setting up a conference call.

three tier architecture A form of client/server architecture that adds a middle tier between the client interface environment and the database management server environment.

TIE lines Leased telephone lines that require no dialing.

time division multiplexing (TDM) A multiplexing technique in which the sharing of a signal is accomplished by dividing available transmission time on a medium among users.

time feasible system One that is installed in a timely fashion that meets organizational needs.

time value of money States that one dollar today is worth more than one dollar promised a year from now because the dollar can be invested now and make interest.

timeout An action that occurs when a transmitting or receiving workstation has not received data or a response in a specified period of time.

token bus A type of local area network that is designed to be a deterministic protocol like token ring but to operate on a bus topology, not a ring.

token management A system that controls who talks when during the current session by passing a software token back and forth.

token ring A local area network that uses the ring topology for the hardware and a round robin protocol for the software.

topologies The physical layout or configuration of a local area network or a wide area network.

Transmission Control Protocol (TCP) The Internet protocol that turns an unreliable sub-network into a reliable network, free from lost and duplicate packets.

transparent bridge An interconnection device designed for CSMA/CD LANs, observes network traffic flow and uses this information to make future decisions regarding frame forwarding.

transport layer A layer of the OSI and Internet models that is concerned with an error-free, end-to-end flow of data.

transposition-based cipher An encryption technique in which the order of the plaintext is not preserved, as it is in substitution-based ciphers.

trees The more complex bus topologies that consist of multiple cable segments that are all interconnected.

Trojan horse A destructive piece of code that hides inside a harmless looking piece of code, such as an e-mail or an application macro.

tunneling protocol The command set that allows an organization to create secure connections using public resources such as the Internet.

twisted pair wire Two or more pairs of single conductor wires that have been twisted around each other.

two tier architecture A client/server system in which the user interface is located in the user's workstation environment and the database management system is located in a server.

unified messaging A telecommunication service that allows users to utilize a single desktop application to send and receive e-mail, voice mail, and fax.

uniform resource locator (URL) An addressing technique that identifies files, web pages, images, or any other type of electronic document that resides on the Internet.

uninstall software A program that works with the user to locate and remove applications that are no longer desired.

Universal Serial Bus (USB) A modern standard for interconnecting modems and other peripheral devices to microcomputers.

unmanaged hub A hub in a local area network that contains little or no intelligence at all and cannot be controlled from a remote location.

unshielded twisted pair (UTP) The most common form of twisted pair, does not wrap any of the wires with a metal foil or braid.

unspecified bit rate (UBR) A class of service offered by ATM that is capable of sending traffic that may experience bursts of data, but there are no promises as to when the data may be sent, and if there are congestion problems, there is no congestion feedback (as ABR provides).

uplink The satellite connection from a groundstation to the satellite.

Usenet A voluntary set of *rules* for passing messages and maintaining newsgroups.

User Datagram Protocol (UDP) A no-frills transport protocol that does not establish connections or watch for datagrams that have existed for too long.

user-network interface   The connection between a user and the network in asynchronous transfer mode (ATM).

V.90 standard   A 56,000 bps dial-up modem standard approved by a standards making organization and not a single company; it is slightly incompatible with both x2 and K56flex.

variable bit rate (VBR)   A class of service offered by ATM that is similar to frame relay service, VBR is used for real time applications, such as compressed interactive video, which is time dependent, and non-real time applications, such as sending e-mail with large, multi-media attachments, which is not time dependent.

very high data rate DSL   A form of digital subscriber line that is very fast format (between 51 and 55 Mbps) over very short distances (less than 300 meters).

very small aperture terminal (VSAT)   A two-way data communications service performed by a satellite system in which the groundstations use non-shared satellite dishes.

Vigenére Cipher   Possibly the earliest example of a polyalphabetic cipher, created by Blaise de Vigenére in 1586.

virtual channel connection (VCC)   Used in asynchronous transfer mode, a logical connection over a virtual path connection.

virtual circuit   A connection through a network that is not a dedicated physical connection, but is a logical connection created by routing tables at each node/router along the connection.

virtual LAN (VLAN)   A technique in which various workstations on a local area network can be configured via software and switches to act as a private segment local area network.

virtual path connection (VPC)   Used in asynchronous transfer mode to support a bundle of VCCs that have the same endpoints.

virtual private network   A data network connection that makes use of the public telecommunication infrastructure, but maintains privacy through the use of a tunneling protocol and security procedures.

virus   A small program that attaches itself to another program or computer file and when loaded or executed causes problems with a computer.

voice network   A type of network that is designed to support standard telephone calls.

voice over Asynchronous Transfer Mode (VToA)   Another technique of packet voice that offer better quality and faster throughput than Internet telephony.

voice over frame relay (VoFR)   Allows the internal telephone systems of companies to be connected using frame relay PVCs.

voice over Internet Protocol (VoIP)   A technology that allows the transmission of telephone calls over the Internet.

voice over packet   A name for telephone calls over the internet.

voice over the Internet   One name for telephone calls over the internet.

weighted network graph   A structure consisting of nodes and edges in which the traversal of an edge has a particular cost associated with it.

white noise   A relatively constant type of noise and is much like the static you hear when a radio is tuned between two stations.

wide area network   Interconnects computers and computer-related equipment in order to perform a given function or functions and typically uses local and long distance telecommunication systems.

wireless topology   A network configuration that uses radio waves for intercommunication.

World Wide Web (WWW)   The collection of resources on the Internet that are accessed via the HTTP protocol.

worm   A special type of virus that manages to store itself in a location that can cause the most damage to the resident computer system.

X.21   An interface standard that was designed to replace the aging RS-232 standard.

xDSL   The generic name for the many forms of digital subscriber line (DSL).

Index

1-persistent algorithm, 210
4B/5B encoding, 42, 221
8.3/125 cable, 79
10Base2, 219
10Base5, 219
10Base5 standard, 219
10BaseT, 219
10Broad36 specification, 219
32-bit binary addresses, 343
56K leased line, 364
62.5/125 cable, 79
100BaseT4, 219
100BaseTX, 219
100VG-AnyLAN, 222
103/133 Series modems, 127
201 Series modems, 127
202 Series modems, 127
208 Series modems, 127
209A Series modems, 127
212A Series modems, 127
1000BaseCX, 219
1000BaseLX, 219
1000BaseSX, 219
1000BaseT, 219
56,000 bps modems, 107, 113

A

access rights, 400–401
Active Directory, 272–273
adaptive algorithm, 308
adaptive routing, 308
ADSL (Asymmetric Digital Subscriber Line), 29, 379
agent, 444
almanac, 334
amplitude, 34
amplitude modulation, 43–44, 107
amplitude shift keying, 43–44
AMPS cellular telephone systems, 90
AM radio, 29
analog amplifiers, 166
analog signals, 3, 61–62
    analog signal transmission and, 52
    broadband coaxial technology, 74
    broadband signals, 198
    coaxial cable, 74
    digital signal transmission and, 47–51
    frequency division multiplexing, 145
    impulse noise, 161–162
    modems, 112–113
    noise, 31–32
        transmitting digital data with, 43–46
        white noise, 161
analog trunks, 361
analytical modeling, 435–436
ANSI (American National Standards Institute), 19, 120
anti-theft devices, 395
antivirus software, 278, 401–402
APIs (application programming interfaces), 270
application layer, 19
    encapsulating packet, 238
    error detection, 167
applications, 278
    baseline, 438
    capacity planning, 436
    connectionless network, 296–297
    connection-oriented, 295–296
    databases, 262, 280
    desktop publishing, 280
    office suites, 262, 281
    presentation software, 281
    project management, 262
    sending data, 179
    stand-alone spreadsheets, 281
    web browsers, 262
    word processing, 262, 281
area code, 361
ARP (Address Recognition Protocol), 319, 333, 340
ARPA (Advance Research Projects Agency), 318
ARPANET, 318
ASCII character set, 57
ASCII string, 60–61
asynchronous connections, 127, 129–131
asynchronous time division multiplexing, 154
AT command set, 126
ATM (asynchronous transfer mode), 4, 158, 180–181, 310, 358, 373–376
AT&T, 118, 355–356
attenuation, 36–38, 70, 165
auditing, 399–400
availability, 441–442

B

backbone, 77
backplane, 245
backup software, 278
balanced transmission, 125
bandwidth, 35, 111–113
banking transactions, 413
baseband coaxial technology, 74
baseband local area networks, 198

baseband signaling, 198
baselines, 437–439
basic service set, 207
basic telephone systems. *See* POTS (plain old telephone system)
baud, 42
Baudot, Emile, 58
Baudot code, 58–59
baud rate, 41–42
beepers, 68, 89–90
Bell interface standards, 127
benchmarking, 435
bindery, 267
BISYNC protocol, 454–457
blacklisting, 109
BOCs (Bell Operating Companies), 361
boot sector virus, 401
bottlenecks, 244
bps (bits per minute) relationship with frequency, 50–51
Bradner, Scott, 143
braided coaxial cable, 75
breakup of AT&T, 361
BRI (basic rate interface), 118, 153–154
bridges, 4, 234–240
    bottlenecks, 240
    in combination with switches, 249
    converting data formats, 234
    dissimilar LANs (local area networks), 234
    as filters, 234–235
    LANs (local area networks), 194
    ports, 245
    remote, 243
    removing CSMA/CD medium access control sublayer information, 238–239
    source-routing, 240–243
    spanning tree, 240–241
    versus switches, 245
    transparent, 236–240
broadband coaxial cable
    analog signals, 74
    frequency division multiplexing, 147
broadband systems, 198–200
broadband wireless systems, 68, 91–92
broadcast networks and OSI model, 215
broadcast subnet, 295
BSD (Berkeley Software Distribution), 274
buffer preallocation, 310
bulk carrier facilities, 84
burst rate, 370
business goals, 424–425
bus networks, 200–201
bus topology, 197–201

C

CA (certificate authority), 412
cable modems, 106, 115–117
cables, proper shielding for, 165

cable telephony, 363
cable television, 114
    assignment of frequencies, 146
    broadband transmission system, 200
cable testers, 445
callback security, 109
capacity planning, 434–437
    example, 446–438
CardBus, 111
CARL (Colorado Alliance of Research Libraries), 320
cascade port, 222
Category 1 twisted pair wire, 70
Category 2 twisted pair wire, 70
Category 3 twisted pair wire, 70
Category 4 twisted pair wire, 70–71
Category 5e twisted pair wire, 71
Category 5 twisted pair wire, 70–71
Category 6 twisted pair wire, 71
Category 7 twisted pair wire, 71
CBR (Constant Bit Rate), 375
CCIE (Cisco Certified Internet Engineer), 440
CCITT (Consultative Committee on International Telephone and Telegraphy), 19
CDMA (code division multiple access), 54, 88
CDPD (cellular digital packet data), 89
CDSL (Consumer CDSL), 379
cells, 86–87, 90
cellular radio systems, 68
cellular telephones, 86, 90, 147
centralized reservation system, 84
centralized routing, 305–306
Centrex (central office exchange service), 362
Centronics parallel interface, 137
Cerf, Vinton, 317
certificates, 411
Chamber, John, 317
channels, 90, 145
character-oriented protocol, 454
chirp signal spread spectrum, 54
choke packet, 310
ciphertext, 404
CIR (committed information rate), 400–401
circuit cards, hot-swappable, 245
circuit switched subnet, 293, 297
Class A IP addresses, 344
Class B IP addresses, 344
Class C IP addresses, 344
Class D IP addresses, 344
CLECs (competitive local exchange carriers), 343
client-server applications, 413
client/server systems, 8, 264–266
CMTS (Cable Modem Termination System), 116
CNA (Novell NetWare Certified Novell Administrator), 440
CNE (Novell NetWare Certified Novell Engineer), 440
CNSE (Certified Network Systems Engineer), 440
CO (central office), 360
coaxial cable, 74–75
codec, 3

collision window, 210
Combiner, 116
Common Name Resolution Protocol, 346
compilers, 280
composite signals, 37
computer auditing, 399–400
computer connections, 133–137
computer networks, 2–4
    See also networks
    baseline, 437–439
    basic configurations, 6–14
    capacity planning, 428–437
    components, 14–15
    local area network-to-local area network, 10
    local area network-to-wide area network, 10–11
    microcomputer-to-Internet, 9
    microcomputer-to-local area network, 8
    microcomputer-to-mainframe computer, 7
    satellite and microwave, 12
    sensor-to-local area network, 11–12
    terminal-to-mainframe computer configurations, 6
    wireless telephone, 13–14
computers, interfacing devices to, 119
computer simulations, 435
Computer Telephony, 357
computer telephony integration, 322
computer terminals, 6
computer terminal-to-mainframe configurations, 6
conducted media, 68, 79
connecting to Internet, 311–313
connectionless applications, 296–297
connectionless protocol, 444
connection-oriented applications, 297–298
connection-oriented protocol, 444
contention-based protocols, 209–211
control center, 443
control characters, 55–56
controlled number of concurrent users license, 282
converting data
    digital to analog signal, 107
    into signals, 39–53
cookies, 326
corporate license, 276
correcting error, 178–180
crash protection software, 278–279
CRC (cyclic redundancy checksum), 169–171, 180–181
CRL (certificate revocation list), 412
crosstalk, 163
    jitter, 164
    proper shielding of cables, 165
    twisted pair wire, 69
cryptography, 404
CS (Certified Specialist), 440
CSE (Certified Solutions Expert), 440
CSMA/CD networks
    collisions, 210–211, 245, 250, 438
    data converted to Internet data frame, 252
    dedicated network segments, 247
    frame format, 216

    IEEE 802.3 standard, 216
    network layer, 252
    non-deterministic protocol, 211
    popularity of, 213
    shared network segments, 246
    switches, 245
    transparent bridges, 236–239
CSU (Channel Service Unit), 115
CSU/DSU (Channel Service Unit/Data Service Unit), 114–115
CTI (Computer Telephony Integration), 358, 380–382
cutthrough architecture and switches, 245
cycle, 425–426
cyclic checksum, 132

D

D-AMPS (Digital-Advanced Mobile Phone Service), 88, 90
data, 3, 30–38
    analog, 31–32, 39
    conversions example, 59–62
    converting into signals, 39–53
    digital, 32–33, 39
    noise, 31–32
    transmitting, 3
database software, 280
data binding, 330
data codes
    ASCII (American National Standard Code for Information Interchange), 57–58
    Baudot code, 58–59
    EBCDIC (Extended Binary Coded Decimal Interchange Code), 55–56
data communications, 2–3
datagram packet switched network, 294, 321
data link connections, 129–133
data link layer, 17
    connections, 106
    error control, 160
    error detection, 160, 166
    frame, 294
    frame relay, 368
data networks, 3
data packets, 328
data transfer formula, 114
data transfer rate, 50–51
data transmission
    errors, 160–165
    noise, 160–165
    rates, 107
    speed, 95
DB-9 9-pin serial port connectors, 136
DB-15 connector, 137
DB-25 connector, 137
DCE (data-circuit terminating equipment), 119–120
decibel, 38
decryption. *See* encryption and decryption
delay distortion, 165

DELETE method, 326
delta modulation, 50–51
demand priority protocol, 214
denial of service attacks, 402–403
dense wavelength division multiplexing, 144, 157–159
Department of Defense, 333
DES (Data Encryption Standard), 408
DES Cracker, 409
desktop publishing, 280
detailed LCD, 430–431
deterministic protocol, 212
devices, 120
DHTML (Dynamic HTML), 327, 329–330
diagnostic tools, 445–446
dial-up modem, 9
Differential Manchester encoding, 41, 60
digital certificates, 411–412
digital data, 3, 32
    noise, 33
    transmitting with analog signals, 43–46
    transmitting with digital signals, 39–42
digital encoding schemes, 39–42
digital equipment, 165
digital repeaters, 166
digital signals, 32, 39, 61–62
    analog data transmission with, 47–51
    data transfer rate, 50
    digital data transmission with, 39–42
    frequency, 50
    impulse noise, 162
    modems, 112–113
    noise, 33
    TDM (time division multiplexing), 147
    timing irregularities, 164–165
    white noise, 161
digital signatures, 409–410
digital subscriber line, 106
digital trunks, 361
Dijkstra's least cost algorithm, 301, 302–304, 309
DIN (Deutsches Institut fur Normung) connector, 135–136
directory security, 263
direct sequence spread spectrum, 54–55
discarding datagram, 336
discount rate, 433
discovery frames, 242–243
distributed algorithm, 308, 309
distributed processing, 188
distributed reservation system, 84
distributed routing, 306–307
DMS100, 118
DNS (Domain Name Server), 341
DNS (Domain Name System), 21, 343
DNS (Domain Naming Service), 272
DOCSIS (Data Over Cable Service Interface Specification), 116
domain names, converting to IP address, 343
dotted-decimal notation, 343
downlink, 83

downshift code, 58
downstream transmissions, 115
draft standard, 18
DSL (digital subscriber line), 4, 114, 117–118, 376–380
DSL.Lite, 379
DSL modems, 118
DSU (Data Service Unit), 115
DTE (data terminal equipment), 119–120
duplex connections, 132
DWDM (dense wavelength division multiplexing), 143
dynamic HTML, 319

E

EBCDIC (Extended Binary Coded Decimal Interchange Code), 55–56
echo, 163–164
echo suppressor, 164
e-commerce, 331–332
EDI (Electronic Data Interchange), 331
effective bandwidth, 35
EIA (Electronic Industries Association), 19, 120
EIA-232E standard, 122, 124
EIA RS-449 standard, 125
electrical testers, 445
electromagnetic interface, 72
electromagnetic interference, 76
    jitter, 164
    proper shielding of cables, 165
electromagnetic noise, 161
Electronic Frontier Foundation, 409
electronic mail. *See* e-mail
electronic mail service, 193
electronic office, 194
electronic security, 335
e-mail, 5, 322–323, 413
e-mail bombing, 402
e-mail daemon, 323
e-mail systems, 322
encapsulation, 25
encryption algorithms, 404–405
encryption and decryption
    ciphertext, 404
    cryptography, 404
    DES (Data Encryption Standard), 408
    DES Cracker, 409
    digital signatures, 409–412
    Electronic Frontier Foundation, 409
    encryption algorithms, 404–405
    Kerberos, 410–412
    key, 404–405
    monoalphabetic substitution-based ciphers, 405
    PGP (Pretty Good Privacy), 410
    polyalphabetic substitution-based ciphers, 405–406
    public key cryptography, 402–402
    S-HTTP (Secure-HTTP), 407
    SSL (secure sockets layer), 408
    transposition-based ciphers, 406–407
Entrust Technologies, 418

e-retailing, 331
error bursts, 168
error control, 160
    doing nothing approach to, 172
    example, 180–181
    Go-back-N ARQ, 175
    returning message, 172–175
    Selective-reject ARQ, 175–178
    sliding window protocols, 174–175
    Stop-and-Wait ARQ, 172–174
error correction, 178–180
error detection, 160
    example, 180–181
    at final destination only, 167
    parity checks, 167–169
    software, 17
    techniques, 166–171
error messages, 172
errors
    attenuation, 165
    correcting, 178
    crosstalk, 163
    delay distortion, 165
    echo, 163–164
    impulse noise, 161–162
    jitter, 164–165
    prevention, 165–166
    white noise, 161
Ethernet, 219, 222
Ethernet card and cable modems, 115
even parity, 167
exchange, 361
exclusive-ORed, 54
expert system virus analysis, 402
extended service set, 207
extensions to FTP, 346
external modem, 110
external security, 395
extranets, 333

F

fallback negotiation, 165
Fast Ethernet, 219
faxes, 109
FCC (Federal Communications Commission), 80,
    113, 146
FDDI (fiber data distributed interface), 220
FDMA (frequency division multiple access), 90
feasibility studies, 431–434
fiber distribution node, 116
fiber exhaust, 157
fiber optic cables
    8.3/125 cable, 79
    62.5/125 cable, 79
    ATM (asynchronous transfer mode), 158
    backbone, 77
    cable modems, 117
    dense wavelength division multiplexing, 157–159

    electromagnetic interference, 76–77
    error-free high data transmission rates, 76
    full duplex, 116
    high bandwidth transmission, 152
    high cost of, 77
    laser, 76
    LED (light-emitting diode), 76
    multi-mode transmission, 79
    noise, 77
    one-way transmission, 77
    photo diode, 76
    photo receptor, 76
    reflection, 78
    refraction, 78
    SDH (synchronous digital hierarchy), 158
    secure transmission media for sensitive data, 404
    single-mode transmission, 78–79
    SONET (synchronous optical network), 158
    wiretapping, 76–77
files, uploading and downloading, 319–320
file security, 263
file server, 187
filters, 234–235
financially feasible system, 431
firewalls, 253, 413–416
FireWire, 127
flooding, 300–305, 309
flow control, 17
FM radio, 29
forward error correction, 178
forward search least cost algorithm, 301
Fourier Analysis, 37
fractional-T1
fragmenting datagram, 336
frame, 129
frame formats, 216
frame relay, 180, 310, 312–313
    availability, 367
    burst rate, 370
    CIR (committed information rate), 370–371
    CRC (cyclic redundancy checksum), 181
    data link layer, 368
    do nothing approach to error control, 172
    error detection, 181
    good security, 367
    high throughput, 377
    interconnections, 369
    versus Internet, 371
    layer 2 protocol, 368
    layers, 181
    low error rates, 367
    PVC (permanent virtual circuit), 367–368
    reasonable pricing, 367
    setup, 368–370
    SVCs (Switched Virtual Circuit), 372–373
    very high transfer speeds, 367
    VoFR (voice over frame relay), 371–372
frames, 17
Free Software Foundation, 274

frequencies, 34–35
    cells, 87
    digital signal, 50
    pseudorandom sequence of, 53
    relationship with bps (bits per second), 50–51
frequency division multiplexing, 88, 144–147, 198
frequency hopping spread spectrum, 54
frequency modulated signal, 60–61
frequency modulation, 44–45
frequency shift keying, 44–45, 60
FTP (file transfer protocol), 21, 319
full duplex connections, 132–133
full duplex data link protocol, 457
full duplex fiber optic line, 116
full duplex switches, 250
full-T1, 151

G

game port, 137
Gates, Bill, 105
gateway, 232
Gaussian noise, 161
generating polynomial, 169
GEO (Geosynchronous earth orbit) satellites, 85–86
GET method, 326
Go-back-N ARQ, 175
GPS (global positioning systems), 85
ground line, 125
guard band, 145
guarding against viruses, 401–402

H

hackers, 393
half-duplex protocol, 454
hamming distance, 178–179
hard disk storage space, 263
hardware, 4
hash, 410
Hayes modems, 126
Hayes Smartmodem, 126
Hayes standards, 126
HDLC (High-level Data Link Control), 462
HDSL (High-bit rate DSL), 379
HEAD method, 326
help desk, 443
HFC (Hybrid Fiber Coax), 363
high data transfer rates, 50, 192
high speed telephone line, 4
historic RFC, 19
home-to-Internet connection, 311–312
hop count, 305
hop limit, 305
horizontal parity, 168–169
hot-swappable circuit cards, 245
HTML (HyperText Markup Language), 319, 326–328
HTML tags, 327–328
HTTP (HyperText Transport Protocol), 21, 312, 319, 326–327

hub polling, 134
hubs, 4, 97, 201–202, 244
    100VG-AnyLAN, 222
    baseband coaxial, 74
human voice, 359
hybrid telephones, 88–89
hyperlinks, 328
Hz (Hertz), 34

I

IAB (Internet Architecture Board), 345
IBM token ring, 214
ICMP (Internet Control Message Protocol), 319, 333, 339–340
ID systems, 397–399
IEA-232 standard, 122–123
IEEE (Institute for Electrical and Electronics Engineers), 120
IEEE 802.3 standard, 218–219
IEEE 802.5 standard, 220
IEEE 802.11 standard, 207, 208
IEEE 802.12 standard, 214, 221
IEEE 802 frame formats, 216–217
IETF (Internet Engineering Task Force), 346, 413
IMAP (Internet Message Access Protocol), 323
impulse noise, 161–162
IMTS (Improved Mobile Telephone Services), 86
information service, 9
infrared transmissions, 90–91
intelligent checksum analysis, 402
interchange circuits, 120
interconnection devices
    bridges, 234–250
    hubs, 244
    routers, 252–253
    switches, 244–240
interface layer, 20, 238
interfaces
    components, 121
    conforming to standards, 120
    EIA-232E standard, 121–124
    EIA RS-449 standard, 125
    FireWire, 127
    Hayes standards, 126
    RS-232 standard, 121–124
    standards, 120–126
    USB (Universal Serial Bus), 128–129
    X.21 standard, 125–126
interfacing, 3
    devices and physical layer, 106
    modems, 119–126
intermodulation distortion, 45
internal modem, 110
Internet, 318
    application services, 333–334
    backdoor entry to data, 417
    connecting to, 311–313
    connectionless packet delivery service, 333
    connections, 9

example, 349–350
versus frame relay, 371
future of, 345–349
hackers, 393
IAB (Internet Architecture Board), 345
IETF (Internet Engineering Task Force), 346
interconnectivity, 394
interface services, 333–334
Internet protocol, 252
IRTF (Internet Research Task Force), 346
ISOC (Internet Society), 345
locating documents, 341–345
network layer, 252
protocols, 319
reliable transport service, 333–334
RFC (Request for Comment), 18–19
routing, 299
W3C (The World Wide Web Consortium), 346
World Wide Web, 325–332
Internet2, 348–349
Internet application layer, 21
Internet draft, 19
Internet Explorer, 336
Internet Fax, 346
Internet layer, 238
Internet model, 19–21
example, 23–25
Internet network layer, 21
Internet protocols
ARP (Address Recognition Protocol), 340
ICMP (Internet Control Message Protocol), 339–340
IP (Internet Protocol), 334–337
PPP (Point-to-Point Protocol), 341
PPTP (Point-to-Point Tunneling Protocol), 341
TCP (Transmission Control Protocol), 338–339
tunneling protocols, 341
UDP (User Datagram Protocol), 340
Internet server software, 279
Internet services
e-mail, 322–323
FTP (file transfer protocol), 319–320
listserv, 334
remote login, 320
streaming audio and video, 325
telephony, 321
Telnet, 320
Usenet, 324
Internet software, 278
Internet standard, 18
Internet transport layer, 21
internetworking, 233
intraLATA, 361
intranets, 332–333
intrusion detection, 397
IP (Internet Protocol), 21, 319, 333
encapsulation process, 334
fields, 347
functions, 336
importance of, 337

IP datagrams, 334–337
IPv6, 347–348
objective, 337
IP addresses, 341, 344–345
IP datagrams, 334–337
IP multicasting, 344–345
IPP (Internet Printing Protocol), 346
IP subnet masking, 345
IPv6, 347–348
IPX (Internet packet exchange), 268
IrDA (Infrared Data Association), 91
ISDN (Integrated Services Digital Network), 4, 114, 117, 152, 365–366
ISDN modems, 117–118
ISDN multiplexing, 152–154
ISDN networks, 117
ISDN switch type protocols, 118
ISO (International Standards Organization), 16, 19, 120
ISOC (Internet Society), 345
isochronous connection, 127
isolated routing, 307
ISPs (Internet Service Providers), 9, 311, 345
ITU (International Telecommunications Union), 120
ITU (International Telephone Union), 113
ITU-T (International Telecommunications Union-Telecommunication Standards Sector), 19

J
JavaScript, 280
JTAPI (Java Telephony API), 382

K
K56flex format, 113
Kerberos, 410–411
key, 404–405

L
lambdas, 143, 157
LANs (local area networks), 3–5, 8, 290, 429
100VG-AnyLAN, 221–222
advantages, 195
bridges, 194, 234–240
broadband, 198
broadcast, 192
connecting, 10
connecting to WANs (wide area networks), 252
costs, 196
databases and applications, 193
data communicating devices, 192
data transfer rates, 195
disadvantages, 196
distributed processing, 194
electronic mail service, 193
electronic office, 194
error rates, 195
Ethernet, 218–219

FDDI (fiber data distributed interface), 220–221
file server, 193
    functions of, 193–195
    hardware, 4
    high data transfer rates, 192
    home office example, 225–226
    hubs, 97
    IBM token ring, 220
    licensing agreements, 196
    logical design, 201
    manager or system administrator, 196
    medium access control protocols, 209–214
    mixing platforms on, 195
    monitoring manufacturing events, 194
    multiple, 232–233
    network servers, 251
    network topologies, 196–206
    NOS (network operating system), 263–264
    passive device, 97
    physical design, 201
    popularity, 231
    print server, 193
    reasons for interconnecting, 232–233
    routers, 194
    shared network, 202
    sharing hardware and software resources, 192, 195
    small geographic area, 192
    small office example, 223–225
    software conflicts, 196
    support devices, 262
    transferring video images and video streams, 194
    very low error rates, 192
    weakest link, 196
    wiring example, 97–98
LAN software example, 284–285
LAN support devices. *See* network support devices
LAP (Link Access Protocol), 464
LAP-B, 464
laptops and modems, 110–111
laser, 76
laser light, 143
LATAs (local exchange carriers), 361
layer 2 protocol, 368, 373
LCD (location connectivity diagram), 430
leaf objects, 269
Learning Ware, 348
lease digital telephone line, 114
leased line services, 364–365
least cost algorithm, 301, 305–306
LEC (local exchange carriers), 362
LED (light-emitting diode), 76
licensing agreements, 281–282
linear projection, 435–436
line disconnect attack, 403
line-of-sight transmission, 81–82
link state routing algorithm, 308–309
Linux, 262, 274–277
listserv, 334
LLC (logical link control) sublayer, 215

LMDS (Local Multipoint Distribution System), 92
local area network tester, 446
local area network-to-local area network
    configuration, 10
local loop, 360
local loop-back testing, 109
Local Multipoint Distribution Service, 295
locating documents, 341–345
logical connection, 22–23
longitudinal parity, 168–169
longitudinal redundancy check, 168–169
loss of signal strength, 36–38

M

macro virus, 401
MAC (medium access control) sublayer, 215–216
MAC sublayer address, 235
mailserv, 324
mainframe operating systems, 263
majordomo, 324
MAN (metropolitan area network), 290
managed hubs, 244
Manchester encoding, 41–42, 131
MAU (multi-station access unit), 205–206
maximum data transfer rate, 51
media
    carrying multiple signals, 144
    cost, 94
    data transmission speed, 95
    dense wavelength division multiplexing, 144
    distance, 95
    environments, 96
    expandability, 95–96
    frequency division multiplexing, 144–147
    maintenance costs, 94
    multiplexing, 144
    observing stated capacities of, 166
    propagation speed, 95
    security, 96
    selection criteria, 94–96
    speed, 95
    statistical time division multiplexing, 144
    time division multiplexing, 144
media converters, 283
medium access control protocols
    contention-based protocol, 209–211
    reservation protocols, 209, 214
    round robin protocols, 209, 211–214
memory security, 263
memory storage space, 263
Message Transfer Agent, 322
metropolitan area networks, 3
MIB (Management Information Base), 444
Microcom Corporation, 108
microcomputer operating systems, 263
microcomputers, 7
microcomputer-to-Internet configuration, 9
microcomputer-to-local area network configuration, 8

microcomputer-to-mainframe computer configuration, 7

Micro-marketing, 331

Microsoft Internet Explorer, 325

"Microsoft Windows NT Server 4.0 *versus* UNIX," 261

microwave systems, 68

MIDI (Musical Instrument Digital Interface), 137

milnet, 318

MIME (Multipurpose Internet Mail Extension), 322

MIME system, 323

miniature card technology, 111

minicomputer operating systems, 263

MMDS (multichannel multipoint distribution service), 68, 92

MNP 1-5 standards, 108

mobile telephones
   824 - 849 MHz range, 90
   869 - 894 MHz range, 90
   AMPS (Advanced Mobile Phone Service), 88
   cells, 86–87, 90
   cellular telephone, 86
   channels, 90
   D-AMPS (Digital-Advanced Mobile Phone Service), 88
   FDMA (frequency division multiple access), 90
   hybrid telephones, 88–89
   IMTS (Improved Mobile Telephone Services), 86
   MIN (mobile ID number), 87–88
   MSAs (Mobile Service Areas), 86
   MTSO (mobile telephone switching office), 87–88
   PCS (Personal Communication Systems), 88
   PCS (personal communication systems), 86
   towers, 87

modem pools, 118–119, 262

modems, 3, 43, 127, 262
   56,000 bps, 107, 113
   alternatives, 114–118
   analog signaling, 112–113
   bandwidth limitations, 111–113
   connection negotiation, 108
   data compression, 108
   data transmission rates, 107
   digital signaling, 112–113
   downstream transmission, 112
   error correction, 108
   external, 110
   fallback negotiation, 108
   fall forward negotiation, 108
   faxes, 109
   interfacing, 3
   internal, 110
   K56flex format, 113
   laptop, 110–111
   local loop-back testing, 109
   MNP 5 compression standard, 108
   operating principles, 107
   PCMCIA (Personal Computer Memory Card International Association) standard, 111
   remote loop-back testing, 109
   security, 109
   standard telephone functions, 108
   testing functions, 109
   V.90 standard, 113
   V-series, 128
   x2 format, 113

Modified Final Judgment, 361

modulation, 43, 107

monoalphabetic substitution-based ciphers, 405

MTBF (mean time between failures), 441

MTSO (mobile telephone switching office), 87–88

MTTR (mean time to repair), 441

multi-level generic scanning, 400

multi-mode transmission, 79

multiple LANs (local area networks), 232–233

multiplexed earth station satellite system, 84

multiplexing, 3, 52, 144
   business example, 160–162
   comparing techniques, 159–160
   dense wavelength division multiplexing, 157–159
   frequency division multiplexing, 144–147
   TDM (time division multiplexing), 147–157

multiplexors, 145, 262

multipoint connections, 133

multitasking operating system, 263

N

Navigator and layers, 330

NDS (Novell Directory Services), 267

NDS tree, 269–270

Netscape Navigator, 325

NetWare, 261, 266–271, 276–277

NetWare 3, 267–268

NetWare 4, 269–280

NetWare 5, 280

network architecture models, 14
   Internet model, 19–21
   logical and physical connections, 22–23
   OSI (Open Systems Interconnection) model, 16–19

network design example, 446–448

network graph, 299–301

Network in a Box, 225–226

network layer, 17, 294

network management protocol, 443

network managers, 439–441

network modeling, 429–431

Network-Network Interface, 373

networks, 2, 4, 429
   See also computer networks
   amount of traffic on each node, 439
   availability, 441–442
   backbone, 77
   bottlenecks, 244
   connected in mesh, 290
   control center, 443
   diagnostic tools, 445–446
   dividing large, 233
   error statistics, 439
   firewalls, 413–416
   generating usable statistics, 441–442

handling congestion, 309–311
improving, 437
limiting use of, 395–396
MTBF (mean time between failures), 441
MTTR (mean time to repair), 441
peak periods of use, 438
reliability, 442
reliable service, 295
security, 394
utilization data, 438
network server licenses, 284
network servers, 251, 263–264
network software
applications, 278, 280–281
Internet server software, 279
Internet software, 278
programming tools, 278, 280
utilities, 278–279
network support devices, 283–284
network topologies
bus topology, 197–201
comparing, 208–209
ring topology, 203–206
star-wired bus topology, 201–203
wireless topology, 206–208
Network World Fusion Web site, 143
newsgroups, 324
NI1, 118
NIC (network interface card), 197, 235
nodes, 4, 292, 298–299, 309–310, 437
noise, 160–165
analog signals, 31–32
Category 1 twisted pair, 70
data transmission rates, 107
digital data, 33
digital signals, 33
fiber optic cable, 77
non-deterministic protocol, 211
non-facilities based CLECs, 363
non-persistent algorithm, 210
Nortel Networks Certification, 440
NOS (network operating systems)
comparison, 277
functions, 264
interacting with Internet, 264
Linux, 262, 274–275
NetWare, 262, 267–271
network support functions, 264
OS/2, 275–276
OS/2x, 262
summary, 276–277
Unix, 262, 273–274
Windows NT, 271–273
NRX-L (Non-Return to Zero-Level) digital encoding
scheme, 39–41
NRZ-I (Non-Return to Zero-Inverted) encoding scheme,
40–41
NSFnet (National Science Foundation), 318
Nyquist's Theorem, 49, 51, 61

O
OC (optical carriers), 152
odd parity, 167
office suites, 281
ohm, 75
OKI Electric Industry Company Web site, 191
OKI MSM9225 chip, 191
one-way pagers, 89
operating systems, 262–263, 282
operationally feasible system, 432
operational security, 395–396
organization chart, 269
OS/2, 275–276
OS/2 Warp, 276
OS-2 Warp 4, 276
OS/2x, 262
OSI (Open Systems Interconnection) model, 16–19,
215–216
OSPF (Open Shortest Path First) protocol, 309
OU (organizational unit), 269
overview LCD, 430

P
packet filter, 415
packets, 294, 305
packet switched subnets, 296–298
packet switched virtual circuit subnet, 298
packet voice, 322, 331
pagers, 68, 89–90
PAM (pulse amplitude modulation), 47
parallel port connector, 137
parallel ports, 137
parasitic virus, 401
parity bit, 130
parity checks, 167–169, 171
passive device, 97, 197
passwords, 303–304, 397–398
Patsuris, Penelope, 317
payback analysis, 432–434
pbs (bits per second), 42
PC Card Hosts, 111
PC Cards, 111
PC Card Software, 111
PC card technology, 111
PCMCIA (Personal Computer Memory Card International Association) standard, 111
PCS (Personal Communication Systems), 68, 88
PDNs (public data networks), 463
period, 34
peripheral devices
cable modems, 106
connecting, 105–106
digital subscriber line, 106
interfacing, 127–129
modems, 106, 107–119
permit system, 310
persistence algorithms, 210
personal computers, 7

PGP (Pretty Good Privacy), 410
phase modulation, 45–46, 107
phases, 36, 425–426
phase shift keying, 45–46
phono jacks, 137
photo diode, 76
photo receptor, 76
physical connection, 22–23
physical damage, 395
physical layer, 17
physical media, 68
ping storm, 402–403
PKI (Public Key Infrastructure), 411–413, 418
plenum, 72
plus code modulation, 47–49
point-to-point connections, 133, 135
polling, 106, 134
polyalphabetic substitution-based ciphers, 405–406
polymorphic virus, 401–402
POP3 (Post Office Protocol), 323
ports, 245
POTS (plain old telephone system), 358
   analog trunks, 361
   cable telephony, 363
   Centrex (central office exchange service), 362
   channels and frequencies, 359
   CLECs (competitive local exchange carriers), 363
   CO (central office), 360
   data transmission speed, 359
   digital trunks, 361
   ILECs (Incumbent Local Exchange Carriers), 363
   interLATA, 361
   intraLATA, 361
   LATAs (local access transport areas), 360–362
   LEC (local exchange carriers), 362
   limitation of telephone signals, 359
   local loop, 360
   Modified Final Judgment, 361
   PBX (private branch exchange), 362
   private lines, 363
   standard telephone line, 361
   subscriber loop, 361
   Telecommunications Act of 1996, 363
   telephone lines, 360–361, 363
   telephone network before and after 1984, 361–363
   telephone networks after 1996, 363–364
   TIE lines, 363
   transmitting human voice, 359
   trunks, 360–361
p-persistent algorithm, 210
PPP (Point-to-Point Protocol), 341
PPTP (Point-to-Point Tunneling Protocol), 341, 349
presentation layer, 19
presentation software, 281
PRI (primary rate interface), 153
primary, 134
printers, 193, 283
print server, 193
private key, 407

private lines, 363
programming environment, 280
programming tools, 278, 280
ProNet terrestrial microwave system, 100
propagation delay, 83
propagation speed, 95
proposed standard, 18
protocol analyzer, 446
protocols, 319
   baselines, 437–438
   hierarchy, 333–334
   stop-and-wait, 172–174
proxy server, 415–416
public key cryptography, 408–409
PUT method, 326
PVC (permanent virtual circuit), 367–368

Q
QoS (quality of service), 253, 311
quadrature amplitude modulation, 46, 107
quadrature phase modulation, 45–46
quantization noise, 112–113
quantizing noise, 49

R
radio waves, 80
RADSL (Rate-Adaptive DSL), 379
RAID (redundant array of independent drives), 251
random passwords, 398
RBOCs (Regional Bell Operating Companies), 361–362
redirection, 267
redundant circuits and switches, 245
reflection, 78
refraction, 78
relationship between frequency and bps (bits per
   second), 50–51
reliability, 442
reliable service, 295
remote access, 279
remote bridges, 243
remote login, 320
remote loop-back testing, 109
remote procedure call, 266
repeater ring topology, 204
repeaters, 70
reservation protocols, 214
reservation system, 84
resolver, 343
resource security, 263
resource sharing access rights, 400
RFC (Request for Comment), 18–19
Rifkin, Stanley Mark, 407
ring topology, 203–206
RIP (Routing Information Protocol), 308
riser, 72
RMON (Remote Network Monitoring), 445
roll-call polling, 134

round robin protocols, 211–214
routers, 4, 10–11, 252–253
    firewall, 253
    LANs (local area networks), 188
    nodes as, 298–299
    as packet filters, 415
    restricting access with, 416
routing
    adaptive, 308
    algorithms, 300
    centralized, 305–306
    Dijkstra's least cost algorithm, 301, 302–304
    distance vector routing algorithm, 308
    distributed, 306–307
    examples, 306
    flooding, 304–305, 309
    forward search least cost algorithm, 301
    Internet, 299
    isolated, 307
    link state routing algorithm, 308–309
    static, 307
RS-422A (X.27) standard, 125
RS-423A (X.26) standard, 125
RSTP (Real Time Streaming Protocol), 325
RTP (Real-Time Protocol), 325
runts, 216

S

sampling rate, 49
satellite and microwave configurations, 12
satellite configurations, 84–85
satellite microwave transmissions, 82–86
satellite systems, 68
Scotia OnLine, 418
scripting languages, 280
SCSI (small computer system interface), 251
SDH (synchronous digital hierarchy), 152–153, 158
SDLC (systems development life cycle)
    acknowledging packets, 459
    address field, 458
    analysis phase, 427–428
    basic packet format, 458
    bit patterns, 457–458
    control field, 458–459
    CRC (cyclic redundancy checksum), 459
    cycle, 425–426
    data, 459
    design phase, 428
    implementation phase, 428
    I (Information) packets, 458
    maintenance phase, 428–429
    M subfield, 460
    network models, 427
    phases, 425–426
    plan, 426–427
    planning phase, 427
    polling and request for initialization example, 460
    primary polling secondary, 461

    S (Supervisory) packets, 458
    S subfield, 459–460
    systems analyst, 426
    U (Unnumbered) packets, 458
SDLC code independent, 457
secondaries, 134
secret key cryptography, 410
security, 394
    access rights, 400–401
    anti-theft devices, 395
    auditing, 399–400
    backdoor entry, 109
    banking example, 418
    basic measures, 394–400
    biometric techniques, 398–399
    blacklisting, 109
    callback security, 109
    commitment to, 416–417
    encryption and decryption techniques, 404–411
    expected level of, 416
    external, 395
    firewalls, 413–416
    guarding against viruses, 401–402
    ID systems, 397–399
    intrusion detection, 397
    media, 96
    modems, 109
    operational, 395–396
    passwords, 397
    physical damage, 395
    PIN (personal identification number), 397
    PKI (Public Key Infrastructure), 411–413
    policy design issues, 416–417
    standard system attacks, 402–404
    surveillance, 396–397
    Windows NT, 272
Security National Bank, 407
selection, 134
Selective-reject ARQ, 175–178
self-clocking, 41
self-clocking signal, 132
sensors, 11–12
sensor-to-local area network connection, 11–12
serial ports, 136
servers, 4, 8, 263–264
server software, 279
services, 263
session layer, 18
Shannon, Claude, 114
Shannon's formula, 51
shared network
    LANs (local area networks), 202
    segments, 246
sharing software and hardware, 195
SHF (super high frequency) microwave frequencies, 92
shift keying, 43
S-HTTP (Secure-HTTP), 407
signals, 3
    amplitude, 34

analog, 31–32, 39
attenuation, 36–38
bandwidth, 35
calculating data transfer rate, 51
composite, 37
conversion examples, 59–62
converting data into, 39–53
coupling between two paths, 163
db (decibel), 37–38
digital, 32–33, 39
frequencies, 34–35
fundamentals, 34–36
greater frequency of, 50
Hz (Hertz), 34
loss of strength, 36–38
maximum data transfer rate, 51
noise, 31–32
period, 34
phase, 36
quantization noise, 112–113
reflective feedback of, 163–164
strength, 38
transmitting, 3
signature-based scanner, 401
simple parity, 167–168
simplex connections, 133
single-bit errors, 168
single-mode transmission, 78–79
single-stranded coaxial cable, 75
single-user earth station satellite system, 85
site license, 282
sliding window protocols, 174–175
slope overload noise, 50–51
smart media card technology, 111
SMTP (Simple Mail Transfer Protocol), 21, 322
smurfing, 402–403
SNMP (Simple Network Management Protocol), 443–445
SNMP manager, 444
software licensing agreements, 262
Solaris 2.x, 274
SONET (synchronous optical network), 4, 152–153, 158
source-routing to bridges, 240–243
spanning tree bridges, 240–241
special control character, 59
spectrum, 35
spike, 161–162
spitterless DSL, 378
splitter, 116
spoofing, 402–403
spread spectrum technology, 53–54
SPX (sequenced packet exchange), 268
SSL (secure sockets layer), 408
stand-alone spreadsheets, 281
standard system attacks, 402–404
standard telephone line, 361
start bit, 130
star topology, 201
star-wired bus topology, 201–203

star-wired ring topology, 205
state information, 332
static routing, 307
station, 291
statistical time division multiplexing, 144, 154–157
statistics, 441–442
stealing passwords, 303–304
stealth virus, 401–402
stop-and-wait protocols, 172–174, 454
STP (shielded twisted pair), 71–72
streaming audio and video, 325
STS (synchronous transport signals), 152
stub, 266
submasking, 341
subnets, 4, 292
    broadcast subnet, 295
    circuit switched subnet, 293
    datagram packet switched subnet, 294
    as graph, 299–301
    nodes, 298–299
    packet switched subnet, 294–295
    types, 293–295
    types and network applications, 297–298
    virtual circuit packet switched subnet, 294–295
subscriber extension, 361
subscriber loop, 361
support devices, 262, 284
surveillance, 396–397
SVCs (Switched Virtual Circuit), 372
switches
    ATM (asynchronous transfer mode), 373
    backplane, 245
    versus bridges, 245
    in combination with bridges, 249
    CSMA/CD networks, 245
    cutthrough architecture, 245
    dedicated network segments, 246
    full duplex, 250
    hot-swappable circuit cards, 245
    isolating traffic patterns, 247–249
    multiple access, 247–249
    ports, 241
    primary goals, 247
    redundant circuits, 245
    shared network segments, 246
    speed, 241
symmetric connection, 377
synchronization points, 18
synchronous connections, 131–132
synchronous time division multiplexing
    demultiplexor, 148–149
    empty time slot, 149–150
    ISDN multiplexing, 152–154
    maintaining synchronization, 150
    T1 multiplexing, 150–152
system attacks, 402–404
systems analyst, 426

T

T1 line, 114, 150–151
T1 lines, 364
T-1 lines, 70
T1 multiplexer, 151
T1 multiplexing, 150–152
tags, 327–328
tap, 197
tape drives, 283
TAPI (Telephony API), 382
TCP (Transmission Control Protocol), 21, 319, 333–334
    datagram format, 338–339
    functions, 238–239
TCP/IP (Transmission Control Protocol/Internet Protocol), 5, 9, 277
TDM (time division multiplexing)
    digital signals, 147
    statistical time division multiplexing, 154–157
    synchronous time division multiplexing, 148–154
TDMA (time division multiple access), 88, 90
technically feasible system, 431
telecommunications, 4
Telecommunications Act of 1996, 363
telecommunications systems example, 382–386
tele-immersion, 348
telephone line conditioning or equalization, 165
telephone lines, 360–361, 363
telephone networks
    before and after 1984, 361–363
    after 1996, 363–364
telephone signal limitations, 359
telephone systems. *See* POTS (plain old telephone system)
telephony, 321, 331
Telnet, 21, 320
terminal emulation, 7
terminal emulation program, 320
terminal-to-mainframe computer connections, 6, 133–135
terminate-and-stay-resident antivirus software, 402
terrestrial microwave transmission, 81–82
testing functions, 109
text-based editor, 326
thermal noise, 161
thick coaxial cable, 75
thin client workstations, 283–284
thin coaxial cable, 75
three tier architecture, 265
throughput and cable modems, 117
TIE lines, 363
time division multiplexing, 88, 144
time feasible system, 432
time-hopped spread spectrum, 54
timeout, 173
time value of money, 433
token bus, 214
token management, 18
token passing protocols, 212
token ring, 212–213, 216–217, 220

top-level domains, 342–343
topologies, 196–206
traffic analyzer, 446
transmitting
    analog data with analog signals, 52
    analog data with digital signals, 47–51
    digital data with analog signals, 43–43
    signals, 3
transmitting data, 3, 160
transparent bridges, 236–240
transport layer, 17–18
    error detection, 167
    transport header information, 238
transposition-based ciphers, 406–407
trees, 199–201
Trojan Horse, 403
Trojan horse, 401
trunks, 360–361
TSAPI (Telephony Services API), 382
tunneling protocols, 341, 349–350
twisted pair cable
    AWG (American Wiring Gauge) number, 72
    Category 1, 70
    Category 2, 70
    Category 4, 70–71
    Category 5, 70–71
    Category 5e, 71
    Category 6, 71
    Category 7, 71
    crosstalk, 69
    electromagnetic interface, 72
    modular connectors, 201
    plenum, 72
    riser, 72
    summary of characteristics, 73
    wire thickness, 72
two-way pagers, 89

U

UDP (User Datagram Protocol), 21, 319, 333–334, 340, 444
unbalanced circuit, 125
uninstall software, 279
Universal Serial Bus interface, 105
Unix, 262, 274, 276–277
unmanaged hubs, 244
uplink, 83
uploading and downloading files, 319–320
UPS (uninterruptable power supplies), 283
upshift code, 58
upstream transmissions, 115
URLs (uniform resource locators), 341–342
USB (Universal Serial Bus), 128–129, 136
Usenet, 324
User Agent, 322
User-Network Interface, 373
utilities, 278–279
UTP (unshielded twisted pair), 71

V

V.90 standard, 113
VBR (Variable Bit Rate), 375
VCC (virtual channel connection), 373
VDSL (Very high data rate DSL), 379
VeriSign, 412
vertical redundancy check, 167–168
Vigenére, Blaise de, 406
Vigenére Cipher, 406
virtual circuit packet, 298
virtual circuit packet switched subnet, 294–295
virtual laboratories, 348
viruses, 401–402, 414
VLAN (virtual LAN), 247
VoFR (Voice over frame relay), 322
VoFR (voice over frame relay), 371–372
voice, 151
voice networks, 3
voice over packet, 331
voice over the Internet, 331
VoIP (voice over Internet Protocol), 321
VPC (virtual path connection), 373
VPN (virtual private networks), 413
    tunneling protocol, 341, 349–350
VPScript, 280
VSAT (Very Small Aperture Terminal), 85, 98–100

W

W3C (The World Wide Web Consortium), 346
WANs (wide area networks), 3–5, 429
    basics, 291–298
    combinations of network applications with subnet
        types, 297–298
    components, 4
    connection-oriented, 295–296
    differences from LANs, 290
    home-to-Internet connection, 311–312
    logical network design, 295–297
    mesh design, 290
    network congestion, 309–311
    nodes, 292
    physical network design, 295
    routing, 298–309
    station, 291
    subnet, 292–295
    work-to-Internet connection, 312–313
Warp Connect, 276
Warp Server, 276
wave division multiplexing, 157
Web browsers, 299, 325
Web pages
    authoring tool, 316
    creating, 327–331
    DHTML (Dynamic HTML), 327, 329–330
    HTML (HyperText Transport Protocol), 327

    publishing software, 279
    XML (eXtensible Markup Language), 327
    XML (eXtensible markup language), 330
Web server, 271
weighted network graph, 300
white noise, 161
Windows 2000, 272–273
Windows NT, 271–273
Windows NT/2000, 276–277
Windows NT version, 271–272
wireless, 3
wireless LANs, 206–208
wireless media action example, 98–100
wireless telephone configurations, 13–14
wireless topology, 206–208
wireless transmissions
    broadband wireless systems, 91–92
    CDPD (cellular digital packet data), 89
    infrared transmissions, 90–91
    mobile telephones, 86–89
    pagers, 89–90
    radio waves, 80
    satellite microwave transmission, 82–86
    summary, 92–93
    terrestrial microwave transmission, 81–82
word processing, 281
workgroups, 400
workstations, 4
    bridges, 242
    discovery frames, 242–243
    modem pools, 118–119
    sending and receiving signals simultaneously, 250
    thin client, 283–284
work-to-Internet connection, 312–313
World Wide Web Distributed Authoring and
        Versioning, 346
worms, 401
WWW (World Wide Web), 319
    cookies, 332
    creating web pages, 327–331
    DHTML (Dynamic HTML), 329–330
    e-commerce, 331–332
    HTML (hypertext transport protocol), 327
    HTTP (hypertext transport protocol), 327
    state information, 332
    XML (eXtensible markup language), 330

X

x2 format, 113
X.25 standard, 463–466
X.509 standard, 411–412

Z

Zimmermann, Philip, 410